Lecture Notes in Physics

Springer
Berlin
Heidelberg
New York
Hong Kong
London
Milan
Paris
Tokyo

The Editorial Policy for Edited Volumes

The series *Lecture Notes in Physics* (LNP), founded in 1969, reports new developments in physics research and teaching - quickly, informally but with a high degree of quality. Manuscripts to be considered for publication are topical volumes consisting of a limited number of contributions, carefully edited and closely related to each other. Each contribution should contain at least partly original and previously unpublished material, be written in a clear, pedagogical style and aimed at a broader readership, especially graduate students and nonspecialist researchers wishing to familiarize themselves with the topic concerned. For this reason, traditional proceedings cannot be considered for this series though volumes to appear in this series are often based on material presented at conferences, workshops and schools.

Acceptance

A project can only be accepted tentatively for publication, by both the editorial board and the publisher, following thorough examination of the material submitted. The book proposal sent to the publisher should consist at least of a preliminary table of contents outlining the structure of the book together with abstracts of all contributions to be included. Final acceptance is issued by the series editor in charge, in consultation with the publisher, only after receiving the complete manuscript. Final acceptance, possibly requiring minor corrections, usually follows the tentative acceptance unless the final manuscript differs significantly from expectations (project outline). In particular, the series editors are entitled to reject individual contributions if they do not meet the high quality standards of this series. The final manuscript must be ready to print, and should include both an informative introduction and a sufficiently detailed subject index.

Contractual Aspects

Publication in LNP is free of charge. There is no formal contract, no royalties are paid, and no bulk orders are required, although special discounts are offered in this case. The volume editors receive jointly 30 free copies for their personal use and are entitled, as are the contributing authors, to purchase Springer books at a reduced rate. The publisher secures the copyright for each volume. As a rule, no reprints of individual contributions can be supplied.

Manuscript Submission

The manuscript in its final and approved version must be submitted in ready to print form. The corresponding electronic source files are also required for the production process, in particular the online version. Technical assistance in compiling the final manuscript can be provided by the publisher's production editor(s), especially with regard to the publisher's own LaTeX macro package which has been specially designed for this series.

LNP Homepage (springerlink.com)

On the LNP homepage you will find:

– The LNP online archive. It contains the full texts (PDF) of all volumes published since 2000. Abstracts, table of contents and prefaces are accessible free of charge to everyone. Information about the availability of printed volumes can be obtained.

– The subscription information. The online archive is free of charge to all subscribers of the printed volumes.

– The editorial contacts, with respect to both scientific and technical matters.

– The author's / editor's instructions.

R. Narayanan D. Schwabe (Eds.)

Interfacial Fluid Dynamics and Transport Processes

Editors

Ranga Narayanan
University of Florida,
Department of Chemical Engeneering,
Bifurcation and Nonlinear Instability Lab.,
Gainesville,
32611-9103 Florida, USA

Dietrich Schwabe
Justus-Liebig-Universität Giessen,
I.Physikalisches Institut,
Heinrich-Buff-Ring 16,
35390 Gießen, Germany

Electronic Supplements to this book are available online at springerlink.com

Cataloging-in-Publication Data applied for

A catalog record for this book is available from the Library of Congress.

Bibliographic information published by Die Deutsche Bibliothek
Die Deutsche Bibliothek lists this publication in the Deutsche Nationalbibliografie; detailed bibliographic data is available in the Internet at http://dnb.ddb.de

ISSN 0075-8450
ISBN 3-540-40583-6 Springer-Verlag Berlin Heidelberg New York

Springer-Verlag Berlin Heidelberg New York
a member of BertelsmannSpringer Science+Business Media GmbH

springer.de

Typesetting: Camera-ready by the authors/editor
Cover design: *design & production*, Heidelberg

Printed on acid-free paper
54/3141/du - 5 4 3 2 1 0

Preface

Springer Verlag has been pleased to bring out this special volume on interfacial fluid dynamics and transport processes. There are seventeen articles and each article is written in a pedagogical manner dealing with relevant research issues and questions. The intended audience is post-doctoral scientists, academicians and graduate students intending to pursue research and it is our hope that this volume will have lasting value.

Several issues arise within the general field of interfacial transport such as the instability of interfacial processes and driven flows. Instabilities occur when there is a sudden change in the structure of a solution as a control parameter is smoothly varied. They are usually accompanied by a change in the patterns in fluid flow or temperature and concentration fields. Transport phenomena related instability at the interface has much of its origin in the seminal works of Rayleigh who in the later part of the 19th century worked on jets, gravitationally unstably stratified fluid layers, and on the first ideas on convection. Some of these ideas were subsequently modified by the work of Marangoni, Block and Pearson on surface tension driven instabilities. Over the years similar concepts have found place in solidification and melting, electrodeposition, and other phase change problems.

Renewed interest in this subject has occurred because of its applications to materials science, biomedical engineering and space processing. Understanding instabilities is very important as this is related to pattern formation during the processing of materials on earth, in space and related to information technology and the bio-sciences. The first five papers of this volume are related to instabilities at the interface. Four papers are presented on topics of interfacial instabilities due to Marangoni flows in multiple layers or with multiple effects such as double diffusion. A fifth paper gives a tutorial presentation on the instability of spreading films. All of these papers discuss several research issues that have gained importance in recent years.

The next eight papers are concerned with flows that are driven by thermocapillary means. Many of these papers are concerned with the technologically important configuration of the liquid bridge, the numerical calculation of the flows in the bridge and the onset of instabilities from a basic flow state. The eighth paper in this part is concerned with the controlled migration of a droplet on account of thermocapillary effects. All of the articles discuss the physics and provide several helpful references for newcomers to the field.

The last four papers of the volume deal with a variety of very interesting problems that are tied to the general theme of this volume and offer an assortment of topics. A paper on electrocapillary flows acquaints the reader with electric fields that are applied tangentially to surfaces and provides experimental evidence of the phenomena that arise. There is a detailed paper on the numerical simulation of a drop undergoing fragmentation, a problem that is of technological importance in emulsification. This is followed by an article on the stability of emulsions under thermocapillary effects. The volume then concludes with a very helpful article that discusses the energetics of phase change between two fluids.

This volume was motivated by the first International Marangoni Association workshop held on September 12, 2001 in the Castle Rauischholzhausen near Marburg in Germany. Despite the unfortunate fateful events that took place the day before, the conference saw the presentation of over 40 papers and posters. A follow-up meeting in Gainesville, FL was held in March 2002 as many U.S. scientists were unable to attend the first meeting. Authors were selected, based on their research interests and presentations to write the articles that are presented in this volume. In short, the reader will get a very good idea of the various facets of interfacial fluid flow and transport from several scholars who have devoted several years of research to this important field.

No meeting can be run successfully without a nominal amount of funding. We are indeed fortunate to have received support from a number of agencies and institutions. We acknowledge with gratitude support from the U.S. National Science Foundation's Chemical and Thermal Systems Directorate via grant number CTS 0109096, NASA's Office of Biological and Physical Research, Physical Sciences Division, Code UG via grant number W-24346, The European Space Agency (ESA), WE-Heraeus-Stiftung through Deutsche Physikalische Gesellschaft (DPG) and the Justus-Liebig-Universitt Giessen as well as the University of Florida's Research Foundation and College of Engineering.

Finally, we would like to thank Ms. G. Hakuba, Ms. S. Thoms and Dr. C. Caron of Springer-Verlag, Heidelberg and Mr. H. Ho of the University of Florida for their patience and assistance in the preparation of this volume.

R. Narayanan
Gainesville, FL, USA
and
D. Schwabe
Gießen, Germany

April 2003

Contents

List of Contributors

T. Azami
NEC Corporation,
34 Miyukigaoka,
Tsukuba 305-8501, Japan
azami@frl.cl.nec.co.jp

R. Balasubramaniam
National Center for Microgravity
Research on Fluids and Combustion,
Mail Stop 110-3,
NASA Glenn Research Center,
Cleveland, OHIO 44135, USA
bala@grc.nasa.gov

S.W. Benintendi
Department of Mechanical and
Aerospace Engineering,
University of Dayton,
Dayton, Ohio 45469, USA

C.P. Benjamin IV
The George W. Woodruff School of
Mechanical Engineering,
Georgia Institute of Technology,
Atlanta, Georgia 30332-0405, USA

Th. Boeck
Laboratoire de Modélisation en
Mécanique,
Université Pierre et Marie Curie,
8 rue du Capitaine Scott,
75015 Paris, France
thomas@lmm.jussieu.fr

L.M. Braverman
Department of Computer Engineering,
Karmiel ORT College,
Karmiel, Israel

C. Chan
Department of Aerospace and
Mechanical Engineering,
College of Engineering and Mines,
The University of Arizona,
P.O. Box 210119,
1130 N. Mountain, RM N614,
Tucson AZ 85721-00119, USA
cholik@email.arizona.edu

C.F. Chen
Department of Aerospace and
Mechanical Engineering,
University of Arizona,
Tucson, AZ 85721, USA
chen@ame.arizona.edu

E. Chénier
Université de Marne-la-Vallée,
Laboratoire d'Etude des Transferts
d'Energie et de Matière,
77454 Marne-la-Vallée
Cedex 2, France
labrosse@azur.epfl.ch

J.M. Davis
Microfluidic Research & Engineering
Laboratory,
Department of Chemical Engineering,
Princeton University,
Princeton, NJ 08544-5263, USA
jeffreyd@princeton.edu

C. Delcarte
Université de Paris-Sud XI,
Laboratoire d'Informatique pour
la Mécanique et les Sciences de
l'Ingénieur,
Bat. 508, B.P.133,
91403 Orsay Cedex, France
delcarte@limsi.fr

B.J. Fischer
Microfluidic Research and Engineering
Laboratory,
Department of Chemical Engineering,
Princeton University,
Princeton, NJ 08544-5263, USA
bfischer@princeton.edu

A.A. Golovin
Department of Engineering Sciences
and Applied Mathematics,
Northwestern University,
Evanston, IL 60208–3100, USA
a-golovin@northwestern.edu

O.A. Gudz
Novosibirsk State University,
Novosibirsk, Russia

T. Hibiya
Tokyo Metropolitan Institute of
Technology,
6-6 Asahigaoka,
Hino 191-0065, Japan
hibiya@mtc.biglobe.ne.jp

D. Johnson
Department of Chemical Engineering,
University of Alabama,
Tuscaloosa, AL 35487-0203, USA
djohnson@coe.eng.ua.edu

C. Karcher
Department of Mechanical
Engineering,
Ilmenau University of Technology,
P.O. Box 100565,
98684 Ilmenau, Germany
christian.karcher
@mb.tu-ilmenau.de

G. Kasperski
Maitre de Conferences, Université
Paris-Sud XI,
LIMSI-CNRS BP133,
91403 Orsay Cedex, France
kaspersk@limsi.fr

R.E. Kelly
Department of Mechanical and
Aerospace Engineering,
University of California,
Los Angeles, CA 90095-1597, USA
rekhome@ucla.edu

H.C. Kuhlmann
ZARM – University of Bremen,
Am Fallturm,
28359 Bremen, Germany
kuhl@zarm.uni-bremen.de

G. Labrosse
Université de Paris-Sud XI,
Laboratoire d'Informatique pour
la Mécanique et les Sciences de
l'Ingénieur,
Bat. 508, B.P.133,
91403 Orsay Cedex, France
Gerard.Labrosse@limsi.fr

J-C. Legros
Université Libre de Bruxelles,
MRC, CP-165/62,
50, av. F.D.Roosevelt,
B-1050 Brussels, Belgium
jclegros@ulb.ac.be

S.P. Lin
Mechanical and Aeronautical Engineering Department
Clarkson University,
Potsdam, NY 13699-5725, USA
gw02@clarkson.edu

D.E. Melnikov
Université Libre de Bruxelles,
MRC, CP-165/62,
50, av. F.D.Roosevelt,
B-1050 Brussels, Belgium
denmel@mrc.ulb.ac.be

M. Mojahed
Université Libre de Bruxelles,
MRC, CP-165/62, 50,
av. F.D.Roosevelt,
B-1050 Brussels, Belgium
mmojahed@ulb.ac.be

S. Nakamura
NEC Corporation,
34 Miyukigaoka,
Tsukuba 305-8501, Japan
nakamusi@frl.cl.nec.co.jp

R. Narayanan
Department of Chemical Engineering,
University of Florida,
Gainesville, Florida 32611 USA
ranga@che.ufl.edu

A.A. Nepomnyashchy
Department of Mathematics,
Technion –
Israel Institute of Technology,
32000 Haifa, Israel
nepom@math.technion.ac.il

C. Nienhüser
ZARM – University of Bremen,
Am Fallturm,
28359 Bremen, Germany

O. Ozen
Department of Chemical Engineering,
University of Florida,
Gainesville, Florida 32611 USA
ozen@che.ufl.edu

A.G. Petrova
Altai State University,
Barnaul, Russia

V.V. Pukhnachov
Lavrentyev Institute of Hydrodynamics,
Novosibirsk, Russia
pukh@hydro.nsc.ru

Y. Renardy
Department of Mathematics,
460 McBryde Hall,
Virginia Tech,
Blacksburg, VA 24061-0123, USA
renardyy@calvin.math.vt.edu

D. Schwabe
I. Physikalisches Institut der Justus-Liebig Universitaet,
Heinrich-Buff-Ring 16,
35392 Giessen, Germany
Dietrich.Schwabe@exp1.physik.uni-giessen.de

V.M. Shevtsova
Université Libre de Bruxelles,
MRC, CP-165/62,
50, av. F.D.Roosevelt,
B-1050 Brussels, Belgium
vshev@ulb.ac.be

B.-C. Sim
Department of Mechanical Engineering,
Hanyang University,
Ansan, Kyunggi-do, 425-791, Korea
sbcsim@templab.hanyang.ac.kr

I.B. Simanovskii
Department of Mathematics,
Technion –
Israel Institute of Technology,
32000 Haifa, Israel

M.K. Smith
The George W. Woodruff School of
Mechanical Engineering,
Georgia Institute of Technology,
Atlanta, GA 30332-0405, USA
marc.smith@me.gatech.edu

M. Sumiji
Tokyo Institute of Technology,
4259 Nagatsuta-cho, Midori-ku,
Yokohama 226-8502, Japan

A. Thess
Department of Mechanical
Engineering,
Ilmenau University of Technology,
P.O. Box 100565,
98684 Ilmenau, Germany
thess@tu-ilmenau.de

S.M. Troian
Microfluidic Research and Engineering
Laboratory,
Department of Chemical Engineering,
Princeton University,
Princeton, NJ 08544-5263, USA
stroian@princeton.edu

O.V. Voinov
Institute of Mechanics of Multiphase
Systems,
Tyumen, Russia

A. Zebib
Department of Mechanical and
Aerospace Engineering,
Rutgers University,
98 Brett Road,
Piscataway, NJ, 08854-8058, USA
zebib@jove.rutgers.edul

E.N. Zhuravleva
Altai State Technological University,
Barnaul, Russia

Large Wavelength Disturbances in Two-Fluid Bénard–Marangoni Convection and Their Control

R.E. Kelly

Department of Mechanical and Aerospace Engineering, University of California, Los Angeles CA 90095-1597, USA

Abstract. The stability of large wavelength, small amplitude disturbances is discussed on the basis of first principles, initially for the case without active control. Special attention is given to the occurrence of discontinuities that can form in the curves of Marangoni number for neutral stability versus wavenumber as the depth ratio of the two layers varies. The use of linear proportional control is then discussed for the two-fluid system, and significant change in the Marangoni number for neutral stability is shown to be possible for such disturbances.

1 Introduction

The use of active control with feedback in controlling interfacial instabilities is very much in an early stage of development. This type of control involves a sensor that detects the deformation of the interface from its original state along with an actuator that is designed to restore the interface to its original configuration, or nearly so, by appropriate modification of the fluid system. For a small aspect ratio system whose behavior is dominated by a few modes, a single sensor and actuator might be sufficient to achieve stabilization. For larger systems involving many modes and for which localized disturbances might first develop, arrays of sensors and actuators are required, although rapid scanning devices might possibly be used instead. The sensors and actuators are related by a control law, which is designed so as to achieve a certain goal. Some further comments about configurations will be made in Sect. 4.

The connection between the sensor and actuator makes this kind of control fundamentally different from passive control, such as obtained by imposing an unchanging thermal [8] or magnetic [15] field or by imposing active control of a predetermined type, such as modulation in time of a background field [14,20]. A brief review of these other types of control has been given by the author [13], along with a discussion of applications of active control with feedback to certain types of interfacial instabilities. Several of these applications concern thermocapillary instability, a topic that has been discussed more generally in several recent review articles [11,26,32] and books [4,16]. The basic mechanism for thermocapillary instability in a gas-liquid system is straightforward and can (almost) be explained without mathematical analysis. Say that we consider an interface between an upper layer of gas and a lower layer of liquid so that the interface is stable from the viewpoint of Rayleigh–Taylor instability [3]. The layers are

R.E. Kelly, Large Wavelength Disturbances in Two-Fluid Bénard–Marangoni Convection and Their Control, Lect. Notes Phys. **628**, 1–20 (2003)
http://www.springerlink.com/

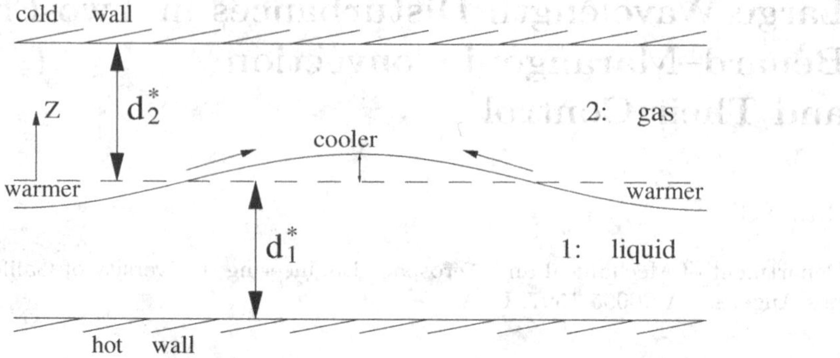

Fig. 1. Schematic showing thermocapillary effect for the case of a deformable interface between a liquid and a gas

horizontal and are taken to be contained between two parallel walls, the lower of which is heated while the upper is cooled (gravity is considered to act downwards). Say that the interface is disturbed so that it has a crest that is closer to the cold wall and a trough that is closer to the warm wall, as shown in Fig. 1. If the surface tension of the interface is assumed to decrease as temperature increases, and vice versa, a traction due to the variation of surface tension will be established along the interface that will tend to reinforce the disturbance in the liquid layer, leading to instability. Actually, as discussed in the pioneering work by Pearson [22], deflection of the interface is not essential for the instability to occur at wavelengths on the order of the liquid depth; any disturbance that leads to local heating of the interface will suffice. However, interfacial displacement is essential for instability of the large wavelength disturbances (i.e., wavelength much greater than the depth) that will be considered in this article. Such disturbances tend to be the most unstable ones either under conditions of microgravity or for very thin fluid layers [31]. The situation is more complicated for the case of two immiscible liquid layers, and so we will discuss in more detail the two-liquid case in the following section. Also, the thermocapillary instability can be aided by buoyancy effects associated with Rayleigh–Bénard (RB) convection. When buoyancy effects are negligible so that the instability is associated mainly with thermocapillary effects, we refer to the resulting convection as being of the Bénard–Marangoni (BM) type. We will concentrate on this particular type of convection, which arises via a bifurcation. However, the basic concept discussed in Sect. 3 is also applicable to other types of interfacial instability [13]. It has been discussed previously within the context of BM convection [1,17,19] for the case of a single liquid layer bounded from above by a motionless gas. However, both analyses [11,28] and experiments [12] have now been done that account for the rich structure of the general two-fluid problem, and so we will start with a brief discussion of that situation.

2 Two-Fluid Bénard–Marangoni Instability

We shall discuss the Bénard–Marangoni problem without any control in order to provide a reference for the later results with control. We first define the basic state whose stability will be investigated.

2.1 The Basic State of Conduction

With reference to Fig. 1, the upper wall at $z^* = d_2^*$ will be considered to be at a fixed temperature $\overline{T}^*_{\mathrm{WU}}$ whereas the lower wall at $z^* = -d_1^*$ will be taken to be either at a constant temperature $\overline{T}^*_{\mathrm{WL}}$ or with a constant heat flux condition so that

$$q^*_{\mathrm{WL}} = -K^*_{01}\frac{d\overline{T}^*_1}{dz^*}\left(z^* = -d_1^*\right) = \text{constant},$$

where K^*_{01} denotes the thermal conductivity in the lower layer and an asterisk is used for the moment to denote dimensional variables that will appear later in nondimensional form. Subscripts $j = 1$ and 2 will refer to quantities associated with the lower and upper layers, respectively. The thermal condition at the lower wall does not affect the temperature profile of the basic state but is important for the stability problem. We will denote the temperature of the interface for the undisturbed convection state as T_0^* and use T_0^* as a reference temperature. For the problem with a fixed temperature at the lower wall, the conduction solution is

$$\overline{T}^*_1(z) = T_0^* - \left(\overline{T}^*_{\mathrm{WL}} - T_0^*\right)\left(z^*/d_1^*\right),$$

$$\overline{T}^*_2(z) = T_0^* - \left(T_0^* - \overline{T}^*_{\mathrm{WU}}\right)\left(z^*/d_2^*\right),$$

where

$$T_0^* = \frac{d\overline{T}^*_{\mathrm{WL}} + K\overline{T}^*_{\mathrm{WU}}}{d+K}$$

and d and K denote the nondimensional ratios

$$d = \frac{d_2^*}{d_1^*},$$

$$K = \frac{K^*_{02}}{K^*_{01}}.$$

For the case of a fixed heat flux at the lower wall, we have

$$\overline{T}^*_1 = T_0^* - \left(\frac{q^*_{\mathrm{WL}}}{K^*_{01}}\right)z^*,$$

$$\overline{T}^*_2 = T_0^* - \left(\frac{q^*_{\mathrm{WL}}}{K^*_{02}}\right)z^*,$$

so

$$T_0^* = \overline{T}^*_{\mathrm{WU}} + \frac{q^*_{\mathrm{WL}}d_2^*}{K^*_{02}}.$$

It is common practice in this problem (e.g., see Sect. 3.5 of [16]) to define a characteristic temperature difference on the basis of the lower layer as

$$\Delta\overline{T}_1^* = \overline{T}_{\mathrm{WL}}^* - T_0^*$$

because, for the case of a gas-liquid system, this definition allows the results to be compared easily to the results holding for a single liquid layer. However, in general the characteristic temperature difference just as well could be defined as

$$\Delta\overline{T}_t^* = \overline{T}_{\mathrm{WL}}^* - \overline{T}_{\mathrm{WU}}^*,$$

and this definition might be more appropriate for the two-liquid problem. The relation between the two characteristic temperature differences is

$$\Delta\overline{T}_t^* = \left(\frac{d}{K}+1\right)\left(\overline{T}_{\mathrm{WL}}^* - T_0^*\right).$$

We shall define an equation of state for the surface tension σ as

$$\sigma^*\left(T^*\right) = \sigma_0^* - \sigma_T^*\left(T^* - T_0^*\right) \tag{1}$$

where $\sigma_0^* = \sigma^*\left(T_0^*\right)$ and $\left(-\sigma_T^*\right)$ denotes $d\sigma^*/dT^*$ evaluated at $T^* = T_0^*$. The interface is considered to be planar in the basic state, and so surface tension does not affect the pressure distribution of the basic state. We will also assume that the fluids are incompressible so that the density $\rho^* = \rho^*\left(T^*\right)$. For small temperature differences, an acceptable equation of state for density is

$$\rho_j^*\left(T^*\right) = \rho_{0j}^*\left[1 - \alpha_j^*\left(T^* - T_0^*\right)\right],$$

where α_j^* is the coefficient of thermal expansion.

2.2 The Governing Linearized Stability Equations and Boundary Conditions

The basic state is now considered to be perturbed by a disturbance consisting of a velocity

$$\boldsymbol{V}_j^*\left(x^*, z^*, t^*\right) = \boldsymbol{i}_{x^*} u_j^*\left(x^*, z^*, t^*\right) + \boldsymbol{i}_{z^*} w_j^*\left(x^*, z^*, t^*\right),$$

where $\boldsymbol{i}$ denotes a unit vector, and a corresponding temperature $\widetilde{T}_j^*\left(x^*, z^*, t^*\right)$, defined by

$$T_j^*\left(x^*, z^*, t^*\right) = \overline{T}_j^*\left(z^*\right) + \widetilde{T}_j^*\left(x^*, z^*, t^*\right).$$

We will consider formulation of the two-dimensional problem in order to minimize the number of equations; the extension to the three-dimensional problem is straightforward. The basic equations are the linearized form of the Oberbeck–Boussinesq equations, which are obtained from the Navier–Stokes and energy equations by assuming that density varies only in the buoyancy term and that other fluid properties (e.g., viscosity) are constant [3,32]. The linearized equations, valid for small-amplitude disturbances, are

$$\frac{\partial \boldsymbol{V}_j^*}{\partial t^*} = -\frac{1}{\rho_{0j}^*}\nabla \tilde{p}_j^* + g\alpha_j^* \widetilde{T}_j^* \boldsymbol{i}_{z^*} + \nu_{0j}^* \nabla^2 \boldsymbol{V}_j^*, \tag{2}$$

$$\frac{\partial \widetilde{T}_j^*}{\partial t^*} + w_j^* \frac{d\overline{T}_j^*}{dz^*} = \kappa_{0j}^* \nabla^2 \widetilde{T}_j^*, \tag{3}$$

and

$$\nabla \cdot \boldsymbol{V}_j^* = 0, \tag{4}$$

where t^* denotes time, $\boldsymbol{i}_{z^*}$ is a unit vector in the z^*-direction, $\tilde{p}_j^*$ is the perturbation pressure, w_j^* is the vertical component of perturbation velocity, g is gravitational acceleration, and ν_{0j}^* and κ_{0j}^* represent reference values of kinematic viscosity and thermal diffusivity, respectively, in each layer. Equations (2) and (3) are the linearized form of the momentum and energy equations, respectively, and (4) is the continuity equation.

By taking the divergence of (2) and using (4), we obtain

$$\nabla^2 p_j^* = \rho_{0j}^* g \alpha_j^* \frac{\partial \widetilde{T}_j^*}{\partial z^*}. \tag{5}$$

If we now operate on (2) by the Laplacian operator, make use of (5) and take the z-component of the resulting equation, we obtain the following equation for $\widetilde{T}_j^*$ and w_j^*:

$$\frac{\partial}{\partial t^*}\nabla^2 w_j^* = \nu_{0j}^* \nabla^4 w_j^* + g\alpha_j^* \nabla_\perp^2 \widetilde{T}_j^*, \tag{6}$$

where $\nabla_\perp^2$ represents the horizontal Laplacian operator. Equations (3) and (6) are the governing equations for w_j^* and $\widetilde{T}_j^*$.

The boundary conditions on $\boldsymbol{v}_j^*$ are the no-slip conditions at the walls, namely,

$$\boldsymbol{v}_1^*(x^*, -d_1^*, t^*) = \boldsymbol{v}_2^*(x^*, d_2^*, t^*) = 0. \tag{7}$$

For the problem with constant temperatures at both walls, we have

$$\widetilde{T}_1^*(x_1^*, -d_1^*, t^*) = \widetilde{T}_2^*(x^*, d_2^*, t^*) = 0, \tag{8}$$

whereas, for the problem when the heat flux is prescribed at $z^* = -d_1^*$, we have

$$\frac{\partial \widetilde{T}_1^*}{\partial z^*}(x^*, -d_1^*, t^*) = 0, \qquad \widetilde{T}_2^*(x^*, d_2^*, t^*) = 0. \tag{9}$$

We also have interfacial boundary conditions applied at the mean level of the interface ($z^* = 0$). First, we have continuity of w_j^* and the kinematic condition which relates w_j^* to the displacement $\eta^*(x^*, t^*)$ of the interface:

$$w_1^*(x^*, 0, t^*) = w_2^*(x^*, 0, t^*) = \frac{\partial \eta^*}{\partial t^*}.$$

Continuity of u_j^* at $z^* = 0$ requires, from (4), that

$$\frac{\partial w_1^*}{\partial z^*} = \frac{\partial w_2^*}{\partial z^*}.$$

We also have continuity of temperature and the normal component of heat flux at the disturbed interface, which give at the linear approximation

$$\frac{d\overline{T}_1^*}{dz^*}\eta^* + \widetilde{T}_1^*(x^*,0,t^*) = \frac{d\overline{T}_2^*}{dz^*}\eta^* + \widetilde{T}_2^*(x^*,0,t^*) \tag{10}$$

and

$$K_{01}^* \frac{\partial \widetilde{T}_1^*}{\partial z^*}(x^*,0,t^*) = K_{02}^* \frac{\partial \widetilde{T}_2^*}{\partial z^*}(x^*,0,t^*). \tag{11}$$

Formulation of the normal and tangential stress conditions is more complicated but is discussed thoroughly in the books referenced, e.g., Chaps. 1 and 2 of [28], and basic research papers, e.g., [29]. The tangential stress condition at $z^* = 0$ contains the variable surface tension term giving rise to BM convection and is, where μ_{0j}^* denotes the dynamic viscosity of each layer,

$$\mu_{01}^*\left(\frac{\partial u_1^*}{\partial z^*} + \frac{\partial w_1^*}{\partial x^*}\right) - \mu_{02}^*\left(\frac{\partial u_2^*}{\partial z^*} + \frac{\partial w_2^*}{\partial x^*}\right) = \frac{\partial \sigma^*}{\partial x^*} \tag{12}$$

or by use of (1) and (4) after taking the x^*-derivative of (12),

$$\mu_{01}^*\left(-\frac{\partial^2 w_1^*}{\partial z^{*2}} + \frac{\partial^2 w_1^*}{\partial x^{*2}}\right) - \mu_{02}^*\left(-\frac{\partial^2 w_2^*}{\partial z^{*2}} + \frac{\partial^2 w_2^*}{\partial x^{*2}}\right)$$
$$= -\sigma_T^*\left(\frac{d\overline{T}_1^*}{dz^*}\frac{\partial^2 \eta^*}{\partial x^{*2}} + \frac{\partial^2 \widetilde{T}_1^*}{\partial x^{*2}}\right). \tag{13}$$

Although the dependence upon temperature of the surface tension is all-important in (13), it is a higher-order effect in the normal stress condition, which is

$$\tilde{p}_1^* - \tilde{p}_2^* + (\rho_{01}^* - \rho_{02}^*)\, g\eta^* + 2\mu_{01}^*\frac{\partial w_1^*}{\partial z^*} - 2\mu_{02}^*\frac{\partial w_2^*}{\partial z^*} = \sigma_0^*\frac{\partial^2 \eta^*}{\partial x^{*2}}$$

applied at $z^* = 0$.

We now nondimensionalize as follows:

$$(x^*, y^*, z^*) = d_1^*(x, y, z), \qquad t^* = \left(\frac{d_1^{*2}}{\kappa_{01}^*}\right)t, \qquad \eta^* = d_1^*\eta,$$

$$\widetilde{T}_j^* = \Delta\overline{T}_{\mathrm{CH}}^*\widetilde{T}_j, \qquad \boldsymbol{V}_j^* = (\kappa_{01}^*/d_1^*)\,\boldsymbol{V}_j, \qquad \tilde{p}_j^* = \left(\frac{\mu_{01}^*\kappa_{01}^*}{d_1^{*2}}\right)\tilde{p}_j,$$

where $\Delta\overline{T}_{\mathrm{CH}}^*$ is a characteristic temperature difference equal to $\Delta\overline{T}_1^*$ for the constant temperature lower wall and $q_{\mathrm{WL}}^* d_1^*/K_{01}^*$ for the constant heat flux case. The following ratios arise:

$$d = \frac{d_2^*}{d_1^*}, \quad K = \frac{K_{02}^*}{K_{01}^*}, \quad \kappa = \frac{\kappa_{02}^*}{\kappa_{01}^*},$$

$$\mu = \frac{\mu_{02}^*}{\mu_{01}^*}, \quad \alpha = \frac{\alpha_2^*}{\alpha_1^*}, \quad \rho = \frac{\rho_{02}^*}{\rho_{01}^*}, \quad \nu = \frac{\nu_{02}^*}{\nu_{01}^*}.$$

The basic state temperatures for the two layers can now be written as

$$\frac{\overline{T}_1^* - T_0^*}{\Delta \overline{T}_{\mathrm{CH}}^*} = -z, \qquad \frac{\overline{T}_2^* - T_0^*}{\Delta \overline{T}_{\mathrm{CH}}^*} = -\frac{z}{K}.$$

The following nondimensional parameters also arise:

$$\mathrm{Ma} = \frac{\sigma_T^* \Delta \overline{T}_{\mathrm{CH}}^* d_1^*}{\mu_{01}^* \kappa_{01}^*}, \qquad \mathrm{Ra} = \frac{g \alpha_1^* \Delta T_{\mathrm{CH}}^* \left(d_1^*\right)^3}{\nu_{01}^* \kappa_{01}^*}, \qquad \mathrm{Pr} = \frac{\nu_{01}^*}{\kappa_{01}^*}$$

where Ma is a Marangoni number, Ra is a Rayleigh number and Pr is a Prandtl number. Note that the ratio

$$\frac{\mathrm{Ra}}{\mathrm{Ma}} = \frac{\rho_{01}^* g \alpha_1^* \left(d_1^*\right)^2}{\sigma_T^*}$$

decreases as d_1^* decreases for fixed fluid properties. We will assume that d_1^* is sufficiently small that $\mathrm{Ra} \ll \mathrm{Ma}$ and so buoyancy effects can be neglected (in a microgravity environment, this is clearly true even without assuming d_1^* small). We are considering a deformable interface and so two other parameters arise, namely,

$$\mathrm{Cr} = \frac{\mu_{01}^* \kappa_{01}^*}{\sigma_0^* d_1^*}, \qquad \mathrm{Bo} = \frac{g\left(\rho_{01}^* - \rho_{02}^*\right)\left(d_1^*\right)^2}{\sigma_0^*}, \tag{14}$$

where Cr is a crispation number and Bo is a Bond number associated with the density difference across the interface. The Bond number decays less rapidly with d_1^* than Ra as d_1^* becomes small. Alternatively, one of the parameters in (14) could be used along with a Galileo number, Ga, defined as

$$\mathrm{Ga} = \frac{\mathrm{Bo}}{\mathrm{Cr}} = \frac{g\left(\rho_{01}^* - \rho_{02}^*\right)\left(d_1^*\right)^3}{\mu_{01}^* \kappa_{01}^*} \tag{15}$$

which does decay as $\left(d_1^*\right)^3$. However, note that

$$\mathrm{Ra} = \mathrm{Ga}\left(\frac{\rho_{01}^*}{\rho_{01}^* - \rho_{02}^*}\right)\left(\alpha_1^* \Delta T_{\mathrm{CH}}^*\right)$$

and so $\mathrm{Ra} \ll \mathrm{Ga}$ for sufficiently small values of $(\alpha_1^* \Delta T_{\mathrm{CH}}^*)$, a condition that is required in order to make the Boussinesq approximation. Because so many parameters enter into the problem, it can be a challenge to appreciate their relative order of magnitude as well as to obtain a clear idea of how the stability results depend on them.

The nondimensional form of the governing equations (6) and (3), after omitting the buoyancy term, is

$$\left(\frac{\nu^{1-j}}{\mathrm{Pr}}\right)\frac{\partial}{\partial t}\nabla^2 w_j = \nabla^4 w_j,$$

$$\kappa^{1-j}\,\frac{\partial \widetilde{T}_j}{\partial t} - (\kappa K)^{1-j}\,w_j = \nabla^2 \widetilde{T}_j,$$

and the interfacial conditions at $z = 0$ are

$$w_1 = w_2 = \frac{\partial \eta}{\partial t}, \qquad \frac{\partial w_1}{\partial z} = \frac{\partial w_2}{\partial z},$$

$$-\eta + \widetilde{T}_1 = -K^{-1}\eta + \widetilde{T}_2, \tag{16}$$

$$\frac{\partial \widetilde{T}_1}{\partial z} = K\frac{\partial \widetilde{T}_2}{\partial z},$$

$$\left(\frac{\partial^2 w_1}{\partial x^2} - \frac{\partial^2 w_1}{\partial z^2}\right) - \mu\left(\frac{\partial^2 w_2}{\partial x^2} - \frac{\partial^2 w_2}{\partial z^2}\right) = -\mathrm{Ma}\left(\frac{\partial^2 \widetilde{T}_1}{\partial x^2} - \frac{\partial^2 \eta}{\partial x^2}\right),$$

$$\mathrm{Cr}\left[\tilde{p}_1 - \tilde{p}_2 + 2\left(\frac{\partial w_1}{\partial z} - \mu\frac{\partial w_2}{\partial z}\right)\right] + \mathrm{Bo}\eta = \frac{\partial^2 \eta}{\partial x^2}, \tag{17}$$

where information about $\tilde{p}_1$ and $\tilde{p}_2$ can be obtained from the nondimensional form of the momentum equations (2).

The wall boundary conditions at $z = -1$ and $z = d$ have the same form as the dimensional equations (7)–(9) and so are not listed.

2.3 Analysis and Results

The standard way to proceed is to assume that a disturbance can be represented in a Fourier manner by a superposition of normal modes of the form

$$\left(w_j, \widetilde{T}_j, \tilde{p}_j, \eta\right) = \left(W_j(z), \Theta(z), \pi_j(z), \eta_0\right) e^{ikx+\lambda t}.$$

where $\lambda = \lambda_r + i\lambda_i$ is the complex growth rate. The stability characteristics of a single mode are determined so that, for instance, the value of Ma for neutral stability ($\lambda_r = 0$), say Ma_N, can be determined as a function of k and the other parameters. As we will indicate, instability can occur in general for $\mathrm{Ma} < 0$ as well as $\mathrm{Ma} > 0$ for sufficiently large $|\mathrm{Ma}|$. By allowing k to vary, the minimum value of $|\mathrm{Ma}_N|$ with respect to k can be determined, and this value yields the critical Marangoni number $|\mathrm{Ma}_{\mathrm{cr}}|$. The corresponding value of wavenumber is the critical wavenumber, k_{cr}.

Even if $\lambda_r = 0$, λ_i can be nonzero, in which case it corresponds to a frequency, and the onset of convection then occurs in an oscillatory manner. If $\lambda_i = 0$, instability occurs in a monotonic manner for $|\mathrm{Ma}| > |\mathrm{Ma}_N|$. Oscillatory onset has been observed [12] for $k \sim 0(1)$ but not yet for $k \ll 1$, which is the case that we will focus on. At any rate, we will assume $\lambda_i = 0$ and concentrate in Sect. 3 on the control of monotonic instability. The governing ordinary differential equations are

$$\left(\frac{\mathrm{d}^2}{dz^2} - k^2\right)^2 W_j = \lambda\left(\frac{\nu^{1-j}}{\mathrm{Pr}}\right)\left(\frac{\mathrm{d}^2}{dz^2} - k^2\right) W_j,$$

and

$$\left(\frac{\mathrm{d}^2}{\mathrm{d}z^2} - k^2\right)\Theta_j = \lambda\kappa^{1-j}\Theta_j - (\kappa K)^{1-j} W_j,$$

to be solved subject at $z = 0$ to the interfacial conditions

$$W_1 = W_2 = \lambda\eta_0, \qquad \frac{\mathrm{d}W_1}{\mathrm{d}z} = \frac{\mathrm{d}W_2}{\mathrm{d}z},$$

$$\Theta_1 - \Theta_2 = \eta_0 \left(1 - \frac{1}{K}\right),$$

$$\frac{\mathrm{d}\Theta_1}{\mathrm{d}z} = K\frac{\mathrm{d}\Theta_2}{\mathrm{d}z},$$

$$\left(\frac{\mathrm{d}^2 W_1}{\mathrm{d}z^2} + k^2 W_1\right) - \mu\left(\frac{\mathrm{d}^2 W_2}{\mathrm{d}z^2} + k^2 W_2\right) = \mathrm{Ma}k^2\left(\eta_0 - \Theta_1\right), \tag{18}$$

and, after expressions for perturbation pressure are substituted into (17),

$$\left(\frac{\mathrm{d}^2}{\mathrm{d}z^2} - 3k^2 - \frac{\lambda}{\mathrm{Pr}}\right)\frac{\mathrm{d}W_1}{\mathrm{d}z} - \mu\left(\frac{\mathrm{d}^2}{\mathrm{d}z^2} - 3k^2 - \frac{\lambda}{\nu\mathrm{Pr}}\right)\frac{\mathrm{d}W_2}{\mathrm{d}z}$$
$$= \frac{k^2}{\mathrm{Cr}}\left(\mathrm{Bo} + k^2\right)\eta_0.$$

The solutions must also satisfy the wall boundary conditions, which are

$$W_1(-1) = \frac{\mathrm{d}W_1}{\mathrm{d}z}(-1) = W_2(d) = \frac{\mathrm{d}W_2}{\mathrm{d}z}(d) = 0,$$

$$\Theta_2(d) = 0 \quad \text{and} \quad \Theta_1(-1) = 0 \quad \text{or} \quad \frac{\mathrm{d}\Theta_1}{\mathrm{d}z}(-1) = 0.$$

If a three-dimensional disturbance had been considered, similar equations with $k^2 = k_x^2 + k_y^2$ would have been obtained due to horizontal isotropy. The general boundary value problem can be solved analytically for $\lambda = 0$ [29], giving rise to a very complicated eigenvalue relation which is usually evaluated numerically, although asymptotic forms are useful for insight. For moderate values of d and not too small values of (Bo/Cr), disturbances with $k \sim 0(1)$ tend to be the most unstable but, for $d \ll 1$ or $d \gg 1$ in a bounded region, large wavelength disturbances with $k \ll 1$ are most unstable, e.g. see Fig. 2 of [31] and Fig. 4a of [12]. We will concentrate on the case $k \ll 1$ in the next section and so obtain results here by expanding in terms of k^2, as follows:

$$\Theta_j = \Theta_{j0} + k^2\Theta_{j1} + \cdots \tag{19}$$

$$W_j = k^2 W_{j1} + k^4 W_{j2} + \cdots \tag{20}$$

$$\eta_0 = \eta_{00} + k^2\eta_{01} + \cdots \tag{21}$$

$$\mathrm{Ma} = \mathrm{Ma}_0 + k^2\mathrm{Ma}_1 + \cdots \tag{22}$$

$$\lambda = k^2\lambda_1 + k^4\lambda_2 + \cdots \tag{23}$$

The expansion for W_j follows from (18), given (19) and (21). For $k^2 \ll 1$, the disturbance velocity is very nearly parallel to the walls and so $W_j \ll 1$. From (23), $k^2 = 0$ corresponds to neutral stability; we wish to determine if $\lambda = 0$ also for some range of $k^2 > 0$. At lowest order in k^2, the governing equations are

$$\Theta''_{j0} = 0,$$

$$\Theta_{10}(0) - \Theta_{20}(0) = \eta_{00}\left(\frac{K-1}{K}\right), \tag{24-a}$$

$$\Theta'_{10}(0) = K\Theta'_{20}(0), \tag{24-b}$$

$$\Theta_{20}(d) = 0, \tag{25-a}$$

$$\Theta_{10}(-1) = 0 \quad \text{or} \quad \Theta'_{10}(-1) = 0, \tag{25-b}$$

where a prime denotes a derivative with respect to z. The choice in (25-b) is between a constant temperature wall condition and a constant heat flux condition. At this order, the problem consists of finding the thermal field associated with a small deflection of the interface. Polynomial solutions exist with the following form:

for $\Theta_{10}(-1) = 0$,

$$\theta_{10} = \eta_{00}\left(\frac{K-1}{K+d}\right)(1+z),$$

$$\theta_{20} = \eta_{00}\left(\frac{K-1}{K(K+d)}\right)(z-d),$$

and for $\theta'_{10}(-1) = 0$,

$$\Theta_{10} = \eta_{00}\left(\frac{K-1}{K}\right),$$

$$\Theta_{20} = 0.$$

With these solutions, we can show from either side of (10) that the disturbed interfacial temperature is 180° out of phase with η, i.e., it is cooler at a maximum of η, in accordance with the earlier discussion.

At $\mathrm{O}(k^2)$, we determine $W_{j1}(z)$ from

$$W''''_{j1} = 0, \tag{26}$$

$$W_{11}(-1) = \frac{\mathrm{d}W_{11}}{\mathrm{d}z}(-1) = W_{21}(d) = \frac{\mathrm{d}W_{21}}{\mathrm{d}z}(d) = 0, \tag{27}$$

$$W_{11}(0) = W_{21}(0) = \lambda_1\eta_{00}, \qquad W'_{11}(0) = W'_{21}(0), \tag{28}$$

$$W''_1(0) - \mu W''_2(0) = \mathrm{Ma}_0\left(\eta_{00} - \Theta_{10}(0)\right), \tag{29}$$

$$W_1'''(0) - \mu W_2'''(0) = \left(\frac{\text{Bo}}{\text{Cr}}\right)\eta_{00}. \tag{30}$$

The growth-rate λ_1 and the value of Ma_0 for neutral stability ($\lambda_1 = 0$), say $\text{Ma}_{0,N}$, are determined at this order. After generating polynomial solutions to (26) and applying the interfacial and wall boundary conditions (27–30), we obtain for the case of a constant temperature lower wall

$$\lambda_1\left\{4\mu(d+1)^2 + d^{-1}(d^2-\mu)^2\right\} = \frac{1}{2}\text{Ma}_0 d(d^2-\mu)\left(\frac{d+1}{d+K}\right) - \frac{1}{3}\left(\frac{\text{Bo}}{\text{Cr}}\right)d^2(d+\mu).$$

First consider the case when $\text{Bo} > 0$ so that both gravity and surface tension are stabilizing. It is then possible to have $\lambda_1 > 0$, i.e. instability, if Ma_0 and $(d^2 - \mu)$ have the same sign. Hence, if $d^2 > \mu$, instability can occur for heating from below (Ma_0 sufficiently positive), but, for $d^2 < \mu$, instability can occur only for heating from above (Ma_0 sufficiently negative). The value $\text{Ma}_{0,N}$ associated with $\lambda_1 = 0$ is

$$\text{Ma}_{0,N} = \frac{2}{3}\left(\frac{\text{Bo}}{\text{Cr}}\right)\left\{\frac{d(d+\mu)(d+K)}{(d+1)(d^2-\mu)}\right\}. \tag{31}$$

The corresponding result for the case of a constant heat flux condition at the lower wall is

$$\text{Ma}_{0,N} = \frac{2}{3}\left(\frac{\text{Bo}}{\text{Cr}}\right)\left\{\frac{d(d+\mu)K}{(d^2-\mu)}\right\}. \tag{32}$$

For situations when disturbances with $k \to 0$ become unstable first, (31) and (32) are then the critical values $\text{Ma}_{0,\text{cr}}$. For both cases, $\text{Ma}_{0,N} \to -\infty$ as $d^2 \to \mu$ from below, and $\text{Ma}_{0,N} \to +\infty$ as $d^2 \to \mu$ from above. Hence, $\text{Ma}_{0,N}$ is a discontinuous function of d due to the fact that it cannot change sign by going through zero (for $\text{Ma} = 0$, only decaying solutions exist with $\text{Bo} > 0$). For an air-water interface at a reference temperature $T_0^* = 25°\text{C}$, $\mu \cong 0.0201$, and so the change in sign occurs when $d \cong 0.142$, a value which would seem to be achievable in an experiment. However, the phenomena has apparently not yet been observed. It can be seen from the dimensional form of the tangential stress condition (12) or (13) that a change in sign of $\text{Ma}_{0,N}$ is expected as d^2 passes the value of μ. If we had made z^* nondimensional in layer 2 on the basis of d_2^*, as appropriate for $d_2^* \ll d_1^*$, the ratio (μ/d^2) would appear on the left-hand side of (13). For $(\mu/d^2) \ll 1$, the circulation in the lower layer is just as described in connection with Fig. 1 with $\text{Ma} > \text{Ma}_{0,N} > 0$, i.e., a clockwise circulation is generated in the lower layer with flow along the interface from the trough to the crest. In particular, hot fluid is being convected upwards from below near the trough, which is destabilizing. Due to coupling of the layers via the shear stress, a counterclockwise circulation occurs in the upper layer which brings cold fluid downwards in the vicinity of the trough which tends to be stabilizing. If this effect is sufficiently strong, i.e., large values of μ/d^2, there is no instability for heating from below, but instability can occur if the direction of heating is switched (see also [11]).

Because $\mathrm{Ma}_{0,N} \sim (\mathrm{Bo}/\mathrm{Cr})$, it will tend to be less than the corresponding value $\mathrm{Ma}_N(k)$ for finite wavenumbers if (Bo/Cr) is relatively small, i.e., the Galileo number is small. As an example, consider the case of an air-water interface when $d \to \infty$. For finite wavenumbers, the minimum value of $\mathrm{Ma}_N(k)$ is not affected significantly by surface deformation and is approximately equal to 80 for a constant temperature wall [22]. Comparing this result to that obtained from (2.62) as $d \to \infty$, we see that large wavelength disturbances will indeed be the critical disturbances if $(\mathrm{Bo}/\mathrm{Cr}) < 120$. From the definition (2.34) of Ga, it seems that this will be true when the liquid layer is very thin [31], or for a microgravity condition, or when the densities of the two fluids are almost identical.

The ratio of (31) to (32) is equal to $(d+K)/(d+1)K$ and so tends to unity as $d \to 0$ and to K^{-1} as $d \to \infty$. Hence, the two predicted values of $\mathrm{Ma}_{0,N}$ differ the most for large depth ratios.

For $\mathrm{Bo} < 0$, corresponding to gravity acting upwards, the term involving Bo is destabilizing and is associated with Rayleigh–Taylor instability [3]. As pointed out by Burgess et al. [2], the large wavelength disturbances can then be stabilized for $d^2 > \mu$ by cooling from below (heating from above) so $\mathrm{Ma}_0 < 0$, whereas, for $d^2 < \mu$, heating from below promotes stability. This is an example of using the thermocapillary effect as a passive control for Rayleigh–Taylor instability.

The above results hold for $k^2 \to 0$, and higher order terms in the expansion (19–23) are required to determine how Ma_N depends on k. The result to $0(k^2)$ for the case of an isothermal lower wall is given in [28] as the ratio of two polynomials:

$$\mathrm{Ma}_N = \frac{240\kappa d(\mathrm{Bo}+k^2)(d+K)(d+\mu)}{360\ \mathrm{Cr}\ \kappa(d+1)(d^2-\mu)+3k^2\mathrm{Bo}\ d^3(\kappa-d^2)}, \tag{33}$$

which reduces to (31) as $k^2 \to 0$. The corresponding result for the fixed heat flux surface at $z=-1$ has been obtained for the author by L.S. de B. Alves as

$$\mathrm{Ma}_N = \frac{240\kappa dK(\mathrm{Bo}+k^2)(d+\mu)}{360\mathrm{Cr}\ \kappa(d^2-\mu)+k^2\mathrm{Bo}d^3(5\kappa-3d^2)}. \tag{34}$$

In both (33) and (34), the terms of $\mathrm{O}(k^2)$ are formally of higher order than the leading terms. However, $\mathrm{Cr} \ll 1$ usually, and the results suggest that the denominator can then have a zero if $d^2 < \mu$ and $d^2 < \kappa$ in (33), or $d^2 < 5\kappa/3$ in (34), for a small but nonzero value of k^2, say, k_d^2, given by

$$k_d^2 = 120\left(\frac{\mathrm{Cr}}{\mathrm{Bo}}\right)\kappa(d+1)(\mu-d^2)/d^3(\kappa-d^2) \tag{35}$$

for the case of an isothermal wall and

$$k_d^2 = 360\left(\frac{\mathrm{Cr}}{\mathrm{Bo}}\right)\kappa(\mu-d^2)/d^3(5\kappa-3d^2) \tag{36}$$

for a constant heat-flux condition. Say that $d^2 < \mu$ so that $\mathrm{Ma}_{0,N} < 0$ and further that $\kappa > d^2$ in (35) or $\kappa > \frac{3}{5}d^2$ in (34) so that real values of k_d exist.

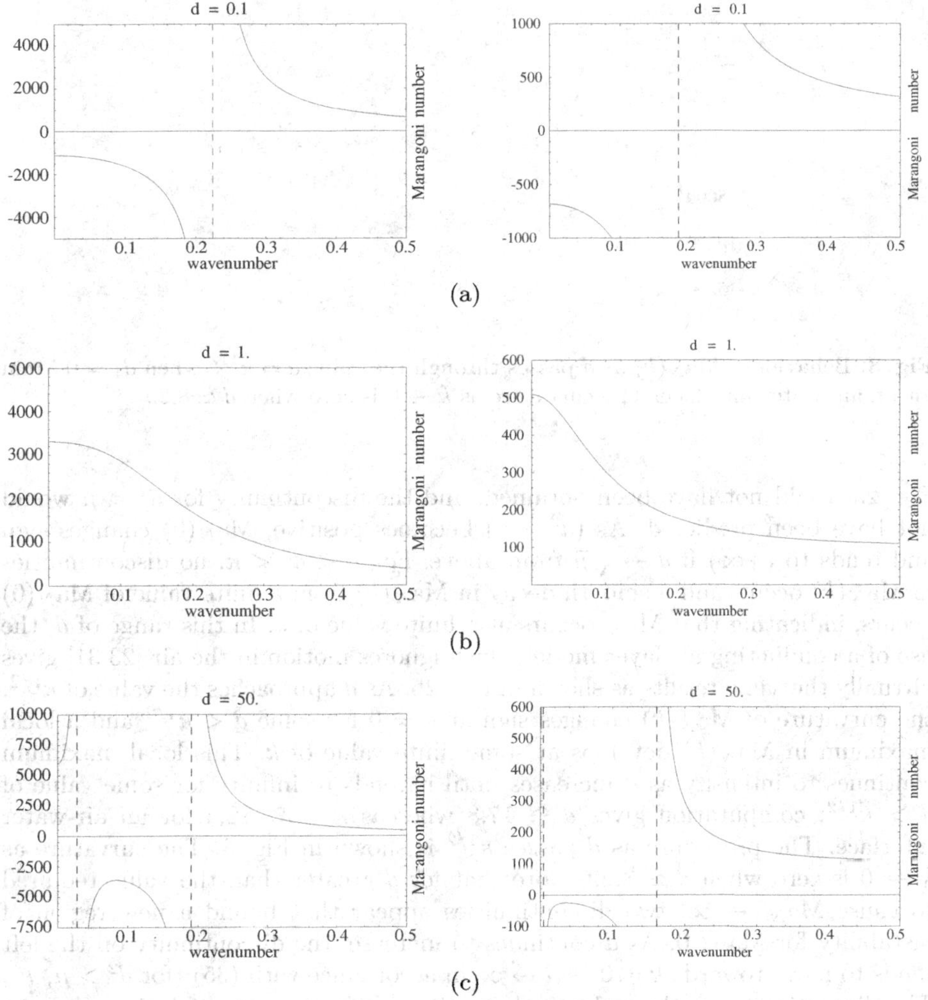

Fig. 2. Marginal stability curves for air over water for prescribed temperature and heat flux at the bottom rigid surface for $Cr = 3.689\,10^{-6}$ and $Bo = 3.38\,10^{-2}$ ($d_1 = 0.5\,mm$)

A plot of $\mathrm{Ma}_N(k)$ for the air-water case with $d = 0.1$ is shown in Fig. 2a, using results obtained numerically by L. Alves from the exact eigenvalue relation for arbitrary k (e.g., Eq. (3.209) of [4]) with $\mathrm{Cr} = 3.689 \times 10^{-6}$, $\mathrm{Bo} = 3.38 \times 10^{-2}$ and $d_1 = 0.5\,\mathrm{mm}$. As shown in Fig. 2a for a constant temperature bottom surface, a discontinuity in $\mathrm{Ma}_N(k)$ occurs at a fixed value of k, across which Ma_N changes sign. As d decreases, the discontinuity moves to the right so that a wider range of $\mathrm{Ma}_N < 0$ occurs. As $d \to \mu^{1/2}$, the discontinuity moves toward $k = 0$ according to (35), and $\mathrm{Ma}_N(0) \to -\infty$. It is important to note that, if we had ignored motion in the air on the basis that $\mu \ll 1$, the solutions with $\mathrm{Ma}_N < 0$ in

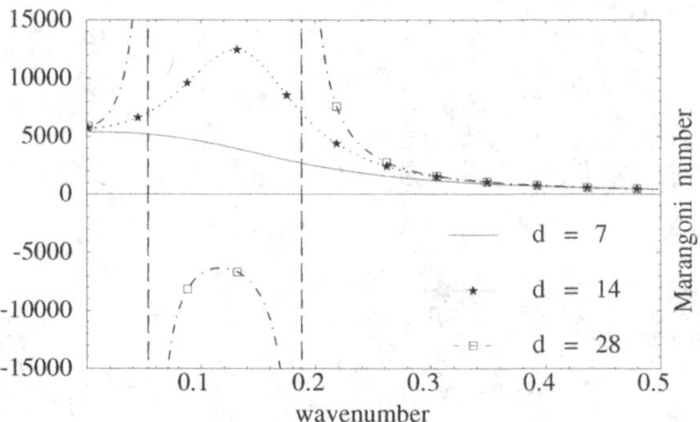

Fig. 3. Behavior of $\mathrm{Ma}_N(k)$ as d passes through the value $d = \kappa^{1/2}$ when $d_1 = 0.5\,\mathrm{mm}$ for an air–water interface. The curvature as $k \to 0$ is zero when $d \simeq 8.29$

Fig. 2a would not have been obtained, and the discontinuity for $d^2 < \mu$ would not have been predicted. As $(d^2 - \mu)$ becomes positive, $\mathrm{Ma}_N(0)$ changes sign and tends to $(+\infty)$ if $d \to \sqrt{\mu}$ from above. For $\mu < d^2 < \kappa$, no discontinuities in $\mathrm{Ma}_N(k)$ occur, and a smooth decay in $\mathrm{Ma}_N(k)$ from a finite value of $\mathrm{Ma}_N(0)$ occurs, indicating that Ma_{cr} occurs at a finite value of k. In this range of d, the use of a conducting air layer model which ignores motion in the air [23,31] gives virtually the same results as shown in Fig. 2b. As d approaches the value of $\kappa^{1/2}$, the curvature of $\mathrm{Ma}_N(k)$ changes sign at $k = 0$ for some $d < \kappa^{1/2}$ and a local maximum in $\mathrm{Ma}_N(k)$ develops at some finite value of k. This local maximum continues to intensify as d increases until it tends to infinity for some value of $d > \kappa^{1/2}$; computation gives $d \cong 17.8$ whereas $\kappa^{1/2} \cong 12.4$ for an air-water interface. The procedure as d passes $\kappa^{1/2}$ is shown in Fig. 3. The curvature as $k \to 0$ is zero when $d \simeq 8.29$. Note that for d greater than the value required to cause $\mathrm{Ma}_N \to \infty$, two discontinuities appear that bound a new region of instability for $\mathrm{Ma} < 0$. As d continues to increase, the discontinuity on the left tends to move towards $k = 0$ as $d \to \infty$ in accordance with (35) (for $d^2 > \mu$, κ). The discontinuity on the right tends to an asymptotic value of k that must be calculated. The situation for $d = 50$ is shown in Fig. 2c, where the region of instability for $\mathrm{Ma} < 0$ is evident. Once a region with $\mathrm{Ma} < 0$ occurs for nonzero values of k, the need for the left discontinuity is obvious. The right discontinuity is required when instability occurs at large values of k for $\mathrm{Ma} > 0$, such as for the air-water interface.

Although the conducting gas layer model [23,31] does not predict the discontinuity occurring for $d^2 < \mu$, it does predict accurately the discontinuities shown in Fig. 2c for $d^2 > \kappa$ as long as $d^2 \gg \mu$. Hence, its use is restricted for a gas-liquid interface only by this condition. In contrast, the traditional one-layer liquid layer bounded by a passive gas [6] is restricted also by the condition $d^2 \ll \kappa$ because this model does not allow for any discontinuities in $\mathrm{Ma}_N(k)$.

The fact that a corridor of stability for a range of k exists in Fig. 2c if $\mathrm{Ma} > 0$ suggests that the oscillatory onset of instability should be considered in this range. For the conducting air layer model, however, Regnier et al. [24] concluded that oscillatory solutions do not exist for the gas-liquid interface if the restrictions imposed by the Boussinesq approximation are imposed. Nonetheless, for more general fluids, oscillatory instability can occur within this corridor of stability for monotonic disturbances. For instance, for the case of the liquids n-hexane and acetonitrile used in the experiment of Juel et al. [12], Engel and Swift [5] found oscillatory solutions to be possible in a similar region [see their Fig. 4]. The predicted frequencies and wavenumbers were found to be in satisfactory agreement with the experimentally observed values.

Before going on, it should be mentioned that the possibility of discontinuities in $\mathrm{Ma}_N(k)$ are discussed on the basis of the large wavelength results in both [4] and [28]. The results presented here in Figs. 2 and 3 were obtained numerically on the basis of the full eigenvalue relation and include novel features such as Fig. 3 that have not previously been discussed. The results help to define more precisely the range of validity for the passive and conducting gas-layer models.

3 Active Control of the Two-Fluid Bénard–Marangoni Instability

We will now discuss the possibility of using active control to stabilize BM convection in a two-fluid system. It will become obvious how modification of the control could be used instead to destabilize the fluid system, if desired. In accordance with the discussion of Sect. 1, actuators and sensors need to be chosen. We will build on the results of Sect. 2 and consider either the lower wall temperature or its heat flux to be controllable on a spatially continuous basis, in a manner similar to the active thermal control used experimentally for RB convection [9,10,30]. Different kinds of sensors could be used, and the ultimate choice will presumably be based on the design of a particular experiment. Attention should be given to the choice, however, because the degree of stabilization can depend on the particular choice of sensor and actuator. In [13], the author argued that sensing of the deflection of the interface is superior to sensing of the temperature of the interface as far as the degree of stabilization is concerned, at least in regard to stabilizing the large wavelength instability (when finite wavelength disturbances are the most unstable ones, deflection of the interface is usually slight, and so sensing of the interfacial temperature is preferable). We will assume that the deflection of the interface can be measured in a continuous manner in order to extend the linear results of [17] to the two-fluid case. However, the choice is not unique, and the continuous measurement of $\eta(x,t)$ is not necessarily required. For a small apparatus, perhaps measurement at a few discrete points might be possible (as long as they are not nodal points). For a large apparatus, however, the initial disturbances are likely to be localized, and so more continuous sampling would be desirable.

We therefore assume that the amplitude η_0 can be measured, and we will relate it to the wall thermal actuator via a control law. At this point, the plentiful results of modern control theory can be used, as has been done for the case of RB convection [18]. Our aim here is merely to extend the results obtained in [17] by means of a linear proportional control law to the two-fluid case, rather than to solve the optimal control problem. For the case of a controlled temperature at the lower wall, we define

$$\Theta_1(-1,k) = \Theta_c(k) = \Theta_{c0} + k^2\Theta_{c1} + \cdots \tag{37}$$

where Θ_{c0}, Θ_{c1}, etc. are to be assigned values that arise from the analysis in order to achieve a certain degree of stabilization. In the present analysis, we will determine only Θ_{c0} but it is important to realize that Θ_c depends in general on k, due to the fact that $\mathrm{Ma}_N = \mathrm{Ma}_N(k)$. Our control law will be

$$\Theta_c(k) = \Lambda(k)\eta_0(k) \tag{38}$$

so that, with $\Lambda > 0$, heating will occur at regions where $\eta > 0$, i.e., where cooling of the interface occurs, and vice versa for $\eta < 0$. The control temperature therefore tends to weaken the mechanism of instability. The function $\Lambda(k)$ is called the gain, and it too can be expanded as

$$\Lambda(k) = \Lambda_0 + k^2\Lambda_1 + \cdots . \tag{39}$$

For the case of heat flux control, the control law is taken as

$$\Theta_c'(k) = -\Gamma(k)\eta_0(k), \tag{40}$$

and so, for $\Gamma > 0$, the heat transfer is increased in regions where $\eta > 0$. We also expand Θ_c' and $\Gamma(k)$ in terms of k, in a manner similar to (37) and (39).

At lowest order, we solve again for $\Theta_{10}(z)$ but with either the boundary condition

$$\Theta_{10}(-1) = \Theta_{c0} = \Lambda_0\eta_{00}$$

or

$$\Theta_{10}'(-1) = \Theta_{c0}' = -\Gamma_0\eta_{00}.$$

The other boundary and interfacial conditions are the same as (24-a,24-b; 25-a). For the case of a controlled temperature at the lower wall, we obtain

$$\Theta_{10}(z) = \Lambda_0\eta_{00}\left\{1 - \left(\frac{K}{K+d}(1+z)\right)\right\} + \eta_{00}\left(\frac{K-1}{K+d}\right)(1+z) \tag{41}$$

and

$$\Theta_{20}(z) = \eta_{00}\left\{\frac{\Lambda_0}{K+d} - \frac{K-1}{K(K+d)}\right\}(d-z). \tag{42}$$

For the case of a controlled heat flux at the lower wall, the results are

$$\Theta_{10}(z) = \eta_{00}\left\{-\Gamma_0\left(z - \frac{d}{K}\right) + \frac{K-1}{K}\right\} \tag{43}$$

and

$$\Theta_{20}(z) = \frac{\eta_{00}\Gamma_0}{K}(d-z). \tag{44}$$

Proceeding to the solution for $W_{j1}(z)$, which is affected only by the revised value of $\Theta_{10}(0)$ in (29), the result for $\mathrm{Ma}_{0,N}$ is

$$\mathrm{Ma}_{0,N} = \frac{2\ \mathrm{Bo}}{3\ \mathrm{Cr}} \left\{ \frac{d(d+\mu)(d+K)}{(d+1-\Lambda_0 d)(d^2-\mu)} \right\} \tag{45}$$

for the case of a controlled wall temperature, and

$$\mathrm{Ma}_{0,N} = \frac{2\ \mathrm{Bo}}{3\ \mathrm{Cr}} \left(\frac{K}{1-\Gamma_0 d} \right) \left\{ \frac{d(d+\mu)}{d^2-\mu} \right\} \tag{46}$$

for a controlled heat flux at $z=-1$. These results can be compared to (31) and (32), respectively. As $\Lambda_0 \to (d+1)/d$ or $\Gamma_0 \to d^{-1}$ from below, $\mathrm{Ma}_{0,N} \to (+\infty)$ for both cases if $d^2 > \mu$ and $\mathrm{Ma}_{0,N} \to (-\infty)$ if $d^2 < \mu$. Hence, significant stabilization can be realized in principle for nonzero Λ_0 or Γ_0. The mechanism for stabilization can be seen from examination of the nondimensional temperature of the displaced interface, given by (16) or, at lowest order, by (24-a) as

$$-\eta_{00} + \Theta_{10}(0) = -\eta_{00} \left(\frac{d+1-\Lambda_0 d}{K+d} \right)$$

for the case of temperature control after use of (41), or as

$$-\eta_{00} + \Theta_{10}(0) = -\eta_{00} \left(\frac{1-\Lambda_0 d}{K} \right)$$

for the case of heat flux control, after use of (43). As expected, heating from below diminishes the amount of cooling due to a positive displacement. Eventually, a condition is reached when the two effects cancel and $|\mathrm{Ma}_{0,N}|$ then goes to infinity. For still larger values of Λ_0 or Γ_0, instability can be induced only by changing the direction of heating, similar to the change associated with the sign of $(d^2-\mu)$. The change in sign of $\mathrm{Ma}_{0,N}$ again occurs on a discontinuous basis as Λ_0 exceeds $(d+1)/d$ or Γ_c exceeds d^{-1}, as shown in Fig. 4 for the case of temperature control with $d=1$.

Due to the fact that only thermal control is used and that the principal effect concerns the interfacial temperature, no alteration in the condition governing the change in sign of $\mathrm{Ma}_{0,N}$ at $d=\mu^{1/2}$ occurs. As discussed earlier, the change in sign is controlled by the shear stress and, because we are considering a Boussinesq fluid, it is not affected directly by thermal control at this order of k^2.

For the case when gravity exerts a destabilizing effect on the interface, i.e., Bo<0, it has already been mentioned that cooling the wall at $z=-1$ can be used as a passive control to stabilize the interface (Ma<0 for $d^2>\mu$). The results given by (45) and (46) suggest that the amount of cooling required can be reduced if we use active control and switch the signs of Λ_0 and Γ_0. In such a situation, the active control would augment the passive control.

4 Discussion

The above results hold, of course, only for disturbances with $k \to 0$. Even for situations where such disturbances are critical for the case without control, large

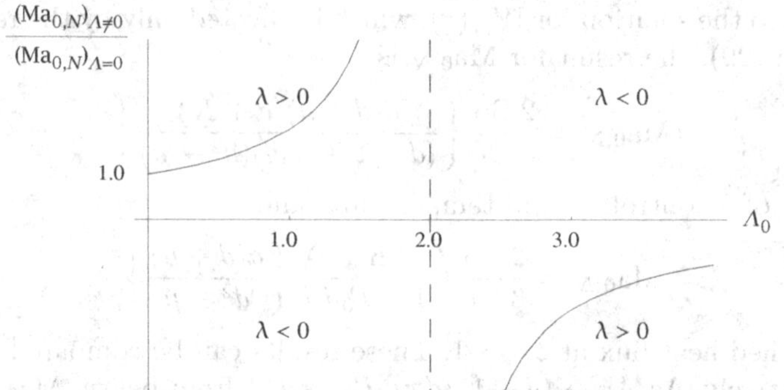

Fig. 4. The value of $|Ma_{0,N}|$ with control relative to the value without control as a function of the lowest order gain Λ_0 for the case of a controlled temperature at the lower wall and $d = 1$

values of $|Ma_{0,N}|$ with control suggest that finite wavelength disturbances should be considered in order to determine Ma_{cr}.

Only the effect of control at the lowest order in k^2 has been determined in the above analysis, which concerns mainly alteration of the interfacial temperature. Significant change in the value of $Ma_N(0)$ was found to be possible, and the reason for the change was discussed. At the next order, the effects of control upon the velocity field and the higher order thermal field would be determined in order to predict further change of Ma_N for $k > 0$. For instance, the possible use of control to shift the discontinuities in $Ma_N(k)$ occurring for finite k, as discussed in Sect. 2, could then be investigated, as well as to influence the bandwidth of unstable disturbances. Information about $\Lambda(k)$ and $\Gamma(k)$ can be used, via the inverse Fourier transform, to say how the control should be distributed in physical space. As long as large wavelength disturbances are of primary concern, an alternate approach is to make use of the slowly varying form of the governing equations based on lubrication theory [21]. Besides allowing the problem to be discussed directly in physical space, these equations allow for discussion of nonlinearity, as associated with the finite amplitude distortion of the interface and subcritical instability. This approach has been used in [17] for a a conducting air layer model in order to explore how control might prevent rupture of a liquid film, as well as to indicate how subcritical instability might be eliminated so that supercritical bifurcation occurs instead.

We now review briefly some recent work concerning active control of interfacial instabilities. An extensive theoretical investigation of the active control of evaporatively driven instabilities in thin, single liquid films has recently been published [7]. Thermocapillary effects are considered along with the destabilizing effects of differential evaporation and vapor recoil. As in the above discussion, knowledge of the thickness of the film is assumed to be known via sensing of some type. However, the heating of the film is achieved internally by absorption of electromagnetic radiation directed at the liquid surface. The radiation consists of a mean part and a variable part that is related to the varying film thickness by a linear control law. Absorption at a solid substrate as well as within the liquid is considered. Stabilization seems to be achieved by exploiting the thermocapillary effect, just as in the above discussion. However, more sophisticated feedback is discussed, such as the use of a time-varying gain.

Active control with feedback has also been used in an experiment [27] concerning another type of thermocapillary convection that is relevant to the growth of single crystals. Liquid with a free surface is contained in a circular annulus. The inner wall of the annulus is heated relative to the outer one, so that a mean thermocapillary flow is generated. Such a mean flow can be expected whenever a component of heating along an interface exists. In the investigation reported in [27], the radial surface flow is toward the outer cylinder. At a critical value of Ma, this convection state develops a time dependence, due to the onset of waves propagating around the annulus. Two cold-wire temperature sensors situated 60° apart were used to detect the oscillations. Local heating of the surface was achieved by a single actuator placed slightly above the liquid's surface whose output is related to the temperature sensed by means of a linear control law. Emphasis was placed on reducing the amplitude of the oscillations in the supercritical regime rather than determining the stability boundary. Suppression was more effective in the weakly nonlinear regime than strongly nonlinear, but an optimal range for the gain was obtained for both regimes. It should be emphasized that suppression was achieved here by use of a relatively simple system partly because a well-defined fundamental wave existed without control.

Another study published since the earlier review by the author [13] concerns suppression of the morphological instability of the melt/crystal interface during directional growth of a binary crystal from its melt [25]. External heating of the melt is assumed to occur near the interface, with an intensity related to the deformation of the interface. On the basis of this linear stability analysis, it would seem that significant stabilization of the interface is possible. As the authors of [25] point out, delay times associated with the actuators might affect their results, and a similar statement could be made of the other theoretical work discussed in this article.

More experimental efforts in the field of active control of interfacial instabilities are certainly desirable, along with closer collaboration between theorists and experimentalists.

Acknowledgements

It is a pleasure to acknowledge past support of the author's research on Bénard-Marangoni convection by the NASA Microgravity Fluid Physics Program. The author is indebted to Leonardo Alves for the numerical computations reported in Sect. 2 and to Arthur Or for collaborating on earlier research on this topic.

References

1. H.H. Bau: Intl. J. Heat Mass Transfer **42**, 1327 (1999)
2. J.M. Burgess, A. Juel, W.D. McCormick, J.B. Swift, H.L. Swinney: Phys. Rev. Lett. **86**, 1203 (2001)
3. S. Chandrasekhar: *Hydrodynamic and Hydromagnetic Instability*, (Clarendon Press, Oxford 1961)
4. P. Colinet, J.C. Legros, M.G. Velarde: *Nonlinear Dynamics of Surface-Tension-Driven Instabilities*, (Wiley-VCH, Berlin 2001)
5. A. Engel, J.B. Swift: Phys. Rev. E **62**, 6540 (2000)
6. G. Gousbet, J. Maguet, C. Roze, R. Darrigo: Phys. Fluids A **2**, 903 (1990)
7. R.O. Grigoriev: Phys. Fluids **14**, 1895 (2002)
8. D.A. Goussis, R.E. Kelly: J. Fluid Mech. **223**, 25 (1991)
9. L.E. Howle: Phys. Fluids **9**, 1861 (1997)
10. L.E. Howle: J. Fluid Mech. **411**, 39 (2000)
11. D. Johnson, R. Narayanan: Chaos **9**, 124 (1999)
12. A. Juel, J.M. Burgess, W.D. McCormick, J.B. Swift, H.L. Swinney: Physica D **143**, 169 (2000)
13. R.E. Kelly: 'Thermocapillary Control with Feedback of Large Wavelength Interfacial Instabilities'. In: *Interfaces for the 21st Century*, ed. by M.K. Smith et al., (Imperial College Press, London 2002)
14. S.P. Lin, L.N. Chen, D.R. Woods: Phys. Fluids **8**, 3247 (1996)
15. T.E. Morthland, J.S. Walker: J. Fluid Mech. **382**, 87 (1999)
16. A.A. Nepomnyashchy, M.G. Velarde, P. Colinet: *Interfacial Phenomena and Convection*, (Chapman and Hall/CRC, Boca Ratón, FL 2002)
17. A.C. Or, R.E. Kelly, L. Cortelezzi, J. L. Speyer: J. Fluid Mech. **387**, 321 (1999)
18. A.C. Or, R.E. Kelly: J. Fluid Mech. **440**, 27 (2001)
19. A.C. Or, L. Cortelezzi, J. L. Speyer: J. Fluid Mech. **437**, 175 (2001)
20. A.C. Or, R.E. Kelly: J. Fluid Mech. **456**, 161 (2002)
21. A. Oron, S.H. Davis, S.G. Bankoff: Rev. Mod. Phys. **69**, 931 (1997)
22. J.R.A. Pearson: J. Fluid Mech. **4**, 489 (1958)
23. C. Pérez-García, B. Echebarría, M. Bestehorn: Phys. Rev. E **57**, 475 (1998)
24. C. Regnier, P.C. Dauby, G. Lebon: Phys. Fluids **12**, 2787 (2000)
25. T.V. Savina, A.A. Nepomnyashchy, S. Brandon, D. R. Lewin, A. A. Golovin: J. Crys. Growth **240**, 292 (2002)
26. M.F. Schatz, G.P. Neitzel: Ann. Rev. Fluid Mech. **33**, 93 (2001)
27. J. Shiomi, G. Amberg, H. Alfredsson: Phys. Rev. E **64**, 031205 (2001)
28. I.B. Simanovskii, A.A. Nepomnyashchy: *Convective Instabilities in Systems with Interface*, (Gordon and Breach Science Publ., Langhorne, PA 1993)
29. K.A. Smith: J. Fluid Mech. **24**, 401 (1966)
30. J. Tang, H.H. Bau: J. Fluid Mech. **363**, 153 (1998)
31. S.J. VanHook, M.F. Schatz, J.B. Swift, W.D. McCormick, H.L. Swinney: J. Fluid Mech. **345**, 45 (1997)
32. M.G. Velarde, A.A. Nepomnyashchy, M. Hennenberg: Adv. Appl. Mech. **37**, 167 (2001)

Convective Instabilities in Layered Systems

A.A. Nepomnyashchy[1], I.B. Simanovskii[1], T. Boeck[2], A.A. Golovin[3], L.M. Braverman[4], and A. Thess[5]

[1] Department of Mathematics, Technion – Israel Institute of Technology, 32000 Haifa, Israel
[2] Laboratoire de Modélisation en Mécanique, Université Pierre et Marie Curie, 8 rue du Capitaine Scott, 75015 Paris, France
[3] Department of Engineering Sciences and Applied Mathematics, Northwestern University, Evanston, IL 60208–3100, USA
[4] Department of Computer Engineering, Karmiel ORT College, Karmiel, Israel
[5] Dept. of Mechanical Engineering, Ilmenau University of Technology, P.O. Box 100565, 98684 Ilmenau, Germany

Abstract. We discuss intrinsic multi-layer convection phenomena which cannot be explained in the framework of the one-layer approach. The physical mechanisms of instabilities are described, and appropriate mathematical models are presented. A special attention is payed to the problem of self-consistency of the mathematical model based on the Boussinesq approximation, which takes into account the deformations of the interface. We consider numerous types of oscillatory Marangoni instabilities, instabilities under a combined action of buoyancy and thermocapillary effect, and anticonvection. Both two-layer systems and three-layer systems are studied.

1 Introduction

The most typical approach to the investigation of Marangoni-Bénard convection and thermocapillary flows is the one-layer approach. When it is applied, the full problem for the fluid motion and for the heat/mass transfer is formulated only in the liquid phase, while the influence of the gas phase is described in a phenomenological way by means of the Biot number.

However, there exist intrinsic two-layer phenomena which are caused by processes in fluids on both sides of the interface. These phenomena cannot be understood without an analysis of the interfacial hydrodynamic and thermal interaction.

As the first example, let us mention the onset of Marangoni convection in a liquid-gas system. The one-layer approach predicts the monotonic Marangoni instability only for heating from the side of the liquid [47]. The two-layer approach reveals the appearance of the monotonic Marangoni instability for both ways of heating, depending on the ratio of layers thicknesses [56].

The two-layer approach can be unavoidable also in the case of the buoyancy convection. It can be found in textbooks [14,27] that the buoyancy-driven convective instability of a mechanical equilibrium state appears only when the heating is from below (in the case of fluids with positive heat expansion coefficients). Strange as it may seem, it is not correct in the presence of an interface between

A.A. Nepomnyashchy et al., Convective Instabilities in Layered Systems, Lect. Notes Phys. **628**, 1–44 (2003)
http://www.springerlink.com/

two fluids. An amazing phenomenon, which is called "anticonvection", has been predicted for some specific two-fluid systems heated from above [15,51,62].

It is known that the stability problem for the mechanical equilibrium state in a system with an interface is not self-adjoint (see, e.g., [52]), thus an oscillatory instability is possible. However, the one-layer approach is unable to reveal several oscillatory instabilities in systems with a non-deformable interface. The first example of the Marangoni oscillatory instability was found by Sternling and Scriven [57] in a two-fluid system (see also [33,48]). The hydrodynamic and thermal interaction between convective motions on both sides of the interface can produce oscillations also in the case of Rayleigh convection [16,45]. A specific kind of oscillations caused by the competition of buoyancy and thermocapillary effect can appear in a two-layer system in the case when the buoyancy convection is excited in the top layer [17].

In the present paper we consider several important types of convective instabilities typical of layered systems. We shall first discuss the basic physical mechanisms of instabilities in a heated fluid layer with a free surface. In Sect. 3 we present basic mathematical models. Section 4 is devoted to the description of some phenomena that occur in two-layer systems: oscillatory Marangoni instability, oscillatory instability due to the combined action of buoyancy and thermocapillary effect, and anticonvection. Effects specific to three-layer systems are discussed in the final Sect. 5.

2 Mechanisms of Convective Instabilities in a Single Layer

We consider a horizontal liquid layer between a hot rigid plate and a cold free surface; the total temperature drop across the layer is Θ. Density, kinematic and dynamic viscosity, heat conductivity, thermal diffusivity and thermal expansion coefficient of the liquid are respectively ρ_1, ν_1, η_1, κ_1, χ_1 and β_1; the mean thickness of the layer is a_1. As the units of length, time, velocity, pressure and temperature we use a_1, a_1^2/ν_1, ν_1/a_1, $\rho_1\nu_1^2/a_1^2$ and Θ, respectively. When the temperature difference Θ is sufficiently small, there is no convection in the fluid. The system keeps the *conductive regime* (mechanical equilibrium state) which is characterized by a constant vertical temperature gradient. When Θ is larger than a certain critical value, this regime becomes *unstable* and the convective motion appears in the fluid.

In the present paper we shall take into account two basic physical effects that produce the convective instability: buoyancy and thermocapillary effect.

The buoyancy is considered typically in the framework of the Boussinesq approximation. It means that the buoyancy force is taken into account in the equation of motion, but the thermal expansion is neglected in the continuity equation. This approximation is based on the assumption that the relative density change, $\varepsilon_\beta \sim \beta_1\Theta \ll 1$, while the non-dimensional parameter characterizing the generation of convection by the buoyancy force, the *Rayleigh number* $R = g\beta_1\Theta a_1^3/\nu_1\chi_1$ (g is acceleration of the gravity), certainly cannot be neglected in studying the convective phenomena, i.e. $R = O(1)$. Also, the thermophysical parameters of the liquid are assumed to be constant.

It is tempting to take into account the deformation h of the free surface caused by the convective flow. This deformation is determined by the balance of the gravity force and the normal stresses on the free surface:

$$\rho_1 g h \sim p - \eta_1 \partial v_n / \partial n,$$

where p is the pressure and v_n is the normal component of velocity. Using the units defined above, we find that the *non-dimensional* deformation of the free surface is proportional to $1/Ga$, where $Ga = ga_1^3/\nu_1^2$ is the *Galileo number*. Because $1/Ga = \varepsilon_\beta P/R$, where $P = \nu_1/\chi_1$ is the *Prandtl number*, we come to the conclusion that the deformation of the free surface caused by convection is a *non-Boussinesq effect*. If this effect is included into consideration while the corrections of the same order $O(\varepsilon_\beta)$ in the continuity equation are neglected, that may lead to *erroneous* results (see [61]).

Another physical effect that may produce convection in a layer with free surface is the thermocapillary effect, i.e. the dependence of the surface tension σ on the temperature T. Typically, the surface tension decreases when the temperature grows (the *normal* thermocapillary effect; $\alpha = -d\sigma/dT > 0$). The *anomalous* thermocapillary effect ($\alpha < 0$) was observed in aqueous alcohol solutions, nematic liquid crystals, binary metallic alloys etc. (see, e.g., [28] and references therein) as well as in some liquid-liquid systems such as 10cS silicone oil–ethylene glycol [3]. The characteristic non-dimensional parameter governing the appearance of the thermocapillary convection is the *Marangoni* number, $M = \alpha \Theta a_1/\eta_1 \chi_1$; $M > 0$ for the normal thermocapillary effect, and $M < 0$ for the anomalous thermocapillary effect.

The ratio of the Rayleigh number to the Marangoni number can be written as $R/M = (a_1/a_c)^2$, where $a_c^2 = \alpha/g\beta_1\rho_1$. In "thick" layers, $a \gg a_c$, the buoyancy mechanism of instability prevails, and the thermocapillary effect can be neglected. In "thin" layers, $a \ll a_c$, the thermocapillary effect plays the dominant role, and the buoyancy is not important.

3 Mathematical Models

We shall present the full mathematical model governing convection in a system of two horizontal layers of immiscible fluids with different physical properties (see Fig. 1a). The system is bounded from above and from below by two rigid plates kept at constant different temperatures (the total temperature drop is Θ). The variables corresponding to the top layer are marked by subscript 1, the variables corresponding to the bottom layer are marked by subscript 2. Density, kinematic and dynamic viscosity, heat conductivity, thermal diffusivity, heat expansion coefficient of the m-th fluid are respectively ρ_m, ν_m, η_m, κ_m, χ_m and β_m; a_m is the thickness of the m-th layer ($m = 1, 2$). The surface tension coefficient σ is a linear function of the temperature T: $\sigma = \sigma_0 - \alpha T$.

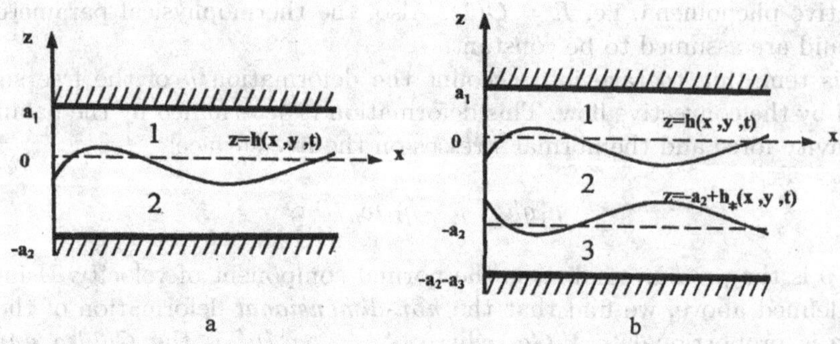

Fig. 1. Basic configuration: **a** two-layer system; **b** three-layer system

Let us introduce the following notation:

$$\rho = \rho_1/\rho_2,\ \nu = \nu_1/\nu_2,\ \eta = \eta_1/\eta_2,\ \kappa = \kappa_1/\kappa_2,\ \chi = \chi_1/\chi_2,\ \beta = \beta_1/\beta_2,\ a = a_2/a_1.$$

As the units of length, time, velocity, pressure and temperature we use the parameters of the top layer: a_1, a_1^2/ν_1, ν_1/a_1, $\rho_1\nu_1^2/a_1^2$ and Θ.

The nonlinear equations of convection *in the framework of the Boussinesq approximation* have the following form [52]:

$$\frac{\partial \mathbf{v}_1}{\partial t} + (\mathbf{v}_1 \cdot \nabla)\,\mathbf{v}_1 = -\nabla p_1 + \nabla^2 \mathbf{v}_1 + G T_1 \mathbf{e}_z,$$
$$\frac{\partial T_1}{\partial t} + \mathbf{v}_1 \cdot \nabla T_1 = \frac{1}{P}\nabla^2 T_1, \tag{1}$$
$$\nabla \cdot \mathbf{v}_1 = 0;$$

$$\frac{\partial \mathbf{v}_2}{\partial t} + (\mathbf{v}_2 \cdot \nabla)\,\mathbf{v}_2 = -\rho\nabla p_2 + \frac{1}{\nu}\nabla^2 \mathbf{v}_2 + \frac{G}{\beta} T_2 \mathbf{e}_z,,$$
$$\frac{\partial T_2}{\partial t} + \mathbf{v}_2 \cdot \nabla T_2 = \frac{1}{\chi P}\nabla^2 T_2, \tag{2}$$
$$\nabla \cdot \mathbf{v}_2 = 0.$$

Here $G = g\beta_1\Theta a_1^3/\nu_1^2$ is the Grashof number (g is the gravity acceleration), $P = \nu_1/\chi_1$ is the Prandtl number of the fluid in the top layer, $\mathbf{e}_z$ is a unit vector directed upwards.

The boundary conditions at the rigid boundaries are:

$$\mathbf{v}_1 = 0,\ T_1 = 0 \ \text{ at } \ z = 1, \tag{3}$$

$$\mathbf{v}_2 = 0,\ T_2 = s \ \text{ at } \ z = -a, \tag{4}$$

with $s = 1$ for heating from below and $s = -1$ for heating from above.

The boundary conditions at the deformable interface $z = h$ include the balance of normal stresses:

$$p_1 - p_2 + W_0 K(1 - \varepsilon_\alpha T_1) + Ga\, \delta h$$
$$= \left[\left(\frac{\partial v_{1i}}{\partial x_k} + \frac{\partial v_{1k}}{\partial x_i} \right) - \eta^{-1} \left(\frac{\partial v_{2i}}{\partial x_k} + \frac{\partial v_{2k}}{\partial x_i} \right) \right] n_i n_k, \tag{5}$$

the balance of tangential stresses:

$$\left[\left(\frac{\partial v_{1i}}{\partial x_k} + \frac{\partial v_{1k}}{\partial x_i} \right) - \eta^{-1} \left(\frac{\partial v_{2i}}{\partial x_k} + \frac{\partial v_{2k}}{\partial x_i} \right) \right] \tau_i^{(l)} n_k - \frac{M}{P} \tau_i^{(l)} \frac{\partial T_1}{\partial x_i} = 0, \ l = 1, 2, \tag{6}$$

the continuity of the velocity field:

$$\mathbf{v}_1 = \mathbf{v}_2, \tag{7}$$

the kinematic equation for the evolution of the interface shape:

$$\frac{\partial h}{\partial t} + v_{1x} \frac{\partial h}{\partial x} + v_{1y} \frac{\partial h}{\partial y} = v_{1z}, \tag{8}$$

the continuity of the temperature field:

$$T_1 = T_2, \tag{9}$$

and the continuity of the heat flux normal components:

$$\left(\frac{\partial T_1}{\partial x_i} - \kappa^{-1} \frac{\partial T_2}{\partial x_i} \right) n_i = 0. \tag{10}$$

Here $W_0 = \sigma_0 a_1 / \eta_1 \nu_1$, $\varepsilon_\alpha = \alpha\theta/\sigma_0$, $\delta = \rho^{-1} - 1$; K is the curvature of the interface, $\mathbf{n}$ is the normal vector and $\boldsymbol{\tau}^{(l)}$, $l = 1, 2$ are the tangential vectors of the interface; p_1, p_2 are the differences between the overall pressure and the hydrostatic pressure. The definitions of the Galileo number Ga and the Marangoni number M have been given in the previous section.

As it was explained above, the system of equations governing convection in the Boussinesq approximation (1), (2) and the system of boundary conditions at the deformable interface (5)–(10) should be used with caution. If the buoyancy is taken into account (i.e. G is not small) and the densities of fluids are not close (i.e. δ is not small), then $Ga = O(\varepsilon_\beta^{-1})$ is large and the shape of the interface is determined in the zeroth order by the balance of hydrostatic and capillary stresses:

$$\delta h + r_c^2 K(1 - \varepsilon_\alpha T) = C, \tag{11}$$

where $r_c = (W_0/Ga)^{1/2} = (\sigma_0/\rho_1 g a_1^2)^{1/2}$ is the dimensionless capillary radius, and the constant C corresponds to the pressure difference across the interface. The zeroth order boundary condition (11) should replace (5) by solving (1)–(2), because these equations themselves are written in the zeroth order in ε_β.

In the case of a closed cavity, the contact angles on the lateral solid walls are taken into account, and generally $h \neq 0$, $C \neq 0$. If the layers are infinite in the horizontal direction, the *zeroth order* shape of the interface is just $h = 0$, and $C = 0$. The boundary conditions on the interface $z = 0$ are:

$$\frac{\partial v_{1x}}{\partial z} - \eta^{-1}\frac{\partial v_{2x}}{\partial z} - \frac{M}{P}\frac{\partial T_1}{\partial x} = 0, \quad \frac{\partial v_{1y}}{\partial z} - \eta^{-1}\frac{\partial v_{2y}}{\partial z} - \frac{M}{P}\frac{\partial T_1}{\partial y} = 0, \tag{12}$$

$$\mathbf{v}_1 = \mathbf{v}_2,\ T_1 = T_2,\ \frac{\partial T_1}{\partial z} - \kappa^{-1}\frac{\partial T_2}{\partial z} = 0. \tag{13}$$

After the calculation of the velocity, pressure and temperature fields in the framework of the Boussinesq approximation (boundary value problem (1), (2), (12), (13), one can find the *correction* of the first-order in ε_β to the shape of the interface $h(x, y, t)$ by solving the equation

$$\delta h + r_c^2(1 - \varepsilon_\alpha T|_{z=0})\left(\frac{\partial^2 h}{\partial x^2} + \frac{\partial^2 h}{\partial y^2}\right) =$$

$$\varepsilon_\beta G^{-1}\left[p_2 - p_1 + 2\left(\frac{\partial v_{1,z}}{\partial z} - \eta^{-1}\frac{\partial v_{2,z}}{\partial z}\right)\right]\Bigg|_{z=0}. \tag{14}$$

Recall that the influence of the interface deformation $h \sim \varepsilon_\beta$ on the convective motion is a non-Boussinesq effect, and it cannot be explored without a consideration of other non-Boussinesq effects that have the same order of smallness.

In the case of the thermocapillary convection, using the system of (1)–(2) (with $G = 0$) together with the boundary conditions (5)–(10) corresponding to a deformable interface is not forbidden. However, if the relative change ε_α of the surface tension coefficient is small, then the coefficient $W_0 = M/\varepsilon_\alpha P$ is large. As one can see from the boundary condition (5), the deformation of the interface produced by the thermocapillary convection has a *small curvature* in the latter case. If the wavelength of the interfacial deformation is not large, the deformation is suppressed by the surface tension, and the boundary conditions (12)–(13) corresponding to a flat interface can be used. However, the *longwave* deformations are not suppressed by the surface tension, and they should be considered using the original boundary conditions (5)–(10).

We shall consider also a model for a *three-layer* fluid system (see Fig. 1b). The third layer of the mean thickness a_3 is added. The parameters of the bottom fluid are denoted by subscript 3. The surface tension coefficient of the lower interface is $\sigma_* = \sigma_{*0} - \alpha_* T$. We define

$$\rho_* = \frac{\rho_1}{\rho_3},\ \nu_* = \frac{\nu_1}{\nu_3},\ \eta_* = \frac{\eta_1}{\eta_3} = \rho_*\nu_*,\ \chi_* = \frac{\chi_1}{\chi_3},$$

$$\kappa_* = \frac{\kappa_1}{\kappa_3},\ \beta_* = \frac{\beta_1}{\beta_3},\ a_* = \frac{a_3}{a_1}, \bar{\alpha} = \frac{\alpha_*}{\alpha}.$$

The boundary value problem for the description of a three-layer system is an extension of the system of (1)–(10) which contains additional equations,

$$\frac{\partial \mathbf{v}_3}{\partial t} + (\mathbf{v}_3 \cdot \nabla)\,\mathbf{v}_3 = -\rho_*\nabla p_3 + \frac{1}{\nu_*}\nabla^2\mathbf{v}_3 + \frac{G}{\beta_*}T_3\mathbf{e}_z,$$

$$\frac{\partial T_3}{\partial t} + \mathbf{v}_3 \cdot \nabla T_3 = \frac{1}{\chi P} \nabla^2 T_3, \tag{15}$$

$$\nabla \cdot \mathbf{v}_3 = 0,$$

and additional boundary conditions,

$$\mathbf{v}_3 = 0,\ T_3 = s \ \text{ at } \ z = -a - a_*. \tag{16}$$

The boundary conditions (4) are replaced by the following set of boundary conditions at $z = -a + h_*$:

$$p_2 - p_3 + W_{*0} K_* (1 - \varepsilon_{\alpha *} T_1) + Ga\, \delta_* h_*$$
$$= \left[\eta^{-1} \left(\frac{\partial v_{2i}}{\partial x_k} + \frac{\partial v_{2k}}{\partial x_i} \right) - \eta_*^{-1} \left(\frac{\partial v_{3i}}{\partial x_k} + \frac{\partial v_{3k}}{\partial x_i} \right) \right] n_{*i} n_{*k}, \tag{17}$$

$$\left[\eta^{-1} \left(\frac{\partial v_{2i}}{\partial x_k} + \frac{\partial v_{2k}}{\partial x_i} \right) - \eta_*^{-1} \left(\frac{\partial v_{3i}}{\partial x_k} + \frac{\partial v_{3k}}{\partial x_i} \right) \right] \tau_{*i}^{(l)} n_{*k} - \frac{\bar{\alpha} M}{P} \tau_{*i}^{(l)} \frac{\partial T_3}{\partial x_i} = 0, \tag{18}$$
$$l = 1, 2,$$

$$\mathbf{v}_2 = \mathbf{v}_3, \tag{19}$$

$$\frac{\partial h_*}{\partial t} + v_{3x} \frac{\partial h_*}{\partial x} + v_{3y} \frac{\partial h_*}{\partial y} = v_{3z}, \tag{20}$$

$$T_2 = T_3, \tag{21}$$

$$\left(\kappa^{-1} \frac{\partial T_2}{\partial x_i} - \kappa_*^{-1} \frac{\partial T_3}{\partial x_i} \right) n_{*i} = 0, \tag{22}$$

where $W_{*0} = \sigma_{*0} a_1 / \eta_1 \nu_1$, $\varepsilon_{\alpha_*} = \alpha_* \theta / \sigma_{*0}$, $\delta_* = \rho_*^{-1} - \rho^{-1}$, K_* is the curvature of the lower interface, $\mathbf{n}_*$ is the normal vector and $\boldsymbol{\tau}_*^{(l)}$ are the tangential vectors of the lower interface.

In the limit of a non-deformable interface, the boundary conditions at $z = -a$ are:

$$\eta^{-1} \frac{\partial v_{2x}}{\partial z} - \eta_*^{-1} \frac{\partial v_{3x}}{\partial z} - \frac{\bar{\alpha} M}{P} \frac{\partial T_3}{\partial x} = 0,\ \eta^{-1} \frac{\partial v_{2y}}{\partial z} - \eta_*^{-1} \frac{\partial v_{3y}}{\partial z} - \frac{\bar{\alpha} M}{P} \frac{\partial T_3}{\partial y} = 0, \tag{23}$$

$$\mathbf{v}_2 = \mathbf{v}_3,\ T_2 = T_3,\ \kappa^{-1} \frac{\partial T_2}{\partial z} - \kappa_*^{-1} \frac{\partial T_3}{\partial z} = 0. \tag{24}$$

4 Two-Layer Systems

The problem (1)–(10) for any choice of parameters has the solution

$$\mathbf{v}_1^0 = \mathbf{v}_2^0 = 0,\ h^0 = 0,\ T_1^0 = -\frac{z-1}{1+\kappa a},\ T_2^0 = -\frac{\kappa z - 1}{1 + \kappa a},$$

$$p_1^0 = -\frac{G}{1+\kappa a} \left(\frac{z^2}{2} - z \right),\ p_2^0 = -\frac{G}{\rho \beta (1 + \kappa a)} \left(\frac{\kappa z^2}{2} - z \right), \tag{25}$$

which corresponds to the mechanical equilibrium state. Its stability can be investigated in the framework of the linear stability theory. The boundary value problem (1)–(10) is linearized near the solution (25). The solutions of the linearized problem are presented as a superposition of normal modes characterized by a wave vector $\mathbf{k} = (k_x, k_y)$ and a complex growth rate $\lambda = \lambda_r + i\lambda_i$:

$$(\tilde{\mathbf{v}}_1(z), \tilde{p}_1(z), \tilde{T}_1(z), \tilde{\mathbf{v}}_2(z), \tilde{p}_2(z), \tilde{T}_2(z)) \exp(ik_x x + ik_y y + \lambda t); \tag{26}$$

in what follows the sign "tilde" will be omitted.

Since the problem is isotropic, the growth rate λ depends only on the wave vector modulus $k = |\mathbf{k}|$ but not on its direction. That is why it is sufficient to consider only two-dimensional disturbances with $\mathbf{k} = (k, 0)$ which do not depend on the coordinate y. Introducing the stream function disturbances

$$v_{mx} = \psi'_m,\ v_{mz} = -ik\psi_m\ (m = 1, 2)$$

where the prime stands for d/dz, and eliminating pressure disturbances in the usual way, we obtain the following boundary eigenvalue problem:

$$-\lambda D\psi_m = -c_m D^2\psi_m + ikGb_m T_m, \tag{27}$$

$$\lambda T_m - ikA_m\psi_m = \frac{d_m}{P} DT_m,\ (m = 1, 2), \tag{28}$$

where $D = d^2/dz^2 - k^2$;

$$z = 1 : \psi_1 = \psi'_1 = T_1 = 0, \tag{29}$$

$$z = -a : \psi_2 = \psi'_2 = T_2 = 0, \tag{30}$$

$$z = 0 : \psi'''_1 - \eta^{-1}\psi'''_2 + [\lambda\delta - 3k^2(1 - \eta^{-1})]\psi'_1 +$$

$$ik\{Ga[\delta + \epsilon_\beta s(1 - \rho^{-1}\beta^{-1})/(1 + \kappa a)] + Wk^2\}h = 0, \tag{31}$$

$$\eta(\psi''_1 + k^2\psi_1) - (\psi''_2 + k^2\psi_2) - \frac{\eta M}{P}\left(T_1 - \frac{s}{1 + \kappa a}h\right) = 0, \tag{32}$$

$$\psi'_1 = \psi'_2,\ \psi_1 = \psi_2 = i\frac{\lambda}{k}h, \tag{33}$$

$$T_1 - T_2 = \frac{s(1 - \kappa)}{1 + \kappa a}h,\ \kappa T'_1 - T'_2 = 0, \tag{34}$$

where $c_1 = b_1 = d_1 = e_1 = 1$, $c_2 = 1/\nu$, $b_2 = 1/\beta$, $d_2 = 1/\chi$, $e_2 = \rho$; $A_1 = dT_1^0/dz = -1/(1 + \kappa a)$, $A_2 = dT_2^0/dz = -\kappa/(1 + \kappa a)$ are the dimensionless temperature gradients, $W = W_0(1 - \epsilon_\alpha s/(1 + \kappa a))$.

The eigenvalue problem (27)–(34) determines the spectrum of the growth rates. Because this problem is not self-adjoint, the eigenvalues can be complex: $\lambda = \lambda_r + i\lambda_i$. The condition $\lambda = 0$ corresponds to the boundary of a monotonic instability, while the conditions $\lambda_r = 0$, $\lambda_i \neq 0$ correspond to the boundary of an oscillatory instability.

If the deformation of the interface h can be neglected, the system of boundary conditions at the interface is simplified:

$$z = 0 : \eta\psi_1'' - \psi_2'' - \frac{ik\eta M}{P}T_1 = 0,\ \psi_1' = \psi_2',\ \psi_1 = \psi_2 = 0, \tag{35}$$

$$T_1 = T_2,\ \kappa T_1' = T_2'. \tag{36}$$

4.1 Marangoni Instabilities

In the present subsection, we consider the case of a pure Marangoni instability ($G = 0$, $M \neq 0$). As explained in Sect. 2, the influence of buoyancy can be ignored in the case when the layer thicknesses are much smaller than $a_c = (\alpha/g\beta_1\rho_1)^{1/2}$, i.e. for thin layers or under microgravity conditions.

Monotonic Instability Modes. Solving the boundary value problem (27)–(34) with $\lambda = 0$, one obtains the monotonic neutral curve which is determined by the following *exact* formula [56]:

$$sM(k) = \frac{8Pk^2(1+\kappa a)(\kappa D_1 + D_2)(\eta B_1 + B_2)}{\kappa[\eta P(\chi E_2 - E_1) - 8k^5(D_1 + D_2)(\eta F_1 - F_2)(Ga\delta + Wk^2)^{-1}}, \tag{37}$$

where

$$D_1 = \frac{C_1}{S_1},\ D_2 = \frac{C_2}{S_2},\ B_1 = \frac{S_1C_1 - k}{S_1^2 - k^2},\ B_2 = \frac{S_2C_2 - k}{S_2^2 - k^2a^2},$$

$$E_1 = \frac{S_1^3 \quad k^3C_1^3}{S_1(S_1^2 - k^2)},\ E_2 = \frac{S_2^3 - k^3a^3C_2^3}{S_2(S_2^2 - k^2a^2)},\ F_1 = \frac{1}{S_1^2 - k^2},\ F_2 = \frac{a^2}{S_2^2 - k^2a^2},$$

$$S_1 = \sinh k,\ C_1 = \cosh k,\ S_2 = \sinh ka,\ C_2 = \cosh ka.$$

The denominator of the expression (37) can change its sign when k is changed. In this case, the monotonic neutral curve is discontinuous, i.e. the Marangoni convection can be excited for *both* ways of heating, from below ($s = 1$) and from above ($s = -1$), but in different intervals of wavenumbers k. The term in the denominator that contains $Ga\delta$ and W is caused by the deformation of the interface. If $Ga\delta \gg 1$ and $W \gg 1$, this term can be dropped except for small k. Thus, one can distinguish between the *short-wave* Marangoni instability which is insensitive to the interface deformation, and the *long-wave* instability which is essentially associated with the deformations of the interface. In the longwave limit, the critical Marangoni number $M_c = M(0) \sim Ga$. Since $Ga = ga_1^3/\nu_1^2$, the deformational type of the monotonic instability is important in the case of very thin layers or under microgravity conditions.

The short-wave monotonic instability produces stationary convective patterns (see [24]). The long-wave instability is typically non-saturable, i.e. it leads to blow-up physically corresponding to the formation of dry spots ([60]). The absence of the saturation is inavoidable in the framework of the one-layer approach. Indeed, let us take a layer with the mean thickness corresponding to the critical value of the Marangoni number, and consider a longwave modulation of its local thickness $h(x,y)$. The local *critical* Marangoni number is proportional to the Galileo number and hence to $h^3(x,y)$. The *actual* local Marangoni number is proportional to $h(x,y)$. Thus, the instability is enhanced in the regions where the thickness decreases. The nonlinear evolution of the interfacial deformation is governed to the leading order by the *Funada equation* [9]

$$\frac{\partial h}{\partial t} = -\nabla_\perp^2 \left[A\nabla_\perp^2 h + B(M - M_c)h + Ch^2\right], \tag{38}$$

where $A > 0$, $B > 0$, C are some constant real coefficients, $\nabla_\perp^2 = \partial^2/\partial x^2 + \partial^2/\partial y^2$. It is well known that this equation, which is called also *Sivashinsky equation* [55], is subject to a blow-up in finite time (see, e.g., [1]). When the two-layer approach is used, the dependence of the critical temperature difference on the thickness of the liquid layer may be *non-monotonic* [52]. If the ratio of the thicknesses of the layers $a = a_c$ corresponds to the minimum value of the critical temperature difference, both the increase and the decrease of the local liquid layer thickness weaken the instability. In the latter case, the coefficient C in (38) vanishes, and the cubic nonlinearity has to be incorporated. If a is close to a_c, the quadratic and the cubic nonlinearities are of the same order, and the problem is governed by the *Cahn-Hilliard equation*,

$$\frac{\partial h}{\partial t} = -\nabla_\perp^2 \left[A\nabla_\perp^2 h + B(M - M_c)h + Ch^2 - Dh^3\right], \tag{39}$$

which is saturable.

Oscillatory Instability Modes. A liquid with a free surface is subject to two oscillatory modes. The first one is the well-known wave of surface deformations, called *capillary-gravity (transverse) wave*, with the dispersion relation

$$\omega^2 = gk + \frac{\sigma}{\rho}k^3 \tag{40}$$

(in the case of an infinitely deep liquid layer). The second oscillatory mode, called *dilational (longitudinal) wave*, appears on a flat surface if it manifests some *elastic* properties [31,32]. The physical origin of the surface elasticity can be the Marangoni effect. For instance, in an infinitely deep liquid layer heated *from above* the dispersion relation of dilational waves is

$$\omega^2 = \frac{1}{\rho}\left(-\frac{d\sigma}{dT}\right)Ak^2, \tag{41}$$

where A is the equilibrium temperature gradient [30].

Typically, both above mentioned wavy modes decay because of the viscosity. However, in both one-layer systems and two-layer systems the waves can be

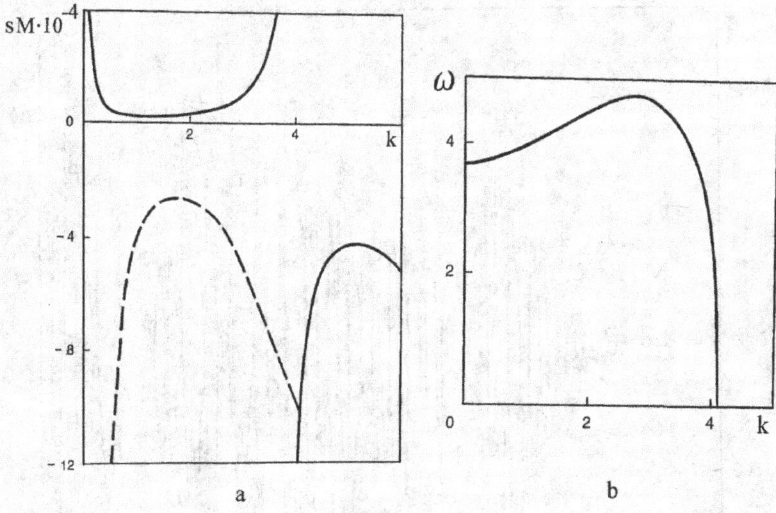

Fig. 2. **a** Neutral stability curves and **b** dependence of the oscillations frequency on the wavenumber for the system n-octane–methanol; $a = 1.6$

spontaneously generated when the frequencies of two different oscillatory modes approach each other [10,30,46]. The instability appears because of a "repulsion" between eigenvalues which is a manifestation of the phenomenon of *avoided crossing* observed in many physical systems (see, e.g., [59]).

However, in two-layer systems longitudinal waves can become unstable without mixing with transverse waves. The crucial point is the difference between viscous and heat diffusion time scales for different fluids, which leads to a certain *time delay* in the counteraction of convective heat fluxes generated by motions in the top and bottom layers. The purely longitudinal oscillatory instability in two-layer fluid systems was discovered by Sternling and Scriven [57] almost simultaneously with the appearance of the Pearson's theory [47] of the monotonic Marangoni instability. Their theory dealt with a system of two layers of *infinite thicknesses*, i.e. it described the *short-wave asymptotics* of the neutral curve ($k \gg 1$). However, the minimum of the neutral curve is usually located at $k \sim 1$, where the criteria for instability are significantly changed ([48,52]). A simple criterion has been obtained for long-wave Marangoni oscillations: they appear if the heating is performed from the side of the fluid with the larger value of the Prandtl number.

The system n-octane-methanol ($\nu = 1.14$, $\eta = 1.02$, $\kappa = 0.698$, $\chi = 0.934$, $P = 7.84$) is an example of the physical system where the two-layer longitudinal oscillations have been predicted ([4]). A typical neutral curve obtained in the absence of interfacial deformations (the limit $Ga \to \infty$) is shown in Fig. 2. One can see that the monotonic instability (solid line) takes place for $s > 0$ (by heating from below), as $k < k_d$, and for $s < 0$ (by heating from above), as

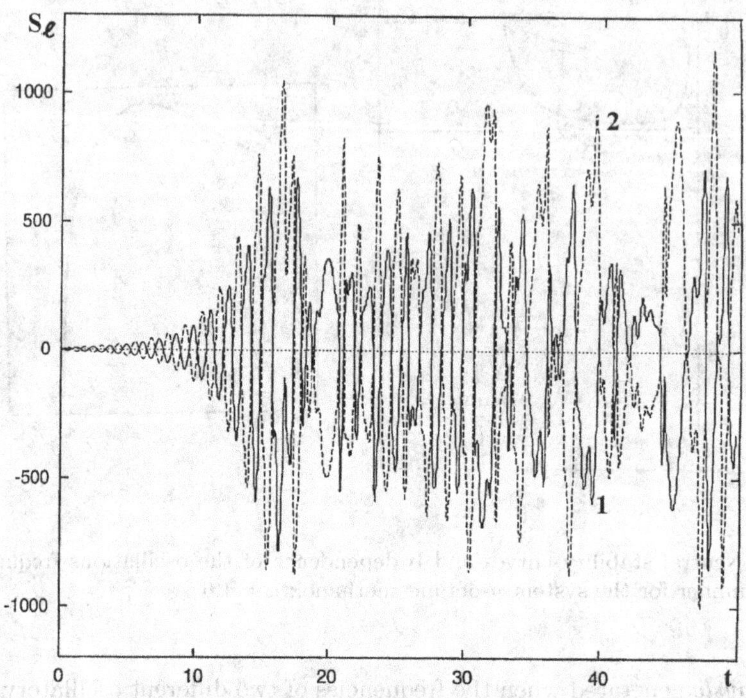

Fig. 3. Time evolution of the integral characteristics $S_1(t) = \int_0^{L/2} dx \int_0^1 dz\psi_1(x,z,t)$ (*solid line*) and $S_2(t) = \int_0^{L/2} dx \int_{-a}^0 dz\psi_2(x,z,t)$ (*dashed line*) for nonlinear standing Marangoni waves in the system *n*-octane–methanol; $a = 1.6$

$k > k_d$; $k_d \approx 3.95$. The absence of the monotonic instability in the long-wave region $k < k_d$ by heating from above is a condition favourable for the appearance of the oscillatory instability. There exists a codimension-2 point (k_*, sM_*) where the frequency of the oscillations tends to zero, and the oscillatory neutral curve (dashed line) terminates in the monotonic one. The weakly nonlinear bifurcation analysis [4] reveals a subcritical instability of the equilibrium state with respect to standing waves, i.e. the instability is not saturated on the level of small-amplitude waves. Our numerical simulations of the fully nonlinear 2D problem, (1)–(10), in this regime exhibit spatially-periodic standing waves with the complicated tempral oscillation (see Fig. 3).

4.2 Combined Rayleigh-Marangoni Convection

If the thicknesses of the layers are comparable with $a_c = \alpha/g\beta_1\rho_1$, the influence of the buoyancy on the excitation of convection cannot be ignored.

In a one-layer system heated from below, the buoyancy bulk forces and the thermocapillary interfacial stresses act in the *same* direction and produce together a stationary instability, provided the thermocapillary effect is normal, i.e. the surface tension coefficient decreases as the temperature increases ($M > 0$) [39]. For two-layer systems, the situation is more intricate. If the flow in the bottom layer is dominant, the actions of buoyancy and thermocapillary effect are similar to those in the one-layer system. If the flow in the top layer is dominant, the buoyancy forces and the thermocapillary stresses act in *opposite* directions. Their competition leads to a stabilization of the stationary instability, as well as to the generation of a specific kind of linear oscillatory instability, which was predicted theoretically [52] and observed in the experiments [24]. In the case of the anomalous thermocapillary effect ($M < 0$), one can obtain an oscillatory instability when the flow in the bottom layer is dominant, and only a stationary instability if the flow in the top layer prevails [3].

A direct three-dimensional simulation of the system of (1)–(2) with the boundary conditions (3)–(4), (12)–(13) under the combined action of buoyancy and anomalous thermocapillary effect ($G > 0$, $M < 0$) has been performed recently for the 10cS silicone oil–ethylene glycol system [2] with the set of parameters $\nu = 0.6493$, $\eta = 0.549$, $\kappa = 0.6194$, $\chi = 1.096$, $\beta = 1.4516$, $P = 94$, $a = 1.8$. In the region of stationary instability, the most typical patterns are a stationary hexagons and stationary rolls. The buoyancy acts in favour of the roll pattern. In the region of an oscillatory instability, the so-called *alternating rolls* turn out to be the typical kind of oscillatory pattern. This pattern is a nonlinear superposition of two systems of standing waves with orthogonal wave vectors. The temporal phase shift between standing waves of different spatial orientations is equal to $T/4$, where T is the full period of oscillations. Thus, one observes some kind of roll patterns that change their orientation with the time interval $T/4$. Recall that the alternating roll pattern is one of the generic wave patterns that appear in the rotationally invariant systems due to a primary oscillatory instability of the spatially homogeneous state [42,50,58].

Some more unusual patterns appear on the boundary between the regions of stationary patterns and alternating rolls. An oscillatory instability of hexagonal pattern leads to the appearance of periodic or chaotic pulsations of hexagons. Another stable non-stationary structure is shown in Fig. 4. A roll pattern is modulated both in transverse and in longitudinal direction, so that one observes a system of elongated rectangles. The wave of modulation moves in the direction parallel to the wavevector of the background system of rolls.

4.3 Anticonvection

In the present subsection we shall describe the amazing phenomenon of "anticonvection" which appears in two-layer systems when heating is from above ($s = -1$). We disregard the thermocapillary effect ($M = 0$), and assume that for both fluids the density decreases with temperature ($G > 0$, $\beta > 0$). Because in both layers the density gradients are directed downward, one could expect that the buoyancy convection is impossible. Nevertheless, a specific kind of convective instability can occur in the system if the thermal diffusivity and the thermal expansion coefficient of the top fluid are much smaller that those

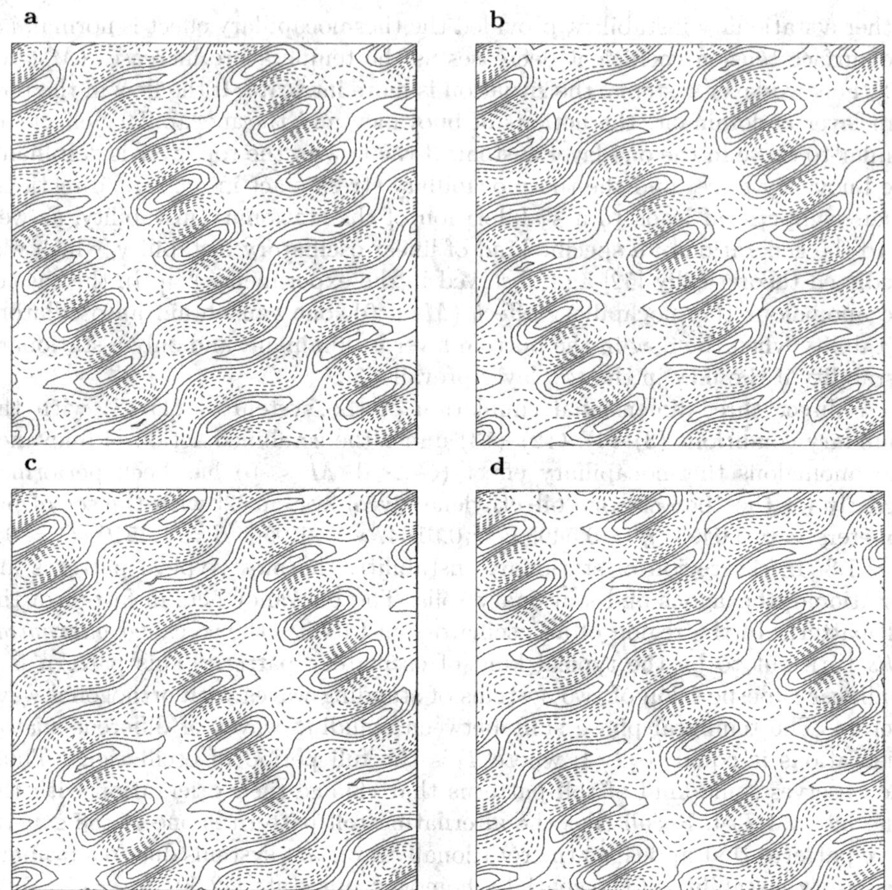

Fig. 4. Modulated rolls, snapshots of isotherms on the interface. $G = 15.5$, $\mathrm{M} = -550$, aspect ratio 25

of the bottom fluid ($\chi \ll 1$, $\beta \ll 1$) [15,51,62]. The physical essence of this instability mechanism is as follows. Let an element of the top fluid move down towards the interface. Due to the low thermal diffusivity of the top fluid this element keeps its temperature higher than that of the surrounding fluid. Since the thermal expansion coefficient of the top fluid is small, this buoyancy force caused by that temperature difference cannot stop the downward motion of the top fluid element. This motion produces a hot spot at the interface which generates an upward convective motion in the bottom, highly buoyant, fluid. Because of the high thermal diffusivity of the bottom fluid, the latter motion does not change the temperature field but it produces tangential stresses at the interface that support the initial velocity disturbance in the top layer. Similar arguments can be used for the explanation of the appearance of anticonvection in the case $\chi \gg 1$, $\beta \gg 1$.

Actually, the conditions for the excitation of the instability described above turned out to be rather restrictive, and in fact only one physical system satisfying these conditions was found (water–mercury). However, the appearance of anticonvection can be essentially simplified in the presence of a constant, spatially uniform heat source or sink at the interface which is applied in such a way that the temperature gradient in one layer is much smaller than in the other one [34,38,40,41]. Assume that a constant heat release rate Q (Q may be positive or negative) is set at the interface $z = 0$. Define $G_Q = g\beta_1 Q a_1^4/\nu_1^2\kappa_1$. Dimensionless temperature gradients in the equilibrium state are

$$A_1 = -\frac{sG + a\kappa G_Q}{G(1+\kappa a)},\ A_2 = \frac{\kappa(sG - G_Q)}{G(1+\kappa a)} \tag{42}$$

($s = -1$ for heating from above). Thus, $A_1 \ll A_2$ if $G_Q \approx G/a\kappa$, $A_1 \gg A_2$ if $G_Q \approx -G$. Let us rewrite the system of (27), (28) for $\lambda = 0$ using the transformation $T_m(z) = P\theta_m(z)$, $\psi_m(z) = \Phi_m(z)/A_1$, $m = 1, 2$:

$$D^2\Phi_1 - ikRA_1\theta_1 = 0;\ D\theta_1 + ik\Phi_1 = 0;$$

$$D^2\Phi_2 - ikRA_1\frac{\nu}{\beta}\theta_2 = 0;\ D\theta_2 + ik\frac{A_2}{A_1}\chi\Phi_2 = 0,$$

where $R = GP$ is the Rayleigh number. One can see that the parameters characterizing the appearance of anticonvection are actually RA_1 and $\chi A_2/A_1$. One can introduce an artificial system without a heat release at the interface, i.e. with temperature gradients $\tilde{A}_1 = 1/(1+\kappa a)$, $\tilde{A}_2 = \kappa/(1+\kappa a)$, but with a renormalized value of the thermal diffusivity ratio $\tilde{\chi}$ satisfying the relation

$$\frac{A_2}{A_1}\chi = \frac{\tilde{A}_2}{\tilde{A}_1}\tilde{\chi}. \tag{43}$$

The critical Rayleigh number for the original system with a heat release at the interface which produces temperature gradients A_1, A_2 is connected with the critical Rayleigh number for that artificial system by the relation

$$A_1 R(A_1, A_2, \chi) = \tilde{A}_1 R(\tilde{A}_1, \tilde{A}_2, \tilde{\chi}). \tag{44}$$

Thus, for any two-fluid system the quantity $\tilde{\chi}$, which determines the existence of anticonvection, can be made very small, if $A_2/A_1 \ll 1$, or very large, if $A_2/A_1 \gg 1$ [37].

Direct three-dimensional simulations of anticonvective flows show that the hexagonal pattern is the most typical one. However, in some cases one observes some kind of the distorted square pattern. In Fig. 5, isolines of the temperature in the plane $z = 0$ are shown for the system 10 cS silicone oil–ethylene glycol with $a = 1$ (the corresponding physical parameters have been given in Sect. 4.2); the temperature is higher in the centers of the convective cells. The anticonvection appears when the heating from above and the interfacial heat release are balanced in such a way that the temperature gradient in the bottom layer is

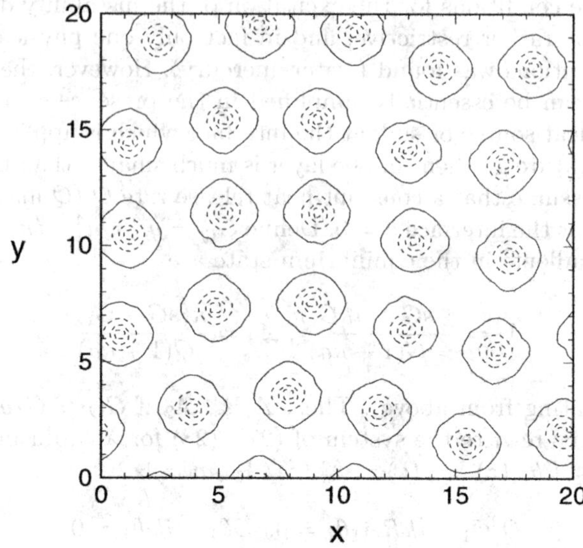

Fig. 5. Isotherms in the plane $z = 0$. $G = 3717$, $G_q = 6000$, $a = 1$

much larger than that in the top layer. Figures 6 and 7 show the isolines of the temperature and vertical velocity in the plane $y = 0$ for the same values of parameters as in Fig. 5. This plane crosses convective cells at different distances from their centers (see Fig. 5), that is why the distortion of the isotherms and the intensity of the flow is so non-uniform. The intensity of the flow in the top layer, where the temperature gradient is relatively small, is much higher than that in the bottom layer. Note that the flow in the bottom layer has a "multi-storey" structure.

4.4 Influence of Lateral Walls

We shall discuss now the influence of lateral walls on the convection patterns.

In the case of infinite layers, the disturbances with the wave-vector modulus $|\mathbf{k}| = k_c$ have the same excitation threshold M_c (or G_c) independently of the orientation of the vector $\mathbf{k}$; the flow pattern is selected due to the nonlinear interaction of disturbances with different directions of $\mathbf{k}$. In a confined container, the set of eigenmodes is discrete, and the flow pattern near the threshold is determined by the "most dangerous" eigenmode. The dependence of the eigenfunctions on the horizontal coordinate can be easily determined in the case of model stress-free and thermally insulating lateral walls. In a rectangular container, the disturbances of the temperature and the vertical velocity are proportional to $\cos k_x x \cos k_y y$ with a quantized set of the admitted wave vectors (k_x, k_y). The wave vector of the most dangerous mode depends on the aspect ratios of the container in a rather complicated way [5]. In a circular container, the same disturbances in the polar coordinates (r, ϕ) are proportional to $\exp(im\phi)J_m(k_{m,n}r)$,

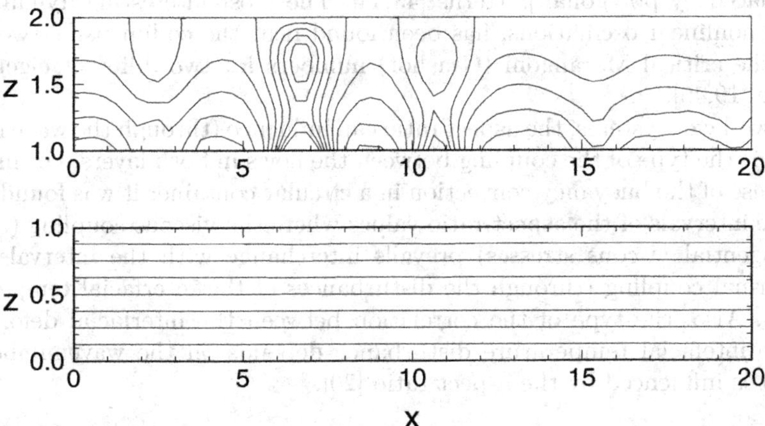

Fig. 6. Isotherms in the plane $y = 0$. $\Delta T_1/\Delta T_2 = 0.004$, where ΔT_m is the difference between temperature values for neighbour isotherms in the m-th fluid

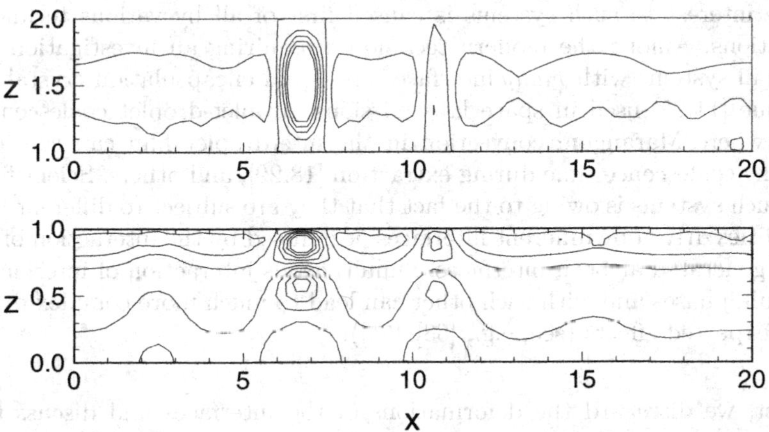

Fig. 7. Isolines of the vertical velocity in the plane $y = 0$. $\Delta v_1/\Delta v_2 = 25$, where Δv_m is the difference between velocity values for neighbour isolines in the m-th fluid

where m is the integer azimutal wave number, J_m is the Bessel function, $k_{m,n}$ is the radial wave number [49].

The theoretical and experimental investigation of the onset of convection in containers with relatively small aspect ratios, carried out for both rectangular [6,8,26] and circular [7,19,63] containers, showed an essential dependence of patterns on the aspect ratios, especially near the instability threshold (see also the review paper [22] and references therein). With the growth of the Marangoni (or Grashof) number the patterns selected by the lateral boundaries (e.g., those with the definite value of the asimutal wavenumber m in circular containers) are replaced by polygonal patterns [43,44]. The most interesting dynamics, including nonlinear oscillations, has been found near the codimension-two points where the critical Marangoni (Grashof) numbers for two different eigenmodes coincide [19,49].

In two-layer systems, the aspect ratio can influence (through the wave number selection) the type of the coupling between the flows in both layers. For instance, in the case of the buoyancy convection in a circular container it was found [20,21] that the intervals of the aspect-ratio values where the viscous coupling (through the tangential viscous stresses) prevails interchange with the intervals where the thermal coupling (through the disturbances of the interfacial temperature) prevails. Also, the type of the correlation between the interfacial deformation and the intefacial temperature disturbance depends on the wavenumber, and hence it is influenced by the aspect ratio [20].

5 Three-Layer Systems

In the previous section we considered convection in systems with a *single* interface. Now we shall discuss the case when a fluid system has *two* interfaces.

The interest to such systems is caused first of all by various technological applications. Among the modern techniques requiring an investigation of convection in systems with *many* interfaces are liquid encapsulation crystal growth technique [11,23] used in space labs missions, droplet-droplet coalescence processes, where Marangoni convection in the interdroplet film can considerably affect the coalescence time during extraction [18,29], and others. Scientific interest in such systems is owing to the fact that they are subject to different kinds of instabilities driven by different interfaces or induced by the interaction of disturbances generated at both interfaces. Simultaneous interaction of interfaces with their bulk phases and with each other can lead to much more complex dynamics and unexpected effects (see, e.g., [53]–[54]).

First, we disregard the deformations of the interfaces and discuss the appearance of *longitudinal* instabilities in the framework of the boundary value problem (1)–(4), (12), (13), (15), (16), (22) and (27). In the present subsection, we shall discuss the case of a pure Marangoni convection ($G = 0$, $M \neq 0$). For the sake of simplicity, we shall consider the *symmetric* case when the fluids 1 and 3 are identical ($\nu_* = \chi_* = \kappa_* = \alpha_* = 1$) and the thicknesses of all the layers are equal ($a = a_* = 1$). Despite the destabilizing density difference across one of the interfaces, the symmetric configuration turned out to be stable under experimental conditions [12], because of the stabilizing action of the interfacial tension.

Note that the geometric symmetry of the system with respect to reflection in the plane $z = -1/2$ does not lead to the corresponding symmetry of eigenfunctions, because the layers 1 and 3 are heated in a different way. The symmetry of the problem manifests itself as the symmetry of the neutral curve: the stability boundary does not depend on the way of heating (i.e. on the sign of parameter

s). The monotonic stability boundary $M = M(k)$ is determined by the analytical formula

$$M^2 = \frac{64k^4(2+\kappa)^2[(1+\kappa)^2C^2-1][(1+\eta)^2(SC-k)^2-(S-kC)^2]}{\eta^2\kappa^2[(1-\chi)^2(S^3-k^3C)^2-\chi^2k^2(kSC+k^2-2S^2)^2]}, \quad (45)$$

where $S = \sinh k$, $C = \cosh k$.

Note that in the case $\chi = 1$ the expression (45) is *negative* for any values of other parameters, thus the monotonic instability *does not exist*. If $3/5 < \chi < 3$, the monotonic neutral curve is located in the region $k > k_*(\chi)$, where $k_*(\chi)$ is determined by the equation

$$\frac{\chi}{|1-\chi|} = f(k_*), \quad (46)$$

where

$$f(k) = \frac{S^3 - k^3C}{k(kSC + k^2 - 2S^2)}. \quad (47)$$

The suppression of the monotonic instability, which is caused by the interaction of disturbances produced by the thermocapillary stresses at both interfaces, provides favourable conditions for the appearance of an *oscillatory* instability [13]. Qualitatively, the oscillations can be described in the following way. First, an intensive Marangoni convection is developed in the top layer and in the middle layer, while the fluid in the bottom layer is almost stagnant. Let us consider a region of *descending* flow in the middle layer. This flow produces a *minimum* of the temperature distribution in a certain point at the lower interface. The thermocapillary stresses at the lower interface generate a new structure around the point of the temperature minimum, which consists of an *ascending* flow in the middle layer and a descending flow in the bottom layer. Thus, a "two-storey" structure appears in the middle layer. Eventually, the new structure ousts the former ones in the middle layer. An intensive ascending motion developing in the middle layer induces a motion in the opposite direction in the top layer. Now, the temperature distribution at the lower interface has a maximum. The changes of velocity and temperature fields described above correspond to one half of the oscillation period. Then a two-storey structure with the opposite sign of the motion appears in the middle layer, etc.

5.1 Combined Rayleigh-Marangoni Convection

If both G and M are different from zero, another type of oscillations can appear which is caused by a competition between the buoyancy and the thermocapillary effect.

In Fig. 8, a stability diagram is shown for the system air–ethylene glycol–Fluorinert FC75 ($\nu = 0.974$, $\nu_* = 18.767$, $\eta = 0.001$, $\eta_* = 0.013$, $\kappa = 0.098$, $\kappa_* = 0.401$, $\chi = 215.098$, $\chi_* = 606.414$, $\beta = 2.62$, $\beta_* = 0.72$, $\bar{a} = 0.080$, $a = a_* = 1$). The boundary of the stability region consists of three fragments. Fragment 1 ($0 < G < 1.5$) corresponds to the boundary of the monotonic Marangoni

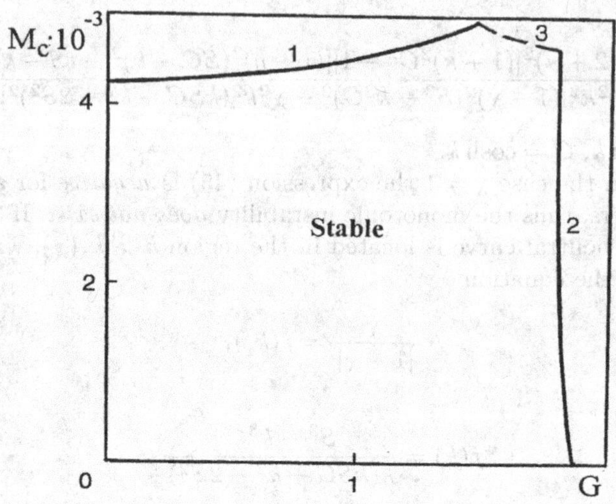

Fig. 8. Stability diagram

instability which is produced mainly near the upper interface, because $\bar{a}$ is rather small. Note that the Marangoni instability is slightly *suppressed* by the buoyancy. Fragment 2 ($1.776 < G < 1.866$) corresponds to the boundary of the monotonic Rayleigh instability which is generated mainly in the bottom layer, because the local Rayleigh number $R_m = g\beta_m A_m^4 a_m/\nu_m \chi_m$ is much higher for the bottom layer ($m = 3$) than for each of other layers. This type of instability is *strengthened* by the thermocapillary effect. The oscillations caused by the competition of both instability mechanisms correspond to fragment 3 ($1.5 < G < 1.776$).

5.2 Marangoni Convection in Systems with Deformable Interfaces

As discussed above, the deformations of interfaces cannot be ignored for *long-wavelength* Marangoni instabilities. The linear stability theory for a fluid system with two interfaces [35] in the long-wave limit predicts the existence of *two* critical Marangoni numbers $M = M_1$ and $M = M_2$ for the monotonic instability. Depending on the parameters of the system, the signs of the critical Marangoni numbers can coincide, but they can also be different, which corresponds to the appearance of the monotonic instability by heating either from below or from above. Also, an *oscillatory* instability boundary may arise, unlike the case of a single interface. The long-wavelength oscillatory instability is characterized by the dispersion relation $\omega \sim k^2$ between the frequency, ω, and the wavenumber.

The nonlinear evolution of the monotonic deformational modes in the case of a three-layer system is similar to that in the case of a two-layer system and leads to the destruction of the layers or to the appearance of a stationary interface deformation. For an oscillatory deformational instability, the weakly nonlinear

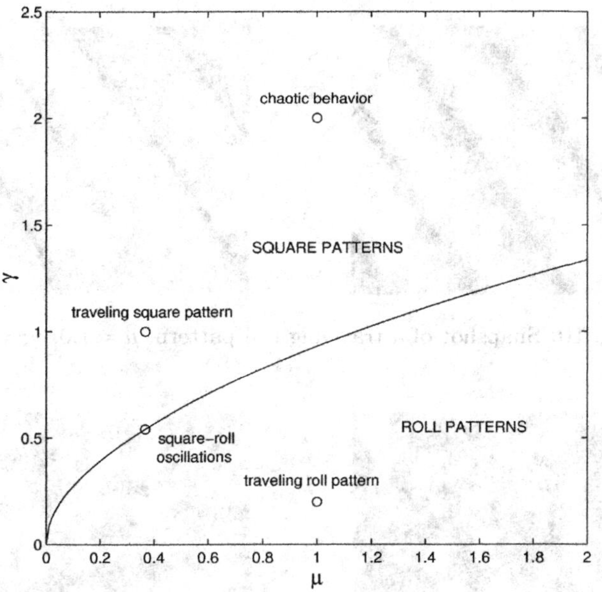

Fig. 9. Parameter regions corresponding to wave patterns with different symmetries. Circles correspond to various patterns found in numerical simulations

theory was developed in the case when the relative thickness of the bottom layer is small in comparison with the thicknesses of the top and the middle layer [25]. After appropriate rescaling the corresponding coupled equations for interfacial deformations H and H_* of both interfaces are:

$$\frac{\partial H}{\partial \tau} + \nabla^2(\nabla^2 H + \mu H + \gamma H^2 - H^3 + H_*) = 0, \tag{48}$$

$$\frac{\partial H_*}{\partial \tau} - \nabla^2 H = 0. \tag{49}$$

We have performed the direct numerical simulations of the system (48), (49). The results of simulations are summarized in Fig. 9. The most typical wavy patterns are traveling rolls (Fig. 10) and traveling squares (Fig. 11). Near the boundary between the regions of above-mentioned patterns, alternating rolls were observed (Fig. 12). Also, some spatially chaotic patterns were found. In the latter flow regime, the deformation of the upper interface displays irregular "spots" of nearly flat interface which split and merge in a chaotic manner (Fig. 13).

Acknowledgements

A.A.N. and I.B.S. acknowledge the support by the Israel Science Foundation.

Fig. 10. Snapshot of a traveling roll pattern, $\mu = 1.0$, $\gamma = 0.2$

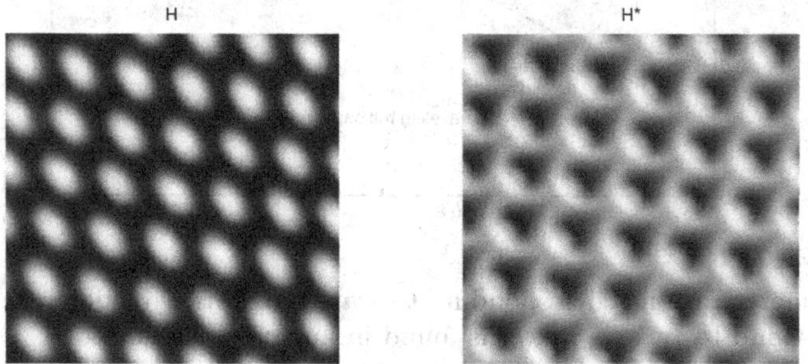

Fig. 11. Snapshot of a traveling square pattern, $\mu = 0.368$, $\gamma = 1.0$

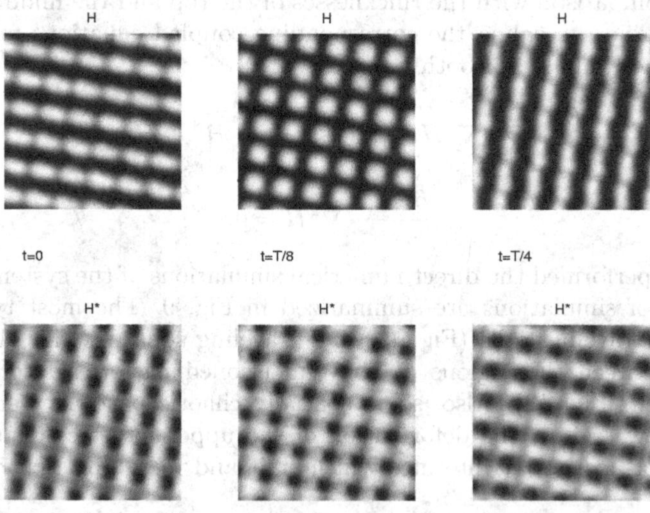

Fig. 12. Alternating rolls, different oscillation phases; $\mu = 0.368$, $\gamma = 0.5415$

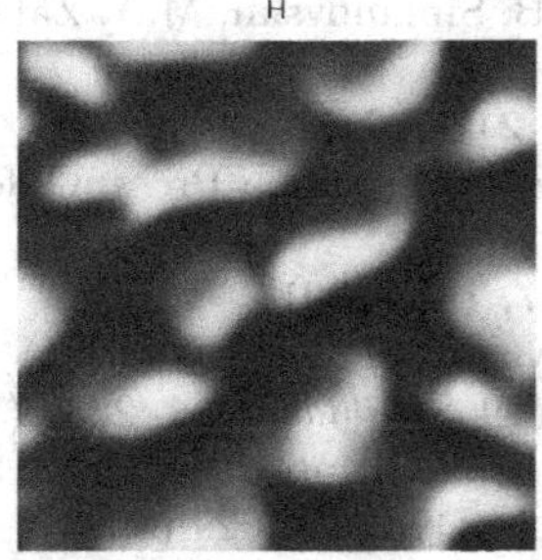

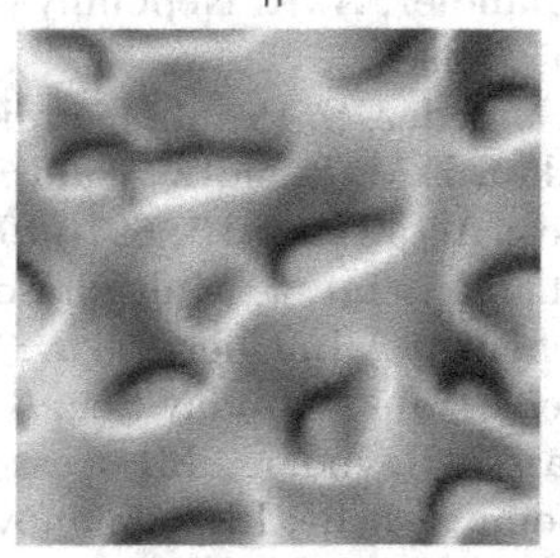

Fig. 13. Snapshot of a chaotic pattern, $\mu = 1.0$, $\gamma = 2.0$

References

1. A.J. Bernoff, A.L. Bertozzi: Physica D **85**, 375 (1995)
2. T. Boeck, A. Nepomnyashchy, I. Simanovskii, A. Golovin, L. Braverman, A. Thess: Phys. Fluids **14**, 3899 (2002)
3. L.M. Braverman, K. Eckert, A.A. Nepomnyashchy, I.B. Simanovskii, A. Thess: Phys. Rev. E **62**, 3619 (2000)
4. P. Colinet, Ph. Géoris, J.C. Legros, G. Lebon: Phys. Rev. E **54**, 514 (1996)
5. P.C. Dauby, G. Lebon, P. Colinet, J.C. Legros: Q. J. Mech. Appl. Math. **46**, 683 (1993)
6. P.C. Dauby, G. Lebon: J. Fluid Mech. **329**, 25 (1996)
7. P.C. Dauby, G. Lebon, E. Bouhi: Phys. Rev. E **56**, 520 (1997)
8. H.A. Dijkstra: Micrograv. Sci. Technol. **VII**, 307 (1995)
9. T. Funada: J. Phys. Soc. Jpn. **56**, 2031 (1987)
10. P.L. Garcia-Ybarra, M.G. Velarde: Phys. Fluids **30**, 1649 (1987)
11. P. Géoris, J.C. Legros: 'Pure thermocapillary convection in a multilayer system: first results from the IML-2 mission'. In: *Materials and Fluids under Low Gravity*, ed. by L. Ratke, H. Walter and B. Feuerbacher, (Springer, Berlin, Heidelberg 1996) pp. 299-311
12. P. Géoris, M. Hennenberg, G. Lebon, J.C. Legros: J. Fluid Mech. **389**, 209 (1999)
13. P. Géoris, M. Hennenberg, I. Simanovskii, A. Nepomnyashchy, I. Wertgeim, J. C. Legros: Phys. Fluids A **5**, 1575 (1993)
14. G.Z. Gershuni, E. M. Zhukhovitsky: *Convective Stability of Incompressible Fluid*, (Keter, Jerusalem 1976)
15. G.Z. Gershuni, E. M. Zhukhovitsky: Izv. AN SSSR, Mekh. Zhidk. Gaza **6**, 28 (1980) (In Russian)
16. G.Z. Gershuni, E. M. Zhukhovitsky: Sov. Phys. Dokl. **27**, 531 (1982)
17. A.Y. Gilev, A.A. Nepomnyashchy, I.B. Simanovskii: Fluid Dyn. **22**, 142 (1987)
18. H. Groothuis, F.G. Zuiderweg: Chem. Engng. Sci. **12**, 288 (1960)
19. D. Johnson, R. Narayanan: Phys. Rev. E **54**, R3102 (1996)
20. D. Johnson, R. Narayanan: Phys. Rev. E **56**, 5462 (1997)
21. D. Johnson, R. Narayanan: Phil. Trans. R. Soc. Lond. A **356**, 885 (1998)
22. D. Johnson, R. Narayanan: Chaos **9**, 124 (1999)
23. E.S. Johnson: J. Crystal Growth **30**, 249 (1975)
24. A. Juel, J.M. Burgess, W.D. McCormick, J.B. Swift, H.L. Swinney: Physica D **143**, 169 (2000)

25. I.L. Kliakhandler, A. A. Nepomnyashchy, I.B. Simanovskii, M.A. Zaks: Phys. Rev. E **58**, 5765 (1998)
26. E.L. Koschmieder, S. Prahl: J. Fluid Mech. **215**, 571 (1990)
27. L.D. Landau, E.M. Lifshitz: *Fluid Mechanics*, (Pergamon Press, Oxford 1987)
28. J.C. Legros: Acta Astr. **13**, 697 (1986)
29. A.M. Leshansky: Int. J. Multiphase Flow **27**, 189 (2001)
30. E.B. Levchenko, A.L. Chernyakov: Sov. Phys. JETP **54**, 102 (1981)
31. J. Lucassen: Trans. Faraday Soc. **64**, 2221 (1968); *ibidem* **64**, 2230 (1968)
32. E.H. Lucassen-Reynders, J. Lucassen: Adv. Colloid Interface Sci. **2**, 347 (1969)
33. A.A. Nepomnyashchy, I.B. Simanovskii: Fluid Dyn. **18**, 629 (1983)
34. A.A. Nepomnyashchy, I.B. Simanovskii: Fluid Dyn. **25**, 340 (1990)
35. A.A. Nepomnyashchy, I.B. Simanovskii: Quarterly J. Mech. and Appl. Math. **50**, 149 (1997)
36. A.A. Nepomnyashchy, I.B. Simanovskii: Phys. Rev. E **59**, 6672 (1999)
37. A.A. Nepomnyashchy, I.B. Simanovskii, L. M. Braverman: Phys. of Fluids **12**, 1129 (2000)
38. A.A. Nepomnyashchy, I.B. Simanovskii: Eur. J. Mech. B/ Fluids **20**, 75 (2001)
39. D.A. Nield: J. Fluid Mech. **19**, 341 (1964)
40. O.V. Perestenko, L.K. Ingel: Izv. Akad. Nauk SSSR, Fiz. Atmosf. Okeana **27**, 408 (1991) (In Russian)
41. O.V. Perestenko, L.K. Ingel: J. Fluid Mech. **287**, 1 (1995)
42. L.M. Pismen: Dyn. Stab. Syst. **1**, 97 (1986)
43. M.L. Ramon, D. Maza, H. L. Mancini: Phys. Rev. E **60**, 4193 (1999)
44. M.L. Ramon, D. Maza, H.L. Mancini, A. Mancho, H. Herrero: Int. J. Bif. and Chaos **11**, 2779 (2001)
45. S. Rasenat, F.H. Busse, I. Rehberg: J. Fluid Mech. **199**, 519 (1989)
46. A.Y. Rednikov, P. Colinet, M.G. Velarde, J.C. Legros: Phys. Rev. E **57**, 2872 (1998)
47. J.R.A. Pearson: J. Fluid Mech. **4**, 489 (1958)
48. J. Reichenbach, H. Linde: J. Coll. Interface Sci. **84**, 433 (1981)
49. S. Rosenblat, G.M. Homsy, S.H. Davis: J. Fluid Mech. **120**, 91 (1982)
50. M. Silber, E. Knobloch: Nonlinearity **4**, 1063 (1991)
51. I.B. Simanovskii: Convective Stability of Two-Layer Systems. PhD Thesis, Leningrad State University, Leningrad (1980) (In Russian)
52. I.B. Simanovskii, A.A. Nepomnyashchy: *Convective Instabilities in Systems with Interface*, (Gordon and Breach, London 1993)
53. I.B. Simanovskii: Physica D **102**, 313 (1997)
54. I.B. Simanovskii: Eur. J. Mech. B/Fluids **19**, 123 (2000)
55. G.I. Sivashinsky: Physica D **8**, 243 (1983)
56. K.A. Smith: J. Fluid Mech. **24**, 401 (1966)
57. C.V. Sternling, L.E. Scriven: AIChE J. **5**, 514 (1959)
58. J.W. Swift: Nonlinearity **1**, 333 (1988)
59. M.S. Triantafyllou, G.S. Triantafyllou: J. Sound Vibr. **150**, 485 (1991)
60. S.J. VanHook, M.F. Schatz, W.D. McCormick, J.B. Swift, H.L. Swinney: Phys. Rev. Lett. **75**, 4397 (1995)
61. M.G. Velarde, A.A. Nepomnyashchy, M. Hennenberg: Adv. Appl. Mech. **37**, 167 (2001)
62. P. Welander: Tellus **16**, 349 (1964)
63. A.A. Zaman, R. Narayanan: J. Coll. Interface Sci. **179**, 151 (1996)

Salt-Finger Instability Generated by Surface-Tension and Buoyancy-Driven Convection in a Stratified Fluid Layer

C.F. Chen and Cho Lik Chan

Department of Aerospace and Mechanical Engineering, The University of Arizona, Tucson, AZ 85721 USA

Abstract. Consider a shallow layer of fluid stratified by a solute gradient with a free surface being heated differentially across the two side walls. Convection due to surface-tension and buoyancy effects will be generated. The cooler, solute-poor fluid sinks along the cold wall and flows toward the hot wall along the bottom of the tank. The warmer, solute-rich fluid returns along the free surface of the layer. The resulting vertical distributions of temperature and solute are conducive to the onset of salt-finger instability, thus generating secondary motion in the primary convection cell. Our recent experiments confirm the existence of such secondary motion. Experiments were conducted in a stratified fluid layer contained in a shallow tank 5 cm wide × 1 cm high × 10 cm long. One side wall (1 × 10 cm) was made of chrome-plated copper and can be maintained at a constant temperature. The opposite side wall was made of Plexiglas for visualization purposes. The fluid was an ethanol-water solution, initially stratified from 96% ethanol at the bottom to 100% ethanol at the top free surface. Experiments were carried out with the copper wall maintained at a temperature lower than the ambient. Secondary motion was first noted at an 0.8°C temperature difference. At a 1.7°C temperature difference, the secondary motion became fully developed. Flow visualization in a longitudinal plane perpendicular to the initial temperature gradient showed that the motion consisted of vortices with axes aligned with the direction of the primary motion. A discussion of the instability mechanism based on a parallel flow model of the actual flow is presented following the presentation of the experimental investigation.

1 Introduction

Recent experiments conducted by Chen and Chen [4] showed that salt-finger convection can be generated by combined capillary and buoyant motion along the free surface in a tank of ethanol-water solution stably stratified by the water concentration. The motion is mainly driven by surface tension gradients due to both thermal and solutal effects caused by a small temperature difference, ~ 1°C, across the tank. Buoyant cellular convection in the form of a vertical array of nearly horizontal cells as observed by Thorpe et al. [9] and Chen et al. [3] are absent because the critical ΔT for onset of such convection is much higher than for the initial density gradient in the fluid. At this low ΔT in the bulk fluid below the top convecting layer, the buoyant convection is in the stable region, and the motion is almost imperceptible due to the stabilizing effect of the solute gradient. When the convection in the top layer is well established, the flow on the free

C.F. Chen and Cho Lik Chan, Salt-Finger Instability Generated by Surface-Tension and Buoyancy-Driven Convection in a Stratified Fluid Layer, Lect. Notes Phys. **628**, 45–58 (2003)
http://www.springerlink.com/

surface is from the hot to the cold wall, with a slow return flow in the opposite direction underneath. The surface streams are relatively solute rich while the return flow is relatively solute poor. This situation is favorable for the onset of salt fingers and, furthermore, the fingers are occurring in the presence of shear. Linden [7] showed, under such circumstances, that finger convection appears as longitudinal rolls aligned in the direction of shear. He generated salt fingers in the presence of shear by a counterflow of sugar solution over a salt solution of greater density. The longitudinal rolls appeared in shadowgraphs as alternate dark and bright bands aligned in the direction of shear. Later, Thangam and Chen [8] generated such instabilities by discharging a heated saline jet on the surface of a long tank of fluid stably stratified by salt at rest. Visualization was obtained by Schlieren photographs, which exhibited the longitudinal structures. Chen and Chen [4] used particle trace photography for visualization in their experiments. They observed finger convection by the presence of longitudinal bands in the direction of shear. They further observed that the onset of finger convection occurred near the cold wall due to solutal capillary effects.

Chan and Chen [1] examined a similar flow in a shallow tank filled with an ethanol-water solution stratified from 96% to 100% ethanol. Capillary-driven convection with salt-finger instabilities was observed. In addition, a two-dimensional numerical simulation for the early times prior to the onset of salt fingers was carried out. Results show that steady convection onsets at the cold wall while oscillatory convection onsets at the hot wall. This is due to the fact that the surface tension of the ethanol-water solution increases with decreasing temperature and increasing water concentration. At the hot wall, the surface tension is reduced due to the elevated temperature causing a flow away from the hot surface. With this flow, the water-rich solution below is pulled up to the free surface, thus increasing the surface tension near the wall. A reverse flow is induced. The oscillatory flow persists for a few cycles, then steady convection away from the hot wall is established.

More recently, Chan et al. [2] revisited the salt-finger instability problem in buoyancy-generated layers in the stratified fluid first observed by Chen and Chen [4]. They were able to determine, using flow visualization, that the secondary flow generated by salt-finger instability is in the form of counter-rotating vortices in the longitudinal plane, which is perpendicular to the imposed temperature gradient. Results of a linear stability analysis of a parallel double-diffusive flow model of the actual flow showed that the instability is in the salt-finger mode under the experimental conditions. The streamlines in the longitudinal plane at onset consist of a horizontal row of counter-rotating vortices similar to those observed in the experiments.

In this paper, we report our recent work on applying some of the techniques discussed in [1] and [2] to the problem of salt-finger instability generated by surface-tension- and buoyancy-driven convection in a stratified fluid layer. The experiments are discussed in Sect. 2; the parallel flow model is present in Sect. 3; and stability is discussed in Sect. 4, followed by conclusions.

2 Experiments

2.1 Apparatus and Procedure

Experiments were carried out in a shallow tank 5 cm wide × 1 cm high × 10 cm long. One of the side walls (10 cm × 1 cm) was made of chrome-plated copper, with passages provided for circulating fluid from a constant-temperature bath. The other side wall was made of Plexiglas for visualization purposes, as were the two end walls and the bottom of the tank (see Fig. 1). The fluid used was an ethanol-water solution; it was chosen because its free surface is not easily contaminated. The concentration of the solution varied from 100 wt% ethanol at the top to 96 wt% at the bottom. This was achieved by injecting through a hypodermic needle three equal fluid layers with ethanol concentrations of 100%, 98%, and 96% successively. The needle was held perpendicularly with the exit just above the bottom boundary so that the lighter fluid was lifted upward by the inflow of heavier fluid. Each layer contained 16.7 cm^3 of fluid, and the injections were performed slowly to minimize mixing. The filled tank was covered to prevent evaporation and was not disturbed for 30 min so that diffusion could smooth out the sharp gradients between the layers. Since we have no means of measuring the concentration of ethanol, the diffusion process was simulated by a one-dimensional calculation, taking into account the effect of variable diffusivity. The concentration distribution 30 min after the start of the diffusion process was essentially linear with height, with non-diffusive effects near the top and bottom. With these results in mind, all experiments were begun 30 min after completion of the filling process.

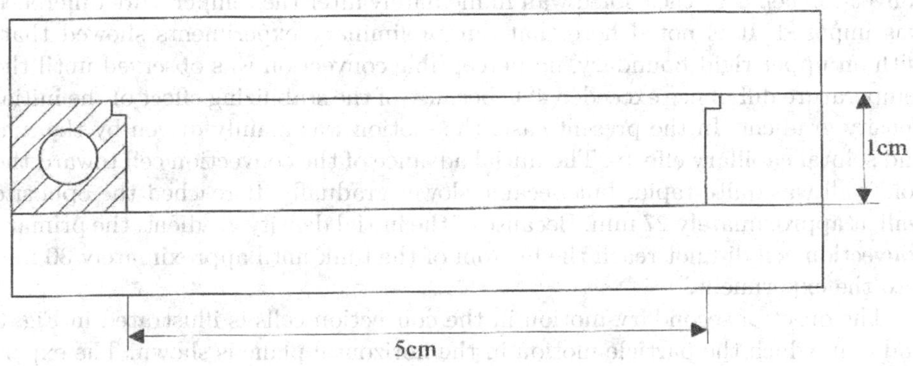

Fig. 1.

Flow visualization was carried out by particle traces with aluminum powder as the tracer. The particles were mixed into the prepared solutions for each layer prior to filling of the tank. A 20-mW He-Ne laser with a cylindrical lens was used to provide a sheet of light through the tank. A CCD video camera was used to image the particle path, which was viewed on a monitor and simultaneously recorded by a VCR. Both vertical and horizontal views were observed. For the horizontal view, the CCD camera was mounted vertically above the tank. The test tank was mounted on a platform capable of vertical motion. In this manner, particle traces could be viewed at any horizontal plane within the 1-cm height of the test tank. Two vertical views were obtained, one in the transverse plane parallel to the imposed temperature gradient and the other in the longitudinal plane perpendicular to the imposed temperature gradient.

2.2 Experimental Results

After a number of preliminary experiments were run to calibrate the system, three experiments were carried out with nearly identical initial conditions to obtain video records of the particle motion in the transverse, horizontal, and longitudinal planes. These conditions were that the initial stratification was from 96% to 100% ethanol, and the temperature of the heat-transfer wall was maintained at $1.7 - 1.9$°C below the initial temperature of the solution. Time-lapse photographs were made of the images displayed on a monitor to record the particle traces.

The motion in the transverse plane is shown in Fig. 2. The experiment was started at 09:50 with an initial temperature difference of 1.7°C. The four photos were taken at 2, 3, 15, and 27 min into the experiment at a 4-sec time exposure. Almost the entire width of the tank, 5 cm, is in view, with the cooling wall on the left and the transparent wall on the right. The onset of the primary convection cell is at the cooled wall immediately after the temperature difference was imposed. It is noted here that our preliminary experiments showed that, with an upper rigid boundary, no perceptible convection was observed until the temperature difference exceeded 4°C because of the stabilizing effect of the initial density gradient. In the present case, the motion was mainly driven by thermal and solutal capillary effects. The initial advance of the convection cell toward the cold wall was quite rapid, but became slower gradually. It reached the opposite wall at approximately 27 min. Because of the initial density gradient, the primary convection cell did not reach the bottom of the tank until approximately 30 min into the experiment.

The onset of secondary motion in the convection cells is illustrated in Fig. 3 and 4, in which the particle motion in the horizontal plane is shown. The experiment was started at 14:41 with an initial temperature difference of 1.9°C. The four photographs were taken at 2, 3, 16, and 26 min into the experiment at an 8-sec time exposure. The horizontal laser sheet was set at the mid-height of the tank. It is seen that even at 2 min, when the primary convection cell only reached half way across the tank, the three-dimensional nature of the motion is clearly

Fig. 2. Video record of particle motion in the transverse plane

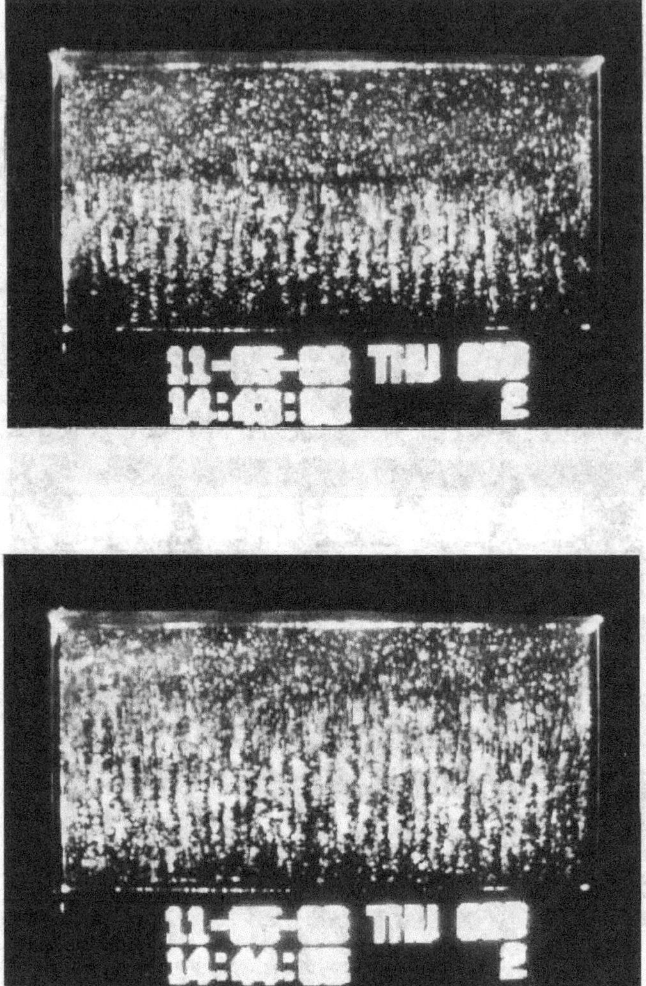

Fig. 3. Record of particle motion in the horizontal plane at 2 and 3 minutes into the experiment

exhibited in the presence of longitudinal rows, where particles are gathered. This flow pattern persisted until 26 min into the experiment.

In order to examine more closely the motion in the longitudinal plane, the vertical laser sheet was broadened to 7-8 mm thick and the centerline of sheet was placed at 1.7 cm from the cold wall, one third of the width. The photos showing the middle 7 cm of the 10-cm length of the tank are presented in Fig. 5. The experiment was started at 11:23 with an initial temperature of 1.7°C. The four photos were taken at 2, 3, 18, and 27 min into the experiment. The photos at 2 and 3 min show the head of the primary convection cell reaching to about

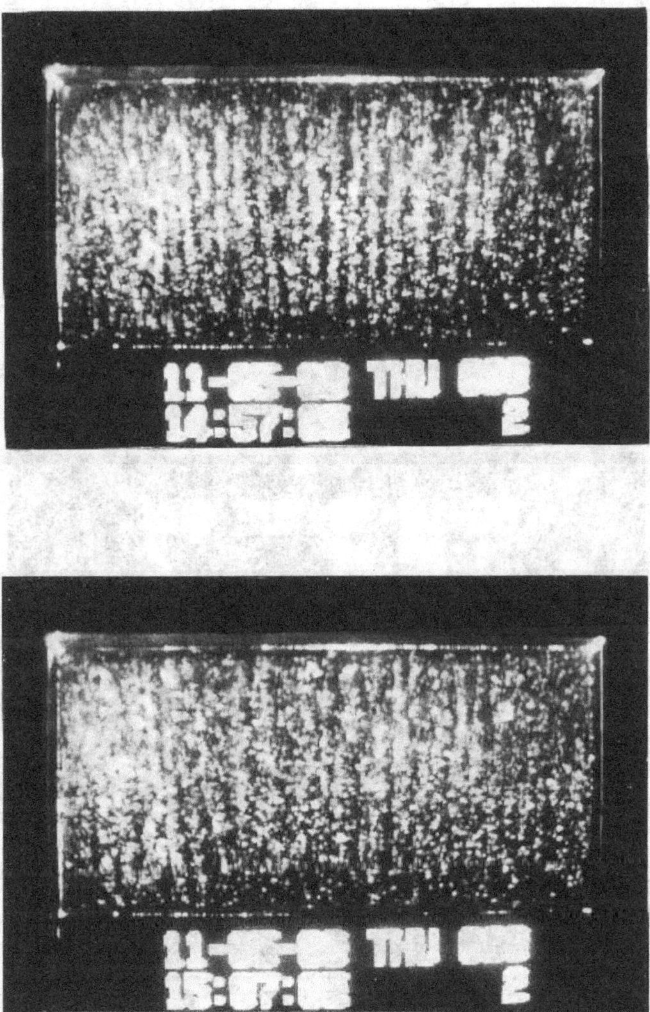

Fig. 4. Record of particle motion in the horizontal plane at 16 and 26 minutes into the experiment

one third of the width of the tank, with particles flowing vertically upward. At 18 min, vortical motion is clearly exhibited, and became well developed at 27 min. Photos taken at later times show deterioration of the vortex structure due to the uncontrolled temperature at the warm wall and the mixing of the solute. Similar structures were observed in buoyancy-driven convection cells in a stratified fluid by us [2], and they were shown to be the manifestation of salt-finger instability.

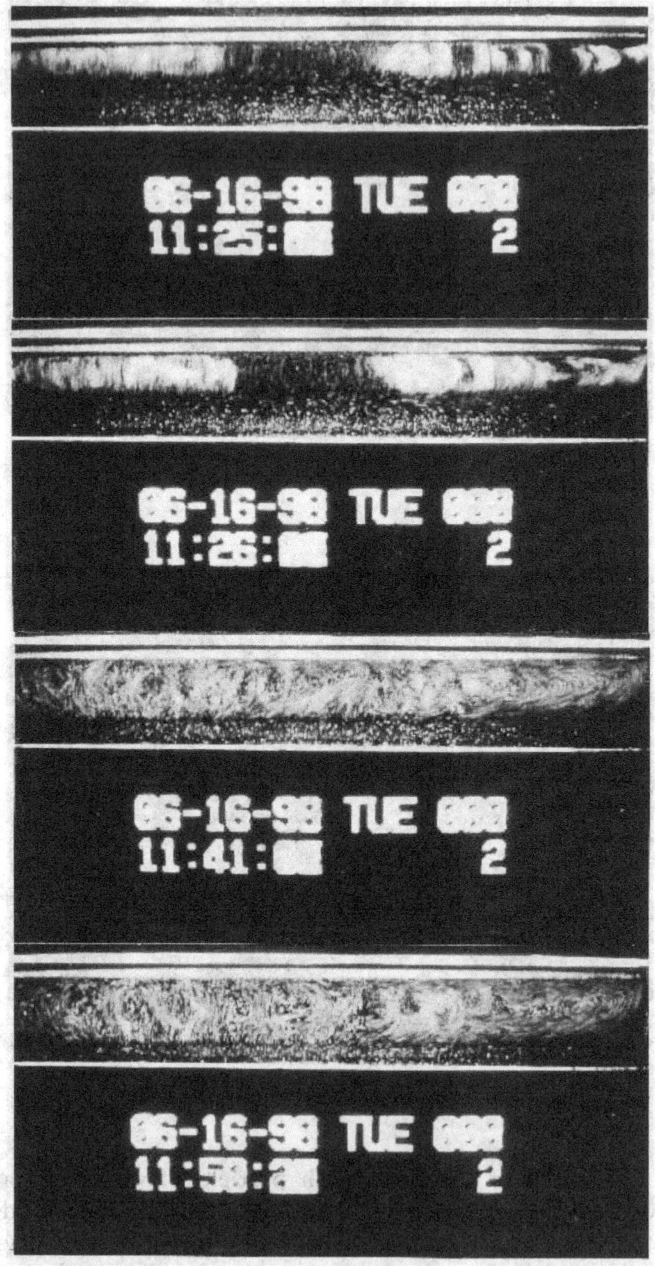

Fig. 5. Motion in the longitudinal plane – the flows in the middle 7 cm of the 10 cm length of the tank are depicted

3 Parallel Flow Model

As we have seen, the flow in the primary convection cell consists of a comparatively warmer and solute-rich fluid flowing above a return flow of cooler and solute-poor fluid. This is remarkably similar to the primary convection flow in layers generated by buoyancy alone in a stratified fluid under sideways heating. Chan et al. [2] carried out a successful linear stability analysis based on a parallel flow model of the actual flow for such a case. It is also noted that parallel flow models have been used successfully by Hart [5], to study convection in a non-rotating atmospheric layer, and by Jeevaraj and Imberger [6], to study the flow in a double-diffusive intrusion. It seems reasonable to expect that a suitable parallel flow model of the present problem can help us to understand the mechanisms that are responsible for the observed instability.

Once the flow in the primary convection cell is established, there is a temperature as well as a solute concentration difference between the two ends. This is because a warmer and solute-rich fluid at the bottom of the layer is being convected upward with the heated fluid at the hot end, and a colder and solute-poor fluid at the top of the layer is being convected downward with the cooled fluid at the cold end. We construct a parallel flow model for the present case as being a fluid layer contained in a shallow tank with imposed constant temperature and solute differences at the two ends. The top surface of the fluid is free. It is assumed that the top and the bottom of the layer are maintained at given linearly varying temperatures and solute distributions between the two end values. The bottom of the layer is rigid while surface tension can act on the free top surface.

To render these ideas into a more concrete form, we consider a long, shallow tank with thickness D, see sketch of the model in Fig. 6. The x-axis is horizontal and the z-axis is vertically upward, with the origin situated at the center of the tank. The flow in the middle portion of the layer away from the ends is nearly horizontal. The velocity components are assumed to be

$$u = u(z), \quad v = w = 0 \tag{1}$$

and the temperature, T, and solute concentration (i.e., water concentration in the present case), S, are linear in x

$$T = T_0 + \gamma x, \quad S = S_0 + \gamma_S x \tag{2}$$

in which γ and γ_S are constant horizontal gradients of T and S, respectively.

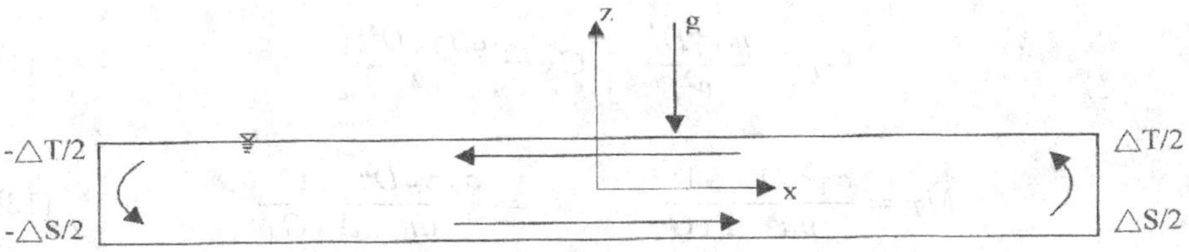

Fig. 6. Sketch of the model

The density ρ and the surface tension λ of the fluid are assumed to be linear in T and S

$$\rho = \rho_0[1 - \alpha(T - T_0) + \beta(S - S_0)] \tag{3}$$

$$\lambda = \lambda_0[1 - \sigma_T(T - T_0) + \sigma_S(S - S_0)] \tag{4}$$

in which α and β are volumetric expansion coefficients and σ_T and σ_S are the surface tension gradients with respect to T and S, respectively. It is noted that, for ethanol-water solutions, $\sigma_T > 0$ and $\sigma_S < 0$. The subscript "0" denotes the reference state. When the parallel flow model is applied to the momentum, energy, and solute conservation equations for an incompressible Boussinesq fluid, we obtain three simple ordinary differential equations. In non-dimensional form, they are

$$\frac{d^3u}{dz^3} = 1 - G_S/G_T \tag{5}$$

$$\frac{d^2T}{dz^2} = G_T \mathrm{Pr}\, u \tag{6}$$

$$\frac{d^2S}{dz^2} = \mathrm{Le}\, G_T \mathrm{Pr}\, u \tag{7}$$

with boundary conditions

$$T = S = 0, \quad \frac{du}{dz} = -(M_T + M_S) = -M, \quad \text{at } z = \frac{1}{2} \tag{8}$$

$$T = S = u = 0, \quad \text{at } z = -\frac{1}{2} \tag{9}$$

$$\int_{-1/2}^{1/2} u(z)\, dz = 0 \tag{10}$$

The thermal and solutal Grashof numbers G_T and G_S, Marangoni numbers M_T and M_S, and the Prandtl and Lewis numbers are defined as

$$G_T = \frac{g\alpha\gamma D^4}{v^2}, \quad G_S = \frac{g\beta\gamma_S D^4}{v^2} \tag{11}$$

$$M_T = \frac{\sigma_T \gamma D^2}{\mu\kappa} \frac{1}{\mathrm{Pr}\, G_T}, \quad M_S = \frac{\sigma_S \gamma_S D^2}{\mu\kappa} \frac{1}{\mathrm{Pr}\, G_T} \tag{12}$$

$$\mathrm{Pr} = \frac{v}{\kappa}, \quad \mathrm{Le} = \frac{\kappa}{\kappa_S} \tag{13}$$

The characteristic length, time, velocity, pressure, temperature, and solute concentrations are D, D^2/v, $g\alpha\gamma D^3/v$, $\rho_0(g\alpha\gamma D^3/v)^2$, γD, and $\gamma_S D$, respectively.

For the basic steady parallel flow, $V_b = (u_b(z), 0, 0)$, $T_b = T_b(x, z)$, and $S_b = (x, z)$, the solutions are

$$u_b = \frac{1}{6}[1 - \frac{G_S}{G_T}]F(z; M) \tag{14}$$

$$T_b = x + \frac{G_T \mathrm{Pr}}{6}[1 - \frac{G_S}{G_T}]G(z; M) \tag{15}$$

$$S_b = x + \frac{\mathrm{Le} G_T \mathrm{Pr}}{6}[1 - \frac{G_S}{G_T}]G(z; M) \tag{16}$$

$$F(z; M) = \frac{z^3}{6} - [\frac{1}{16} + \frac{3M}{4}]z^2 - [\frac{1}{16} + \frac{M}{4}]z + [\frac{1}{192} + \frac{M}{16}] \tag{17}$$

$$G(z; M) = \frac{z^5}{120} - \frac{1}{12}[\frac{1}{16} + \frac{3M}{4}]z^4 - \frac{1}{6}[\frac{1}{16} + \frac{M}{4}]z^3 + \frac{1}{2}[\frac{1}{192} + \frac{M}{16}]z^2 + [\frac{1}{480} + \frac{M}{96}]z - [\frac{1}{3072} + \frac{M}{256}] \tag{18}$$

The scaled velocity $F(z; M)$ is shown in Fig. 7a for $M = 0$ and 0.1. For both cases, flow is toward the hot wall in the lower half of the tank and toward the cold wall in the upper half of the tank. In the absence of surface tension, $M = 0$, $du/dz = 0$ at the free surface. When $M = 0.1$, the effect of surface tension is to increase the return velocity toward the cold wall. The scaled temperature $G(z)$ is shown in Fig. 7b. It is seen that the effect of surface tension is to increase the difference in the temperatures of the hot and cold streams, thus increasing the heat transfer to the cold wall as well as to the top surface.

4 Discussion

Equations (15) and (16) indicate that the scaled temperature and solute concentration distributions across the tank are proportional to each other. In Fig. 7c, we present these two distributions with an arbitrary constant of proportionality of 0.7 to show the regions of potential double-diffusive instability. In the top and the bottom regions, where both the both temperature and solute concentration decrease with height. This is the situation of heating a stably stratified layer from below and, when the adverse temperature gradient exceeds the critical, instability onsets in the oscillatory mode. This is known as diffusive instability. In the comparatively thick middle region, both T and S increase with height and, when the destabilizing solute gradient exceeds the critical value, instability

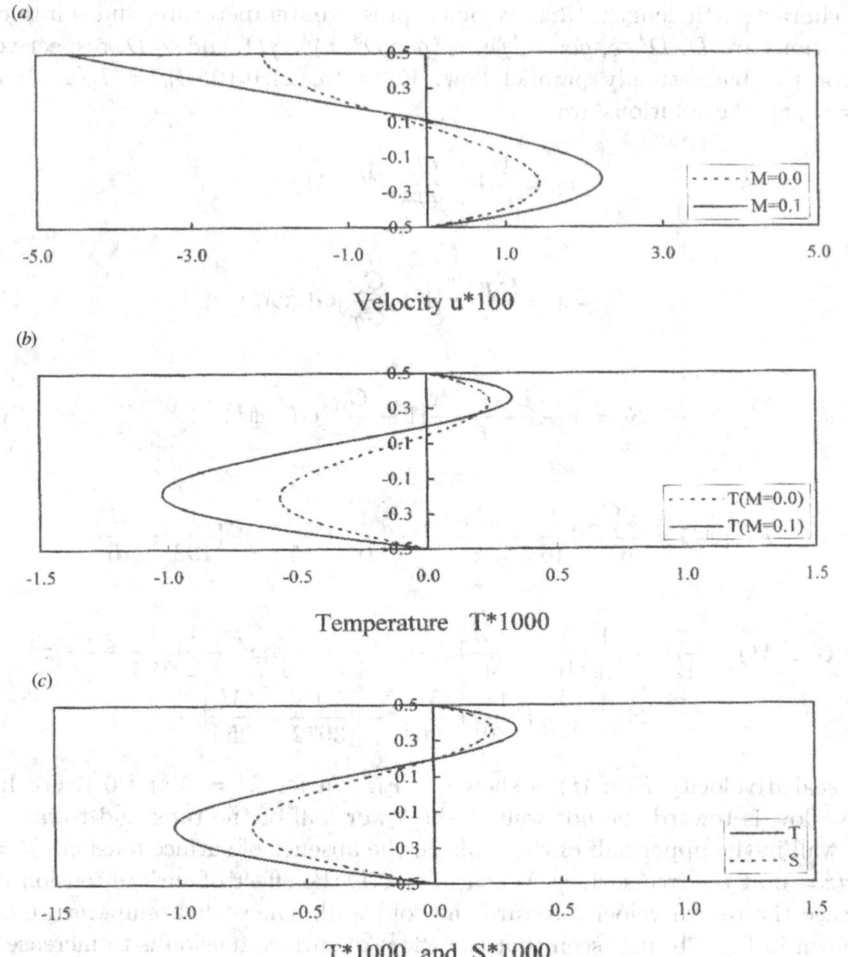

Fig. 7. Scaled velocity and temperature distributions

onsets in the steady convection mode. In most laboratory experiments, an unstable state is easily encountered, and it is usually in the highly supercritical state. Under such conditions, the instability appears as long, narrow, vertical cells, aptly named salt fingers. Such instability is known as salt-finger instability, even though at the marginal state the convection cells are nearly square.

It is seen that, within the flow field, the potential for both diffusive and salt-finger instability exists. For given lateral temperature and solute concentration differences between the two ends, which mode of instability is the most critical one depends on the values of the thermal and solutal Grashof and Marangoni numbers. If the diffusive mode is the most critical one, there will be two rows of generally smaller convection cells confined to the top and bottom layers. If the salt-finger mode is the most critical one, a row of larger convection cells will appear in the middle portion of the tank.

We have considered [2] the parallel flow model with rigid-rigid boundary conditions for a double-diffusive flow generated in a long shallow tank by imposed lateral temperature and solute concentration differences. We found that the velocity, temperature, and the solute distributions in the tank are exactly antisymmetric with respect to the z-axis. Similar to the present case, there are three regions of potential instability, diffusive instability in the top and bottom layers, and salt-finger instability in the middle portion of the tank. We have also carried out a linear stability analysis considering disturbances in the (x, z) plane and the (y, z) plane separately. The marginal stability boundary is shown in the (Gr, Gr_S) plane for given values of Prandtl and Lewis numbers. It is noted that the Marangoni number is not a relevant parameter since there is no surface-tension effect in the problem. For a given pair of values of Gr and Gr_S, one can determine whether the flow is stable, and the mode of instability when the flow is unstable. For the experimental values of the Grashof numbers in which longitudinal vortices were observed, the most critical mode as predicted by stability theory is the salt-finger mode with counter-rotating vortices in the middle of the tank. The predictions compare well with experimental observations. It is conjectured that, for the present case, the salt-finger mode is likely the most critical, and the instability would appear as a row of counter-rotating vortices, as shown in Fig. 4.

5 Conclusions

We have carried out an experimental and analytical study of the secondary flow generated by lateral heating of a shallow, stably stratified fluid layer. It was found that, almost immediately after the lateral temperature difference was applied, a primary convection cell started at the cold end due to the combined effects of thermal and solutal capillarity. As the convection cell advanced toward the warm end, a secondary motion was generated. By using flow visualization techniques, we observed that the particle traces in the longitudinal plane formed a row of vortices whose axes were aligned in the direction of the temperature gradient.

By using a parallel flow model, we determined a simple basic flow that has approximately the same characteristics of the experimental flow field. The temperature and the solute distributions across the tank exhibited regions of potential double-diffusive instability, either in the diffusive mode or in the salt-finger mode. Based on the earlier experiences of Chan et al. [2], it is conjectured that the secondary flow observed in the experiments is the consequence of salt-finger instability that is occurring in the middle portion of the tank.

References

1. C.L. Chan, C.F. Chen: Int. J. Heat Mass Transf. **42**, 2143 (1999)
2. C.L. Chan, W.-Y. Chen, C.F. Chen: J. Fluid Mech. **455**, 1 (2002)
3. C.F. Chen, D.G. Briggs, R.A. Wirtz: Int. J. Heat Mass Transf. **14**, 57 (1971)
4. C.F. Chen, F. Chen: J. Fluid Mech. **352**, 161 (1997)
5. J.E. Hart: J. Atmos. Sci. **29**, 687 (1972)
6. C.G. Jeevaraj, J. Imberger: J. Fluid Mech. **222**, 565 (1991)
7. P.F. Linden: Geophys. Fluid Dyn. **6**, 1 (1974)
8. S. Thangam, C.F. Chen: Geophys. Astrophys. Fluid Dyn. **18**, 111 (1981)
9. S.A. Thorpe, P.K. Hutt., R. Soulsby: J. Fluid Mech. **38**, 375 (1969)

Observations on Interfacial Convection in Multiple Layers without and with Evaporation

O. Ozen[1], D. Johnson[2], and R. Narayanan[1]

[1] Department of Chemical Engineering, University of Florida, Gainesville, Florida 32611, USA
[2] Department of Chemical Engineering, University of Alabama, Tuscaloosa, Alabama, 35487, USA

Abstract. A tutorial is presented on convective instabilities that arise from thermocapillary and buoyancy effects in the presence of evaporation. It focuses on the effect of multiple layers on the nature of the instabilities. A comprehensive explanation of the physics of this type of convection is accompanied by results from calculations and their interpretation.

1 Introduction

This article is about the influence of evaporation on the Rayleigh-Marangoni-Bénard (RMB) problem, a classical problem in fluid mechanics that is concerned with pattern formation due to convection. The RMB problem in the absence of evaporation became noticed via the experiments of Bénard [1] who applied a temperature gradient across a vertical layer of spermaceti and observed that hexagonal cells formed in the liquid layer. Rayleigh [15] gave the first analysis of the instability in the absence of interfacial tension gradients and deflecting surfaces. He attributed the pattern formation in Bénard experiments entirely to buoyancy driven convection. However, Block [2] later identified that the mechanism responsible for the convection in Bénard's experiments was due primarily to interfacial tension gradient-driven convection.

This type of convection has been called Marangoni convection by Scriven and Sternling [17] because it is connected to a qualitative explanation of the phenomenon given much earlier by Marangoni. Applications where convective phenomena play a part are plentiful in chemical engineering, ranging from crystal growth and the coating of films to the casting of alloys and the space processing of glasses and semiconductors. Consequently much analytical, computational and experimental work has been done on this problem and continue to fill the literature wherein the effect of magnetic fields, rotation, porous media and non-Newtonian constitutive equations have been considered. We do not intend to give a review of the problem. Instead we have chosen to write a brief explanation of the problem focusing only on the physics of convection in single and multiple fluid layers and how they are affected by evaporation. The next section deals with the physics of convection without evaporation. It is followed by a section concerned with evaporative instabilities on its own. The article then goes on to give a brief explanation of some of the special features of evaporative interaction

O. Ozen, D. Johnson, and R. Narayanan, Observations on Interfacial Convection in Multiple Layers without and with Evaporation, Lect. Notes Phys. **628**, 59–77 (2003)
http://www.springerlink.com/

with convection followed by a section on the mathematical aspects of the problem. It concludes with some of our thoughts on the scope for future work. The section on physics is a commentary on the convection problem and the interested reader would do well to first read the relevant chapter in the classical treatise by Chandrasekhar [3].

2 The Physics of Buoyancy and Interfacial-Gradient-Driven Convection in the Absence of Evaporation

Before introducing evaporation we shall discuss the pure buoyancy and pure interfacial tension gradient convection problems.

2.1 Pure Rayleigh Convection or Buoyancy-Driven Convection

Buoyancy-driven convection, often referred to as natural convection or Rayleigh convection, occurs when a fluid is subject to a temperature gradient in a gravitational field and when there is a variation of density with respect to temperature. Imagine a layer of liquid bounded vertically by two horizontal rigid plates, with the lower plate at a temperature higher than the upper one. As density typically decreases with an increase in temperature, the fluid near the top plate is heavier than the fluid at the bottom plate, creating a top heavy arrangement, one which is gravitationally unstable. For sufficiently small temperature differences, the fluid simply conducts heat from the lower plate to the upper plate, creating a linear temperature field across the fluid. Now imagine a mechanical perturbation being imparted to an element of fluid so that it is projected downward. As the density of the "fluid packet" is greater than its environment it will proceed downwards. Due to mass conservation, fluid from below will be displaced and will move upwards. This motion will continue unimpeded unless the fluid's kinematic viscosity and thermal diffusivity are high enough so that the mechanical perturbations and thermal perturbations die out quickly and the original mechanically quiescent, thermally conductive state is re-established. Under steady conditions there is a balance that is established between these effects, represented by the dimensionless Rayleigh number, which arises from the scaling of the modeling equations. The Rayleigh number is given by

$$Ra = \frac{\alpha g \Delta T L^3}{\nu \kappa}$$

Here, α is the negative thermal expansion coefficient and is usually positive, g is the magnitude of gravity, ΔT is the vertical temperature difference across the fluid layer and is taken to be positive when the fluid is heated from below, L is the depth of the fluid, ν is its kinematic viscosity, and κ is its thermal diffusivity. If the temperature difference is increased beyond what will be referred to as the critical temperature difference, then the gravitational instability overcomes the viscous and thermal damping effects and the fluid is set into motion, causing buoyancy-driven convection.

2.2 Pure Marangoni Convection or Interfacial Tension Gradient-Driven Convection

Interfacial tension gradient-driven convection or Marangoni convection is unlike buoyancy-driven convection and can occur in a fluid even in the absence of a gravitational field. Imagine a layer of fluid which is bounded below by a rigid plate and whose upper surface is in contact with a passive gas (Fig. 1a). A passive gas is not one without density or viscosity. It is merely a gas wherein one tacitly assumes that thermal or mechanical perturbations cannot occur. Pretend that above the passive gas, there is another rigid plate. For the sake of consistency with the earlier argument, let the lower plate be at a temperature higher than the upper plate's temperature. Now, imagine that the interface between the lower liquid and the passive gas is momentarily disturbed. The regions of the interface that are pushed up experience a cooler temperature. Likewise, the regions of the interface that are pushed down increase in temperature. Typically, interfacial tension decreases with an increase in temperature. Therefore, the regions of the interface which are pushed up increase in interfacial tension, while the regions of the interface which are pushed down, decrease in interfacial tension. When the fluid is pulled along the interface, warmer fluid from the bulk replaces the fluid at the interface near 'x' enhancing the interfacial tension gradient-driven flow. If the temperature difference across the liquid is sufficiently small, then the thermal diffusivity of the fluid will conduct away the heat faster than the disturbance can amplify it. The dynamic viscosity will also resist the flow causing the interface to become flat again and the interfacial tension to become constant. As was the case in buoyancy-driven convection, there exists a critical temperature difference when the interfacial tension gradient-driven flow is not dampened by the thermal diffusivity or viscosity, and the fluid is set into motion. Scaling of the equations that model the interfacial momentum balance lead to a dimensionless group that exhibits the balance between dissipative effects of the viscosity and thermal

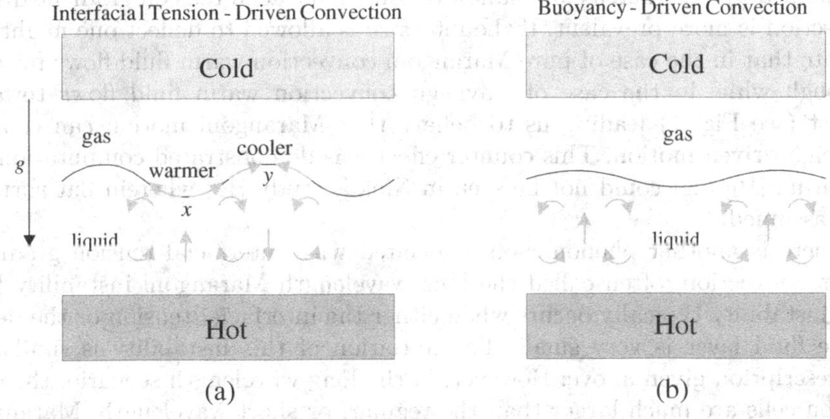

Fig. 1. The physics of pure convection

diffusivity and the promoting effects of the interfacial tension gradient. This dimensionless group is called the Marangoni number, Ma and is given by:

$$Ma = \frac{\gamma_T \Delta T L}{\mu \kappa}$$

Here γ_T is the negative of the gradient of the interfacial tension with respect to temperature, and μ is the dynamic viscosity of the fluid.

It is noteworthy that while the above argument assumes the interface to be deflectable it is not necessary for the interface to undulate in order for Marangoni convection to take place. To see this a little more clearly, look at Fig. 1a, but pretend that the interface is flat. Points 'x' and 'y' will therefore be on the same horizontal level. Imagine an element of fluid projected downward from the interface at point 'y'. Fluid nearby must replace the lost fluid at 'y' and this fluid comes from the region near 'x'. Consequently hot fluid from below moves to 'x'. If the interfacial tension decreases with temperature, as is the case in most substances, then fluid flows from warm surface regions to cold surface regions. Therefore, fluid moves from 'x' to 'y' and the convection continues unless large thermal diffusivity and viscosity damp out the disturbances. This means, theoretically speaking, that Marangoni convection can be analyzed assuming a flat free surface i.e. a surface with an infinite interfacial tension. This fact was observed by Pearson [14] who gave the first theoretical analysis of Marangoni convection using a methodology similar to that used by Rayleigh for the case of buoyancy driven convection.

The extent of either Rayleigh convection or Marangoni convection is primarily a function of the fluid depths. By examining the definitions of Rayleigh and Marangoni numbers we notice that Rayleigh convection is proportional to the cube of the fluid depth while Marangoni convection is directly proportional to it. This ought to come as no surprise as Rayleigh convection arises from body forces whereas Marangoni convection arises from surface forces. From these scaling arguments, we can conclude that for deep fluids, buoyancy-driven convection is more prevalent, whereas for shallow depths, interfacial tension gradient-driven convection is more prevalent. If the interface is allowed to deflect one might anticipate that in the case of pure Marangoni convection warm fluid flows towards a trough while in the case of Rayleigh convection warm fluid flows towards a crest (see Fig. 1) leading us to believe that Marangoni motion can counter Rayleigh driven motion. This counter effect was demonstrated computationally by Sarma [16] and could not be seen in Nield's study [12] wherein flat surfaces were assumed.

There is another phenomenon associated with interfacial tension gradient-driven convection, often called the long wavelength Marangoni instability [13]. This instability typically occurs when either the interfacial tension or the depth of the fluid layer is very small. The initiation of this instability is similar to the description given above. However, in the long wavelength scenario, the convection cells are much larger than the regular, or short wavelength, Marangoni convection. As the convection propagates, it causes large scale deformation in the interface which can actually cause the interface to rupture; that is the interface deforms to such an extent that it comes in contact with the lower plate. This phenomenon occurs in the drying of films and in coating processes.

The Marangoni- Rayleigh interaction takes on interesting proportions when multiple fluid layers are considered and it is to this task that we now turn.

2.3 Physical Effects of Multiple Fluid Layers

Johnson and Narayanan [8,9] give complete details of the different types of convection in multiple stacked fluid layers and only a brief review will be given here. Imagine now a less dense, immiscible layer of fluid above a lower layer of liquid. Here the lower layer is bounded below by a rigid, conducting plate and another rigid, conducting plate bounds the upper layer. Once again let the temperature of the lower plate be greater than the upper plate.

In two vertically stacked, immiscible fluids, there are five different ways in which the convection in the two fluids can couple (Fig. 2). The first way is called the lower dragging mode, where the lower layer becomes unstable to buoyancy-driven convection and drags the upper fluid (Fig. 2a). The reverse of this situation is given in Fig. 2d where the upper fluid becomes unstable to buoyancy-driven convection and drags the lower fluid. This cannot occur if the upper fluid is a gas. At certain fluid depths, buoyancy forces may be equally unstable in both fluid layers, where one of two scenarios may occur. The first scenario is called mechanical or viscous coupling (Fig. 2b), and is characterized by the rolls in each fluid layer rotating in opposite directions. It is important to note here that each fluid layer buoyantly convects independently and consequently the interface becomes an isotherm in the case of pure viscous coupling. The second scenario is called thermal coupling (Fig. 2c) and can be distinguished by rolls in each layer rotating in the same direction. In order to maintain no slip at the interface, there must exist a line of vanishing velocity. A situation can occur when at particular fluid depths where the system tries to couple both mechanically and thermally. As both cannot occur simultaneously, the system oscillates between these two states at the onset of convection. This was first observed numerically by Gershuni and Zhukovitskii [7] and later shown experimentally by Degen et al. [5]

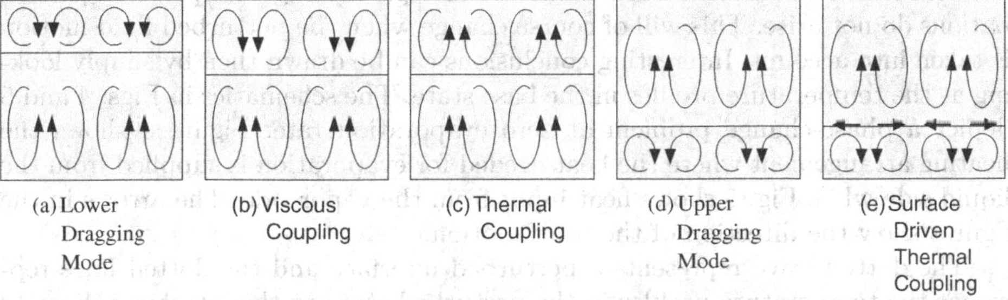

Fig. 2. Coupling between multiple fluid layers

The last figure (Fig. 2e) is an example of what be may called surface-driven thermal coupling. This typically occurs in a liquid-gas system where buoyancy convection is predominant in the gas layer. The convecting gas simultaneously creates a non-uniform temperature profile across the liquid-gas interface immediately causing either Marangoni or buoyancy-driven convection in the lower layer [10]. Notice that the convection in the lower layer is now generated purely by horizontal temperature gradients at the interface and not by viscous dragging.

A completely different type of instability is also possible in two layers of liquids being heated from above. It is called anti convection and discussed in a companion paper in this volume by Nepomnyaschyy et.al. This is of little interest in problems where evaporation also occurs.

Until now we have discussed the physics of convection in the absence of evaporation. Evaporation is by itself an interesting phenomena which on its own merit leads to interfacial instability. We first turn to an explanation of evaporation in the absence of convection and then describe the role of convection on evaporative instability. This is followed by a section on modelling with an interpretation of results.

3 Physics of Evaporative Instability without Convection

To get going lets consider an evaporating liquid in contact with its own vapor. Imagine a liquid vapor bilayer heated from the liquid side and cooled from the vapor side. The only input variables in this problem are the fluid depths, the temperatures of the hot and cold plate and the pressure at the top cold plate. The base state is one where we have a flat interface and it is the stability of this state that is in question. The liquid and vapor depths may be assumed fixed by suitably adjusting the liquid feed and vapor removal rates. In particular, it is possible to adjust the upper plate pressure in such a way that the evaporation or condensation rate is zero and for convenience we will present our results in this framework. To reveal the physics describing this problem we proceed stepwise in constructing the physical model.

First, imagine that the fluid dynamics is left out. Figure 3 shows a constant rate of feed and efflux in the base state and this may even be zero. By imagining that the fluid mechanics is omitted we are also pretending that pressure perturbations do not arise. This will of course change when the perturbed fluid motion is taken into account. Interesting conclusions can be drawn then by simply looking at the temperature profiles in the base state. The schematics in Figs. 4 and 5 depict a phase-change problem at zero evaporation rate. Figure 4 shows the heating arrangement where the heat needed for evaporation is supplied from the liquid side while Fig. 5 shows heat input from the vapor side. The arrows in the figures show the direction of the front motion.

The dotted wave represents a perturbed interface and the dotted lines represent the temperature profiles in the perturbed state at the interface. We note that the temperature at the perturbed interface is the same as the unperturbed state because the temperature is assumed to be in equilibrium with the pressure

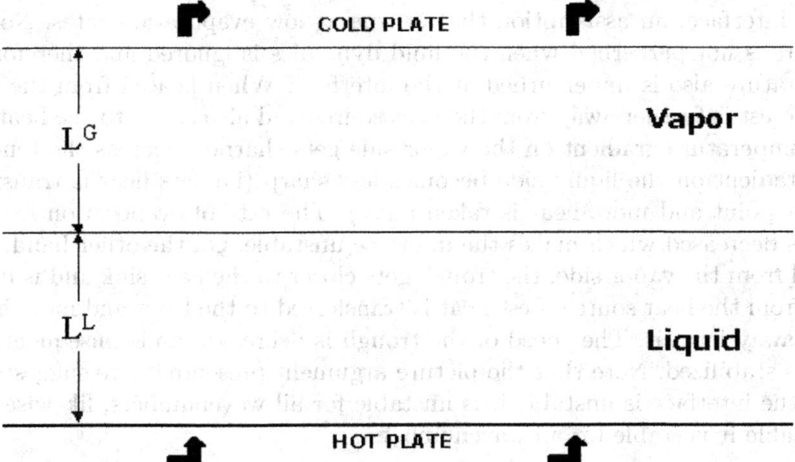

Fig. 3. The physical schematic for an evoporating bi-layer

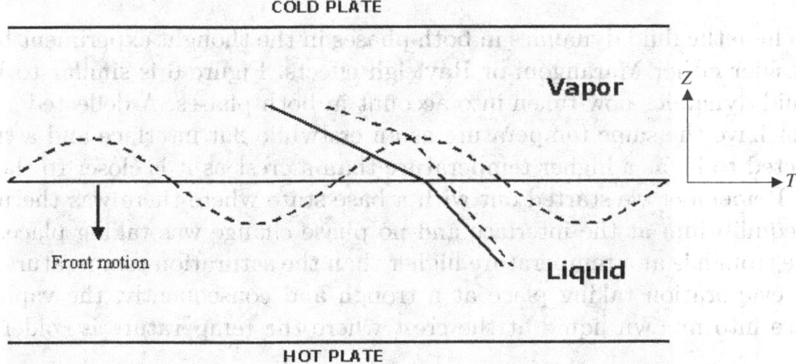

Fig. 4. The physics of evaporation when heat is supplied from the liquid side

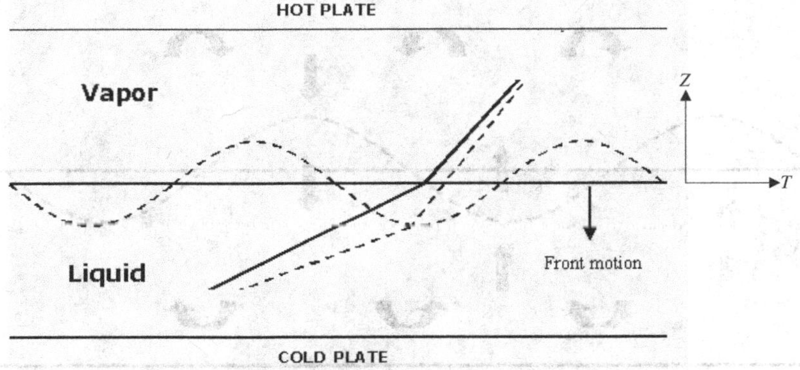

Fig. 5. The physics of evaporation when heat is supplied from the vapor side. Fluid motion is ignored

at the interface, an assumption that is true at low evaporation rates. Now the pressure is not perturbed when the fluid dynamics is ignored and therefore the temperature also is unperturbed at the interface. When heated from the liquid side a crest is further away from the heat source and also closer to the heat sink. The temperature gradient on the vapor side gets sharper whereas the temperature gradient on the liquid side becomes less sharp (i.e. less heat is transferred to that point and more heat is taken away). The rate of evaporation from the crest is decreased which makes the interface unstable. On the other hand, when heated from the vapor side, the trough gets closer to the heat sink and is further away from the heat source. Less heat is transferred to the front and more heat is taken away from it. The speed of the trough is decreased and consequently the front is stabilized. Note that the picture argument presented here suggests that when the interface is unstable it is unstable for all wavenumbers; likewise when it is stable it is stable for all wavenumbers.

4 The Physics of Evaporation with Convection

Now include the fluid dynamics in both phases in the thought experiment but do not consider either Marangoni or Rayleigh effects. Figure 6 is similar to Fig. 4 with fluid dynamics now taken into account in both phases. A deflected surface does not have the same temperature as an erstwhile flat interface and a trough is expected to be at a higher temperature than a crest as it is closer to the heat source. Remember we started out with a base state where there was thermodynamic equilibrium at the interface and no phase change was taking place. Now that the trough is at a temperature higher than the saturation temperature there will be evaporation taking place at a trough and consequently, the vapor will condense into its own liquid at the crest where the temperature is colder than

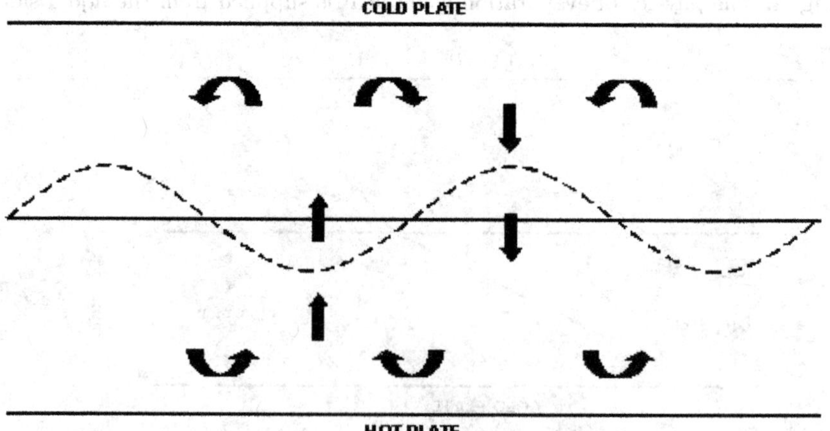

Fig. 6. Evaporation with fluid flow

the saturation temperature when perturbed. There will be upward flow at the troughs and downward flow at the crests as indicated in Fig. 6. Again, instability occurs at every wavenumber. As in other phase-change instability problems it is surface tension that comes in to stabilize the situation as it does so strongly at small wavelengths. But interfacial tension comes attended by the complication that it depends upon temperature and this is a complication that we will account for in the mathematical model along with the reality that buoyancy also must influence the physics. We will discuss these additional but important features when the results of the calculations are given.

Much of the energy required for evaporation may be expected to go into the kinetic energy of the vapor which has a high velocity because of its low density. Evaporation can therefore be expected to offer considerable stabilization to the interfacial gradient instability problem. Now the behavior with wave number of the evaporation instability problem is quite different than the Marangoni problem at low wave numbers and a model and calculations are required to offer insight into the combined physics, and it is to this that we now proceed.

5 The Model

The reader is referred to Fig. 3 where some of the nomenclature is defined. The superscripts L and G stand for liquid and vapor, respectively. The equations that model the physics in unscaled form are given by the Navier Stokes equations in each phase i.e.,

$$\rho^L\left(\frac{\partial \boldsymbol{v}^L}{\partial t}+\boldsymbol{v}^L\cdot\nabla\boldsymbol{v}^L\right)=-\nabla P^L+\rho^L\boldsymbol{g}+\mu^L\nabla^2\boldsymbol{v}^L$$

and

$$\rho^G\left(\frac{\partial \boldsymbol{v}^G}{\partial t}+\boldsymbol{v}^G\cdot\nabla\boldsymbol{v}^G\right)=-\nabla P^G+\rho^G\boldsymbol{g}+\mu^G\nabla^2\boldsymbol{v}^G$$

the energy equations in each phase i.e.,

$$\frac{\partial T^L}{\partial t}+\boldsymbol{v}^L\cdot\nabla T^L=K^L\nabla^2T^L$$

and

$$\frac{\partial T^G}{\partial t}+\boldsymbol{v}^G\cdot\nabla T^G=K^G\nabla^2T^G$$

and the continuity equation in each phase assuming that the fluids are incompressible where $\boldsymbol{v}$, P and T are velocity, pressure, and temperature fields respectively. Further, it is assumed that the bottom and top plate temperatures are kept constant and therefore $T^L(-L^L)=T_{cold}$ and $T^G(L^G)=T_{hot}$ hold. In the equations, ρ, μ and K are the density, dynamic viscosity and thermal diffusivity respectively.

When we assume that the liquid and its vapor are in equilibrium and no phase change is taking place across the interface then we need not feed and remove any fluid through the bottom and top plates. Therefore $v_z^L(-L^L) = 0$ and $v_z^G(L^G) = 0$ hold. The no-slip condition applies along the top and bottom plates, and gives rise to

$$v_x^L(-L^L) = 0 \text{ and } v_x^G(L^G) = 0$$

The interfacial mass balance equation is given by

$$\rho^L(\boldsymbol{v}^L - \boldsymbol{u}) \cdot \boldsymbol{n} = \rho^G(\boldsymbol{v}^G - \boldsymbol{u}) \cdot \boldsymbol{n}$$

where $\boldsymbol{n}$ is the normal and $\boldsymbol{u} \cdot \boldsymbol{n}$ is the interface speed.

At the interface, the tangential components of velocities of both fluids are equal to each other, and they are written as $\boldsymbol{v}^L \cdot \boldsymbol{t} = \boldsymbol{v}^G \cdot \boldsymbol{t}$ where t is the unit tangent vector.

The interfacial tension and its gradient with respect to the temperature at the interface come into the picture through the force balance there. By taking the dot product of the stress balance with the normal and tangential vectors separately, we will get 2 equations, i.e., the normal and tangential stress balance equations. The force balance is given by

$$\left(\rho^L \boldsymbol{v}^L(\boldsymbol{v}^L - \boldsymbol{u}) - \boldsymbol{T}^L\right) \cdot \boldsymbol{n} + \gamma 2H\boldsymbol{n} + \nabla_s \gamma \boldsymbol{t} = \left(\rho^G \boldsymbol{v}^G(\boldsymbol{v}^G - \boldsymbol{u}) - \boldsymbol{T}^G\right) \cdot \boldsymbol{n}$$

where T is the total stress, γ is the interfacial tension and $2H$ is the surface mean curvature.

And, then there is the interfacial energy balance.

$$\left[\rho^L\left(\hat{H}^L + \frac{1}{2}(\boldsymbol{v}^L - \boldsymbol{u})^2\right)(\boldsymbol{v}^L - \boldsymbol{u}) + \boldsymbol{q}^L - \boldsymbol{S}^L \cdot (\boldsymbol{v}^L - \boldsymbol{u})\right] \cdot \boldsymbol{n}$$

$$= \left[\rho^G\left(\hat{H}^G + \frac{1}{2}(\boldsymbol{v}^G - \boldsymbol{u})^2\right)(\boldsymbol{v}^G - \boldsymbol{u}) + \boldsymbol{q}^G - \boldsymbol{S}^G \cdot (\boldsymbol{v}^G - \boldsymbol{u})\right] \cdot \boldsymbol{n}$$

Here $\hat{H}$ is the enthalpy per unit mass, $\boldsymbol{S}$ is the extra stress, and q is the heat flux.

At this point, we have only 11 boundary conditions, and we will complete the boundary conditions with our choice of equilibrium conditions for the temperature and pressure at the interface. When the evaporation rate is very small we may assume that there is thermodynamic equilibrium for the vapor at the phase change boundary and the temperatures of both fluids are equal to each other at the interface, that is

$$P^G = f(T^G) \text{ and } T^L = T^G$$

To study the stability of the interface, arbitrary disturbances are applied and, for a given set of input variables, the growth and decay time constants of these disturbances or equivalently the critical temperature difference across both layers that result in marginal stability are determined. The plan is to linearize the above equations about a known base state and to find the onset of the interfacial instability from the perturbed model. Hereon, the variables of the base state are denoted with the subscript 0, and the variables of the perturbed state with the subscript 1. Thus, the temperature, when perturbed, is described by

$$T = T_0 + \epsilon T_1 + O(\epsilon^2)$$

where ϵ is a small perturbation parameter representing the deviation from the base state. We can further expand T_1 using a normal mode expansion as

$$T_1 = \hat{T}_1(z)\ e^{\sigma t} e^{iwx}$$

Here, σ is the inverse time constant while w is a wave number associated with the given perturbation. A wave number arises because the system is infinite in lateral extent. The same expansion is used for both components of velocity and pressure in both phases. The perturbation equations are then brought into dimensionless form. We have used the following scales: length, L^L; velocity, $\overline{v}$, $\frac{K^L}{L^L}$; time, $\frac{L^L}{\overline{v}}$; pressures in the liquid and in the vapor are scaled as $\rho^L \overline{v}^2$ and $\rho^G \overline{v}^2$, respectively.

5.1 The Base State Solution and the Perturbed Equations

In the base state there is no flow in either phase if the evaporation rate is set equal to zero, thus $\boldsymbol{v}_0^L = \boldsymbol{v}_0^G = \boldsymbol{0}$. The temperature profiles in the liquid and its vapor in dimensionless form become

$$T_0^L = -\frac{k^G}{k^L\delta + k^G} z + \frac{k^L\delta}{k^L\delta + k^G} \quad \text{and} \quad T_0^G = -\frac{k^L}{k^L\delta + k^G} z + \frac{k^L\delta}{k^L\delta + k^G}$$

where $\delta = \frac{L^G}{L^L}$.

After we perturb the domain and boundary equations we get the following equations in the domain in dimensionless form:

$$Pr^L \left(\frac{\partial^2}{\partial z^2} - w^2 \right) \hat{v}_{x1}^L - iw\hat{P}_1^L = \sigma \hat{v}_{x1}^L$$

$$Pr^L \left(\frac{\partial^2}{\partial z^2} - w^2 \right) \hat{v}_{z1}^L + Pr^L Ra^L \hat{T}_1^L - \frac{\partial \hat{P}_1^L}{\partial z} = \sigma \hat{v}_{z1}^L$$

$$\left(\frac{\partial^2}{\partial z^2} - w^2 \right) \hat{T}_1^L - \frac{dT_0^L}{dz} \hat{v}_{z1}^L = \sigma \hat{T}_1^L$$

$$iw\hat{v}_{x1}^L + \frac{\partial \hat{v}_{z1}^L}{\partial z} = 0$$

$$\frac{\nu^L}{\nu^G} Pr^L \left(\frac{\partial^2}{\partial z^2} - w^2 \right) \hat{v}^G_{x1} - iw\hat{P}^G_1 = \sigma \hat{v}^G_{x1}$$

$$\frac{\nu^L}{\nu^G} Pr^L \left(\frac{\partial^2}{\partial z^2} - w^2 \right) \hat{v}^G_{z1} + \frac{\nu^L}{\mu^G} Pr^L Ra^G \hat{T}^G_1 - \frac{\partial \hat{P}^G_1}{\partial z} = \sigma \hat{v}^G_{z1}$$

$$\left(\frac{\partial^2}{\partial z^2} - w^2 \right) \hat{T}^G_1 - \frac{K^L}{K^G} \frac{dT^G_0}{dz} \hat{v}^G_{z1} = \sigma \frac{K^L}{K^G} \hat{T}^G_1$$

$$iw\hat{v}^G_{x1} + \frac{\partial \hat{v}^G_{z1}}{\partial z} = 0$$

Here Pr^L and Ra^L stand for the dimensionless liquid Prandtl and liquid Rayleigh numbers respectively. Pr^L is defined as $\frac{\nu^L}{K^L}$, that is, as the ratio of the liquid kinematic viscosity to the liquid thermal diffusivity, and Ra^L is defined as $\frac{\alpha^L g \Delta T}{\nu^L K^L}(L^L)^3$ where α^L is the negative liquid thermal expansion coefficient and ΔT is the total temperature difference applied across both fluid layers.

At the lower plate, at $z = -1$, the liquid is subjected to $\hat{v}^L_{z1} = 0$, $\hat{v}^L_{x1} = 0$ and $\hat{T}^L_1 = 0$ while the conditions at the top plate become $\hat{v}^G_{z1} = 0$, $\hat{v}^G_{x1} = 0$ and $\hat{T}^G_1 = 0$.

At the interface the interfacial mass balance turns into

$$\hat{v}^L_{z1} - \frac{\rho^G}{\rho^L} \hat{v}^G_{z1} = \sigma \hat{Z}_1$$

And the continuity of temperature and the no-slip conditions become

$$\hat{T}^L_1 + \frac{dT^L_0}{dz} \hat{Z}_1 = \hat{T}^G_1 + \frac{dT^G_0}{dz} \hat{Z}_1 \text{ and } \hat{v}^L_{x1} = \hat{v}^G_{x1}$$

The normal and tangential stress balances in dimensionless form are

$$2 \left(\frac{\mu^G}{\mu^L} \frac{\partial \hat{v}^G_{z1}}{\partial z} - \frac{\partial \hat{v}^L_{z1}}{\partial z} \right) + \frac{1}{Pr^L} \left(\hat{P}^L_1 - \frac{\rho^G}{\rho^L} \hat{P}^G_1 \right)$$

$$+ \left(\left(\frac{\rho^G}{\rho^L} - 1 \right) G^L + Ra^L \left(1 - \frac{\rho^G \alpha^G}{\rho^L \alpha^L} \right) T^L_0|_{z=0} - w^2 Ca \right) \hat{Z}_1 = 0$$

and

$$\frac{\mu^G}{\mu^L} \left(\frac{\partial \hat{v}^G_{x1}}{\partial z} + iw\hat{v}^G_{z1} \right) - \left(\frac{\partial \hat{v}^L_{x1}}{\partial z} + iw\hat{v}^L_{z1} \right) + iwMa \left(\hat{T}^L_1 + \frac{dT^L_0}{dz} \hat{Z}_1 \right) = 0$$

where $G^L = \dfrac{g(L^L)^3}{\nu^L K^L}$ and $Ca = \dfrac{\gamma L^L}{\mu^L K^L}$ are the liquid Galileo and Capillary numbers respectively and where Marangoni number is defined as, $Ma = \dfrac{(\gamma_T)\Delta T}{\mu^L K^L} L^L$.

The interfacial energy balance is

$$\hat{v}^L_{z1} - E\left(\frac{k^G}{k^L}\frac{\partial \hat{T}^G_1}{\partial z} - \frac{\partial \hat{T}^L_1}{\partial z}\right) = \sigma \hat{Z}_1$$

where E stands for the Evaporation number and is given by $E = \dfrac{C^L_p \Delta T}{h_{fg}}$ and where C^L_p and h_{fg} are the liquid heat capacity and the latent heat of evaporation respectively. The equilibrium condition at the interface is

$$\Pi\left(\hat{T}^G_1 + \frac{dT^G_0}{dz}\hat{Z}_1\right) - \hat{P}^G_1 - Pr^L G^L \hat{Z}_1 = 0$$

where $\Pi = \dfrac{h_{fg}\Delta T(L^L)^2}{(K^L)^2 T_0}$ is a dimensionless parameter from the linearized Clapeyron equation and where T_0 is the interface temperature in degrees Kelvin.

5.2 Results and Discussion

A Chebyshev Spectral Tau method is used to solve the resulting eigenvalue problem to determine the critical temperature difference. We have represented the results for several cases by way of graphs. The calculations are done using the physical properties of water and water vapor at its saturation temperature under 1 atm, i.e. at $T_s = 100°$C. The thermophysical properties used in the calculations are given in Table 1.

For each case of interest we present graphs of the ratio of the critical temperature difference for that case to the critical temperature difference of another comparable case against the wavenumber. Table 2 defines the different ratios that are plotted. The subscripts pc, Ma, g, Ra stand for phase change, Marangoni convection, gravity, and buoyancy-driven convection, respectively. For example ΔT_{pc-Ma}, implies the critical temperature difference for the case where there is phase change and the Marangoni convection and R^1 therefore means the ratio of this critical temperature difference to the critical value for the case of pure Marangoni convection.

A fourth case is presented first and it is done so in tabular form for reasons that will soon become obvious. This is the case where the critical temperature difference for a pure phase change problem is compared with the critical temperature difference for the case when the Marangoni effect is added to the phase change problem. The values are tabulated in Table 3 and the reader can draw the conclusion for himself that the addition of the Marangoni effect does little to change the evaporative instability.

Table 3 does not tell the whole story. To see what the interaction of phase change is with the Marangoni problem we must also look at Fig. 7 and observe

Table 1. Thermophysical properties of liquid and vapor water at 100°C

Liquid	Vapor
$\rho^L = 960 \frac{\text{kg}}{\text{m}^3}$	$\rho^G = 0.6 \frac{\text{kg}}{\text{m}^3}$
$\mu^L = 2.9 \times 10^{-4} \frac{\text{kg}}{\text{m sec}}$	$\mu^G = 1.3 \times 10^{-5} \frac{\text{kg}}{\text{m sec}}$
$k^L = 6.8 \times 10^{-1} \frac{\text{J}}{\text{m sec °C}}$	$k^G = 2.5 \times 10^{-2} \frac{\text{J}}{\text{m sec °C}}$
$K^L = 1.7 \times 10^{-7} \frac{\text{m}^2}{\text{sec}}$	$K^G = 2.0 \times 10^{-5} \frac{\text{m}^2}{\text{sec}}$
$\alpha^L = 6.0 \times 10^{-4} \frac{1}{\text{°C}}$	$\alpha^G = 6.0 \times 10^{-3} \frac{1}{\text{°C}}$
$h_{fg} = 2.3 \times 10^6 \frac{\text{J}}{\text{kg}}$	$\gamma = 5.8 \times 10^{-2} \frac{\text{N}}{\text{m}}$
$\gamma_T = 2 \times 10^{-4} \frac{\text{N}}{\text{m°C}}$	

Table 2. Definition of R^1, R^2 and R^3

R^1	$\frac{\Delta T_{pc-Ma}}{\Delta T_{Ma}}$
R^2	$\frac{\Delta T_{pc-Ma-g}}{\Delta T_{pc-Ma}}$
R^3	$\frac{\Delta T_{pc-Ma-Ra}}{\Delta T_{pc-Ma}}$

that the pure Marangoni problem is considerably more unstable than the combined Marangoni and phase change problem. The stability offered by the phase change mechanism is not only attributed to the phase change itself but also due to an active vapor layer that transfers heat as well as momentum. The higher critical temperature difference indicates that there is more energy required in a phase change problem to cause the instability. This extra energy might appear to

Table 3. Comparison of critical temperature differences

Wavenumber, w	ΔT_{pc}(°C)	ΔT_{pc-Ma}(°C)
0.1	0.03	0.03
0.5	4.35	4.35
1.0	6.05	6.04
1.5	7.39	7.39
2.0	9.43	9.43

arise from the phase change but that may be misleading because evaporation in one region of the interface is associated by a companion condensation somewhere else at the interface. The real reason for the extra energy appears to be the large vapor flow that arises on account of the low vapor phase density. The flow in the vapor in a pure Marangoni problem is usually weak as mentioned earlier; on the other hand the vapor flow in a phase change problem is very strong. The big difference between the densities of both phases creates a big jump in the velocity at the interface while the liquid is being transformed into a vapor.[1] The results shown are for the case when the depth of the liquid layer is 10 times the depth of the vapor layer. Our calculations reveal that the wavenumber at which the plot intersects the unity line can be shifted to smaller wavenumbers by simply making the vapor layer deeper indicating that the vapor layer is playing an important role in the stability of the phase change problem.

In drawing Fig. 7 we have taken two similar systems, i.e. two systems where a liquid is underlying its own vapor at zero gravity, and we apply a temperature gradient across both layers. In one case, we pretend that there is no phase change across the interface between the liquid and the vapor. The flow can only be caused by surface tension gradients across the interface. In the other case, though, we will pretend that there is phase change across the interface. And to make a fair comparison, since there is initially no flow in the pure Marangoni problem, we will adjust the pressure on the vapor side in the second system in order to set the phase change rate to zero, as a result phase change is only possible

[1] Now the large vapor rate is often associated with the vapor recoil effect . This effect is of negligible consequence on the forces even though the kinetic energy associated with it is not.

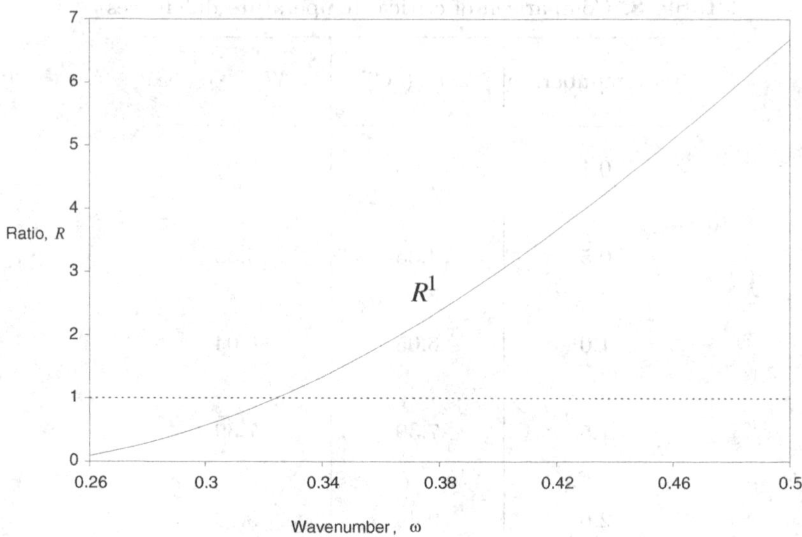

Fig. 7. R^1 vs. ω

through perturbations at the interface. Now we have two systems with no flow in their basic states. They are the same two systems except for one difference, the phase change. Given a perturbation, as expected by earlier arguments, the system where the phase change is allowed turns out to be more stable for large enough wavenumbers. We have plotted the ratio, R^1, versus wavenumber, w, in Fig. 7.

In many physical situations, gravity plays a role. Gravity comes with the burden of buoyancy-driven convection, however the effects of Rayleigh convection can be avoided in applications with very thin liquid and vapor layers. Gravity simply pulls the perturbed interface back to its original position. In other words, gravity plays a stabilizing effect on all wave numbers. However, its effect on the stability of the interface is more obvious for small wave numbers since the stabilizing effect of the interfacial tension is dominant over the gravity for large wavenumbers, and that explains why the critical temperature difference is higher for small wave numbers when the gravity is turned on. This is depicted by the curve labeled R^2 in Fig. 8. Working with thin fluid layers will be difficult in a lab environment. Thus, eventually we have to account for the buoyancy effects on the problem. For that reason Rayleigh convection is now introduced into the problem. It is well known for the Rayleigh problem in the absence of phase change that buoyancy driven convection can easily be stabilized for very small and very large wave numbers. For small wave numbers, the perturbations die out quickly because of the absence of transverse thermal gradients whereas for large wave numbers we observe the strong action of thermal and momentum diffusion. Now look at the plots of R^2 and R^3 versus wavenumber in Fig. 8.

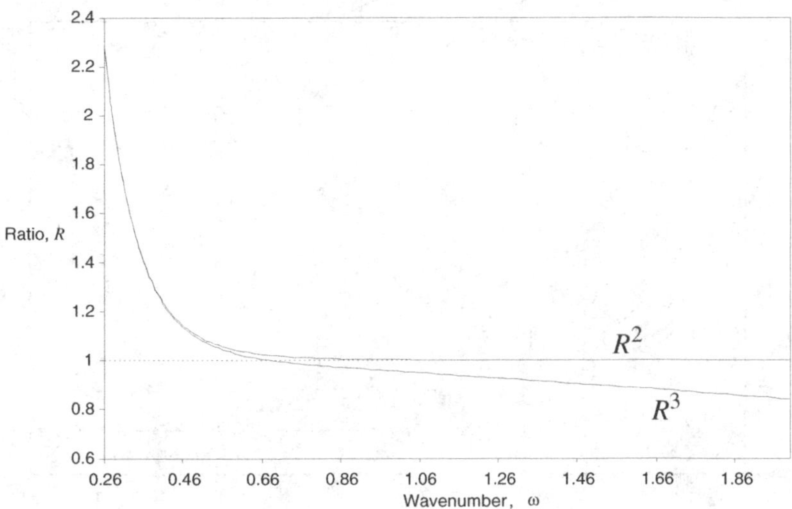

Fig. 8. R^2 and R^3 vs. ω

Table 4. Definition of R^4, R^5, and R^6

R^4	$\dfrac{\Delta T_{\text{critical}} \text{of a system with } \delta = 0.2}{\Delta T_{\text{critical}} \text{of a system with } \delta = 0.1}$
R^5	$\dfrac{\Delta T_{\text{critical}} \text{of a system with } \delta = 0.3}{\Delta T_{\text{critical}} \text{of a system with } \delta = 0.1}$
R^6	$\dfrac{\Delta T_{\text{critical}} \text{of a system with } \delta = 0.4}{\Delta T_{\text{critical}} \text{of a system with } \delta = 0.1}$

Both curves merge for small wavenumbers. In the absence of buoyancy-driven convection, represented by R^2, the stability offered by gravity is swept away by the stability offered through the phase change mechanism so the curve merges with the unity line whereas R^3 diverges to become less stable because of the destabilizing character of buoyancy-driven convection. For large wavenumbers, R^3 merges with the unity line (not shown in figure 8).

In order to show that active vapor layers play a role in the stability of a bilayer phase change problem we vary the depth of the vapor layer and calculate the critical temperature difference for a phase change problem with Marangoni and buoyancy effects. Plots are provided in Fig. 9 where the wavenumber is graphed against the ratio of critical temperature difference for systems with different values of δ. The ratios are defined in Table 4.

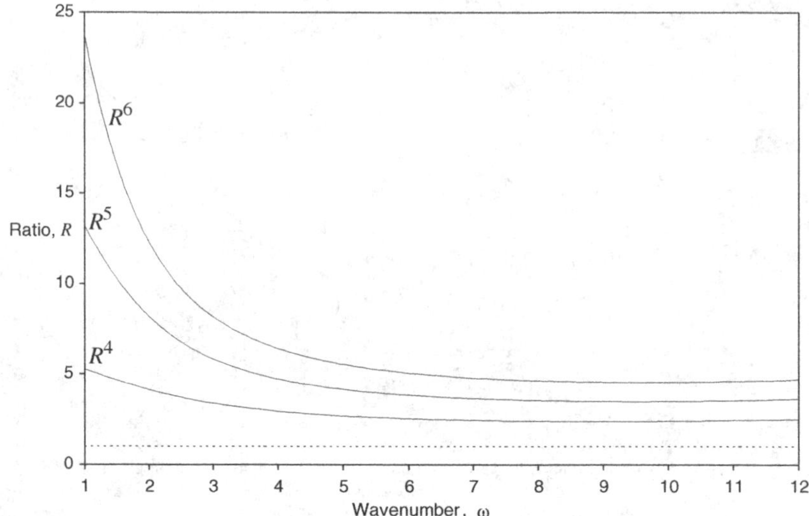

Fig. 9. R^4 through R^6 vs. ω

Figure 9 shows us that increasing the vapor depth actually stabilizes the interface. But, as we go to slightly higher wavenumbers we see the stabilizing effect of the deep vapor layers because the phase change mechanism is overshadowed by the destabilizing effect of buoyancy. The deeper the vapor layer becomes the more unstable it is due to Rayleigh convection which is scaled as the cube of the depth. However, we can still observe the important role the vapor depth is playing in the problem.

6 Scope for Future Work

There are some key issues that remain to be tackled in the convection problem when evaporation is considered. Some of these are computational, others are experimental. For example, we do not yet clearly understand how evaporation affects the change in the coupling mechanism between the fluid layers as the aspect ratio changes. Problems to do with the experimental determination of oscillatory flows that occur when liquid-gas bilayers with evaporation are heated from above will be of relevance to the drying of films. Finally the nonlinear interaction of modes near codimension two points and the rotational modes that can occur in the post onset region of convection are currently a topic of research and deserve attention as well. The problems that we have mentioned comprise a small fraction of the vast array of possible research topics that are still associated with this classical problem. They have been singled out for special attention because of their association with fluid layer interaction.

References

1. H. Bénard: Rev. Gen. Sciences Pure Appl. **11**, 1261 (1900)
2. M. Block: Nature **178**, 650 (1956)
3. S. Chandrasekhar: *Hydrodynamic and Hydromagnetic Stability*, (Oxford University Press, 1961)
4. P.C. Dauby, G. Lebon, E. Bouhy: Phys. Rev. E **56**, 520 (1997)
5. M. Degen, P. Colovas, C. Andereck: Phys Rev E **57**, 6647 (1998)
6. G.Z. Gershuni, E.M. Zhukhovitskii: Izv. Akad. SSSR, Mekh. Gaza **6**, 28 (1981)
7. G.Z. Gershuni, E.M. Zhukhovitskii: Sov. Phys. Dokl. **27**, 531 (1982)
8. D. Johnson, R. Narayanan: Phys. Rev. E **56**, 5462 (1997)
9. D. Johnson, R. Narayanan: Phil. Tran. R. Soc. Lond. A **356** 885 (1998)
10. D. Johnson, R. Narayanan, P.C. Dauby: *The Effect of Air on the Pattern Formation in Liquid-Air Bilayer Convection – How Passive is Air?*. In: "Fluid at Interfaces", eds. W. Shyy and R. Narayanan, (Cambridge Univ. Press, 1998)
11. E.L. Koschmieder, S.A. Prahl: J. Fluid Mech. **215**, 571 (1990)
12. D.A. Nield: J. Fluid Mech. **19**, 1635 (1964)
13. A. Oron, S.H. Davis, S.G. Bankoff: Rev. Mod. Physics **69**(3), 931 (1997)
14. J.R.A. Pearson: J. Fluid Mech. **4**, 489 (1958)
15. L. Rayleigh: Phil. Mag. **32**(6), 529 (1916)
16. G.S.R. Sarma: J. Thermophys. and Heat Transfer,**1**(2), 129 (1987)
17. L.E. Scriven, C.V. Sternling: AIChE J, **5**(4), 514 (1959)
18. A.A. Zaman, R. Narayanan: J. Colloid and Interface Sci. **179**, 151 (1996)

A General Approach to the Linear Stability of Thin Spreading Films

Jeffrey M. Davis, Benjamin J. Fischer, and Sandra M. Troian

Microfluidic Research & Engineering Laboratory (MREL), Department of Chemical Engineering, Princeton University, Princeton NJ 08544-5263, USA

Abstract. Marangoni and thermocapillary driven systems represent two classes of flows in which the variation in surface tension at the gas-liquid interface can generate spontaneously spreading films. This article considers the linear stability of such flows within the lubrication approximation. Since the base state or unperturbed solutions in both cases involve spatially inhomogeneous profiles, the linearized disturbance operators are non-normal and the stability analysis must therefore be generalized beyond a simple modal decomposition. The utility of this type of analysis is first demonstrated for autonomous operators by an example involving thermocapillary spreading subject to a constant thermal gradient. Extension of this non-modal analysis to systems involving non-autonomous operators is demonstrated by an example of Marangoni spreading induced by film contact with an insoluble surfactant monolayer.

1 Introduction

The growing focus on microscale flow phenomena and their extension to microfluidic devices has generated renewed interest in interfacial hydrodynamics and especially in free surface lubrication flows, *i.e.*, flows for which the height to length ratio is exceedingly small. In this limit, liquid systems can sustain an enormous surface to volume ratio. Forces arising from van der Waals interactions, capillarity, thermocapillarity or Marangoni stresses, all of which are usually neglected in large scale flows, dominate the spreading behavior.

Coating flows constitute an important branch of lubrication hydrodynamics in which thin liquid films are made to coat a dry or prewetted substrate by use of body or shear forces. Gravity, centrifugation, or an external gas stream provide the driving forces for falling [18], spin-coated [12,31] or so-called "blown" films [25]. Spreading can also be induced through modulation of the surface tension, which decreases with temperature or surfactant concentration. Gradients in surface tension generate shear stresses at a liquid-vapor interface to produce thermocapillary [26] or Marangoni driven flow [41].

In recent years, numerous experiments have shown that, in many cases, there exists some parameter range for which the liquid film develops instabilities that ultimately destroy the film uniformity. Films driven by gravity, centrifugation or a constant shear stress are observed to develop rivulets at the spreading front as shown in Fig. 1. Multiple unstable fronts can develop in spreading induced by gradients in surfactant concentration, as demonstrated by the images of the cellular and fractal-like instabilities in Fig. 2. The use of surfactants typically

J.M. Davis, B.J. Fischer, and S.M. Troian, A General Approach to the Linear Stability of Thin Spreading Films, Lect. Notes Phys. **628**, 79–106 (2003)
http://www.springerlink.com/

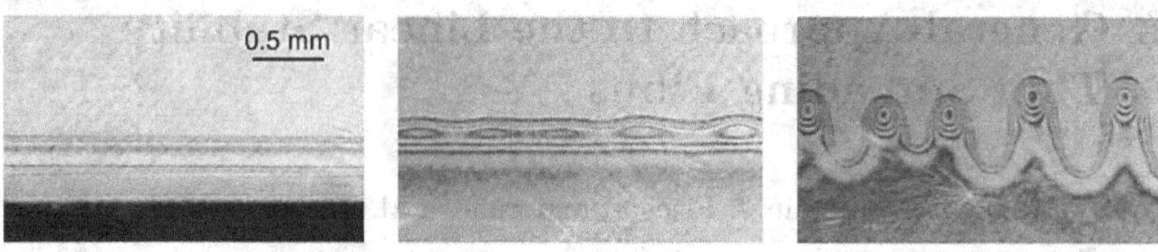

Fig. 1. Thermocapillary spreading of a thin silicone oil film (polydimethyl siloxane) on a differentially heated silicon wafer for $\tau = 0.79$ dyn/cm^2. Spreading proceeds from warmer to cooler regions of the substrate. Optical interference fringes indicate a film thickness $h_c \approx 0.56\mu$m. The time interval between images is 10 min.[23]

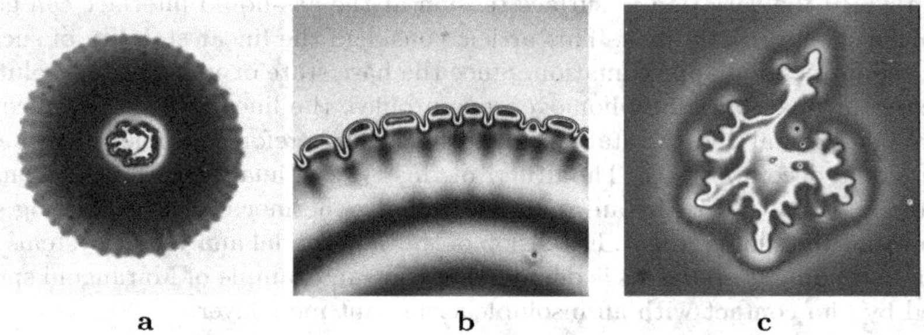

Fig. 2. **a** Spreading of a 5 cSt silicone oil (PDMS) droplet on a silicon substrate prewetted with a $0.5 \pm 0.1\mu m$ film of pure glycerol 3388 sec after deposition (field of view=9.9mm). **b** Magnified view of a cellular pattern which develops at the border of the spreading droplet (field of view=8.1 mm). **c** Magnified view of a dendritic pattern at the drop center (field of view=2.5 mm). The images do not correspond to the same droplet. Courtesy of Dr. A.A. Darhuber, MREL, Princeton University

results in the ramified structures shown in Fig. 3, which are produced by repeated branching and tip-splitting of moving fronts [42].

This article focuses on thin Newtonian films driven to spread cross a smooth, homogeneous surface through modulation of the liquid surface tension γ, although the analysis can be extended to any of the other flows mentioned above. This modulation, which can be enforced via thermal or concentration gradients, creates shear stresses $\boldsymbol{\tau} = \nabla\gamma$ at the gas-liquid interface that drive liquid from regions of low to high surface tension. In what follows, the linear stability of two systems will be considered: (i) thermocapillary spreading in which a constant shear stress is applied to a liquid film by differentially heating the supporting substrate and (ii) Marangoni spreading in which a non-uniform distribution of insoluble surfactant creates a non-constant shear stress that drives the spreading process.

Within the lubrication approximation [6], the simplified Navier-Stokes equations can be integrated, and the velocity in the film can be found directly from the liquid height profile. The kinematic boundary condition, $\boldsymbol{v} \cdot \hat{n} = D\tilde{h}/Dt$, where D/Dt denotes the material derivative and $\boldsymbol{v}$ the fluid velocity, dictates that the normal component of the surface velocity equal the speed of the gas-

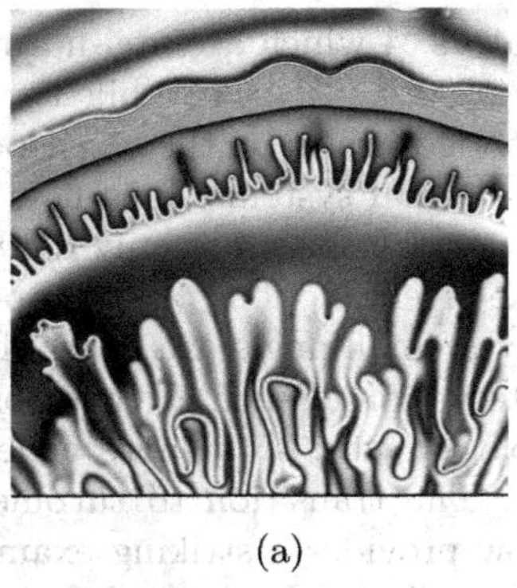

(a)

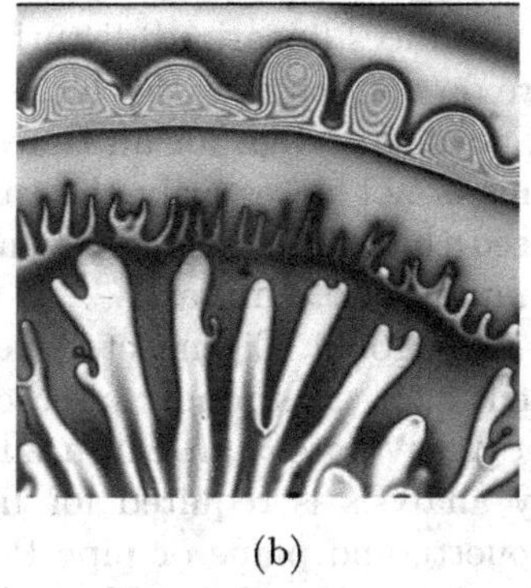

(b)

Fig. 3. Spreading of a glycerol droplet containing sodium dodecyl sulfate (surfactant) on a silicon wafer prewetted with a uniform glycerol layer formed by spin coating at 2000 rpm for 50s at 42% relative humidity. The images show two representative patterns **a** 92 min and **b** 192 min after droplet deposition (field of view = 9.9 mm). The spreading proceeds toward the top of each panel. The black and white contrast is caused by thin film optical interference fringes. Courtesy of Dr. A.A. Darhuber, MREL, Princeton University

liquid interface. Integrating the incompressible form of the continuity equation by parts and using this kinematic condition yields an equation for the film height, $\tilde{h}(\boldsymbol{x},t)$ [32]:

$$\frac{\partial \tilde{h}}{\partial \tilde{t}} + \nabla \cdot \int_0^{\tilde{h}} \boldsymbol{v} \, \mathrm{d}z = \frac{\partial \tilde{h}}{\partial \tilde{t}} + \nabla \cdot \left[\frac{\tilde{h}^2 \nabla \gamma}{2\mu} - \frac{\tilde{h}^3}{3\mu} \nabla p \right] = 0, \tag{1}$$

where $\hat{z}$ is directed normal to the solid substrate. The term proportional to $\nabla\gamma$ describes the contribution to the liquid flux from thermocapillary or Marangoni stresses. The pressure gradient in the third term $\nabla p \equiv \nabla(-\gamma\nabla^2\tilde{h} + A_0\tilde{h}^{-3})$ derives from capillary forces due to interfacial curvature for $|\nabla\tilde{h}|^2 \ll 1$ and attractive van der Waals interactions where A_o is the Hamaker constant. In thermocapillary driven systems, the viscosity μ of the liquid can vary spatially but variations in liquid density are typically much smaller and can be ignored. Other terms accounting for hydrostatic or streamwise gravitational acceleration, or terms arising from boundary conditions used to remove the well known stress singularity at a moving contact line [17], can easily be incorporated into (1). For example, the governing equation in Sect. 2 contains an additional term reflecting a slip boundary condition at the liquid-solid interface. For systems involving surfactant transport, the equation for $\tilde{h}(\boldsymbol{x},t)$ must be coupled to a second equation describing the convection and diffusion of surfactant at the air-liquid interface. This coupling gives rise to a space- and time-dependent shear stress as discussed in Sect. 3.

Depending on the forces used to drive the spreading, the base state solutions, $\tilde{h}(\boldsymbol{x},t)$, assume shapes ranging from constant traveling waves, to self-similar profiles to complex time-dependent waveforms. In all cases, this spatial inhomogeneity produces linearized disturbance operators $\mathcal{A}$ which are non-normal and therefore do not commute with their adjoint, *i.e.* $\mathcal{A}\mathcal{A}^{\dagger} \neq \mathcal{A}^{\dagger}\mathcal{A}$. While the stability of a normal system for all times t is strictly governed by the eigenspectrum of $\mathcal{A}$, this is not necessarily the case for non-normal systems. It is now widely recognized that the modal spectrum for non-normal operators only determines the asymptotic stability as $t \to \infty$ because the eigenfunctions of such operators are not orthogonal. Eigenvectors separated by a small angle are nearly linearly dependent and can strongly interact. This interaction of such eigenvectors with different decay rates can cause large transient growth, so a more generalized stability analysis is required for finite times. The transition to turbulence in both Couette and plane or pipe Poiseuille flow provides a striking example for which traditional modal analysis fails [38]. Farrell and Ioannou [7,8] have developed a rigorous and generalized stability theory for both autonomous and non-autonomous operators. Implementation of this method to non-autonomous systems with non-trivial time dependence, however, can be computationally prohibitive.

1.1 Linear Stability Theory – Modal Approach

The equations governing the evolution of infinitesimal disturbances are obtained by linearizing the relevant interface equations. Discretization of the linearized system produces a set of equations that can be cast in operator form

$$\frac{\mathrm{d}\boldsymbol{G}}{\mathrm{d}t} = \mathbf{A}\boldsymbol{G}, \tag{2}$$

where $\mathbf{A}(t)$ denotes the linearized disturbance operator and $\boldsymbol{G}(t)$ is a vector that represents the state of the system at time t. The traditional approach to stability proceeds by diagonalizing $\mathbf{A}(t)$ into a matrix whose diagonal elements contain the rank ordered eigenvalues, which is equivalent to assuming an exponential time dependence for $\boldsymbol{G}$. If all the eigenvalues have non-positive real part, then the flow is stable. If the real part of any eigenvalue is positive, the flow is unstable. The eigenvalue with largest, positive real part corresponds to the most dangerous or fastest growing mode whose wavelength can be directly compared with experiment. For normal operators, this procedure yields extremely accurate predictions, as in the Bénard problem [2]. For highly non-normal operators, this method can fail to predict instability altogether, as in the transition to turbulence in bounded shear flows [35]. This failure is due to the fact that the matrix used to diagonalize $\mathbf{A}(t)$ is not unitary and therefore cannot be ignored.

1.2 Generalized Linear Stability Theory

The general solution to (2) is given by

$$\boldsymbol{G}(t) = \boldsymbol{\Phi}_{[t,t_0]}\boldsymbol{G}(t_0), \tag{3}$$

where the propagator or matricant $\mathbf{\Phi}_{[t,t_0]}$, which maps the state of the system at time t_0 to its state at time t, obeys the semigroup property $\mathbf{\Phi}_{[t,s]}\mathbf{\Phi}_{[s,t_0]} = \mathbf{\Phi}_{[t,t_0]}$ and satisfies the matrix differential equation [8]

$$\frac{\mathrm{d}\mathbf{\Phi}_{[t,t_0]}}{\mathrm{d}t} = \mathbf{A}(t)\mathbf{\Phi}_{[t,t_0]}, \quad \mathbf{\Phi}_{[t_0,t_0]} = \mathbf{I}. \tag{4}$$

The matricant can be expressed as a multiplicative integral which is formally defined as [13]

$$\mathbf{\Phi}_{[t,t_0]} = \widehat{\int_{t_0}^{t}} [\mathbf{I} + \mathbf{A}(s)ds] \equiv \lim_{\delta t\to 0} \prod_{j=1}^{n} [\mathbf{I} + \mathbf{A}(t_j)\delta t_j], \tag{5}$$

with $n\ \delta t = (t - t_0)$. If the values of the matrix function $\mathbf{A}(t)$ commute, such that $[\mathbf{A}(t_1), \mathbf{A}(t_2)] = 0\ \forall\ t_1, t_2\ \in\ (t_0, t)$, then the propagator reduces to the matrix

$$\mathbf{\Phi}_{[t,t_0]} = \exp\left[\int_{t_0}^{t} \mathbf{A}(s)ds\right]. \tag{6}$$

In general, however, (6) is not a solution to (4), as verified by differentiation [13]. For the case of autonomous operators, $\mathbf{A}$ is independent of t, and (6) reduces further to

$$\mathbf{\Phi}_{[t,t_0]} = \exp\left[\mathbf{A}(t - t_0)\right]. \tag{7}$$

The amplification ratio, σ, of an arbitrary initial perturbation, $\boldsymbol{G}(t_0) \neq \mathbf{0}$, over the time interval $[t_0, t]$ is given by

$$\sigma^2 = \frac{(\boldsymbol{G}(t), \boldsymbol{G}(t))}{(\boldsymbol{G}(t_0), \boldsymbol{G}(t_0))} = \frac{(\mathbf{\Phi}_{[t,t_0]}\boldsymbol{G}(t_0), \mathbf{\Phi}_{[t,t_0]}\boldsymbol{G}(t_0))}{(\boldsymbol{G}(t_0), \boldsymbol{G}(t_0))}$$

$$= \frac{\left(\mathbf{\Phi}^{\dagger}_{[t_0,t]}\mathbf{\Phi}_{[t,t_0]}\boldsymbol{G}(t_0), \boldsymbol{G}(t_0)\right)}{(\boldsymbol{G}(t_0), \boldsymbol{G}(t_0))}, \tag{8}$$

where the inner product is computed in the Euclidean norm $\|\cdot\| = (\cdot,\cdot)^{1/2}$ and the adjoint linear operator, $\mathbf{\Phi}^{\dagger}$, is defined by $(\boldsymbol{u}, \mathbf{\Phi}\boldsymbol{v}) = (\mathbf{\Phi}^{\dagger}\boldsymbol{u}, \boldsymbol{v})$ for vectors $\boldsymbol{u}$ and $\boldsymbol{v}$. It then follows that the maximum amplification of a disturbance during the time interval $[t_0, t]$ is given by the square root of the maximum eigenvalue of $\mathbf{\Phi}^{\dagger}_{[t_0,t]}\mathbf{\Phi}_{[t,t_0]}$. The structure that undergoes maximum amplification is the corresponding eigenvector of the composite operator [8].

The base states for thin films driven by a constant shear stress reduce to traveling waves of constant speed. In the frame of reference defined by the traveling wave, $\mathbf{A}$ is *autonomous* and the propagator $\mathbf{\Phi}_{[t,0]}$ reduces to the form (7). Section 2 provides a rigorous formulation of the stability behavior and optimal perturbations for thermocapillary driven spreading. By contrast, the spreading of thin films driven by Marangoni stresses derived from a finite surfactant source gives rise to base states whose shapes are space- and time-dependent. The operator $\mathbf{A}$ is therefore *non-autonomous*. Evaluation of (5) for time-dependent base states, which must be numerically determined, is computationally challenging. The alternative but less-general approach described in Sect. 3, which reduces the relevant system of equations to an initial value problem, nonetheless allows physical insight into the mechanisms promoting instability. Although this approach prevents identification of so-called optimal perturbations, it offers the flexibility of directly probing critical regions of the flow which are highly susceptible to disturbances.

2 Autonomous Operator – Thermocapillary Spreading

Consider a thin incompressible Newtonian film spreading along a horizontal substrate heated by a constant temperature gradient dT/dx. The liquid film is assumed to be sufficiently thin that hydrostatic pressure is negligible. For small Péclet and Biot numbers, the temperature of the air-liquid interface is identical to the substrate thermal profile. Since the surface tension of a liquid decreases linearly (to a first approximation) with increasing temperature [1], the applied thermal gradient produces a constant shear stress $\tau = d\gamma/dx = (d\gamma/dT) \cdot (dT/dx)$, which drives liquid from warmer to cooler regions [25]. The stress singularity that would otherwise occur at the moving contact line is removed [3] by use of the Greenspan slip condition [16]. Other models can be used to relieve the singularity in thermally driven films including a uniform precursor layer [21,22] and a structured van der Waals film [4]. The Greenspan slip model has also been applied to falling films [5].

There exists an inner region at the front of the spreading film of characteristic length $l = h_c/(3Ca)^{1/3}$, which is obtained by balancing the capillary, thermocapillary and viscous forces controlling the flow [40]. The capillary number is defined by $Ca = \mu U_c/\gamma_o$, where μ is the liquid viscosity (assumed constant for short migration distances), γ_o is the reference surface tension, and $U_c \equiv h_c\tau/2\mu$ is the characteristic flow speed generated by thermocapillary forces. The quantity h_c denotes the characteristic film thickness far from the contact line which can be determined from matching the film thickness to the outer region.

The dimensionless form of (1), extended to include slip at the moving contact line, reduces to

$$h_t - \left(h^2\right)_\chi + \nabla \cdot \left[\left(h^3 + \alpha h\right) \nabla \nabla^2 h\right] = 0, \tag{9}$$

where α is the dimensionless slip coefficient, $\nabla \equiv \hat{\chi}\partial_\chi + \hat{\zeta}\partial_\zeta$, and subscripts denote partial differentiation with respect to χ, ζ, or t. The stretched variables in (9) are defined by $\chi = -x/l$, $\zeta = y/l$, $h = \tilde{h}/h_c$, and $t = \tilde{t}/(l/U_c)$. The term $(h^2)_\chi$ arises from the applied thermocapillary stresses, and the term $\nabla \cdot (h^3 \nabla \nabla^2 h)$ arises from the capillary pressure induced by surface curvature.

The spreading can be viewed from a reference frame moving at constant speed v_o where $\xi = \chi + v_o t$. The position $\xi = \xi_{CL}(\zeta, t)$ denotes the location of the contact line. For unperturbed flow, this (static) location is given by $\xi = 0$. The boundary conditions used to solve the transformed equation

$$h_t + v_o h_\xi - \left(h^2\right)_\xi + \nabla \cdot \left[\left(h^3 + \alpha h\right) \nabla \nabla^2 h\right] = 0, \tag{10}$$

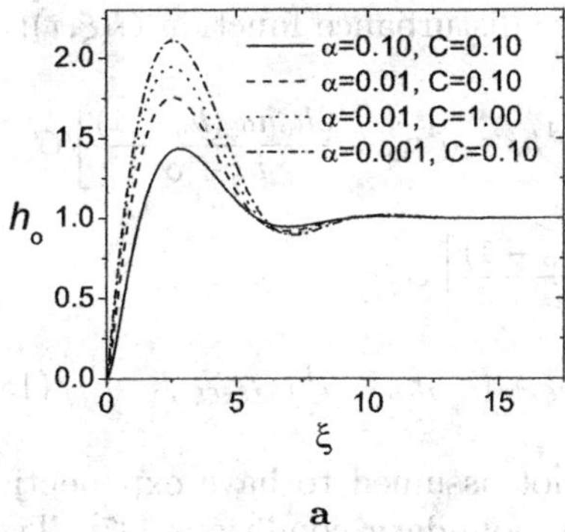

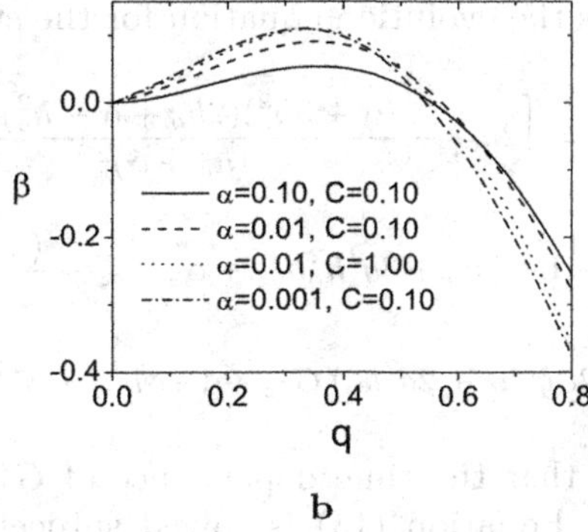

a **b**

Fig. 4. Numerical solution of **a** the dimensionless, steady state profile, $h_o(\xi)$, and **b** the dispersion curves, $\beta(q)$, obtained from a modal analysis

(where $\nabla \equiv \hat{\xi}\partial_\xi + \hat{\zeta}\partial_\zeta$) are $h(\xi \leq \xi_{CL}) = 0$; $h_\xi(\xi = \xi_{CL}) = \mathsf{C}$, which prescribes the contact angle at the moving front; and $h \to 1$ and $h_{\xi\xi\xi} \to 0$ as $\xi \to +\infty$. The latter two constraints predicate a flat and constant film thickness in matching to the other region.

2.1 Steady Traveling Wave Solutions

The third order equation governing the evolution of the steady base state $h(\xi, \zeta, t) = h_o(\xi)$ is found from integration of (10) to be

$$h_{o\xi\xi\xi} = \frac{h_o - 1}{h_o^2 + \alpha}. \tag{11}$$

The above boundary conditions determine that the wave speed is $v_o = 1$ and that the integration constant vanishes. Equation (11) is converted to a system of first order equations which are solved using a standard shooting method for stiff ODEs. Solutions for different values of the slip coefficient ($0.001 < \alpha < 0.10$) and contact slope ($0.1 < \mathsf{C} < 1.0$) are shown in Fig. 4a. Smaller values of C have only a small influence on the height profile and are not shown. The maximum amplitude of the capillary ridge that develops behind the moving contact line increases with decreasing slip or increasing contact slope.

2.2 Linear Stability of Traveling Waves

Profiles with large capillary ridges as shown in Fig. 4a can undergo sinusoidal fingering instabilities with well defined (dimensionless) wavenumber q (scaled by l) in the transverse direction ($\hat{\zeta}$). Substitution of perturbed waveforms $h(\xi, \zeta, t) = h_o(\xi) + \varepsilon h_1(\xi, \zeta, t)$ with $\varepsilon \ll 1$ into (10), where

$$h_1(\xi, \zeta, t) = G(\xi, t)\exp(iq\zeta), \tag{12}$$

yields the evolution equation for the streamwise disturbance function $G(\xi,t)$:

$$\frac{\partial G}{\partial t}=\left[2h_{o\xi}-\frac{(\alpha+3h_o^2)(2h_o+\alpha-h_o^2)h_{o\xi}}{(h_o^2+\alpha)^2}-\alpha q^4h_o-q^4h_o^3-\frac{6h_oh_{o\xi}(h_o-1)}{h_o^2+\alpha}\right]G$$
$$+\left[-1+2h_o+3q^2h_o^2h_{o\xi}+\alpha q^2h_{o\xi}-\frac{(\alpha+3h_o^2)(h_o-1)}{h_o^2+\alpha}\right]G_\xi$$
$$+\left(2\alpha q^2h_o+2q^2h_o^3\right)G_{\xi\xi}+\left(-\alpha h_{o\xi}-3h_o^2h_{o\xi}\right)G_{\xi\xi\xi}+\left(-\alpha h_o-h_o^3\right)G_{\xi\xi\xi\xi}\,. \tag{13}$$

Note that the time dependence of $G(\xi,t)$ is not assumed to have exponential form. Equation (13) is solved subject to four boundary conditions [37]. Two conditions demand that the disturbance decays upon approach to the outer region, *i.e.* $G(\xi\to+\infty)=0$ and $G_\xi(\xi\to+\infty)=0$. The third condition follows from the combined Taylor expansions of h and h_ξ about $\xi=\xi_{CL}$:

$$h(\xi_{CL})\approx h_o(0)+\varepsilon G(0)+\xi_{CL}h_{o\xi}(0)\quad\text{and}$$
$$h_\xi(\xi_{CL})\approx h_{o\xi}(0)+\varepsilon G_\xi(0)+\xi_{CL}h_{o\xi\xi}(0), \tag{14}$$

where terms of order $\varepsilon\xi_{CL}$ are neglected. Combining these results with the boundary conditions $h(\xi_{CL})=h_o(0)=0$ and $h_\xi(\xi_{CL})=h_{o\xi}(0)=\mathsf{C}$ yields one boundary condition for the disturbance at $\xi=0$:

$$h_{o\xi\xi}G-\mathsf{C}G_\xi=0. \tag{15}$$

The second condition at $\xi=0$ is obtained by evaluating (13) at the contact line (where $h_o=0$) and using (15):

$$G_t-\left(\mathsf{C}+\alpha q^2h_{o\xi\xi}\right)G+\alpha\mathsf{C}G_{\xi\xi\xi}=0. \tag{16}$$

2.3 Asymptotic Behavior

The asymptotic stability of a non-normal system in the limit $t\to\infty$ is determined from the modal spectrum [7]. In this limit, solutions to (13) can be further specified according to $G(\xi,t)=H(\xi)\exp(\beta t)$, where β denotes the disturbance growth rate. The numerical solutions were obtained by discretizing (13) using a central difference scheme and using a standard QR algorithm for calculating the relevant eigenvalues and eigenfunctions. The dispersion curves, $\beta(q)$, corresponding to the four base state height profiles shown in Fig. 4a are plotted in Fig. 4b. There exists a band of unstable wavenumbers in the range $0<q\lesssim0.5$ with maximum growth rate at $q_{\max}\approx0.35$. The growth rate with $\alpha=0.01$ and $\mathsf{C}=1.00$ is nearly as large as that obtained with $\alpha=0.001$ and $\mathsf{C}=0.10$. The value $\beta(q_{\max})$ increases as the slip coefficient decreases or the contact slope increases. This behavior confirms that the higher the capillary ridge, the more unstable the flow.

2.4 Optimal Amplification Ratio

Equation (13) can be represented in the form (2), where the autonomous matrix $\mathbf{A}$ contains the elements obtained from the discretization of the linearized equation. The matrix is real, square, banded, nondefective and non-normal. The formal solution to (2) then reduces to the operator exponential acting on the initial condition $\boldsymbol{G}_o$:

$$\boldsymbol{G}(t) = \exp(t\mathbf{A})\boldsymbol{G}_o. \tag{17}$$

It follows that the maximum possible amplification over time t is given by

$$\sigma_{\max}(t) \equiv \sup_{\boldsymbol{G}_o \neq 0} \frac{\|\boldsymbol{G}\|}{\|\boldsymbol{G}_o\|} = \|\exp(t\mathbf{A})\|. \tag{18}$$

This result also follows from (8) where the greatest amplification of any initial perturbation $\boldsymbol{G}_o$ over time t is given by the square root of the maximum eigenvalue [7] of $e^{t\mathbf{A}^\dagger} e^{t\mathbf{A}}$, namely $[\lambda_{\max}(e^{t\mathbf{A}^\dagger} e^{t\mathbf{A}})]^{1/2} = \|e^{t\mathbf{A}}\|$. Any non-defective matrix $\mathbf{A}$ can be decomposed according to the similarity transformation

$$\mathbf{A} = \mathbf{S}\mathbf{\Lambda}\mathbf{S}^{-1}, \tag{19}$$

where $\mathbf{S}$ is the matrix whose columns are the normalized eigenvectors of $\mathbf{A}$ in order of growth rate and $\mathbf{\Lambda}$ is the diagonal matrix of the associated eigenvalues [15]. This identity can be used to establish bounds on $\|e^{t\mathbf{A}}\|$:

$$\exp(\lambda_{\max}t) \leq \|\exp(t\mathbf{A})\| = \|\mathbf{S}\exp(t\mathbf{\Lambda})\mathbf{S}^{-1}\| \leq \|\mathbf{S}\|\|\mathbf{S}^{-1}\|\exp(\lambda_{\max}t), \tag{20}$$

where $\lambda_{\max}$ is the leading entry of $\mathbf{\Lambda}$. For a normal operator $\mathbf{A}$, $\mathbf{S}$ is unitary, and both the lower and upper bounds on $\|\exp(t\mathbf{A})\|$ equal $\exp(\lambda_{\max}t)$ $\forall\, t$. The eigenvalue with largest real part is therefore physically determinant since the growth rate of any disturbance is bounded above by $\lambda_{\max}$, the spectral abscissa of $\mathbf{A}$, which forms the leading entry in $\mathbf{\Lambda}$. For a non-normal operator, the eigenvectors are not orthogonal and the norm of $\mathbf{S}$ and its inverse can be much larger than unity. For highly non-normal systems, several orders of transient amplification can induce nonlinear effects, thereby invalidating the results of modal analysis. The transient behavior of solutions to (13) is therefore determined by examining the time dependence of $\|\exp(t\mathbf{A})\|$. The optimal initial condition that attains the maximum amplification at time t is determined as part of the analysis, which obviates the need to specify an initial form for the perturbation. In these studies, the number of grid points used in discretizing $\mathbf{A}$ ranged from 1600 to 4500; the matrix norms and exponentials were calculated with MATLAB 5.3 [30].

Figure 5 depicts the temporal evolution of $\ln\|\exp(t\mathbf{A})\|$ for selected wavenumbers q. The curves represent the envelopes maximized over all initial conditions of the amplification of individual initial conditions. Initially the system experiences a very small level of transient amplification (small bump near $t = 0$) followed by a brief plateau. By time $t = 5$, the curves with $q \neq 0$ rapidly approach a straight line whose slope equals the eigenvalue predicted from modal analysis. The insignificant level of transient amplification and the rapid convergence to the

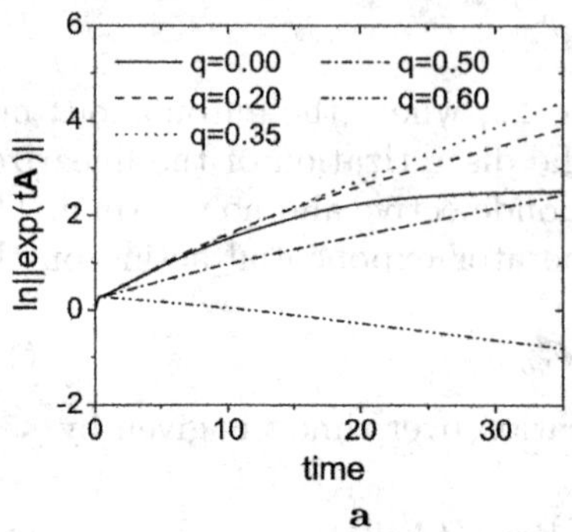

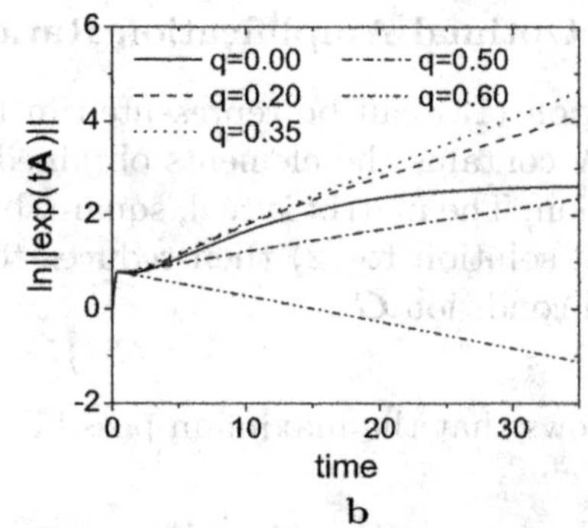

Fig. 5. Maximum possible amplification of disturbances within a time interval t for the height profiles shown in Fig. 1. Parameter values are **a** $\alpha = 0.01$, $\mathsf{C} = 1.0$ and **b** $\alpha = 0.001$, $\mathsf{C} = 0.1$

relevant eigenvalue explains the excellent agreement between experimental measurements of the fingering wavelength and the predictions of the most unstable wavelength from modal theory [21]. The ordering of the wavenumbers according to the degree of amplification level is also identical to the predictions obtained from the eigenspectrum of **A**. The small degree of transient amplification can be traced to the rather small degree of non-normality of the governing operator, as discussed in Sect. 2.5.

Short Time Behavior. The behavior of the disturbance growth rate in the limit $t \to 0$ is found by expanding the matrix $e^{t\mathbf{A}^\dagger} e^{t\mathbf{A}}$ in (8) in a Taylor series:

$$\lim_{t \to 0} \frac{\mathrm{d}}{\mathrm{d}t} \|e^{t\mathbf{A}}\| = \lambda_{\max} \left(\frac{(\mathbf{A} + \mathbf{A}^\dagger)}{2} \right). \tag{21}$$

Transient growth occurs when the maximum eigenvalue of the Hermitian part of **A** is positive. The structure that experiences the most amplification at early times [7] is the eigenvector associated with the maximum eigenvalue of $(\mathbf{A} + \mathbf{A}^\dagger)/2$. Comparison between this eigenvalue and the slopes of the curves in Fig. 5 as $t \to 0$ provides a check on the numerical computations.

Long Time Behavior. Since $\exp(t\mathbf{A}) = \mathbf{S}\exp(\mathbf{\Lambda}t)\mathbf{S}^{-1}$, the maximum amplification in the limit $t \to \infty$ is dominated exponentially by the first column of $\mathbf{S}$ and the first row of $\mathbf{S}^{-1}$ with amplification factor $\exp[\Re e(\lambda_{\max})t]$. Schwartz's inequality reveals that the normalized initialcondition that produces the maximum growth over time t is the complex conjugate of the first row of $\mathbf{S}^{-1}$, namely $[(\mathbf{S}^{-1})_{r1}]^\dagger$, or the first column of $(\mathbf{S}^\dagger)^{-1}$. Let $\mathbf{S}_L$ denote the matrix whose rows consist of the complex conjugates of the normalized left eigenvectors of **A** (the eigenvectors of $\mathbf{A}^\dagger$). By definition, $\mathbf{S}_L\mathbf{S} = \mathbf{I}$, so $\mathbf{S}_L = \mathbf{S}^{-1}$. This relationship implies that the optimal initial condition that undergoes the most amplification as $t \to \infty$ is the first row of $\bar{\mathbf{S}}_L$ (where the overbar denotes the complex conjugate), which is the leading eigenvector of $\mathbf{A}^\dagger$. The spectral abscissa of $\mathbf{A}$, denoted by $\alpha(\mathbf{A})$, is equal to the growth abscissa, $\gamma(\mathbf{A})$, also known as the Lyapunov exponent [39]:

$$\alpha(\mathbf{A}) \equiv \sup_{z\in\Lambda(\mathbf{A})} \Re e(z) \equiv \Re e[\lambda_{\max}(\mathbf{A})] = \gamma(\mathbf{A}) \equiv \lim_{t\to\infty} t^{-1} \ln \|e^{t\mathbf{A}}\| . \quad (22)$$

As shown in Fig. 5, this asymptotic limit is approached quite early in the spreading process. Evaluation of the time for onset of fingering in various thermocapillary experiments indicates that this asymptotic limit is reached well before the instability is observed [23].

2.5 Pseudospectra

An eigenvalue of a matrix $\mathbf{A}$ is a number $z \in \mathbb{C}$ such that $z\mathbf{I} - \mathbf{A}$ is singular, where $\mathbf{I}$ denotes the identity matrix. Determining the magnitude of the resolvent, $(z\mathbf{I} - \mathbf{A})^{-1}$, for a range of $z \in \mathbb{C}$ provides useful information on the behavior of $\mathbf{A}$ (and the operator of which it is the discrete representation) that cannot be determined by merely computing the eigenvalues. For each $\varepsilon \geq 0$, the ε-pseudospectrum of $\mathbf{A}$ is defined as [39]

$$\Lambda_\varepsilon(\mathbf{A}) \quad = \quad \{z \in \mathbb{C} : \|(z\mathbf{I} - \mathbf{A})^{-1}\| \geq \varepsilon^{-1}\}. \quad (23)$$

If $\mathbf{A}$ is normal then $\Lambda_\varepsilon(\mathbf{A})$ is the union of discs formed by the set of points in $\mathbb{C}$ within a distance ε of the spectrum of $\mathbf{A}$, $\Lambda(\mathbf{A})$. The ε-pseudospectrum may be much larger if $\mathbf{A}$ is non-normal. Examination of plots of $\Lambda_\varepsilon(\mathbf{A})$ gives an indication of the extent of non-normality of a matrix and thus of the physical relevance of its eigenvalues.

The pseudospectra of an operator or matrix can also be used to calculate bounds on $\|\exp(t\mathbf{A})\|$. For each $\varepsilon \geq 0$, the ε-pseudospectral abscissa of $\mathbf{A}$ is defined by

$$\alpha_\varepsilon(\mathbf{A}) \quad = \quad \sup_{z\in\Lambda_\varepsilon(\mathbf{A})} \Re e(z). \quad (24)$$

A lower bound on the norm of the matrix exponential for $\varepsilon > 0$, derived from the Laplace transform, is given by [39]

$$\sup_{t\geq 0} \|\exp(t\mathbf{A})\| \geq \varepsilon^{-1}\alpha_\varepsilon(\mathbf{A}). \quad (25)$$

This bound is relevant if the spectrum is confined to the left halfplane since a lower bound on the transient growth can quickly be found by determining how far the pseudospectra extend into the right halfplane. If the spectrum extends into the right halfplane then this bound is not useful since the growth becomes infinite as $t \to \infty$.

An analogue of the Cauchy integral formula in complex analysis, the Dunford-Taylor integral, provides a representation of any analytic function f of an operator (or matrix) as the integral around an appropriate contour in the complex plane [24]:

$$f(\mathbf{A}) \quad = \quad \frac{1}{2\pi i}\int_\Gamma (z\mathbf{I} - \mathbf{A})^{-1} f(z)dz, \quad (26)$$

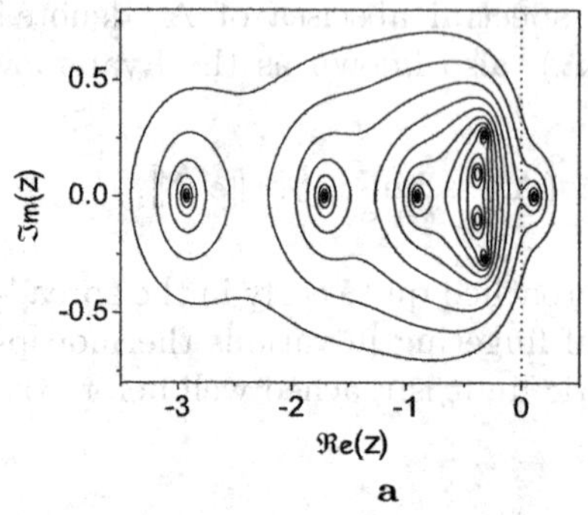

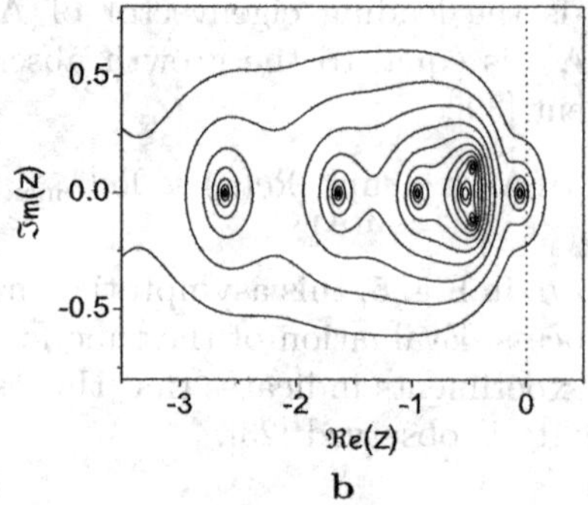

Fig. 6. Plots of the ε-pseudospectra for the base state with $\alpha = 0.001$ and $\mathsf{C} = 0.1$ for wavenumbers **a** $q = 0.35$ and **b** $q = 0.60$. The contours correspond to $\varepsilon = 10^{-1}, 10^{-1.25}, ..., 10^{-3.5}$. The *dotted vertical line* separates the stable and unstable halves of the complex plane

where Γ is any contour enclosing the spectrum of $\mathbf{A}$. Choosing the contour to be the boundary $\partial\Lambda_\varepsilon(\mathbf{A})$ of $\Lambda_\varepsilon(\mathbf{A})$ for some $\varepsilon > 0$ produces an upper bound for the norm of the matrix exponential:

$$\| \exp(t\mathbf{A}) \| \leq \frac{1}{2\pi\varepsilon} \int_{\partial\Lambda_\varepsilon(\mathbf{A})} \exp[t\ \Re e(z)]\ |dz|. \tag{27}$$

In practice, this bound is difficult to compute accurately.

Contours of the ε-pseudospectrum were calculated using the Pseudospectra GUI [43] for MATLAB. Boundaries of the ε-pseudospectrum for wavenumbers $q = 0.35$ and $q = 0.60$ for $\alpha = 0.001$ and $\mathsf{C} = 0.1$ are shown in Fig. 6 for a region near the unstable half of the complex plane. The contours shown correspond to values of $\varepsilon = 10^{-1}$, $10^{-1.25}$,..., $10^{-3.5}$. The abscissa and ordinate denote $\Re e(z)$ and $\Im m(z)$, respectively. The curves exhibit only mild non-normality since each contour exceeds the spectrum of $\mathbf{A}$ by an amount only slightly larger than ε, and the eigenvalue with largest real part appears robust. The non-modal amplification is primarily associated with the pairs of complex conjugate eigenvalues near $\Re e(\lambda) \approx -0.5$. The extent of non-normality is even less for larger values of the slip coefficient, which is consistent with the plots of $\ln \| \exp(t\mathbf{A}) \|$ vs. t shown in Fig. 5. Although the magnitude of the contact slope (which affects the height of the capillary ridge) influences the eigenvalues found from modal analysis, it has a negligible effect on the level of transient amplification achieved before the modal growth rate is established. Because the contours $\varepsilon \geq 10^{-1.5}$ for $q = 0.60$ extend more than a distance ε beyond the leading eigenvalue and into the unstable section of the complex plane, a small level of transient amplification is expected for this asymptotically stable wavenumber. Applying (25) produces a lower bound on amplification of approximately 1.7, while the actual maximum attained amplification is 2.2, as shown in Fig. 5.

2.6 Optimal Perturbations – SVD

The optimal initial perturbation and its corresponding evolved state at time t are directly computed from the singular value decomposition of $\exp(t\mathbf{A})$ according to [7]

$$\exp(t\mathbf{A}) = \mathbf{U\Sigma V}^{\dagger}, \tag{28}$$

where the columns of the unitary matrix $\mathbf{V}$ represent the complete set of initial states and the columns of the unitary matrix $\mathbf{U}$ are orthonormal basis vectors that span the range space of final states. The elements, σ_{i}, of the diagonal matrix $\mathbf{\Sigma}$ describe the growth realized by each initial state as it is transformed by the propagator into the corresponding final state. Note that the singular value decomposition must be calculated for each time at which $\mathbf{U}$, $\mathbf{\Sigma}$, and $\mathbf{V}$ are sought. For an initial perturbation, $\boldsymbol{G}_o = \sum a_{\mathrm{i}}\boldsymbol{V}_{\mathrm{i}}$ with $(\sum |a_{\mathrm{i}}|^2)^{1/2} = 1$, applied at time $t = 0$, the corresponding evolved state at time t is $\boldsymbol{G}(t) = \sum a_{\mathrm{i}}\sigma_{\mathrm{i}}\boldsymbol{U}_{\mathrm{i}} = \exp(t\mathbf{A})\boldsymbol{G}_o$. The vectors $\boldsymbol{V}_{\mathrm{i}}$ are ordered by growth, and the optimal perturbation, $\boldsymbol{V}_{\mathrm{opt}}(t)$, is the initial condition that undergoes the maximum amplification during the interval t. This maximum amplification is denoted by $\sigma_{\max} \equiv \|\exp(t\mathbf{A})\|$ and is given by the leading entry in $\mathbf{\Sigma}$.

The maximum possible amplification at any time is attained by the optimal initial disturbance calculated for that time. The normalized evolved state, $\boldsymbol{U}_{\mathrm{opt}}$, corresponding to $\boldsymbol{V}_{\mathrm{opt}}$ evolves into the leading eigenfunction of $\mathbf{A}$, $\boldsymbol{H}(\xi)$, as $t \to \infty$. The optimal initial disturbance, $\boldsymbol{V}_{\mathrm{opt}}(t \to \infty)$, in this long time limit (at which the most unstable mode dominates) asymptotes to $\boldsymbol{H}^{\dagger}(\xi)$, the eigenvector of the adjoint linearized operator, which is the initial condition that optimally excites the most unstable mode, $\boldsymbol{H}(\xi)$.

The excitation for the most asymptotically unstable wavenumber, $q = 0.35$, is plotted in Fig. 7. The disturbance applied at $t = 0$ that elicits the largest

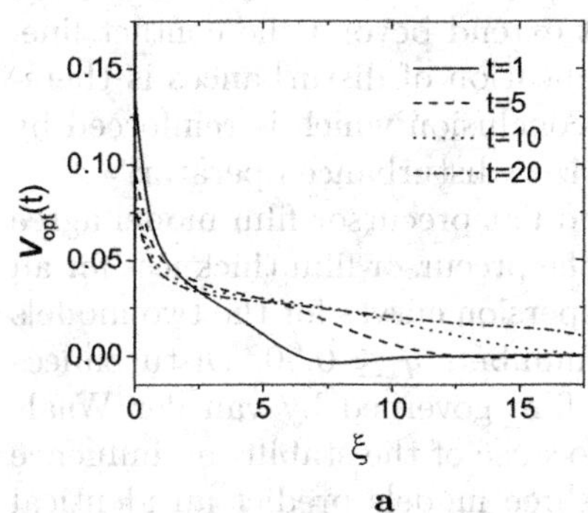

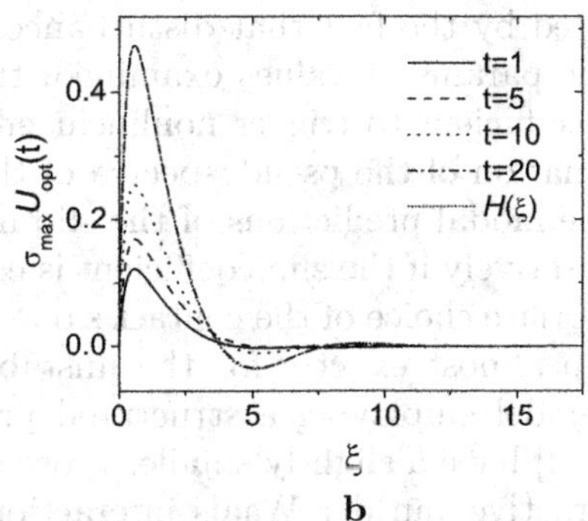

Fig. 7. **a** Optimal initial disturbance, $\boldsymbol{V}_{\mathrm{opt}}$, and **b** the evolved state, $\sigma_{\max}\boldsymbol{U}_{\mathrm{opt}}$, after time t for a disturbance of wavenumber $q = 0.35$ applied to the base state with $\alpha = 0.10$ and $\mathsf{C} = 0.10$. Each initial disturbance is normalized to unit magnitude. The magnitude of the corresponding evolved state is equal to the amplification attained by the initial disturbance at time t. The evolved state at $t = 20$ cannot be distinguished from the leading eigenfunction, $H(\xi)$

response at $t = 1$ is localized at the contact line. The initial disturbances that undergo the most amplification at later times broaden to encompass much of the capillary ridge but retain maxima at the contact line. Although initially focused near the forward portion of the capillary ridge, the system's response to these perturbations broadens at later times to encompass more of the ridge. By a dimensionless time $t = 15$, this response to the optimal disturbance is nearly indistinguishable from the modal eigenfunction, which explains the excellent agreement between the shape of the eigenfunction and the structure of the experimentally observed instability shortly after onset [21].

2.7 Summary of Thermocapillary Spreading Problem

Because thermally driven liquid films have spatially dependent base states, the linearized disturbance operator, **A**, is non-normal. The upper bound on disturbance amplification must therefore be determined from the norm of the matrix exponential, $\| \exp(t\mathbf{A}) \|$, rather than from the eigenvalue of **A** with largest real part, which determines the stability for all time only for normal operators. The transient growth analysis yields several noteworthy results. The ranking of disturbances of wavenumber q from largest to smallest level of disturbance amplification corresponds exactly to the asymptotic results from the modal theory. There is a smooth, rapid transition from the non-modal behavior to the asymptotic results obtained from the eigenspectrum of **A**. The optimal disturbances for both asymptotically stable and unstable flows initially exhibit a strong peak at the contact line. Disturbances that induce instability rapidly broaden to encompass the entire capillary ridge and the corresponding evolved states rapidly asymptote toward the requisite eigenfunction. The slip boundary condition generates less transient amplification than the use of boundary conditions which predicate a flat [21] or van der Waals precursor film [4] ahead of the contact line, even for very small values of the slip coefficient. This smaller transient growth is caused by the fact that disturbances cannot extend beyond the contact line. For the parameter values examined, the amplification of disturbances is therefore insufficient to trigger nonlinear effects, a conclusion which is reinforced by examination of the pseudospectra of the linearized disturbance operator.

The modal predictions of the slip model and flat precursor film model agree quantitatively if the slip coefficient is equal to the precursor film thickness for an appropriate choice of the contact slope. The dispersion curves for the two models overlap almost exactly for the unstable wavenumbers $q \leq 0.50$. Disturbances in a model employing a structured precursor film governed by van der Waals forces [4] have a slightly smaller growth rate because of the stabilizing influence of attractive van der Waals interactions. All three models predict an identical wavelength for the most unstable disturbance.

The insignificant level of transient amplification and the insensitivity of the asymptotic results to the specific characteristics of the precursor region, combined with the mild non-normality of the linearized disturbance operator, explain the excellent agreement between theory and experiment [23]. Unlike the linearized disturbance operator for plane Poiseuille flow, in which the angle between the eigenfunctions decreases exponentially as the Reynolds number increases [33], the angle between the eigenfunctions of the leading eigenvalues in thermally driven films is relatively large (63° in the system studied). This is likely due to the fact that the spatial inhomogeneity of the base state is confined to the capillary ridge. As a result, surface tension, which plays such a critical role in free surface flows, dampens the oscillatory, subdominant modes before significant energy can be transferred to the leading eigenvector. Little transient amplification occurs since the modes interact only weakly.

3 Non-autonomous Operator: Marangoni Spreading from a Finite Surfactant Source

The reduction of (5) to (7) for autonomous operators, which allows straightforward computation of the singular value decomposition of $\| \exp(t\mathbf{A}) \|$, is no longer valid for non-autonomous operators. For the Marangoni problem discussed below, there is no reference frame in which the base states can be rendered time-independent. As a result, the disturbance operator $\mathbf{A}$ is non-autonomous. While the singular value decomposition of (5) can still be used to identify the complete set of optimal perturbations ordered by the level of growth realized over a given time interval, numerical implementation of this approach for non-autonomous systems can be formidable, especially for large matrices. In addition, the long time dynamics is not governed by the fastest growing mode but by the Lyapunov vector growing at the mean rate of the first Lyapunov exponent [8]:

$$\lambda = \limsup_{t \to \infty} t^{-1} \ln(\| \mathbf{\Phi}(t) \|). \tag{29}$$

This exponent is analogous to the spectral abscissa in autonomous systems, where $\lambda > 0$ defines asymptotically unstable flow. Information about the associated Lyapunov vector is often much more difficult to obtain than the analogous exercise for autonomous operators which simply reduces to computing the eigenvalues and eigenvectors of $\mathbf{A}$.

Given these difficulties, the Marangoni problem is posed as an initial value problem. Disturbances are applied to selected regions of the flow and their amplification ratio, which is normalized by the temporal behavior of the evolving base state, is monitored in time. Despite this restriction to a limited set of initial conditions, the system of equations is shown to exhibit large transient growth as disturbances ahead of the spreading front convect past the leading edge. Although all the disturbances investigated decay away as $t \to \infty$, the substantial amplification at intermediate times may signal the presence of convective instabilities.

Even for normal operators, care must be taken in defining the stability criterion for time-dependent base states. Shen [36] first introduced the concept of "momentary stability" to describe the situation which prevails at a given instant when the kinetic energy of disturbances diminishes at a faster rate than the kinetic energy of the base state. Likewise, he designated a system "momentarily unstable" if the kinetic energy contained in the disturbance diminishes at a slower rate than the kinetic energy of the base state. The ratio of the relative energy of the disturbance, $E_d(t)$, to that of the reference (base) state, $E_b(t)$, provides a convenient measure of disturbance growth at time t. Normalizing the energy of the base states and disturbances by the corresponding values at some reference time t_0 leads to the amplification ratio

$$R = \left[\frac{E_d(t)}{E_d(t_o)}\right] / \left[\frac{E_b(t)}{E_b(t_o)}\right] , \tag{30}$$

which surveys the intensification or dissipation of the relative input energy for a given time interval. While this definition falls far short of providing the relevant information contained in the first Lyapunov vector and exponent for non-autonomous systems, investigation of the system's response to selected disturbances help pinpoint the critical features of the spreading profile associated with large transient growth.

3.1 Surfactant Driven Flow

Consider a thin incompressible Newtonian liquid film of constant viscosity μ and initial height h_c partially coated with an insoluble surfactant monolayer of length L_c as shown in Fig. 8a. The initial gradient of the spreading pressure, $\Pi = \gamma_o - \gamma_m$ (see Fig. 8b), induces a shear stress $\tau = d\gamma/dx = (d\gamma/d\Gamma)(d\Gamma/dx)$ at the air-liquid interface where Γ is the surface surfactant concentration. Because surface tension is a decreasing function of surfactant concentration, the positive shear stress drives surfactant and liquid toward regions of higher surface tension (uncoated regions). For a finite monolayer source, the driving force $\Pi/L(t)$, where Π is constant, weakens as the monolayer spreads.

The surfactant monolayer is incorporated into the model through the stress boundary condition and the accompanying equation of motion for the surface concentration. This monolayer, however, is regarded as infinitesimally thin such that its surface viscosity and density can be ignored. Furthermore, since the flow is unbounded in the streamwise x-direction, it is assumed that the molecules

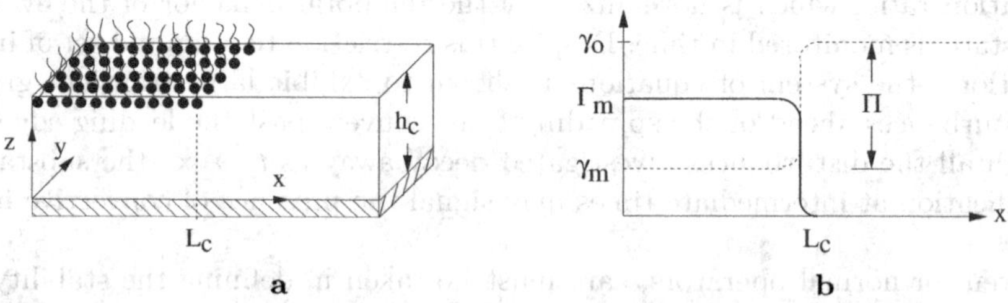

Fig. 8. The initial configuration for Marangoni spreading in rectilinear geometry. The liquid layer has viscosity μ, density ϱ, and initial uniform thickness h_c. The initial surfactant monolayer extends a distance L_c with surface tension γ_m and surface concentration Γ_m. The maximum spreading pressure is defined by $\Pi = \gamma_o - \gamma_m$, where the surface tension of the uncoated film is γ_o

comprising the monolayer behave as an ideal gas. The dimensionless equation of state relating the surface tension to the molecular concentration is assumed to be $\gamma = 1 - \Gamma$. Non-linear equations of state have also been studied [9,14].

Within the lubrication approximation [6] and for vanishing Bond number $\varrho g {h_c}^2/\gamma_m$, the dimensionless evolution equations for h and Γ become [20,29]

$$\frac{\partial h}{\partial t} + \nabla \cdot \left(-\frac{h^2}{2} \nabla \Gamma + \frac{\mathcal{C}}{3} h^3 \nabla^3 h \right) = 0 \,, \tag{31a}$$

$$\frac{\partial \Gamma}{\partial t} + \nabla \cdot \left(-\Gamma h \nabla \Gamma + \frac{\mathcal{C}}{2} \Gamma h^2 \nabla^3 h - \frac{1}{Pe_s} \nabla \Gamma \right) = 0 \,. \tag{31b}$$

where the film height is normalized by the initial film thickness h_c, the surfactant concentration by the initial concentration Γ_m, and the streamwise (x) and transverse (y) coordinates by the initial monolayer extent L_c. The dimensionless group $\mathcal{C} = (h_c/L_c)^2 \gamma_m / \Pi$ represents the ratio between the capillary force, which prefers minimal surface area, and the spreading pressure, which is responsible for the increase in surface area. This group can also be written as $\mathcal{C} = (h_c/L_c)^3/Ca$ where $Ca = \mu U_c/\gamma_m$ is the usual capillary number scaled on the characteristic speed, $U_c = (h_c/L_c)\Pi/\mu$, set by the Marangoni stresses. This characteristic speed establishes the convective time scale L_c/U_c used to normalize the dimensional time. The Péclet number $Pe_s = U_c L_c / D_s = \Pi h_c / \mu D_s$ where D_s is the surfactant diffusion coefficient along the surface, represents the ratio between the convective surfactant flux and the diffusive flux. For large Péclet numbers, the diffusive contribution is negligible and the last term in (31b) can be omitted altogether. Returning to (31a), the second and third terms describe the liquid flux contribution from Marangoni stresses and pressure gradients due to surface curvature, respectively. Equation (31b) represents the convective-diffusive transport of surfactant at the air-liquid interface. The second and third terms couple the surfactant concentration to the surface velocity of the spreading film.

3.2 Base State Flow Profiles

In the absence of a constant shear stress, there is no reference frame that renders the spreading process time-independent. It is possible, however, to find a self-similar solution in the limit of infinite capillary number ($\mathcal{C} \to 0$) and infinite Péclet number. In this limit, only Marangoni stresses drive the flow. A straightforward scaling analysis for spreading in rectilinear geometry [20], as in Fig. 8a, reveals that the leading edge of the monolayer advances in time as $t^{1/3}$. The slopes of the film thickness and concentration profiles, however, contain discontinuities which create difficulties in formulating the linear stability analysis [27]. Inclusion of capillarity and surface diffusion destroys the exact self-similar nature of the solutions; however, within the range of parameter values used in this study, the asymptotic profiles approach self-similar form. It is therefore useful to transform the original variables accordingly:

$$\xi = \frac{x}{t^{1/3}} \,, \quad g_o(\xi, t) = t^{1/3} \Gamma_o(x,t) \ \text{and} \ h_o(\xi, t) = H_o(x,t) \,. \tag{32}$$

The subscript "o" is used to denote the solutions corresponding to the base or unperturbed states. The additional time dependence in the scaling for Γ_o is dictated by the constraint that the total surfactant mass, $\int_0^{L(t)} \Gamma_o(x,t)dx$, remain constant. The dimensionless equations describing the evolution of the film height and surfactant concentration in a frame advancing as $L(t) \sim t^{1/3}$ are given by

$$t\frac{\partial h_o}{\partial t} = \frac{1}{3}\xi h_{o\xi} + \frac{1}{2}\left(h_o^2 g_{o\xi}\right)_\xi - \frac{\mathcal{C}}{3t^{1/3}}\left(h_o^3 h_{o\xi\xi\xi}\right)_\xi , \tag{33a}$$

$$t\frac{\partial g_o}{\partial t} = \frac{1}{3}(\xi g_o)_\xi + \left(g_o h_o g_{o\xi}\right)_\xi - \frac{\mathcal{C}}{2t^{1/3}}\left(g_o h_o^2 h_{o\xi\xi\xi}\right)_\xi + \frac{t^{1/3}}{Pe_s} g_{o\xi\xi} . \tag{33b}$$

The boundary conditions used to solve these equations correspond to symmetry and no-flux of liquid and surfactant at the origin $\xi = 0$ and decay conditions far downstream where the liquid layer is quiescent and free of surfactant:

$$h_{o\xi}(0,t) = 0 , \quad h_{o\xi\xi\xi}(0,t) = 0 \text{ and } g_{o\xi}(0,t) = 0 , \tag{34a}$$

$$h_o(\infty,t) = 1 , \quad h_{o\xi}(\infty,t) = 0 \text{ and } g_o(\infty,t) = 0 . \tag{34b}$$

The initial conditions defined at $t = 1$ correspond to an initially flat liquid layer coated with a surfactant monolayer whose concentration is relatively flat and smoothly decays to zero near the point ξ_o. These two conditions are given by

$$\begin{aligned} h_o(\xi,1) &= 1 \\ g_o(\xi,1) &= g_o{}^{\max} \left[1 - \tanh\left(A(\xi - \xi_o)\right)\right] . \end{aligned} \tag{35}$$

This study was restricted to parameter values $g_o{}^{\max} = 0.5$, $A = 10$, $\xi_o = 0.5$, $\mathcal{C} = 10^{-5}$ and $Pe_s = 5000$. The initial surfactant distribution therefore has an inflection point at ξ_o=0.5 and vanishes at about $\xi = 0.75$, as shown in Fig. 9b. Equations (33a) and (33b) were solved using the method of lines [34], which implements second-order centered differences for the spatial derivatives and a fully implicit Gear's method for the time integration [19]. The number of grid points used in the computations varied between 301 and 751.

Figure 9 depicts the evolution of $h_o(\xi)$ and $g_o(\xi)$ for times $1.0 \leq t \leq 5.0$. As the monolayer spreads along the liquid layer, it shears the underlying liquid film, producing a sharp ridge at the advancing front. Unlike the thermocapillary problem discussed earlier, this ridge is mainly caused by the Marangoni stress which pulls liquid from left to right causing severe thinning near the initial perimeter of surfactant deposition and thickening at the moving front. Capillary forces smooth points of extreme surface curvature, which exist at the apex and thinned portions of the film profile. When capillary and diffusive forces are omitted from the equations, the film thickness (and concentration profile) assumes the shape of a linearly increasing (decreasing) ramp. Behind the apex of the Marangoni ridge, there develops a distinctive linear segment in both h_o and g_o (indicative of the self-similar behavior) which expands in time. As time evolves, $h_o(\xi)$ and $g_o(\xi)$ approach self-similar forms since capillary and diffusive forces weaken considerably with respect to the main driving force.

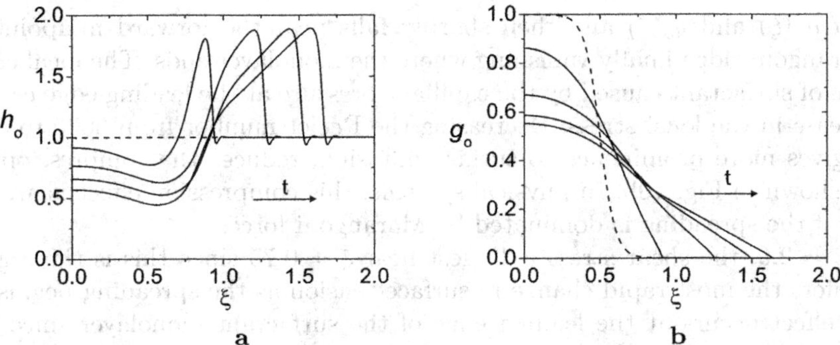

Fig. 9. Evolution of the base state **a** film thickness h_o and **b** surfactant concentration g_o with $Pe_s = 5000$ and $\mathcal{C} = 10^{-5}$ for times $t = 1.0, 1.1, 1.5, 2.6$ and 5.0

Surfactant Compression Effect. It is interesting to examine separately the Marangoni and capillary contributions to the base state surface velocity profile as shown in Fig. 10a for $t = 2.6$. The Marangoni stress causes a sharp increase in the surface speed at the point where the moving film meets the quiescent layer. This increase in speed gives rise to the increase in the film height at the advancing front. The surface tension, acting through the capillary pressure gradient, opposes this motion, causing a negative surface velocity. The combined effect smooths the sharpest features of the velocity profile and introduces a small capillary oscillation just ahead of the monolayer decay point at $\xi \approx 1.5$ to give an overall negative flow speed in this region. This net negative contribution causes a local accumulation of surfactant at the leading edge, which is more easily visualized by examining the gradient in surfactant concentration shown in Fig. 10b. The curve $g_{o\xi}$ decays monotonically away from $\xi = 0$ to reach a minimum at $\xi \approx 0.75$. The gradient plateaus in the region corresponding to the linear por-

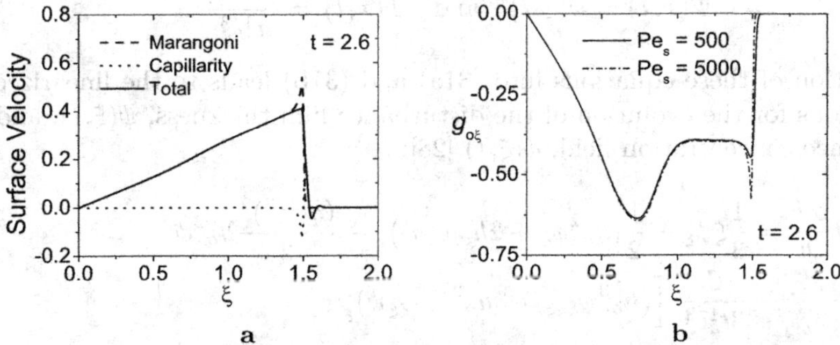

Fig. 10. **a** Marangoni and capillary contributions to the base state surface velocity profile, along with their sum, at time $t = 2.6$ for $Pe_s = 5000$ and $\mathcal{C} = 10^{-5}$. The surface velocity becomes negative just ahead of the step profile. **b** A comparison of $g_{o\xi}$ at time $t = 2.6$ for $Pe_s = 500,\ 5000$ with $\mathcal{C} = 10^{-5}$. The concentration gradient $g_{o\xi}$ undergoes a sharp change at $\xi \approx 1.5$

tions of $h_o(\xi)$ and $g_o(\xi)$ and then sharply falls near the forward midpoint of the Marangoni ridge finally vanishing where the monolayer ends. The local compression of surfactant caused by the capillary pressure at the leading edge causes an increase in the local stress. Decreasing the Péclet number from 5000 to 500, which gives more prominence to surface diffusion, reduces the compression effect as shown in Fig. 10b. In physical systems, this compression effect is always present if the spreading is dominated by Marangoni forces.

At $t = 2.6$, the shear stress is largest near $\xi = 0.75$ since this is the region that suffers the most rapid change in surface tension as the spreading begins. A similar effect occurs at the leading edge of the surfactant monolayer since the local accumulation of surfactant causes a local enhancement in the concentration gradient. It is expected that these two critical regions, which experience the largest shear stress, are particularly vulnerable to disturbances. This issue is discussed further in Sect. 3.3.

3.3 Linear Stability of Time Dependent Base State Profiles

The stability of infinitesimal disturbances to the film thickness and surfactant concentration with sinusoidal variations in the transverse direction ($\hat{y}$) are investigated next. The perturbed waveforms are defined by $h(x,y,t) = h_o(x,t) + \varepsilon \widetilde{H}(x,y,t)$ and $\Gamma(x,y,t) = \Gamma_0(x,t) + \varepsilon\widetilde{\Gamma}(x,y,t)$ ($\varepsilon \ll 1$), where

$$(\widetilde{H}, \widetilde{\Gamma})(x,y,t) = (\Psi, \Phi)(x,t) e^{iqy} \,. \tag{36}$$

The dimensionless wavenumber q, normalized by L_c, is associated with the transverse corrugations, which are denoted by a tilde. With this choice, disturbances to the film height and surfactant concentration are applied in phase. Other choices are possible. The Fourier amplitudes, Ψ and Φ, are further rescaled by transforming to the moving reference frame of the base states as in (32):

$$\Psi(x,t) = \psi(\xi,t) \quad \text{and} \quad \Phi(x,t) = \frac{\phi(\xi,t)}{t^{1/3}} \,. \tag{37}$$

Substitution of these equations into (31a) and (31b) leads to the linearized set of equations for the evolution of the disturbance film thickness, $\psi(\xi,t)$, and the disturbance concentration field, $\phi(\xi,t)$ [28]:

$$\begin{aligned} t\frac{\partial\psi}{\partial t} &= \frac{1}{3}\xi\psi_\xi + \frac{1}{2}\left({h_o}^2\phi_\xi + 2h_o g_{o\xi}\psi\right)_\xi - \frac{(qt^{1/3})^2}{2}{h_o}^2\phi \\ &- \frac{\mathcal{C}}{3t^{1/3}}\Big[\left({h_o}^3\psi_{\xi\xi\xi} + 3{h_o}^2 h_{o\xi\xi\xi}\psi\right)_\xi \\ &- (qt^{1/3})^2\left(({h_o}^3)_\xi\psi_\xi + 2{h_o}^3\psi_{\xi\xi}\right) + (qt^{1/3})^4{h_o}^3\psi\Big] \,, \end{aligned} \tag{38a}$$

$$\begin{aligned} t\frac{\partial\phi}{\partial t} &= \frac{1}{3}(\xi\phi)_\xi + \left(g_o g_{o\xi}\psi + h_o g_{o\xi}\phi + h_o g_o \phi_\xi\right)_\xi - (qt^{1/3})^2 h_o g_o \phi \\ &- \frac{\mathcal{C}}{2t^{1/3}}\Big[\left(g_o {h_o}^2\psi_{\xi\xi\xi} + 2g_o h_o h_{o\xi\xi\xi}\psi + {h_o}^2 h_{o\xi\xi\xi}\phi\right)_\xi \end{aligned}$$

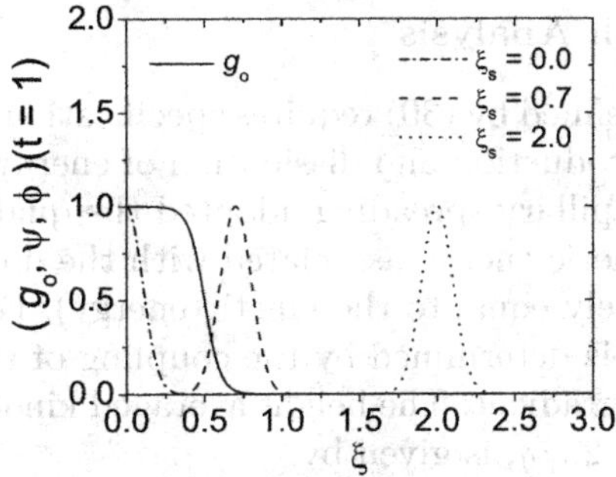

Fig. 11. Three locations of the initial perturbations, $\psi(t=1)$ and $\phi(t=1)$, relative to the initial base state concentration profile $g_o(t=1)$

$$- \left(qt^{1/3}\right)^2 \left(({g_o h_o}^2)_\xi \psi_\xi + 2 g_o {h_o}^2 \psi_{\xi\xi} \right) + \left(qt^{1/3}\right)^4 g_o {h_o}^2 \psi \Big] + \frac{t^{1/3}}{Pe_s} \left[\phi_{\xi\xi} - \left(qt^{1/3}\right)^2 \phi \right] . \tag{38b}$$

Similar boundary conditions used for solving the base state equations are applied to the disturbance equations, namely:

$$\psi_\xi(0,t) = 0 \; , \quad \psi_{\xi\xi\xi}(0,t) = 0 \; , \quad \text{and} \quad \phi_\xi(0,t) = 0 \; , \tag{39a}$$

$$\psi(\infty,t) = 0 \; , \quad \psi_\xi(\infty,t) = 0 \; , \quad \text{and} \quad \phi(\infty,t) = 0 \; . \tag{39b}$$

The initial conditions chosen for this study represent highly localized Gaussian functions that are positioned either well ahead of the initial monolayer front ($\xi_s = 2.0$), at the base of the initial monolayer perimeter ($\xi_s = 0.7$), or at the origin ($\xi_s = 0.0$). These disturbances are described by:

$$\psi(\xi,1) = \phi(\xi,1) = e^{-\mathrm{B}\,(\xi-\xi_s)^2} \; . \tag{40}$$

The amplitudes $\psi(\xi,1)$ and $\phi(\xi,1)$ are set to unity since (38a) and (38b) are linear equations. In this study, $B = 50$. Figure 11 shows three locations of the Gaussian perturbations chosen for this study.

Equations (33a), (33b), (38a) and (38b) are solved simultaneously using the method of lines as described earlier. Equations (33a) and (33b) are discretized and cast into the operator form (2) by defining state vectors $\boldsymbol{G} = [\psi \; \phi]$. Since there exists both an implicit and explicit time dependence in (38a) and (38b), the linearized operator $\mathbf{A}$ is non-autonomous; the generalized stability analysis described in Sect. 1.2 can therefore be invoked. An alternative, less general approach is used below for computational expedience. Gaussian distributed perturbations in the film thickness and surfactant concentration are positioned at key points in the spreading film. This targeted approach uncovers at least one region of the flow which is particularly vulnerable to disturbance growth.

3.4 Transient Growth Analysis

The amplification ratio defined by (30) requires specification of the quantity to be used in monitoring the production and dissipation of energy. Previous studies of centrifugal and thermocapillary spreading adopted the quantity, $h^2(\boldsymbol{x},t)$, which is proportional to the kinetic energy associated with the dominant driving force [21,22,37] (but not precisely equal to the kinetic energy). For Marangoni driven spreading the flow speed is determined by the coupling of the film height to the surfactant concentration gradient. The height averaged kinetic energy of the flow per unit wavelength, $\lambda = 2\pi/q$, is given by:

$$E_b \equiv \frac{1}{2\lambda}\int_0^\lambda \int_0^\infty |\langle \boldsymbol{v_o}\rangle|^2 d\xi dy \quad \text{and} \quad E_d \equiv \frac{1}{2\lambda}\int_0^\lambda \int_0^\infty |\langle \widetilde{\boldsymbol{v}}\rangle|^2 d\xi dy \,. \tag{41}$$

where the subscripts b and d denote the base state and disturbance, respectively. The magnitude of the base state velocity vector is given by $|\langle \boldsymbol{v_o}\rangle|$ and that of the disturbance velocity field by $|\langle \widetilde{\boldsymbol{v}}\rangle|$, with height averaged quantities denoted by angular brackets. The height-averaged base state velocity fields in the streamwise and transverse directions (the vertical component is negligible within the lubrication approximation) are given by

$$\begin{aligned} \langle u_o\rangle &= -\frac{1}{2t^{2/3}} h_o g_{o\xi} + \frac{\mathcal{C}}{3t} {h_o}^2 h_{o\xi\xi\xi} \,, \\ \langle w_o\rangle &= 0 \,, \end{aligned} \tag{42}$$

respectively, while those of the averaged disturbance velocities are given by

$$\begin{aligned} \langle \widetilde{u}\rangle &= \left[-\frac{1}{2t^{2/3}} \left(h_o\phi_\xi + g_{o\xi}\psi\right) + \frac{\mathcal{C}}{3t} h_o \left(h_o\psi_{\xi\xi\xi} + 2h_{o\xi\xi\xi}\psi - t^{2/3}q^2 h_o\psi_\xi\right)\right] e^{iqy} \,, \\ \langle \widetilde{w}\rangle &= \left[-\frac{1}{2t^{1/3}} q h_o\phi + \frac{\mathcal{C}}{3t^{2/3}} q {h_o}^2 \left(\psi_{\xi\xi} - t^{2/3}q^2\psi\right)\right] i e^{iqy} \,. \end{aligned} \tag{43}$$

3.5 Mechanism for Large Transient Growth

Depending on the initial location ξ_s of the applied disturbances, the system undergoes different levels of amplification as shown in Fig. 12. When disturbances are applied at the origin (well inside the monolayer), the amplification ratio R rapidly decays from one to zero (Fig. 12a). The decay is more rapid for larger wavenumbers. The shear stress near the origin is very small and flow is not vulnerable to disturbance amplification.

For disturbances applied further downstream at $\xi_s = 0.7$, near the base of the initial concentration decay point, the system undergoes a brief period of small transient growth. Figure 12b shows that the smallest wavenumbers experience the largest amplification. The mode with $q = 0$ achieves an overall amplification ratio of about 25 and slowly decays toward zero as $t \to \infty$. Despite this short lived boost, disturbances with $q \neq 0$ dissipate their energy by $t \approx 1.5$. Additional studies of the rate of change $R^{-1}dR/dt$ has led to the following interpretation

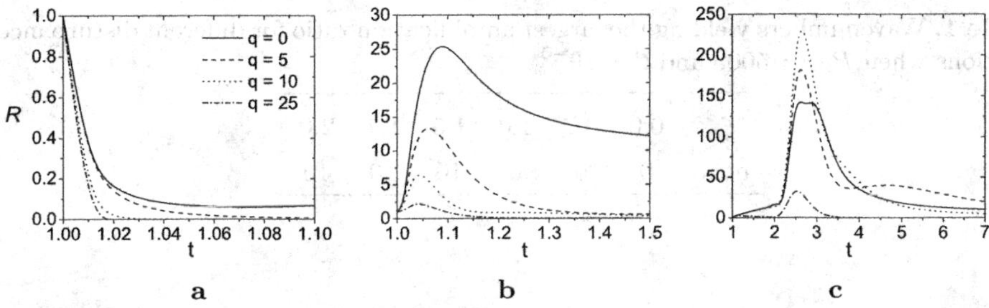

Fig. 12. Time evolution of the amplification ratio for **a** $\xi_s = 0.0$, **b** $\xi_s = 0.7$, and **c** $\xi_s = 2.0$ with disturbance wavenumbers in the range $0 \leq q \leq 25$ for $Pe_s = 5000$ and $\mathcal{C} = 10^{-5}$

of the flow [9]. When a disturbance is first applied to the spreading film, the system counteracts the disturbance flow by establishing Marangoni and capillary pressure gradients in the transverse direction ($\hat{y}$) that drive liquid and surfactant away from the disturbance to produce a momentary stabilizing response. The system overshoots this response, however, giving rise to an enhanced streamwise flow that produces the maximum in R shown for each curve. Eventually, the Marangoni stresses die away and therefore so does the driving force for spreading. In the limit $t \to \infty$ all disturbances decay to zero.

Significant amplification occurs when the initial disturbance is applied well ahead of the initial monolayer at ξ_s=2.0 (see Fig. 12c). In this case, the base state has sufficient time to develop a sizeable Marangoni ridge, which enhances the mobility of the disturbance when the two meet. The amplification therefore occurs at a later time in the spreading process (cf. Fig. 12b,c). Not only are the amplification ratios almost an order of magnitude larger but the duration of the enhanced response is also much longer. Because the speed of the advancing front decays in time as $dL/dt \sim t^{-2/3}$, the disturbance has a longer residence time in the vicinity of the Marangoni ridge than is the case with disturbances localized further upstream. The mode with the largest overall R also switches from $q = 0$ to $q = 10$ with successively smaller enhancement for $q = 5, 0$ and 25 in that order. There is an additional interesting difference between Figs. 12b and 12c. While in Fig. 12b the small q modes attain their maximum amplification ratio at later times than the large q disturbances, Fig. 12c shows that all the wavenumbers resonate at approximately the same time.

A recent study has revealed that the location of applied disturbances has a significant effect on which mode undergoes the strongest amplification [9] as shown in Table 1. Specifically, the $q = 0$ mode undergoes the largest transient growth when disturbances are localized at or behind the initial concentration distribution. By contrast, disturbances placed farther downstream selectively promote the $q = 10$ mode.

Table 1. Wavenumbers yielding the largest amplification ratio for different disturbance locations when $Pe_s = 5000$ and $C = 10^{-5}$

ξ_s	0.0	0.7	1.0	1.5	1.7	2.0
$q_{\max}$	0	0	0	10	10	10

3.6 Summary of Marangoni Driven Spreading

The disturbance functions, $\psi(\xi,t)$ and $\phi(\xi,t)$, corresponding to $q = 10$ and $\xi_s = 2.0$ are plotted in Fig. 13 for times $t = 1.0, 2.6$ and 8.0. In comparing these curves to the base state profiles shown in Figs. 9 and 10b, it is evident that the largest transient response occurs when the disturbances migrate to the point where the concentration gradient undergoes the most rapid change (*i.e.* the region of surfactant compression discussed earlier), which is also a region of high curvature in film thickness. For the case shown, the maximum amplification occurs at $\xi = 1.5$ for $t = 2.6$. A second region of large shear stress at $\xi = 0.7$ is not as vulnerable to disturbances since R, $\psi(\xi,t)$ and $\phi(\xi,t)$ rapidly decay for $t > 3.0$ for $\xi_s = 2.0$. It is likely that these disturbances are stabilized by the much slower base flow speed in this region. A comparison of Figs. 9a, 13a, and 10b indicates that as the disturbance functions advect through the Marangoni ridge into the linear portion of the height profile (where the concentration gradient is constant), the amplification ratio approaches zero. Increasing the level of surface diffusion by decreasing Pe_s leads to a substantial decrease in amplification. For example, disturbances with $q = 10$ initially localized well ahead of the spreading front suffer a reduction in the maximum value of R from approximately 230 to 15 as Pe_s is decreased from 5000 to 100. Values of Pe_s in experiments are estimated to be closer to 5000.

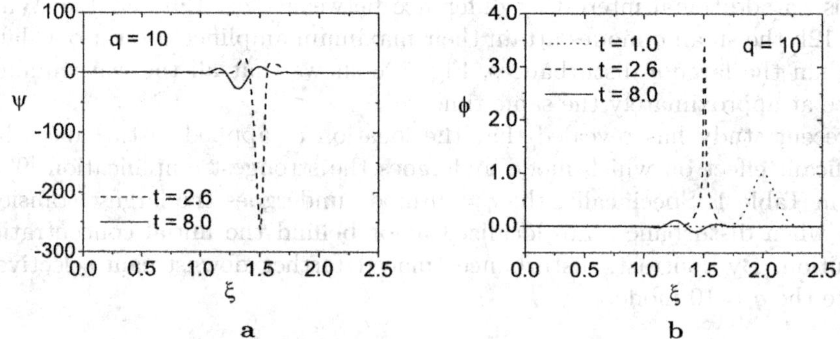

Fig. 13. Solutions for the disturbances in **a** film thickness, ψ, and **b** surfactant concentration, ϕ, for times ranging from $1.0 \leq t \leq 8.0$ with $q = 10$ and $\xi_s = 2.0$. The amplitude of the function $\psi(\xi, t = 1)$ in **a** is too small to be visible

Additional studies have shown that the decay rate of disturbances with $q > 10$ is even more rapid. For $q = 25$, the disturbance functions die away completely before they are convected into the linear regions of the base state profiles. As shown in Fig. 12c, disturbances with $q = 0$ also undergo significant amplification. Interestingly, the functions $\psi(\xi, t)$ and $\phi(\xi, t)$ assume shapes almost identical to $h_{o\xi}$ and $g_{o\xi}$. This equivalence cannot be derived analytically from (38a) and (38b). The equivalence between the disturbance eigenfunction $G(\xi)$ and the first spatial derivative of the base state $h_{\xi}(\xi)$ can be derived analytically for the time-independent equations governing gravitationally driven [40] or thermocapillary driven films [21].

The analysis presented above, which reduces the original stability analysis of a non-autonomous, non-normal system to a simpler initial value problem prevents identification of the optimal perturbations. It nonetheless provides valuable insight into possible mechanisms for instability. Because both the perturbations and the base states evolve in time, however, the energy growth of perturbations must be measured relative to that of the evolving base states. For a finite monolayer spreading over a thin liquid film, the analysis identifies the leading edge of the spreading front, where $h_{o\xi\xi}$ and $g_{o\xi\xi}$ are particularly large, as the region most susceptible to disturbance amplification.

Table (1) indicates that disturbances with $q = 10$ undergo the largest amplification if placed ahead of the initial surfactant monolayer. This placement allows development of a significant Marangoni ridge that amplifies disturbances upon contact. This prediction of a selective wavenumber is encouraging because it is one aspect of the instability that can be measured and compared with experiment. Experimental evidence, however, has not demonstrated whether the observed fingering patterns are an asymptotic instability or a nonlinear instability triggered by nonlinear mechanisms. The fact that all the localized disturbances investigated vanish as $t \to \infty$ predicts that the system is asymptotically stable. This asymptotic decay, however, is likely due to the fact that in these studies, the overall mass of surfactant contained in the source region is rather small. Current studies [10,11] clearly show that surfactant distribution from a more massive source than considered in Sect. 3.2 (modeled either as a large and fixed concentration at the origin or as a time-dependent release) generates more pronounced film thinning behind the advancing front with a consequent bottleneck for surface transport. This bottleneck leads to rapid disturbance localization and sufficient sustained amplification to produce asymptotic instability.

4 Conclusion

The presence of a deformable free surface in thin films driven to spread by body or shear forces gives rise to base states that are spatially nonuniform. This non-uniformity produces linearized disturbance operators that are non-normal and an eigenspectrum that may not be physically determinant at finite times. In this article, two examples of free surface shear flows are investigated, namely thermocapillary and Marangoni driven spreading. The first involves driving the coating flow with a constant shear stress that is induced by a streamwise linear temperature profile. The second involves spreading driven by a non-constant shear stress that is caused by an insoluble surfactant monolayer with non-uniform distribution. The associated disturbance operator for the thermocapillary system is autonomous but non-normal. This example is therefore used to demonstrate a more generalized, rigorous non-modal approach to linear stability for free surface flows. Calculations of the maximum disturbance amplification and the pseudospectra for this system, however, reveal weak non-normality and transient growth such that the modal growth rate is rapidly recovered. Subdominant modes contribute little energy to the leading eigenvector because their oscillatory behavior is rapidly damped by surface tension. Generalization of these results to numerous other lubrication flows involving surface tension and a constant driving force may suggest similarly weak non-normality and transient growth.

The base states corresponding to a finite monolayer spreading on a thin viscous film do not support constant traveling wave solutions. The film thickness and concentration profiles are rather complex waveforms, each with distinctive features that advance at different rates. Of particular interest are two regions of the flow profile where the gradient in shear stress and film curvature is large - namely, the region just ahead of the decay point of the initial surfactant distribution and a region at the leading edge of the spreading film where capillary forces produce surfactant compression. The disturbance operator for this system is non-autonomous and highly non-normal. Although the formalism for generalized linear stability outlined in Sect. 1.2 is applicable to non-autonomous operators, the disturbance propagator is difficult to compute accurately because the evolving base states and disturbance functions must be numerically evaluated in time. This universal approach to linear stability is therefore not applied to the Marangoni spreading problem. The analysis is instead reduced to an initial value problem to provide information about the physical mechanisms driving the observed instability. Initial disturbances are localized to various positions along the spreading profile to help determine the source of the transient disturbance growth. Disturbances initially localized well ahead of the surfactant monolayer undergo amplification by over two orders of magnitude (for the parameters used in this study). This amplification is traced to the leading edge of the Marangoni ridge where the second derivative of the surfactant concentration is particularly large, as is the curvature of the liquid surface.

An interesting avenue for further investigation is the application of successive excitations to the base states by sequential perturbations ahead of the advancing front. Such perturbations could maintain a large disturbance amplification ratio, increasing the potential for non-linear effects. A full nonlinear analysis of the response to such initial perturbations would be needed to ascertain if these effects can produce the requisite destabilization. Although the energy analysis has isolated a region in the flow that is most susceptible to disturbances and an associated preferred wavelength for large transient growth, direct comparison to experiment remains elusive since it is difficult to excite infinitesimal disturbances of specified shape. In contrast to thermocapillary spreading, little quantitative experimental data exists for Marangoni spreading. Additional theoretical and experiment work is required to understand fully the intricate arterial patterns observed experimentally.

Acknowledgment

We kindly acknowledge funding from the National Science Foundation (CTS and DMR), Unilever Research US and NASA Fluid Physics in support of our ongoing studies of thin film flows and hydrodynamic instabilities. JMD also wishes to thank the Department of Defense for a National Defense Science and Engineering Graduate Fellowship.

References

1. A.W. Adamson: *Physical Chemistry of Surfaces*, (John Wiley & Sons, New York 1990)
2. S. Chandrasekhar: *Hydrodynamic and Hydromagnetic Stability*, (Dover Publications, Inc., New York 1961)
3. J.M. Davis, S.M. Troian: "Influence of Boundary Slip on the Optimal Excitations in Thermally Driven Spreading," in review for Phys. Rev. E (2002)
4. J.M. Davis, S.M. Troian: Phys. Rev. E **67**, 016308:1-9 (2003)
5. J.M. Davis, S.M. Troian: Phys. Fluids **15**, 1344 (2003)
6. W.M. Deen: *Analysis of Transport Phenomena*, (Oxford University Press, New York 1998)
7. B.F. Farrell, P.J. Ioannou: J. Atmos. Sci. **53**, 2025 (1996)
8. B.F. Farrell, P.J. Ioannou: J. Atmos. Sci. **53**, 2041 (1996)
9. B.J. Fischer, S.M. Troian: Phys. Rev. E **67**, 016309:1-11 (2003)
10. B.J. Fischer, S.M. Troian: "Thinning and Disturbance Growth in Liquid Films Mobilized by Continuous Surfactant Delivery", in press for Phys. Fluids (2003)
11. B.J. Fischer, S.M. Troian: "The Linear Stability of Thin Films Subject to a Sharp Change in Shear Stress", submitted to Langmuir (2003)
12. N. Fraysse, G.M. Homsy: Phys. Fluids **6**, 1491 (1994)
13. F.R. Gantmacher: *The Theory of Matrices*, vol. 2. (Chelsea Publishing Company, New York 1964)
14. D.P. Gaver, J.B. Grotberg: J. Fluid Mech. **213**, 127 (1990)
15. G.H. Golub, C.F. Van Loan: *Matrix Computations*, 2nd ed. (Johns Hopkins University Press, Baltimore, MD 1990)
16. H. Greenspan: J. Fluid Mech. **84**, 125 (1978)
17. C. Huh, L.E. Scriven: J. Colloid Interface Sci. **35**, 85 (1971)
18. H.E. Huppert: Nature **300**, 427 (1982)
19. A.C. Hindmarsh: 'ODEPACK - A systemized collection of ODE solvers'. In: *Scientific Computing*, ed. by R. S. Stepleman, (North-Holland, Amsterdam 1983) pp. 55
20. O.E. Jensen, J.B. Grotberg: J. Fluid Mech. **240**, 259 (1992)
21. D.E. Kataoka, S.M. Troian: J. Colloid Interface Sci. **192**, 350 (1997)
22. D.E. Kataoka, S.M. Troian: J. Colloid Interface Sci. **203**, 335 (1998)
23. D.E. Kataoka: The Spreading Behavior of Thermally Driven Liquid Films. Ph.D. Thesis, Princeton University, Princeton (1999)
24. T. Kato: *Perturbation Theory for Linear Operators*, (Springer-Verlag, New York 1966)

25. V.G. Levich: *Physicochemical Hydrodynamics*, (Prentice-Hall, Englewood Cliffs 1962)
26. V. Ludviksson, E.N. Lightfoot: AIChE Journal **17**, 1166 (1971)
27. O.K. Matar, S.M. Troian: Phys. Fluids **10**, 1234 (1998)
28. O.K. Matar: The Dynamic Behavior of an Insoluble Surfactant Monolayer Spreading on a Thin Liquid Film. Ph.D. Thesis, Princeton University, Princeton, NJ (1998)
29. O.K. Matar, S.M. Troian: Phys. Fluids **11**, 3232 (1999)
30. *MATLAB 5.3*, (The MathWorks Inc., Natick, MA, 1999)
31. F. Melo, J.F. Joanny, S. Fauve: Phys. Rev. Lett. **63**, 1958 (1989)
32. A. Oron, S.H. Davis, S.G. Bankoff: Rev. Mod. Phys. **69**, 931 (1997)
33. S.C. Reddy, P.J. Schmid, D.S. Henningson: SIAM J. Appl. Math. **53**, 15 (1993)
34. W.E. Schiesser: *The Numerical Method of Lines*, (Academic Press, San Diego 1991)
35. P.J. Schmid, D.S. Henningson: *Stability and Transition in Shear Flows*, (Springer-Verlag, New York 2001)
36. S.F. Shen: J. Aerospace Sci. **28**, 397 (1961)
37. M.A. Spaid, G.M. Homsy: Phys. Fluids **8**, 460 (1996)
38. L.N. Trefethen, A.E. Trefethen, S.C. Reddy, T.A. Driscoll: Science **261**, 578 (1993)
39. L.N. Trefethen: SIAM Rev. **39**, 383 (1997)
40. S.M. Troian, E. Herbolzheimer, S.A. Safran, J. F. Joanny: Europhys. Lett. **10**, 25 (1989)
41. S.M. Troian, X.L. Wu, S. A. Safran: Phys. Rev. Lett. **62**, 1496 (1989)
42. S.M. Troian, E. Herbolzheimer, S.A. Safran: Phys. Rev. Lett. **65**, 333 (1990)
43. T.G. Wright, L.N. Trefethen: SIAM J. Sci. Comp. **23**, 591 (2001)

Thermocapillary Convection in Cylindrical Geometries

Bok-Cheol Sim[1], Abdelfattah Zebib[2], and Dietrich Schwabe[3]

[1] Department of Mechanical Engineering, Hanyang University, Ansan, Kyunggi-do 425-791, Korea
[2] Department of Mechanical and Aerospace Engineering, Rutgers, The State University of New Jersey, Piscataway, NJ 08855-8058, USA
[3] 1. Physics Institute, University of Giessen, Heinrich-Buff-Ring 16, 35392 Giessen, Germany

Abstract. Thermocapillary convection in two types of cylindrical geometries is studied by three-dimensional numerical simulations: an open cylindrical annulus heated from the outside wall and a liquid bridge. The non-deformable free surfaces are either flat or curved as determined by the fluid volume, V, and the Young-Laplace equation. Convection is steady and axisymmetric at sufficiently low values of the Reynolds number, Re, with either flat or curved surfaces. For the parameter ranges considered, it is found that only steady convection is possible at any Re in strictly axisymmetric computations. Transition to oscillatory three-dimensional motions occurs as Re increases beyond a critical value dependent on the aspect ratio, the Prandtl number, and V. Good agreement with available experiments is achieved in all cases.

1 Introduction

Thermocapillary convection is a surface tension driven flow due to a temperature gradient along a free surface. The temperature gradient results in a surface tension variation which causes a free surface shear stress and generates fluid motion in the liquid pool. It is well known that thermocapillary convection is steady and axisymmetric when the temperature difference between two side walls of an open cavity or two disks of a liquid bridge is sufficiently small. The axisymmetric flow undergoes a transition to oscillatory time-dependent, three-dimensional flow as the temperature difference increases beyond a critical value.

In order to obtain homogeneous solids from crystal melt [12], the influence of buoyancy and thermocapillary forces needs to be understood. These forces cause convection in the melt during crystal growth on Earth. However, in a microgravity environment(about $10^{-5}g_0$ in space experiments [13], where g_0 is the normal earth gravitational aceleration), thermocapillary forces dominate and may drive unsteady convection. Unsteady convection is responsible for striations which are bands of different concentrations in the crystal. Therefore, understanding transition to oscillatory flows is important to material processing in space. The most popular methods for growing crystals are the Czochralski and floating zone [11] techniques. Thermocapillary effects in the two techniques were summarized by Schwabe [12].

B.-C. Sim, A. Zebib, and D. Schwabe, Thermocapillary Convection in Cylindrical Geometries, Lect. Notes Phys. **628**, 107–129 (2003)
http://www.springerlink.com/

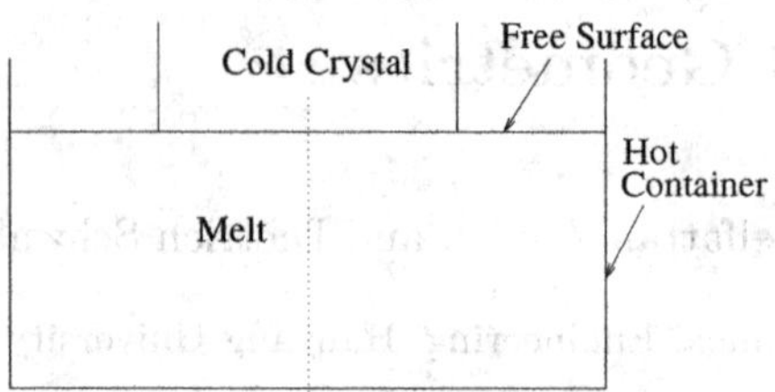

1.1: Czochralski technique

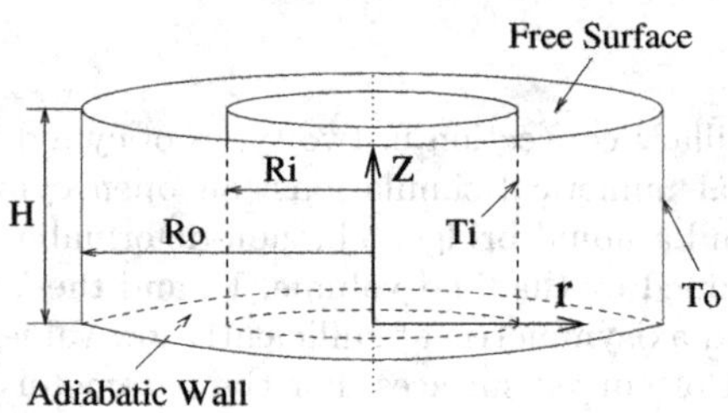

1.2: Physical system

Fig. 1. Physical system. The aspect ratio $Ar = (Ro - Ri)/H$ of the annular gap can be changed between 8 and 1 by adjusting H with a movable bottom

Numerous experimental and numerical studies of surface tension driven flows are available. However, three-dimensional numerical models in an open cylinder or annulus have not been reported even with non-deformable flat surfaces. A few 3D numerical simulations have been performed in a liquid bridge with a non-deformable flat surface. In the present study, thermocapillary convection in two types of cylindrical geometries is investigated by three-dimensional numerical simulations:

1. An open cylindrical annulus heated from the outside wall as shown in Fig. 1.2, constituting a model for the Czochralski crystal growth system shown in Fig. 1.1.
2. A liquid bridge in a half-zone model of the float-zone crystal-growth process as shown in Fig. 20.

For an open annulus with a flat surface, we report the influence of the aspect ratio on critical Reynolds numbers and frequencies, and pattern of convection. In order to study the effect of the free surface shape on thermocapillary convection, the non-deformable free surface in the liquid bridge is either flat or curved as determined by the liquid volume and the Young-Laplace equation. Boundary conditions at the curved interface are first derived, and the influence of the free surface shape on convection is studied. The numerical predictions in each problem are compared with available experiments.

2 Open Annulus Heated from the Outside Wall

The physical system considered is that in the microgravity experiment MAGIA [13] and is shown in Fig. 1.2. It is a cylindrical annulus with inner and outer radii, *Ri* and *Ro*, which is filled with an incompressible, Newtonian fluid of Prandtl number=6.84 to a height H. The aspect ratio, Ar, is defined as $(Ro - Ri)/H$, and the values of Ar are assumed 1, 2.5, 3.33, and 8 (which correspond to different values of H with Ri and Ro fixed, $Ri/Ro = 0.5$). The vertical inside and outside walls have cold and hot temperatures, $Ti = T_{cold}$ and $To = T_{hot}$, respectively. The bottom is an adiabatic solid wall. The horizontal free surface is assumed nondeformable and has convective heat loss to the surroundings with the ambient temperature, $T^*_\infty (= T_{cold})$. The surface tension is assumed a linear function of temperature,

$$\sigma = \sigma_r - \gamma(T^* - T^*_r), \tag{1}$$

where $\gamma = -\partial\sigma/\partial T^*$, and subscript r and superscript $*$ denote respectively a reference state and a dimensional quantity.

Assuming zero gravity, the nondimensional governing equations are as follows:

$$\nabla \cdot \boldsymbol{v} = 0, \tag{2}$$

$$Re\left(\frac{\partial \boldsymbol{v}}{\partial t} + \nabla \cdot (\boldsymbol{v}\boldsymbol{v})\right) = -\nabla P + \nabla^2 \boldsymbol{v}, \tag{3}$$

$$Ma\left(\frac{\partial T}{\partial t} + \nabla \cdot (\boldsymbol{v}T)\right) = \nabla^2 T, \tag{4}$$

where $\boldsymbol{v}$ is the nondimensional velocity vector, and P and T are the nondimensional pressure and temperature. Re is the Reynolds number, Pr is the Prandtl number, and Ma is the Marangoni number defined by

$$Re = \gamma\frac{\Delta T H}{\nu\mu}, Pr = \frac{\nu}{\alpha}, Ma = Pr \cdot Re, \tag{5}$$

where ν, μ, and α are kinematic viscosity, dynamic viscosity, and thermal diffusivity respectively. The length, temperature, velocity, pressure, and time are normalized with respect to H, ΔT, $\frac{\gamma\Delta T}{\mu}$, $\frac{\gamma\Delta T}{H}$, and $\frac{\mu H}{\gamma\Delta T}$, respectively, where $\Delta T = T_{hot} - T_{cold}$.

The velocities in the r, z, and θ directions of a cylindrical coordinate system are u, v, and w, respectively. With the free surface assumed nondeformable, the boundary conditions become

$$\frac{\partial u}{\partial z} + \frac{\partial T}{\partial r} = 0,\ v = 0,\ \frac{\partial w}{\partial z} + \frac{1}{r}\frac{\partial T}{\partial \theta} = 0,\ \frac{\partial T}{\partial z} = -BiT,\ at\ z = 1, \tag{6}$$

$$u = 0,\ v = 0,\ w = 0,\ \frac{\partial T}{\partial z} = 0,\quad at\ z = 0, \tag{7}$$

$$u = 0,\ v = 0,\ w = 0,\ T = 0,\quad at\ r = Ri/H, \tag{8}$$

$$u = 0,\ v = 0,\ w = 0,\ T = 1,\quad at\ r = Ro/H. \tag{9}$$

The Biot number in (6) is given by $Bi = hH/k$ where h is a heat transfer coefficient to the surroundings at the cold wall temperature, and k is the thermal conductivity of the liquid.

2.1 Numerical Aspects

The governing equations (2)–(4) and (6)–(9) are solved by a finite volume scheme with second order accuracy in space and with implicit method in time. The SIMPLER algorithm of Patankar [9] is used to handle the pressure coupling. Nonuniform grids (grid-stretching factor=1.1) are constructed with finer meshes in the regions under the free surface and near the bottom and side walls where boundary layers develop. The azimuthal direction has uniform grids in all cases. All computations are started with either zero or steady state initial conditions at lower Re. In order to examine grid dependence, critical Reynolds numbers, Re_c, are computed with various grids in each aspect ratio. Convergence criteria for a steady state are $|s^{n+1} - s^n| \leq 10^{-10}$ and $|s^{n+1} - s^n|/|s^{n+1}| \leq 10^{-4}$, where s is any of the variables (u, v, w, T) at all points and n is the time marching level. In addition, time histories of velocities and temperatures at the mid-point of the free surface, computed with various grids and time steps, are compared. Table 1 shows Re_c found using different grids with Ar = 1, 2.5, 3.33 and 8, respectively. Re is varied in increments of 10 in order to estimate Re_c. These steps ($\Delta Re = 10$) in Re are less than 3 percent of the reported Re_c.

The numerical code is also validated by comparison of steady surface temperature distribution with experimental results [2]. The parameters for the simulation are $Pr = 97$, $Ar = 0.889(Ri/H = 0.111,\ Ro/H = 1)$ and $Re = 510$, and temperatures in the inside and outside wall are $Ti = T_{hot}$ and $To = T_{cold}$, respectively. The numerical results with $Bi = 2$ are in good agreement with those from experiments as shown in Fig. 2.

Table 1. Grid refinement studies with $Bi = 0$ and various Ar

Ar	Grid numbers($r \times z \times \theta$)	Re_c	Wavenumber	Ar	Grid numbers	Re_c	wavenumber
1	51 × 51 × 21	740	5	2.5	51 × 41 × 31	490	
	61 × 61 × 31	740	5		61 × 51 × 41	490	9
	71 × 71 × 41	740	5		71 × 61 × 51	490	9
3.33	61 × 51 × 41	490	13	8	61 × 41 × 51	600	24
	61 × 51 × 61	490	12		71 × 51 × 56	570	24
	71 × 61 × 51	490	12		71 × 51 × 100	560	20
	71 × 61 × 81	490	12		71 × 51 × 112	560	20

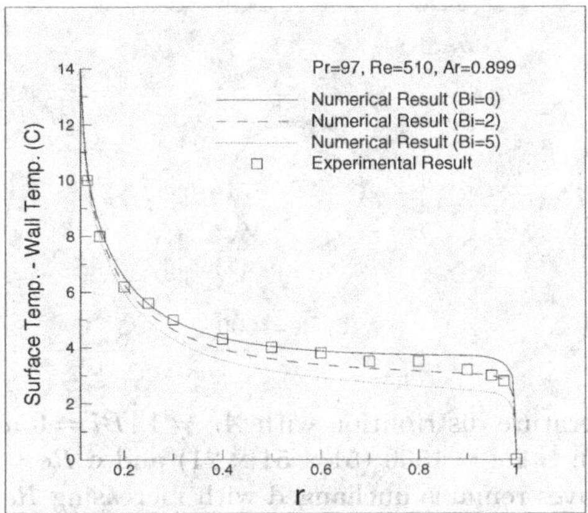

Fig. 2. Surface temperature distribution with $Pr = 97$, $Re = 510$, $Ar = 0.889(Ri/H = 0.111, Ro/H = 1)$, $Ti = T_{hot}$ and $To = T_{cold}$. The numerical results with $Bi = 2$ are in good agreement with experiments [2]

2.2 Results and Discussion

Azimuthal Rotating Waves: $Ar = 1$, 2.5 and 3.33, $Bi = 0$. Re_c for onset of oscillations with $Ar = 1$ is about 740. At $Re = 740$, the flow is steady and the isotherms on the free surface are just circular lines, i.e. axisymmetric. Figure 3 shows the time history of the temperature near the mid-point of the free surface with various supercritical *Re*. Figure 3a is computed from zero initial conditions, and Figs. 3b,c are computed from the same steady state initial conditions ($Re = 700$). As expected, starting from the same initial conditions oscillations begin earlier in time at the higher *Re*. While the amplitudes of temperature oscillations are larger at the higher *Re*, the mean temperature at these points decreases with increasing *Re*.

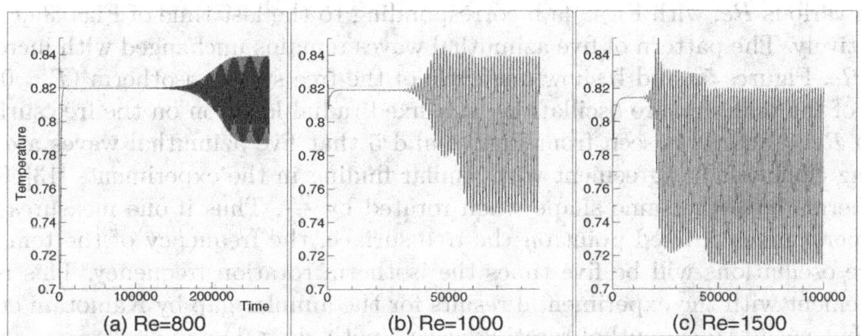

Fig. 3. Time history of the temperature near the mid-point of the free surface with $Ar = 1$, $Bi = 0$ and **a** $Re = 800$, **b** $Re = 1000$ and **c** $Re = 1500$. Same initial conditions are employed in **b** and **c**. The oscillations begin earlier in time at the higher *Re* as expected

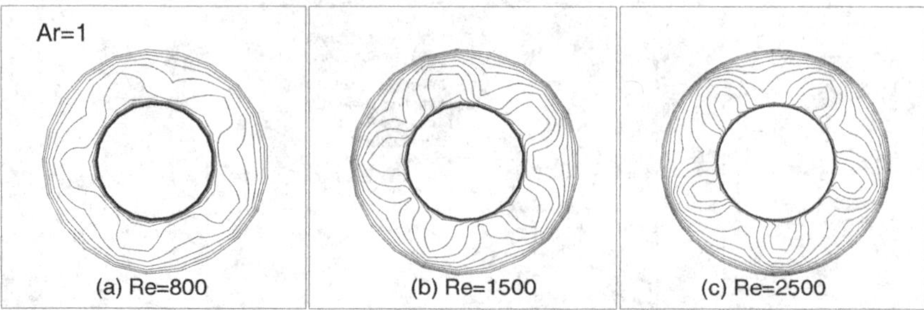

Fig. 4. Surface temperature distribution with $Ar = 1$, $Bi = 0$ and **a** $Re = 800$ (grid: $51(r) \times 51(z) \times 21(\theta)$), **b** $Re = 1500$ ($51 \times 51 \times 21$) and **c** $Re = 2500$ ($71 \times 71 \times 41$). The pattern of five waves remains unchanged with increasing Re

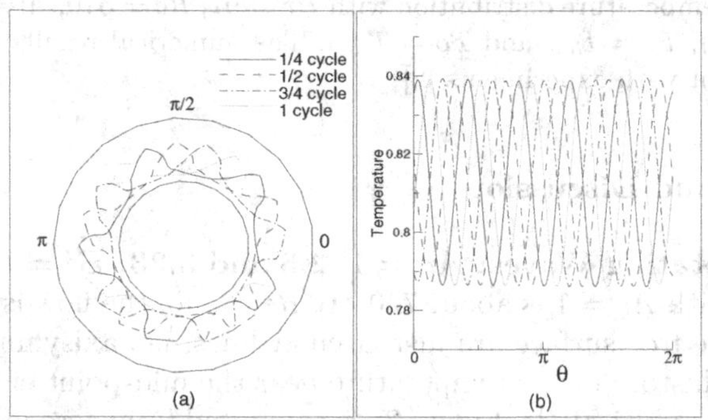

Fig. 5. One cycle of the free surface **a** isotherm ($T = 0.8$), and **b** temperature oscillations at the free surface and $r = 1.47$ with $Re = 800$, $Ar = 1$ and $Bi = 0$. Five azimuthal waves are travelling clockwise

Figure 4 shows instantaneous temperature distributions on the free surface with various Re, with Figs. 4a,b corresponding to the last time of Figs. 3a,c, respectively. The pattern of five azimuthal waves remains unchanged with increasing Re. Figures 5a and b show one cycle of the free surface isotherm ($T = 0.8$), and of the temperature oscillations at a fixed radial location on the free surface with $Re = 800$. It is seen from Figs. 4 and 5 that five azimuthal waves are rotating clockwise in agreement with similar finding in the experiments [13]. The isotherms have the same shape when rotated by $\frac{2\pi}{5}$. Thus if one measures the temperature at a fixed point on the free surface, the frequency of the temperature oscillations will be five times the isotherm rotation frequency. This is in agreement with the experimental results for the annular gap by Kamotani et al. [3], but with two azimuthal rotating waves (with $Ar < 1$).

Re_c with $Ar = 2.5$ and 3.33 are found to be about 490. The corresponding temperature distributions on the free surface are shown in Figs. 6 and 7. While nine azimuthal waves near the critical region appear on the free surface with $Ar = 2.5$, ten azimuthal waves are observed at $Re = 800$ as shown in Fig. 6.

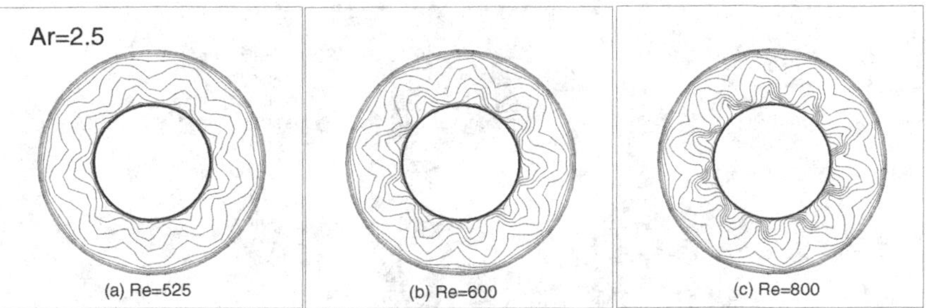

Fig. 6. Temperature distribution on the free surface with $Ar = 2.5$, $Bi = 0$ and **a** $Re = 525$, **b** $Re = 600$, and **c** $Re = 800$ (grid: $61 \times 51 \times 41$). The pattern of nine azimuthal waves in **a** changes as Re increases to ten waves in **c**

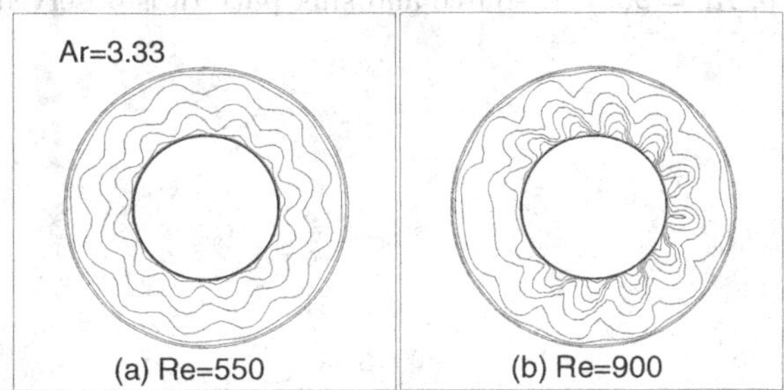

Fig. 7. Temperature distribution on the free surface with $Ar = 3.33$, $Bi = 0$ and **a** $Re = 550$ and **b** $Re = 900$ (grid: $71 \times 61 \times 81$). **a** is a rotating wave and **b** is a wave pattern travelling from a source into opposite directions to a sink

This is in reasonable agreement with the results from the experiments [13], where 11 azimuthal waves are observed at $Re = 2.5Re_c$. With $Ar = 3.33$, the twelve azimuthal waves are rotating clockwise near the critical region in Fig. 7a. It is evident that the number of azimuthal rotating waves near the critical region increases with increasing Ar.

Figures 8 and 9 show temperature fluctuations (deviation from the time-averaged mean temperature at each position) at the free surface and one cycle of temperature fluctuations at fixed radial locations on the free surface corresponding to Fig. 7b. The source and sink at the free surface are observed near $\theta = \pi$ and 0, respectively. The waves travel both clockwise and counterclockwise from $\theta = \pi$ to $\theta = 0$. The fluctuations are stronger near the inside, cold wall as shown in Fig. 9.

Azimuthal Pulsating Waves: $Ar = 8$ and $Bi = 0$ A mesh of $71 \times 51 \times 112$ is necessary to resolve the increased number of azimuthal wavetrains. Re_c with Ar=8 is about 560. Figure 10 gives temperature distributions on the free surface at various Re, while temperature fluctuations on the free surface are shown in

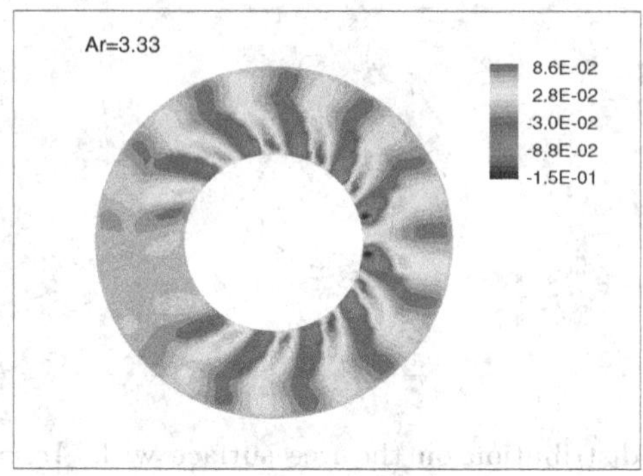

Fig. 8. Snapshot of free surface temperature fluctuations corresponding to Fig. 7b ($Ar = 3.33$, $Bi = 0$, $Re = 900$). A source and sink pattern is observed near $\theta = \pi$ and 0, respectively

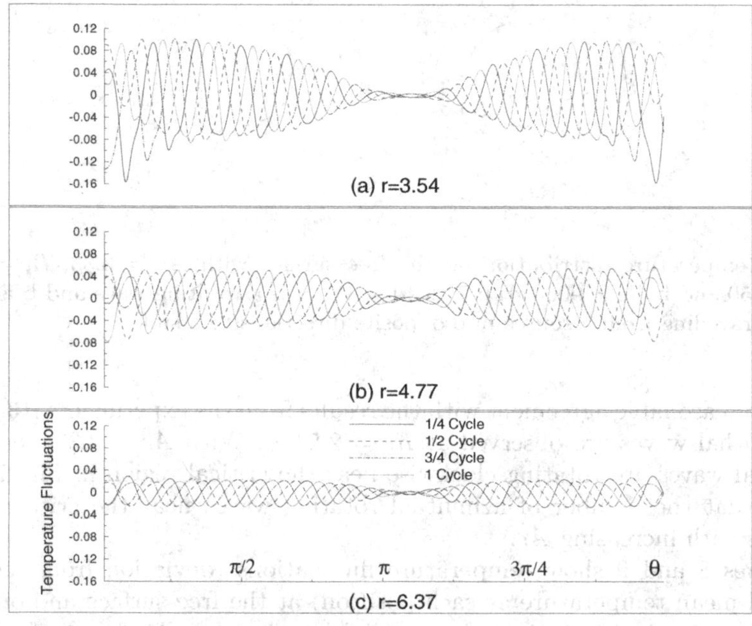

Fig. 9. One cycle of temperature fluctuations at the free surface and **a** $r = 3.54$, **b** $r = 4.77$ and **c** $r = 6.37$ corresponding to Fig. 8 ($Ar = 3.33$, $Bi = 0$, $Re = 900$). The fluctuations decrease with increasing r towards the hot side

Figs. 11 and 12. Twenty azimuthal wavetrains are found on the free surface, and the waves are pulsating as shown in Fig. 12. The inside wall looks like the source of the waves: the waves are generated at the inside cold wall and travel to the outside hot wall as shown in Fig. 13. This is in good agreement with the hydrothermal waves in the infinite layer model [21] and the rectangular cavity

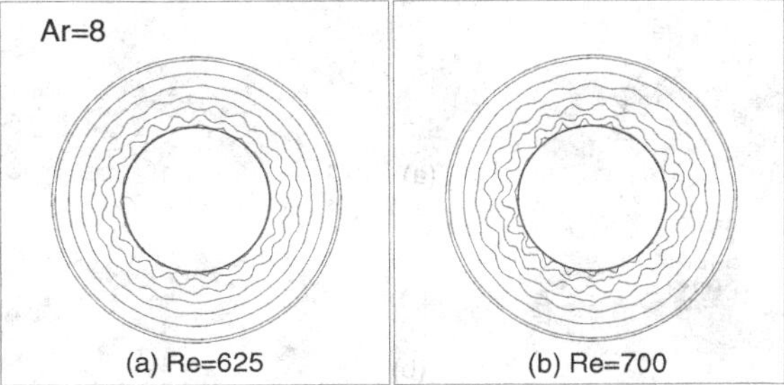

Fig. 10. Temperature distributions on the free surface with $Ar = 8$, $Bi = 0$ and **a** $Re = 625$ and **b** $Re = 700$ (grid: $71 \times 51 \times 112$). The inside cold wall looks like the source of the waves

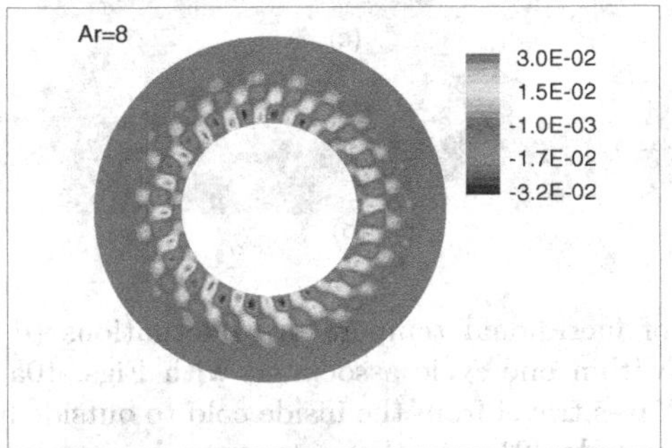

Fig. 11. Snapshot of free surface temperature fluctuations corresponding to Fig. 10a ($Ar = 8$, $Bi = 0$, $Re = 625$) indicating a pulsating pattern

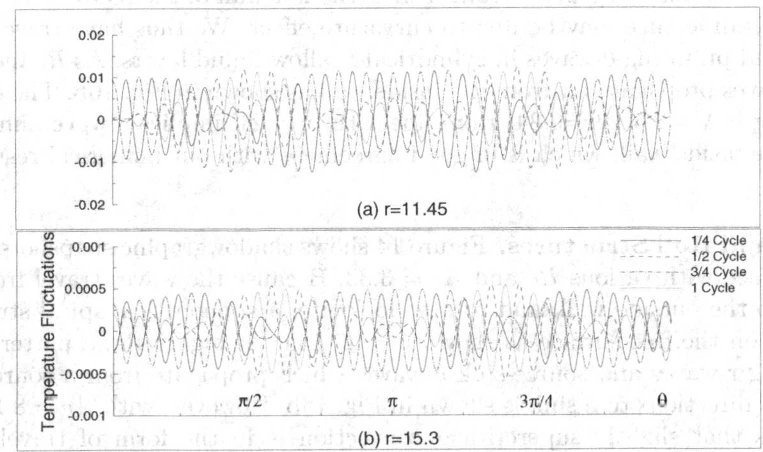

Fig. 12. One cycle of free surface temperature fluctuations at **a** $r = 11.45$ and **b** $r = 15.3$ corresponding to Fig. 11 ($Ar = 8$, $Bi = 0$, $Re = 625$) confirming the standing wave

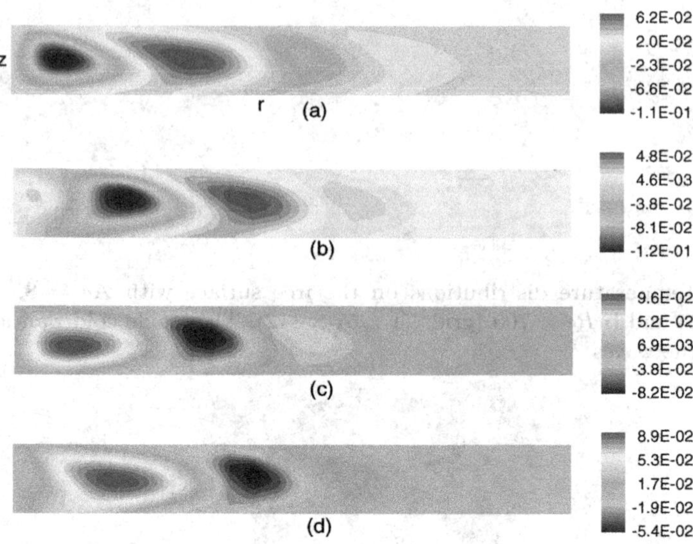

Fig. 13. Snapshots of meridional temperature fluctuations ($\theta = 0$) at four evenly distibuted instances within one cycle associated with Figs. 10a, 11 and 12 ($Ar = 8$, $Bi = 0$, $Re = 625$). Waves travel from the inside cold to outside hot walls in agreement with the infinite layer model [21] and the rectangular cavity simulations [24]

simulations [24]. However, while the oscillations are stronger near the cold wall in the open annulus, they were stronger near the hot wall of the rectangular cavity [24]. This difference may be due to curvature effect. We thus have travelling r-waves and pulsating θ-waves in cylindrical-shallow liquid layers. As Re increases these waves propagate far from the inside wall as shown in Fig. 10b. The critical wavelength $\lambda = 2.5$ (Pr=6.84) from linear theory [21] implies a wave number of 20 at the inside wall, which is in good agreement with our numerical result.

Spiral and Roll Structures. Figure 14 shows shadowgraphic snapshots at the free surface with various Re and $Ar = 3.33$. Because the waves travel from the inside to the outside wall, and rotate clockwise, a supercritical spiral structure appears on the free surface as shown in Fig. 14a. At higher Re, a pattern with travelling r-waves and source-sink θ-waves which propagate from a source into opposite directions to a sink is shown in Fig. 14b. However, with $Ar = 8$ Fig. 15 indicates that slightly supercritical convection is in the form of travelling r-waves and pulsating source-sink θ-waves. These two kinds of spiral patterns are in agreement with experimental results [1].

Figures 16a–d shows streamlines at Re_c with various Ar and $Bi = 0$. The flows are steady and axisymmetric. A single-roll structure is observed at each Ar. The calculated flow structure with $Ar = 1$ in Fig. 16a resembles the observations made on an open rectangular cavity from microgravity experiments [7] as shown

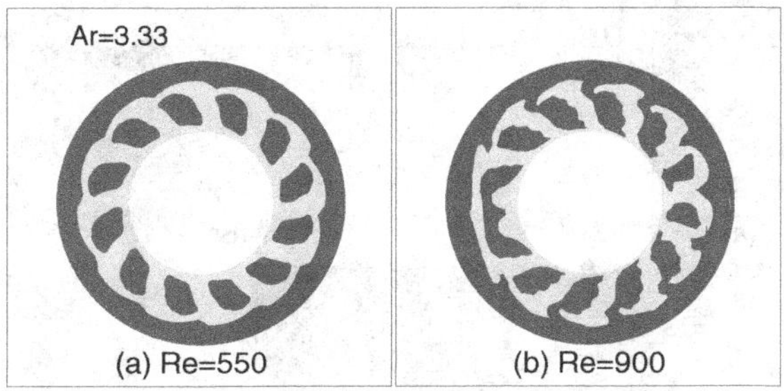

Fig. 14. Shadowgraphic snapshots (contours of $\nabla^2 T$) at the free surface corresponding to Fig. 7 ($Ar = 3.33$, $Bi = 0$, **a** $Re = 550$ and **b** $Re = 900$). **a** shows azimuthal clockwise rotating waves, while **b** indicates azimuthal waves with a source and sink near respectively $\theta = \pi$ and 0

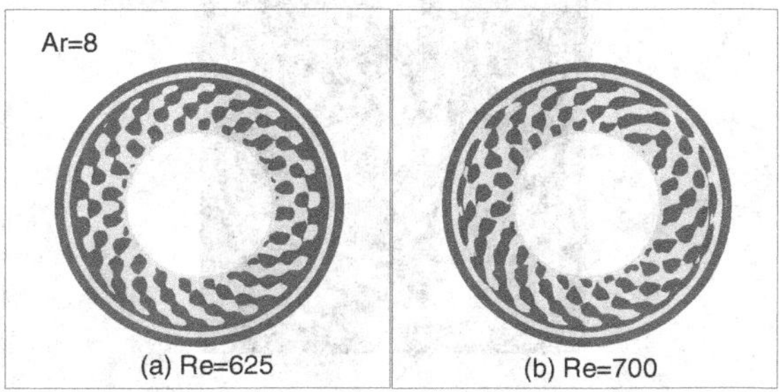

Fig. 15. Shadowgraphic snapshots (contours of $\nabla^2 T$) at the free surface corresponding to Fig. 10 ($Ar = 8$, $Bi = 0$, **a** $Re = 625$ and **b** $Re = 700$). Both **a** and **b** show azimuthal pulsating waves with a source and sink near respectively $\theta = 0$ and π

in Fig. 16e. Figure 17 shows snapshots of meridional streamlines at θ=0 with various Ar and supercritical Re. In steady state, only the single-roll structure is available with Ar=1, 2.5, 3.33 and 8. However, just above critical, two and three rolls are observed with Ar=3.33 and 8, respectively. Figure 10 of [13] with $Ar = 8$ is in perfect agreement. The number of rolls increases with increasing Ar. We can expect the multi-roll structure to appear beyond Re_c in the case of shallow liquid layers. The axisymmetric results are very different from those of two-dimensional rectagular cavities reported by Xu and Zebib [24], where a critical Ar exists, the multi-structure appears at subcritical Re, and the flow can be stable with multi-structure in restabilized region (highly supercritical Re). However, the structures of three-dimensional states in shallow cylindrical and rectangular cavities have in common travelling multi-cells from the cold to hot corners with a standing, pulsating pattern in the third direction.

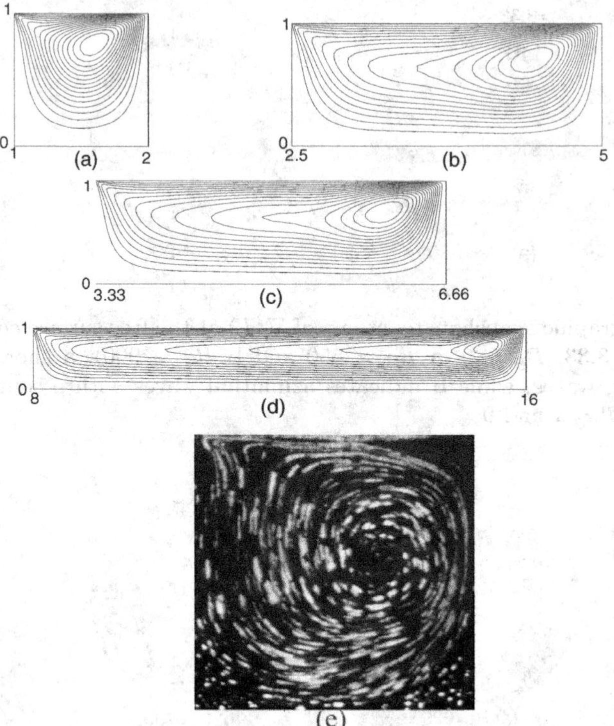

Fig. 16. Streamlines at the critical steady state with $Bi = 0$, varous Re_c and Ar [**a** 740 and 1, **b** 490 and 2.5, **c** 490 and 3.33, and **d** 560 and 8] and **e** streak lines in an open rectangular cavity from microgravity experiments ($20mm \times 20mm \times 20mm, Pr = 7, \Delta T = 9K (Re \approx Re_c)$)[7]. Figures are shown at different scales. A single-roll structure occurs at the steady state

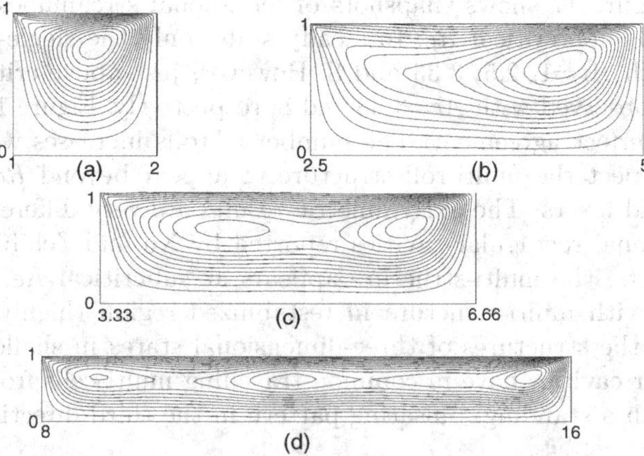

Fig. 17. Snapshots of meridional streamlines at $\theta = 0$ with $Bi = 0$, various supercritical Re and Ar: **a** 800 and 1, **b** 525 and 2.5, **c** 550 and 3.33, and **d** 625 and 8. Multi-cells occur near transition in shallow liquid layers with $Ar \geq 3.33$

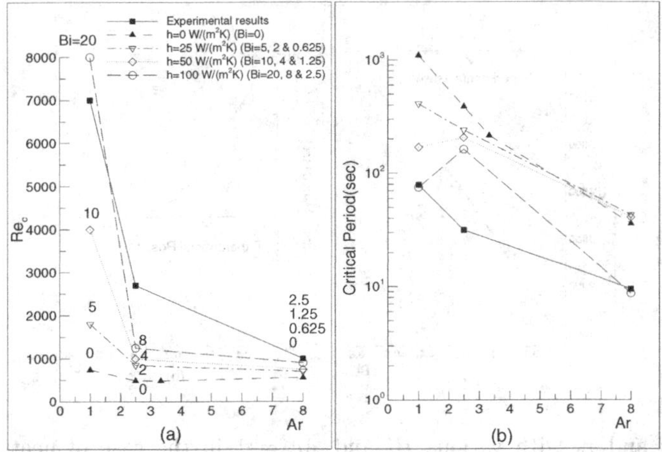

Fig. 18. Re_c and τ_c corresponding to various Ar at various Bi cooling the free surface. Heat loss from the free surface stabilizes the flow, and Re_c increases with increasing Bi at fixed Ar. τ_c decreases with Ar (Bi) at fixed Bi (Ar). Note that the curves here are for constant h so that Bi decreases along these curves with increasing Ar

Critical Reynolds Numbers and Frequencies. Figure 18 shows the effect of Bi on Re_c and the critical dimensional-period, τ_c. The numerical results with $Bi = 0$ are in good qualitative but not in good quantitative agreement with the experiments [13]. The value of h is 25 to 250 $W/(m^2K)$ for gases in forced convection. Because of evaporation, the value will be increased substantially. It can be seen that heat loss from the free surface stabilizes the flow, and Re_c increases with increasing Bi. τ_c decreases with increasing Bi. It is observed that better comparison with experiments is achieved at the larger values of Bi.

In the experiments [13], it is argued that the free surface is effectively heated by the surroundings. This is modelled here assuming $T_\infty = \frac{(T^*_\infty - T_i)}{\Delta T} = \frac{\gamma H}{\nu\mu}\frac{(T^*_\infty - T_i)}{Re}$. Thus, the last boundary condition in (6) at $z = 1$ is replaced by

$$\frac{\partial T}{\partial z} = -Bi\left(T - \frac{6 \times 10^6 H}{5.02 Re}\right), \tag{10}$$

with H in meters. This is the extreme case where T^*_∞ is 6^oC higher than T_i. Figure 19 shows the variation of Re_c and τ_c with Bi and $Ar = 1$. Re_c and τ_c are very sensitive to Bi due to the constant high temperature of the surroundings, and they approach the experimental results [13] with increasing Bi.

2.3 Conclusions

Oscillatory thermocapillary convection in open cylindrical containers is investigated numerically to document its stability characteristics. Five, nine and twelve azimuthal waves near the critical region are found rotating clockwise on the free

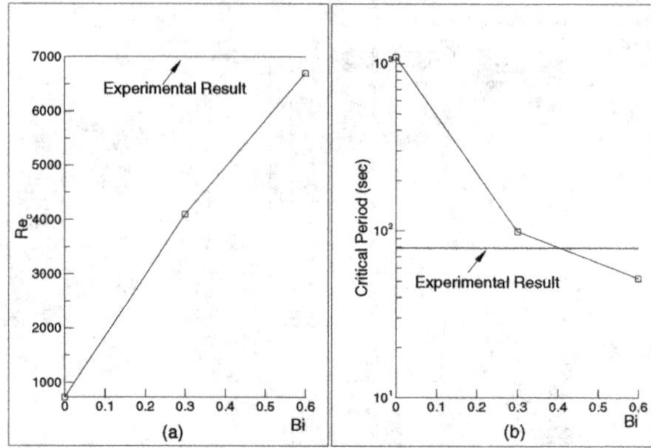

Fig. 19. Re_c and τ_c with various Bi and $Ar = 1$ in the case of heating from the surroundings to the free surface. The heating stabilizes the flow, and Re_c increases with increasing Bi

surface with, respectively, $Ar = 1$, 2.5 and 3.33, while twenty azimuthal pulsating waves are observed on the free surface with $Ar = 8$. While only a single-roll structure is observed up to Re_c with each Ar, the multi-roll structure appears beyond Re_c in shallow liquid layers with $Ar = 3.33$ and 8. In general, the number of azimuthal waves and multicells increases with increasing Ar, and τ_c decreases with increasing Ar.

Both heat loss from the free surface or heating from the surroundings to the free surface stabilize the flow, and either inclusion is necessary to achieve better quantitative agreement with the experiments. Re_c increases with increasing Bi while τ_c shows the opposite trend.

3 Liquid Bridge with a Curved Surface

The physical system considered is a liquid bridge with either flat or curved surface as shown in Fig. 20. $Ar(= R/H)$ of 1 and Pr of 1 and 4 are used to validate the numerical predictions with linear theory, and $Ar = 0.714$ and $Pr = 27$ to compare with experiments. The two upper and lower disks have nondimensional temperatures, $T_h = 1$ and $T_c = 0$. The surface tension is assumed a linear function of temperature as shown in (1).

The nondimensional governing equations are (2)–(4). Re, Pr, Ma and scales for normalization are the same as those from Sect. 2. The boundary conditions become

$$u = 0,\ v = 0,\ w = 0,\ T = 0,\ at\ z = 0, \tag{11}$$

$$u = 0,\ v = 0,\ w = 0,\ T = 1,\ at\ z = 1. \tag{12}$$

The derivation of the boundary conditions and interface equation appears in Appendix. The non-dimensionalized position of the free surface is described by

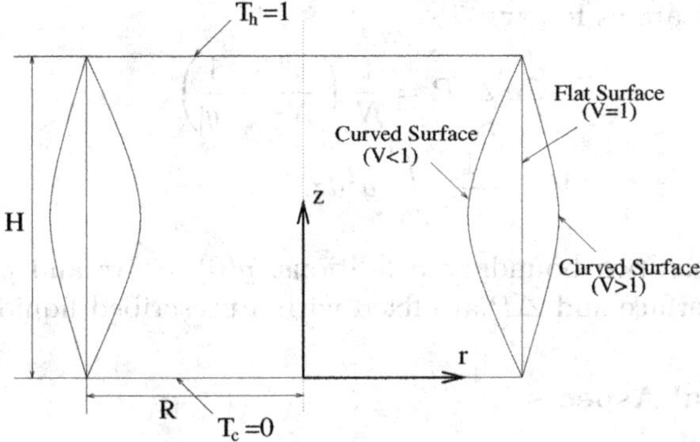

Fig. 20. Physical system

a function $r = g(z)$. Thermal, kinematic and tangential stress balance boundary conditions at the interface are

$$-\frac{1}{N}\left(\frac{\partial T}{\partial r} - g'\frac{\partial T}{\partial z}\right) = BiT, \tag{13}$$

$$u = g'v, \tag{14}$$

$$(1 - g'^2)\left(\frac{\partial v}{\partial r} + \frac{\partial u}{\partial z}\right) + 2g'\left(\frac{\partial u}{\partial r} - \frac{\partial v}{\partial z}\right) = -SN\left(g'\frac{\partial T}{\partial r} + \frac{\partial T}{\partial z}\right), \tag{15}$$

$$\frac{\partial w}{\partial r} - \frac{w}{r} + \frac{1}{r}\frac{\partial u}{\partial \theta} - g'\left(\frac{\partial w}{\partial z} + \frac{1}{r}\frac{\partial v}{\partial \theta}\right) = \frac{-SN}{r}\frac{\partial T}{\partial \theta}, \tag{16}$$

where $N = (1 + g'^2)^{1/2}$ and $g' = dg/dz$. It is known that isotherms near hot and cold walls are compressed in high Pr flow [25] and the driving force near the cold wall is much less effective for the overall flow than that near the hot wall [6]. The peak of the surface velocity near the cold wall causes convergence problems in numerical simulations. Thus, in some of our numerical experiments, we use regularization (damping) near the cold wall only by changing the driving thermocapillary forces in (15) and (16). The regularization factor [14], S, in the lower 5 percent of the free surface is defined as

$$S = 0.25[1 - cos(20\pi z)]^2 \quad at\ z \leq 0.05, \tag{17}$$

$$S = 1 \quad otherwise. \tag{18}$$

The effect of this regularization on the critical Re and frequency will be discussed.

The location of the interface, $g(z)$, is determined by the normal stress balance. The deviation of the free surface shape from the static meniscus depends on the Capillary number, $Ca = \gamma\Delta T/\sigma_o$. When $Ca \ll 1$, the dynamic surface deformation can be neglected [8], and the normal stress balance equation simplifies to the Young-Laplace equation. The interface and liquid volume equations in a state of rest are as follows:

$$-Ca \bigtriangleup P = \frac{1}{N}\left(\frac{g''}{N^2} - \frac{1}{g}\right), \tag{19}$$

$$V = \frac{1}{Ar^2}\int_0^1 g^2 dz. \tag{20}$$

Equation (19) has two boundary conditions, $g(0) = Ar$ and $g(1) = Ar$. The shape of the interface and ΔP are fixed with a prescribed liquid volume.

3.1 Numerical Aspects

In order to solve the problem with a curved surface, the governing equations are transformed from the physical domain(r, z, θ) into a rectangular computational domain(ξ, η, ζ).

$$\xi = r/g(z) \tag{21}$$

$$\eta = z \tag{22}$$

$$\zeta = \theta \tag{23}$$

The governing equations transformed into the computational domain are

$$\frac{1}{\xi}\frac{\partial \xi u}{\partial \xi} - \xi g' \frac{\partial v}{\partial \xi} + g\frac{\partial v}{\partial \eta} + \frac{1}{\xi}\frac{\partial w}{\partial \zeta} = 0, \tag{24}$$

$$Re\left[\frac{\partial u}{\partial t} + \frac{1}{\xi g}\frac{\partial \xi u^2}{\partial \xi} - \frac{\xi g'}{g}\frac{\partial uv}{\partial \xi} + \frac{\partial uv}{\partial \eta} + \frac{1}{\xi g}\frac{\partial uw}{\partial \zeta} - \frac{w^2}{\xi g}\right]$$
$$= -\frac{1}{g}\frac{\partial p}{\partial \xi} - \frac{u}{(g\xi)^2} + \nabla^2 u - \frac{2}{(g\xi)^2}\frac{\partial w}{\partial \zeta}, \tag{25}$$

$$Re\left[\frac{\partial v}{\partial t} + \frac{1}{\xi g}\frac{\partial \xi uv}{\partial \xi} - \frac{\xi g'}{g}\frac{\partial v^2}{\partial \xi} + \frac{\partial v^2}{\partial \eta} + \frac{1}{\xi g}\frac{\partial vw}{\partial \zeta}\right] = -\frac{\partial p}{\partial \eta} + \frac{\xi g'}{g}\frac{\partial p}{\partial \xi} + \nabla^2 v, \tag{26}$$

$$Re\left[\frac{\partial w}{\partial t} + \frac{1}{\xi g}\frac{\partial \xi uw}{\partial \xi} - \frac{\xi g'}{g}\frac{\partial vw}{\partial \xi} + \frac{\partial vw}{\partial \eta} + \frac{1}{\xi g}\frac{\partial w^2}{\partial \zeta} + \frac{uw}{\xi g}\right]$$
$$= -\frac{1}{\xi g}\frac{\partial p}{\partial \zeta} - \frac{w}{(g\xi)^2} + \nabla^2 w + \frac{2}{(g\xi)^2}\frac{\partial u}{\partial \zeta}, \tag{27}$$

$$Pr Re\left[\frac{\partial T}{\partial t} + \frac{1}{\xi g}\frac{\partial \xi uT}{\partial \xi} - \frac{\xi g'}{g}\frac{\partial vT}{\partial \xi} + \frac{\partial vT}{\partial \eta} + \frac{1}{\xi g}\frac{\partial wT}{\partial \zeta}\right] = \nabla^2 T, \tag{28}$$

$$\nabla^2 = \frac{1}{g^2}\left[\frac{1}{\xi} + \xi g'^2\right]\frac{\partial}{\partial \xi}\left(\xi\frac{\partial}{\partial \xi}\right) - \xi\frac{g''g - g'^2}{g^2}\frac{\partial}{\partial \xi}$$
$$-\frac{2\xi g'}{g}\frac{\partial^2}{\partial \eta \partial \xi} + \frac{\partial^2}{\partial \eta^2} + \frac{1}{(\xi g)^2}\frac{\partial^2}{\partial \zeta^2} \tag{29}$$

The transformed boundary conditions become

$$\text{At } \eta = 0, \quad T = 0, \; u = 0, \; v = 0, \; w = 0, \tag{30}$$

$$\text{At } \eta = 1, \quad T = 1, \; u = 0, \; v = 0, \; w = 0. \tag{31}$$

At the interface($\xi = 1$),

$$\left(\frac{1}{g} + \frac{g'^2}{g}\right)\frac{\partial T}{\partial \xi} - g'\frac{\partial T}{\partial \eta} = -NBiT, \tag{32}$$

$$u = g'v \tag{33}$$

$$\left(\frac{1+g'^2}{g}\right)\frac{\partial v}{\partial \xi} - 2g'\frac{\partial v}{\partial \eta} + \left(\frac{g'+g'^3}{g}\right)\frac{\partial u}{\partial \xi} + (1-g'^2)\frac{\partial u}{\partial \eta} = -SN\frac{\partial T}{\partial \eta}, \tag{34}$$

$$(1+g'^2)\frac{\partial w}{\partial \xi} - gg'\left(\frac{\partial w}{\partial \eta} + \frac{1}{g}\frac{\partial v}{\partial \zeta}\right) - w + \frac{\partial u}{\partial \zeta} = -SN\frac{\partial T}{\partial \zeta}. \tag{35}$$

First, the shape($g(z)$) of the interface with described liquid volume is determined by (19) and (20). The transformed governing equations (24)–(28) and boundary conditions equations (30)–(35) are solved by a finite volume method employing a SIMPLER algorithm. Details of the numerical solver are given in Sect. 2.1. In order to examine grid dependence in axisymmetric models, surface velocity and temperature distributions are computed with various grids and liquid volumes, and a mesh of $51(r)\times 51(z)$ is used. In the three-dimensional model, Re_c are computed with $Bi = 0$ and various grids and Pr. Table 2 shows Re_c found using different grids with $Bi = 0$, and various Pr and V. Re is varied in increments of 10 in order to estimate Re_c. These steps in Re are less than 5 percent of the reported Re_c. A mesh of $41(r) \times 51(z) \times 40(\theta)$ is used for $V \leq 1$, and $51 \times 61 \times 50$ for $V = 1.139$.

Table 2. Grid refinement studies of 3D convection with $Bi = 0$

Ar	V	Pr	Grid ($r \times z \times \theta$)	Re_c	
1	1	1	31 ×31×30	2,650	no regularization
			41 ×41×40	2,580	no regularization
			51 ×51×50	2,570	no regularization
			31 ×31×30	2,550	regularization
			41 ×41×40	2,520	regularization
			51 ×51×50	2,520	regularization
		4	31 ×31×30	1,060	regularization
			41 ×41×40	1,010	regularization
			51 ×51×50	970	regularization
0.714	1	27	31 ×41×30	220	regularization
			41 ×51×40	210	regularization
			51 ×61×50	220	regularization
	0.755		41 ×51×40	220	regularization
			51 ×61×50	220	regularization

Table 3. Comparison of stream function minima($\psi_{min} \times 10^2$) in a 2D liquid bridge ($Bi = 0.3$, $Ar = 1$ and $V = 1$)

Re	Pr	Sumner et al. [22]	Present result without regularization	Present result with regularization
100	10	-0.4221	-0.4217	-0.4183
10	100	-0.4205	-0.4202	-0.4199

Table 4. Comparison with linear stability analysis for Pr=1 and 4 with $Bi = 0$, $Ar = 1$ and $V = 1$

Pr	Present results (regularization) Re_c, ω_c, m	Present results (no regularization) Re_c, ω_c, m	Linear stability [23] (regularization) Re_c, ω_c, m	Linear stability [4] (no regularization) Re_c, ω_c, m
1	2520, 58.0, 2	2580, 58.7, 2	2539, 63.2, 2	2551, 65.0, 2
4	1010, 25.1, 2		1047, 27.9, 2	1002, 28.5, 2

The axisymmetric code is validated by comparison of the minimum values of the stream functions with those from Sumner et al. [22] in Table 3. Table 4 shows Re_c, ω_c and m from the three-dimensional simulations and linear theories [23,4], where ω_c and m are nondimensional critical angular frequency($= 2\pi f H^2/\nu$) and wavenumber. The effect of regularization on the stream function minima, Re_c, and ω_c is less than 5 percent as shown in Tables 3 and 4. Thus, the regularization is used to save computing time in all reported two- and three-dimensional simulations. More details on regularization can be found in [14,23].

3.2 Results and Discussion

Axisymmetric Thermocapillary Convection with $Bi = 0$, $Ar = 0.714$, $Pr = 27$ and Various V. We have investigated thermocapillary convection with various V up to $Re = 1,000$, and have found no axisymmetric oscillatory states in liquid bridges with either flat or curved surfaces. We thus conclude that only azimuthal waves can generate oscillations in this model. While oscillatory thermocapillary convection in a rectangular cavity can be investigated in two-dimensional simulations [24], it can not be realized in an open cylinder with a uniform heat flux [18,19] or in a liquid bridge. Oscillatory buoyant-thermocapillary convection was reported in axisymmetric simulations by Shevtsova and Legros [14], but no oscillations of pure thermocapillary convection was found in their axisymmetric simulations. The details of the axisymmetric results appear in [19,20].

Three-Dimensional Thermocapillary Convection with $Bi = 0$, $Ar = 0.714$ and $Pr = 27$. Re_c for onset of oscillations with a flat cylindrical surface is about 210. Figures 21a and b show temperature fluctuations in the section $z = 0.53$ and at the free surface for $Re = 230$ with $V = 1$ and 1.138($Re_c = 210$), respectively. Figure 22 shows temperature signals from four numerical

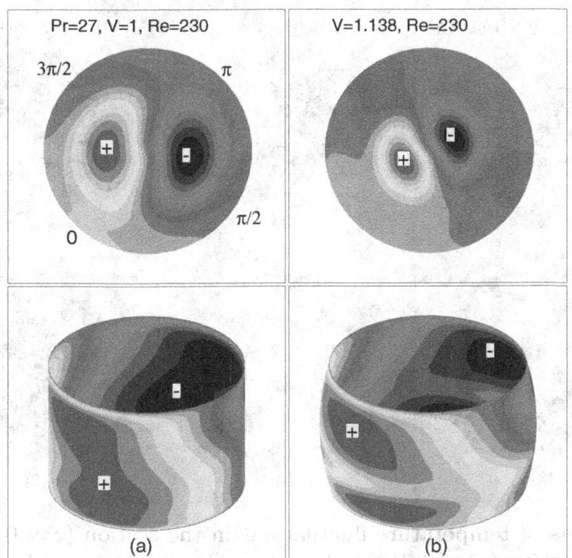

Fig. 21. Snapshots of temperature fluctuations in the section ($z = 0.53$) and the free surface with $Bi = 0$, $Ar = 0.714$ and various V. A pair of hot and cold spots ($m = 1$) is observed

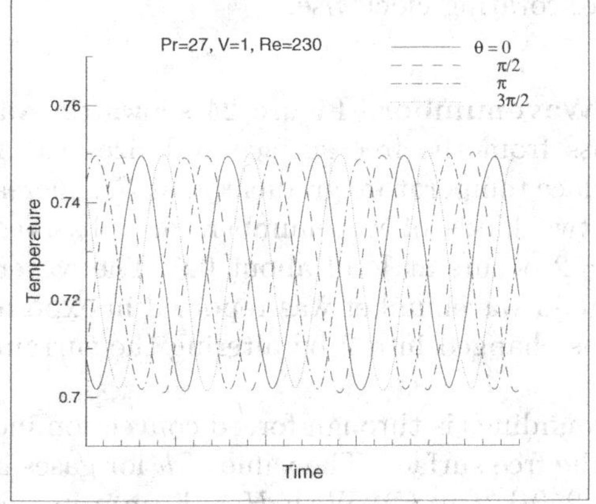

Fig. 22. Temperature oscillations at four fixed points ($r = 1, z = 0.53, \theta = 0, \pi/2, \pi, 3\pi/2$) corresponding to Fig. 21a. A rotating pattern is indicated

thermocouples at different azimuthal positions. The temperature fluctuations consist of a hot and a cold spot rotating clockwise, i.e $m = 1$. This rotating pattern with $m = 1$ remains unchanged with increasing Re beyond the critical value.

Re_c was found to decrease with increasing V in the case of concave surfaces. With $V = 0.874$, 0.755 and 0.647, Re_c are found to be about 190, 220 and 250, respectively. Figure 23 shows temperature fluctuations in the section $z = 0.53$

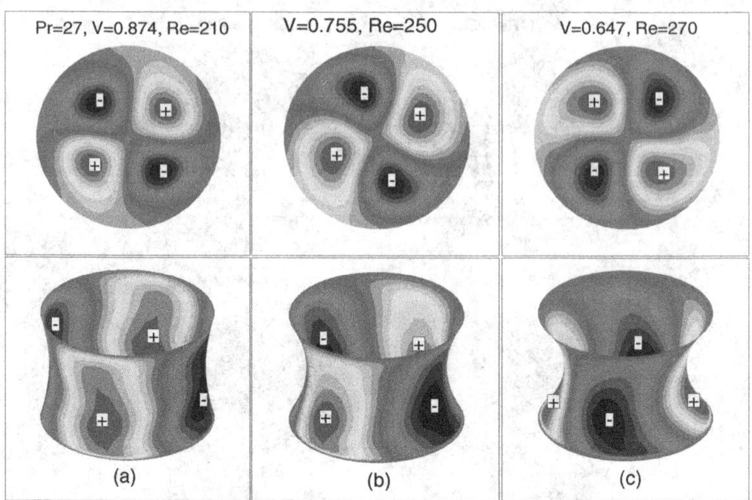

Fig. 23. Snapshots of temperature fluctuations in the section ($z = 0.53$) and the free surface with $Bi = 0$, $Ar = 0.714$ and various V. Two pairs of hot and cold spots ($m = 2$) are observed

and at the free surface with various V. Two pairs of hot and cold spots, i.e. a wavenumber of 2, are rotating clockwise.

Critical *Re* and Wavenumber. Figure 24 shows the variation of Re_c and m with V. Heat loss from the free surface stabilizes the flow as it tends to decrease the free surface temperature gradient, and Re_c increases with increasing Bi. With $Bi = 0$, two kinds of wavenumbers are observed, and a transition from $m = 1$ to $m = 2$ occurs at V of about 0.9. The wavenumber can change with Bi. This change in wavenumber was reported in experiments [15], where a wavenumber of 1 was changed into 2 by altering the surrounding conditions of a liquid bridge.

Heat loss to surroundings is through forced convection induced by the action of shear stresses at the free surface. The value of h for gases in forced convection is in the range $25 - 250W/(m^2K)$. With $H = 1.4mm$ in experiments [6], Bi is in the range $0.32 - 3.2$. In the three-dimensional numerical simulations with a cylindrical surface and $Pr = 7$ of [5], Bi of 6.4 was necessary to compare with the experimental results of [10]. Thus, our $Bi = 1$(a much smaller bridge) is a reasonable assumption to compare with the experimental results of [6]. In addition, our previous work [13,16,17] showed that the inclusion of surface heat loss was necessary to achieve better agreement with experiments. When our numerical results with $Bi = 1$ are compared with those from normal gravity experiments [6], the wavenumber of 1 and the rotating mode are in good agreement. With $V = 1$, transition to $m = 1$ from $m = 2$ occurs at $Bi = 0.5$. Re_c is about 400 and $m = 1$ with $V = 1$ and $Bi = 0.5$. With $Bi > 0$, it is found that two different branches can exist in the stability diagram ($V - Re_c$). The most stable range of

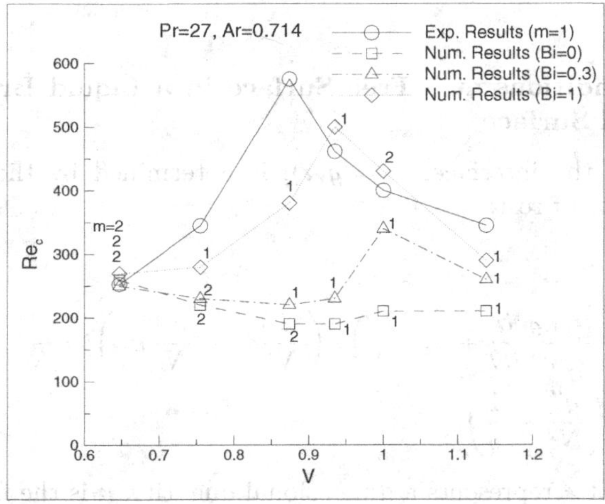

Fig. 24. Variations of Re_c and m with Bi where the experimental results are from [6]. Two different branches are observed in the stability diagram with $Bi > 0$

V with $Bi = 1$ is near 0.94 which is not in good agreement with experimental result [6], $V = 0.87$. However, in the experiments by Sumner et al. [22] at a higher Pr, the most stable range was near $V = 0.95$. Other experiments [15] showed that the range was very close to $V = 1$. Although the experiments were performed with small rods, the difference between numerical and experimental results may be due to gravity-induced bridge deformation since the largest static Bond number($Bd = \rho g H^2/\sigma$) in the experiments [6] was 2.

3.3 Conclusions

Two- and three-dimensional thermocapillary convection in a cylindrical liquid bridge is investigated numerically to document its stability characteristics. Two-dimensional simulations with cylindrical or curved free surfaces predict steady convection even at very high Re. It is found that only azimuthal waves can generate oscillations in the model. The stream function minima and the maximum surface velocity and temperature decrease with increasing Re.

Oscillatory thermocapillary convection is possible only in three-dimensional calculations. This is consistent with experiments. With $Bi = 0$ and either a flat or a convex surface, waves with the wavenumber of 1 are rotating clockwise. The pattern remains unchanged with increasing Re beyond the critical value. The wavenumber of 2 is observed with $Bi = 0$ and a concave surface, and the waves are also rotating clockwise. Heat loss from the free surface stabilizes the flow, and Re_c increases with increasing Bi. The wavenumber changes with Bi. With $Bi > 0$, it is found that two different branches can exist in the stability diagram ($V - Re_c$). The results with $Bi = 1$ are in reasonable agreement with available normal gravity experiments although these seem to favor a single azimuthal wave.

Appendix

Boundary Conditions at a Free Surface in a Liquid Bridge with a Curved Surface

The location of the interface, $r = g(z)$, is determined by the normal stress balance in a state of rest:

$$\frac{\triangle P^*}{\sigma} = \kappa = \nabla \cdot \boldsymbol{n}$$

$$\frac{\triangle P^*}{\sigma} = \left(\boldsymbol{e_r}\frac{\partial}{\partial r^*} + \frac{\boldsymbol{e_\theta}}{r^*}\frac{\partial}{\partial \theta} + \boldsymbol{e_z}\frac{\partial}{\partial z^*}\right)\cdot\left(\frac{1}{N^*}\boldsymbol{e_r} - \frac{g^{*\prime}}{N^*}\boldsymbol{e_z}\right) = \frac{1}{N^*}\left(\frac{1}{g^*} - \frac{g^{*\prime\prime}}{N^{*2}}\right)$$

$$-Ca \triangle P = \frac{1}{N}\left(\frac{g''}{N^2} - \frac{1}{g}\right) \tag{36}$$

where superscript $*$ represents a dimensional quantity, κ is the free surface curvature, $N^* = \left(1 + g^{*\prime 2}\right)^{1/2}$, $^{*\prime} = \frac{d}{dz^*}$ and $\boldsymbol{n}(= \frac{1}{N^*}\boldsymbol{e_r} - \frac{g^{*\prime}}{N^*}\boldsymbol{e_z})$ is the unit normal vector to the free surface.

The thermal boundary condition at the free surface with $T_\infty{}^* = T_c{}^*$ is

$$-k\frac{\partial T^*}{\partial n} = -\frac{k}{N^*}\left(\frac{\partial T^*}{\partial r^*} - g^{*\prime}\frac{\partial T^*}{\partial z^*}\right) = h(T^* - T^*{}_\infty)$$

$$-\frac{1}{N}\left(\frac{\partial T}{\partial r} - g'\frac{\partial T}{\partial z}\right) = BiT \tag{37}$$

The kinematic boundary condition at the free surface is

$$\boldsymbol{v_n} = \boldsymbol{v}\cdot\boldsymbol{n} = 0$$

$$u = g'v \tag{38}$$

Two tangential stress balance equations at the free surface become

$$\boldsymbol{\tau_1}\cdot\underline{\underline{S}}\cdot\boldsymbol{n} = \boldsymbol{\tau_1}\cdot\nabla\sigma$$

$$\frac{1}{N^{*2}}\left(g^{*\prime}S_{rr} + S_{zr} - g^{*\prime 2}S_{rz} - g^{*\prime}S_{zz}\right) = -\frac{\gamma}{N^*}\left(g^{*\prime}\frac{\partial T^*}{\partial r^*} + \frac{\partial T^*}{\partial Z^*}\right)$$

$$\frac{\mu}{N^{*2}}\left[2g^{*\prime}\left(\frac{\partial u^*}{\partial r^*} - \frac{\partial v^*}{\partial z^*}\right) + (1 - g^{*\prime 2})\left(\frac{\partial v^*}{\partial r^*} + \frac{\partial u^*}{\partial z^*}\right)\right] = \frac{-\gamma}{N^*}\left(g^{*\prime}\frac{\partial T^*}{\partial r^*} + \frac{\partial T^*}{\partial Z^*}\right)$$

$$2g'\left(\frac{\partial u}{\partial r} - \frac{\partial v}{\partial z}\right) + (1 - g'^2)\left(\frac{\partial v}{\partial r} + \frac{\partial u}{\partial z}\right) = -N\left(g'\frac{\partial T}{\partial r} + \frac{\partial T}{\partial z}\right) \tag{39}$$

$$\boldsymbol{\tau_2}\cdot\underline{\underline{S}}\cdot\boldsymbol{n} = \boldsymbol{\tau_2}\cdot\nabla\sigma$$

$$\frac{S_{\theta r}}{N^*} - \frac{g^{*\prime}}{N^*}S_{\theta z} = \frac{-\gamma}{r^*}\frac{\partial T^*}{\partial \theta}$$

$$\frac{\mu}{N^*}\left(r^*\frac{\partial}{\partial r^*}\left(\frac{w^*}{r^*}\right) + \frac{1}{r^*}\frac{\partial u^*}{\partial \theta}\right) - \frac{\mu g'}{N^*}\left(\frac{\partial w^*}{\partial z^*} + \frac{1}{r^*}\frac{\partial v^*}{\partial \theta}\right) = \frac{-\gamma}{r^*}\frac{\partial T^*}{\partial \theta}$$

$$\frac{\partial w}{\partial r} - \frac{w}{r} + \frac{1}{r}\frac{\partial u}{\partial \theta} - g'\left(\frac{\partial w}{\partial z} + \frac{1}{r}\frac{\partial v}{\partial \theta}\right) = \frac{-N}{r}\frac{\partial T}{\partial \theta} \tag{40}$$

where $\boldsymbol{\tau_1} = \frac{g^{*\prime}}{N^*}\boldsymbol{e_r} + \frac{1}{N^*}\boldsymbol{e_z}$ and $\boldsymbol{\tau_2} = \boldsymbol{e_\theta}$ are two tangential unit vectors to the free surface. $\underline{\underline{S}}$ is the viscous stress tensor.

References

1. N. Garnier, A. Chiffaudel: Eur. Phys.J. **B19**, 87 (2001)
2. Y. Kamotani, S. Ostrach, A. Pline: J. Heat Transfer **117**, 611 (1995)
3. Y. Kamotani, S. Ostrach, J. Masud: J. Fluid Mech. **410**, 211 (2000)
4. M. Levenstam, G. Amberg, C. Winkler: Phys. Fluids **13**, 807 (2001)
5. J. Leypoldt, H. Kuhlmann, H. Rath: J. Fluid Mech. **414**, 285 (2000)
6. J. Masud, Y. Kamotani, S. Ostrach: J. Thermophysics Heat Transfer **11**, 105 (1997)
7. J. Metzger, D. Schwabe, A. Cramer, A. Scharmann: Final Report of Sounding Rocket Experimnets in Fluid Science and Material Science, Texus 28-30, MASER 5&MAXUS1(ESA SP-1132, **4**, 60-71 October (1994)
8. M. Mundrane, A. Zebib: J. Thermophysics Heat Transfer, **9**, 795 (1995)
9. S. Patankar: *Numerical heat transfer and fluid flow*, (Mcgraw-hill book company 1980)
10. F. Preisser, D. Schwabe, A. Scharmann: J. Flud Mech. **126**, 545 (1983)
11. P. Rudolph: Prog. Crystal Growth and Charact. **29**, 275 (1994)
12. D. Schwabe: Physico-Chemical Hydrodynamics **2**, 263 (1981)
13. D. Schwabe, A. Zebib, B.-C. Sim: J. Fluid Mech. (in press) (2003)
14. V. Shevtsova, J. Legros: Phys. Fluids **10**, 1621 (1998)
15. V. Shevtsova, M. Mojahed, J. Legros: Acta Astronautica **44**, 625 (1999)
16. B.-C. Sim, A. Zebib, D. Schwabe: J. Fluid Mech. (in press) (2003)
17. B.-C. Sim, A. Zebib: Phys. Fluids **14**, 225 (2002)
18. B.-C. Sim, A. Zebib: Int. J. Heat Mass Transfer **45**, 4983 (2002)
19. B.-C. Sim: Thermocapillary convection in cylindrical geometries. Ph. D. dissertation, Rutgers University (2002)
20. B.-C. Sim, A. Zebib: J. Thermophysics Heat Transfer **16**, 553 (2002)
21. M. Smith, S. Davis: J. Fluid Mech. **132**, 119 (1983)
22. L. Sumner, G. Neitzel, J.-P. Fontaine, P. Dell'Aversana: Phys. Fluids **13**, 107 (2001)
23. M. Wanschura, V. Shevtsova, H. Kuhlmann, H. Rath: Phys. Fluids **7**, 912 (1995)
24. J. Xu, A. Zebib: J. Fluid Mech. **364**, 187 (1998)
25. A. Zebib, G. Homsy, E. Meiburg: Phys. Fluids **28**, 3467 (1985)

Surface Tension Driven Flow of Molten Silicon: Its Instability and the Effect of Oxygen

Taketoshi Hibiya[1,2,3], Takeshi Azami[2], Masanobu Sumiji[3], and Shin Nakamura[2]

[1] Tokyo Metropolitan Institute of Technology, 6-6 Asahigaoka, Hino 191-0065, Japan
[2] NEC Corporation, 34 Miyukigaoka, Tsukuba 305-8501, Japan
[3] Tokyo Institute of Technology, 4259 Nagatsuta-cho, Midori-ku, Yokohama 226-8502, Japan

Abstract. Surface-tension-driven flow of molten silicon, which is one of mechanisms of heat and mass transfer during crystal growth, was investigated by using a liquid-bridge configuration under microgravity and on earth. Using microgravity is a convenient way to study surface-tension-driven flow, because buoyancy flow can be suppressed so that only surface-tension-driven flow can be distinguished. In the liquid-bridge configuration, which corresponds to floating-zone growth, flow instability and its three-dimensional structure were investigated through measurement of temperature-oscillation, flow visualization, optical pyrometry of the melt surface, observation of oscillation of the melt/crystal interface, and observation of surface oscillation by phase-shift interferometry. Azimuthal wave number m for instability structure depends on the aspect ratio of the bridge, Γ, which is defined as the ratio of height h to radius r.

Surface-tension-driven flow was found to be affected by oxygen partial pressure of the ambient atmosphere, which corresponds to concentration of oxygen in Si melt. This is very important finding, because for the Czochralski growth system, oxygen dissolves into melt from a crucible wall made of SiO_2. It was also found that surface tension and its temperature coefficient strongly depend on oxygen partial pressure. Oxygen partial pressure at a Si melt surface can be deduced experimentally and theoretically by measuring the oxygen partial pressure at the inlet of the gas flow system.

A cellular pattern was observed at a surface of 20 cm deep Czochralski melt, whereas we found a hydrothermal wave at a surface of 8-mm-thick thin melt. Observed patterns are discussed in light of driving force of surface-tension-driven flow in the Czochralski melt.

1 Introduction

Silicon crystals with high quality and large diameter are required for highly developed IT (information technology) society. Currently, 200- and 300-mm-diameter wafers are used for LSI (large-scale-integration) production, and the diameter is expected to increase up to 450 mm by 2015. However, as the diameter increases, to further improve the productivity of LSI chip fabrication, control of defects in the crystals becomes more important. This means that to improve the quality of silicon single crystals, the heat and mass-transfer processes involved in crystal growth must be fully understood and controlled, because silicon-crystal growth is governed by the heat and mass transfer at the crystal/melt interface in the case of both the floating-zone and the Czochralski methods. Both these

T. Hibiya et al., Surface Tension Driven Flow of Molten Silicon: Its Instability and the Effect of Oxygen, Lect. Notes Phys. **628**, 131–155 (2003)
http://www.springerlink.com/

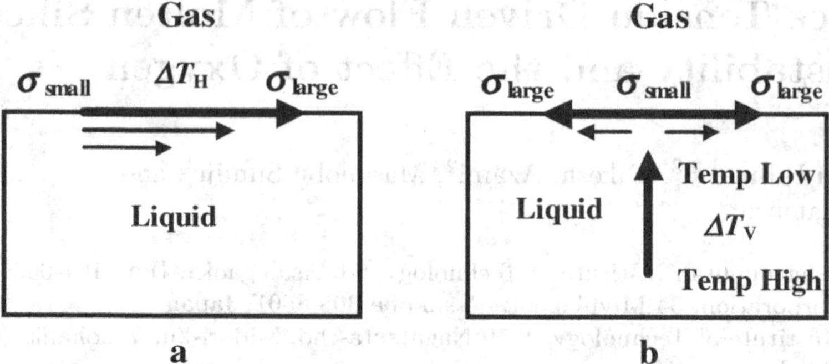

Fig. 1. Mechanism of surface-tension-driven flow: **a** thermocapillary effect, **b** Marangoni effect in the classical definition

methods involve as transport mechanism buoyancy flow, surface-tension-driven-flow, forced flow and diffusion.

Surface-tension-driven flow is the most poorly understood of the above transport phenomena, because the earth's gravitational field makes it difficult to distinguish it from buoyancy flow. The existence of surface-tension-driven-flow in molten silicon was confirmed for the first time during the floating-zone crystal growth experiment under a microgravity condition [10]. From the viewpoint of the geometry of surface, two kinds of cryltal growth systems must be considered. One is the surface-tension-driven-flow on the surface of a liquid-bridge configuration, which corresponds to that in floating-zone crystal growth. The other occurs on a flat, horizontal surface and corresponds to that in a Czochralski growth system, where a silicon melt is sustained in a SiO_2 crucible.

Driving force for surface-tension-driven flow is the gradient in surface tension. Surface-tension-driven flow is categorized into two cases, i.e., surface-tension-driven flow with or without large enough temperature gradient perpendicular to the free surface. In the first case, the driving force is the difference in surface tension along the melt surface caused by a temperature gradient or a concentration gradient in the adsorbate or solute, as shown in Fig. 1a. Surface-tension-driven flow due to a temperature gradient is called thermocapillary flow, and flow due to a concentration gradient in the adsorbate or solute is called solutocapillary flow. Surface-tension-driven flow observed in the floating-zone case is considered as thermocapillary flow. The second case is a flow driven by surface tension gradient due to temperature fluctuation at the surface of the melt with large enough temperature difference vertical to the melt surface, (larger than critical one) as shown in Fig. 1b. If the flow starts at the melt surface, flow takes place from inside of the melt to the surface to compensate the lost volume due to surface flow. Due to large vertical temperature gradient in the melt, the melt with higher temperature than that at the surface is supplied to the surface, so that temperature at the plume point at the surface becomes higher than that in the surrounding portion. This temperature rise drives circular flow continuously and forms a cellular pattern on the surface, i.e., Marangoni-Bénard pattern. This mechanism is called the Marangoni effect in a classical definition [32]. In the case of the Czochralski melt system, thermocapillary effect (due to a temperature difference between the crucible wall and the growing crystal), solutocapillary effect (due to oxygen concentration difference), and Marangoni effect in the classical definition (due to the vertical temperature difference in the thermal inverse layer) are nonlinearly coupled, as discussed later.

Eyer et al. [10] found that dopant striations are formed in silicon single crystals grown by a floating-zone (FZ) method even under microgravity and concluded that an oscillatory surface-tension-driven flow must exist. Since then studies on the surface-tension-driven flow of molten silicon in a liquid-bridge configuration have been intensified both experimentally and numerically. Cröll et al. [7] showed that dopant boron was distributed not homogeneously along the growth direction of a silicon crystal, but it changed gradually in the growth direction. This suggests that transport in the melt was governed not by diffusion but by surface-tension-driven flow. Cröll et al. [8] also experimentally determined the critical Marangoni number for transition from stationary flow to oscillatory flow, Ma_{c2}, as 100–200 under microgravity, where Marangoni number is determined as

$$Ma = (|\partial\sigma/\partial T|\Delta Th)/(\mu\kappa) \qquad (1)$$

where $|\partial\sigma/\partial T|$ is the temperature coefficient of surface tension, ΔT is the temperature difference in the system, h is the characteristic length of the system, μ and κ are viscosity and thermal diffusivity of molten silicon, respectively (μ $=7.0\times10^{-4}$Pa·s and κ $=2.1\times10^{-5}$·m^2·s^{-1}). Although microgravity is the ideal environment to study surface-tension-driven flow, it can be studied on earth if the buoyancy effect is minimized by, for example, using a very short liquid bridge or a very shallow melt. For medium- and high-Pr-number liquids, lots of studies have been carried out one earth and under microgravity. Many numerical studies were also performed. Kuhlmann reviewed these studies [19]. The results were useful for studies on the surface-tension-driven flow of low-Pr-number fluids, e.g. the dependence of the azimuthal wave number on the aspect ratio Γ. However, in the case of low-Pr-number fluids, doing so is experimentally difficult because of their metallic characteristics, i.e., high melting temperature, opaqueness under visible light, and high chemical reactivity. Besides the studies on molten silicon, there have only been a limited number of studies on Marangoni flow of low Pr-number melts such as molybdenum [16], mercury [12] and tin [46,51].

A numerical approach is useful to overcome the above-described difficulties. Levenstam et al. [21], Leypoldt et al. [22], and Imaishi et al. [15] numerically modeled a half-zone liquid bridge configuration, which corresponds to a half of a floating-zone melt with a unidirectional temperature gradient. They reported that an asymmetrical temperature field is caused by the instability of hydrodynamic origin and that the azimuthal wave number of instability depends on the aspect ratio of a liquid bridge Γ, whereas hydrothermal wave appears for high-Pr-number fluids [22]. Rupp et al. [38] and Kaiser et al. [17] numerically investigated a full-zone configuration in the case of real floating-zone crystal growth. However, it is difficult to calculate the flow, particularly in the case of a high Marangoni number, because of the characteristics of low-Pr-number fluids, whose flow mode evolves from axisymmetric stationary to three-dimensional oscillatory at a small Marangoni number [39]. Therefore, calculations have been limited to flow with a small Marangoni number. On the other hand, due to experimental difficulty, experiments could only be carried out on high-Marangoni-number flow. Experimental and numerical approaches have complementary roles, so as to study the surface-tension-driven-flow of a low-Pr-number fluid such as molten silicon. Recently crystal growers have recognized the existence of surface-tension-driven flow at a molten silicon surface, and the effect of this flow has been taken account in numerical modeling [35,47].

In the present study, flow instability and the effect of oxygen partial pressure on the Marangoni flow of molten silicon in a half-zone configuration were experimentally investigated in terms of temperature oscillation in a liquid bridge, flow visualization, surface oscillation, oscillation of the melt/crystal interface position, and the effect of oxygen partial pressure of an ambient atmosphere on surface-tension-driven flow. The surface-tension-driven flow on a flat surface melt, which corresponds to the Czochralski melt system, was also investigated.

2 Experimental Setup

The experimental set-up, shown in Figs. 2, 3a, and 3b, consists of a molten silicon bridge between the upper and lower carbon rods in a mono-ellipsoidal infrared mirror furnace. Temperature at the upper carbon/melt interface was set higher than that at the lower interface in order to form a half-zone configuration with a unidirectional temperature gradient. This configuration ensured that the phenomena to be studied were simplified. To investigate the flow instability structure, temperature field was monitored by fine thermocouples and flow field

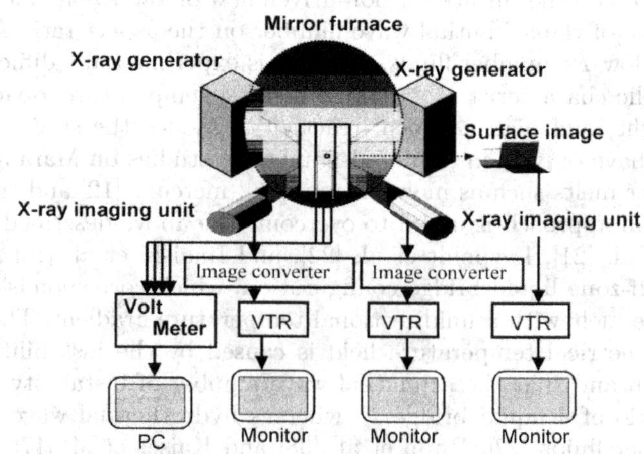

Fig. 2. Mirror furnace equipped with X-ray observation system

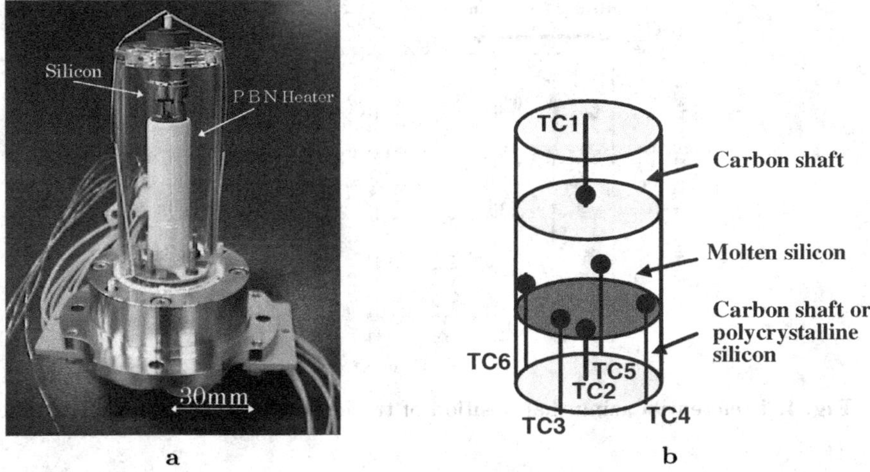

Fig. 3. a Sample assembly for a molten silicon bridge; **b** sketch of the assembly

was observed by an X-ray visualization system with tracer particles made of 450-μm-diameter carbon-coated ZrO_2particles under microgravity and on earth [26]. Two thermocouples (TC1 and TC2) were set near the carbon rod/silicon melt interface to estimate the temperature difference at the melt surface, which was determined empirically as half of the measured temperature difference between TC1 and TC2. Four fine thermocouples (TC3-TC-6) were set 90° -apart each other along the azimuthal direction 2 mm above the lower interface.

Besides the temperature- and the flow-field observation, a non-invasive diagnostics of the surface behavior of the molten silicon bridge was performed by several optical techniques mainly on earth. Phase-shift interferometry using an argon laser with a wavelength of 488 nm was used to distinguish the oscillation at the melt surface due to surface-tension-driven-flow instability from that due to resonant oscillation [31]. Temperature oscillation at the molten silicon surface was also observed by non-contact optical pyrometry using a CCD camera [1]. Moreover, a laser microscope was used to observe the oscillation of the position of the melt/crystal interface, so as to correlate this oscillation with fluctuation of asymmetric temperature field due to flow field instability [43]. Surface tension and its temperature coefficient were measured by a sessile drop method in an atmosphere with carefully controlled oxygen partial pressures [25]. A silicon single crystal was grown by a floating-zone method to observe the effect of oxygen partial pressure on the formation of dopant growth striations [45].

3 Modes of Flow Instability

3.1 Phase Relationship of Temperature Oscillation

Temperature oscillation across the liquid bridge (with Marangoni number ranging from 1000 to over 10000) was measured under microgravity and on earth [27]. In

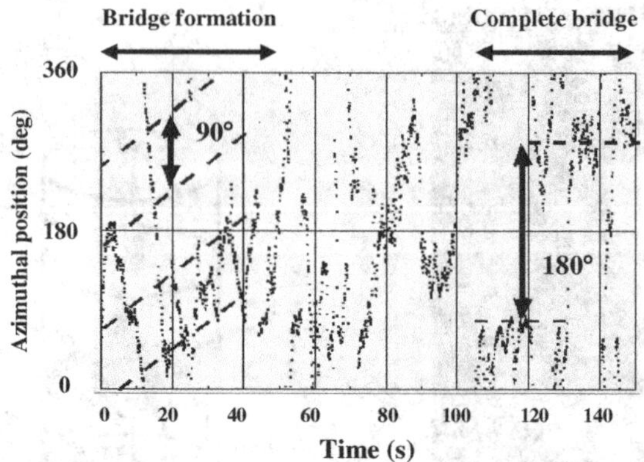

Fig. 4. Preferential azimuthal position of tracer particles as a function of time

this range, the Marangoni flow was oscillatory with single or multiple frequencies. Analysis of the phase relationship of the temperature oscillation measured by four thermocouples showed that there exists non-axisymmetric temperature field within a liquid bridge and it fluctuates with certain frequency. Temperature of non-axisymmetric temperature field is cooler than that at the surface of a liquid bridge, because a liquid bridge is heated from outside by radiation from an infrared lamp. The azimuthal wave number of instability, m, depends on the aspect ratio of the liquid bridge Γas follows:

$$m \approx 2/\Gamma \tag{2}$$

$$\Gamma = h/r \tag{3}$$

where h is the height of the liquid bridge and r is its radius. This dependence of wave number on aspect ratio agrees with that numerically predicted [48].

The aspect-ratio dependence of the wave number of instability(m) was also investigated through a flow-visualization experiment under microgravity. Figure 4 shows experimental results on flow visualization during a parabolic flight of a jet aircraft (a Gulf Stream II) [28]. The molten silicon bridge was 8.5 mm high and 10 mm in diameter (aspect ratio Γ=1.7). The ordinate shows the azimuthal position of the bridge where a tracer particle prefers to stay, and the abscissa shows observation time. In this experiment, we used a neutral-buoyancy tracer particle made of ZrO_2 particle coated with SiO_2 and carbon. Since the density of the tracer particle well matched with that of molten silicon, the trajectory could be observed successfully not only under the microgravity condition but also even in the 2G phases before and after the parabolic flight.

The tracer particle seems to stay at the position 90° -apart each other during the bridge formation period, whereas it stays at the position 180° -apart each other after the bridge was completely formed. This means that oscillatory motion with two-fold symmetry existed during the bridge formation period (short bridge), and that one-fold symmetry was formed after the bridge was completed (8.5-mm-high bridge). These observations agree well with the idea that wave number m is inversely proportional to aspect ratio Γ as shown by (2).

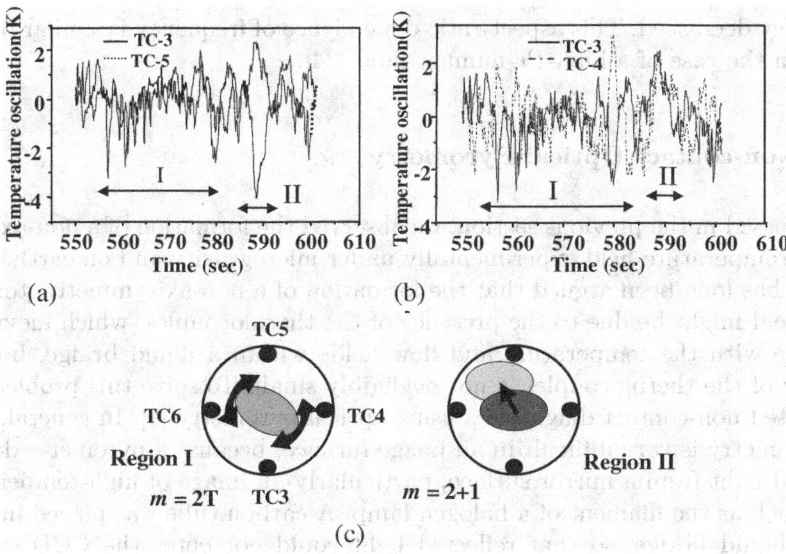

Fig. 5. Non-axisimmetric temperature distribution and its oscillation. Temperature oscillation measured by **a** TC-3 and TC-5, and **b** TC-3 and TC-4. **c** Models of non-axisymmetric temperature distribution for $m = 2\text{T}$ (*left*) and $m = 2 + 1$ (*right*)

The above-mentioned dependence of phase relationship on aspect ratio was also systematically examined on earth (i.e., under normal gravity). Figure 5 shows an example of the temperature oscillation measured by four thermocouples within a molten silicon bridge with an aspect ratio of 1.4 (7 mm high). In region I of the elapsed measurement time, temperature oscillation measured by thermocouples TC-3 and TC-5 (180° -apart) has an in-phase relationship (Fig. 5a), whereas that measured by thermocouples TC-3 and TC-4 (90° -apart) has an anti-phase relationship (Fig. 5b). However, in region II, the oscillation has a larger amplitude than that in Region I, and the oscillation measured by thermocouples TC-3 and TC-5 has an anti-phase relationship, but that measured by thermocouples TC-3 and TC-5 has an in-phase relationship.

In region I, the phase relation of temperature oscillation can be explained by the twist mode fluctuation of two-fold symmetric temperature filed ($m = 2\text{T}$), as predicted by Imaishi et al. [15] (see Fig. 5c). In this mode two-fold symmetric temperature field goes back and forth, as shown by arrows in the left side sketch of Fig. 3c. However, in region II, instability shows one-fold symmetric motion of the two-fold symmetric temperature field ($m = 2 + 1$), which crosses the center part of the bridge, as predicted by Leypold et al. [22]. Dold et al. [9] also observed this behavior experimentally.

In the present work, for $\Gamma = 0.6$, $m = 2\text{T}$ mode existed and $m = 3$ mode was plausibly observed. For $\Gamma = 1.0$ and 1.4, $m = 2\text{T}$ and $m = 2 + 1$ modes were observed. For $\Gamma = 2.0$, $m = 1$ mode was observed. It was also found that the amplitude and main frequency of temperature oscillation depend on aspect ratio; oscillation amplitude increased with increasing aspect ratio, but oscillation frequency decreased. This aspect-ratio dependence of frequency is similar to that found in the case of a high-Pr-number fluid [34].

3.2 Non-contact Optical Pyrometry

As discussed in the previous section, we observed the formation of a non-axisymmetric temperature field experimentally under microgravity and on earth. However, it has long been argued that the formation of a non-axisymmetric temperature field might be due to the presence of the thermocouples, which inevitably interfere with the temperature and flow fields within a liquid bridge, because the size of the thermocouples is not negligibly small. To solve this problem, we attempted non-contact diagnostics using optical pyrometry [1]. In general, optical pyrometry is very difficult in an image furnace, because a pyrometer detects reflected light from a mirror surface, particularly an image of high-temperature parts such as the filament of a halogen lamp. A carbon tube was placed in front of the liquid bridge, so that reflected light could not enter the CCD camera (Fig. 6). A silicon-melt bridge was formed between the upper and lower carbon rods. The melt height was 5 or 10 mm, and the temperature difference between the upper and lower interfaces was set at 7.5, 24.5, or 100 K.

Using the CCD camera, we could observe oscillations in pixel brightness regardless of the existence of the thermocouples. This observation suggests that the formation and oscillation of a non-axisymmetric temperature field within a liquid bridge are not due to the existence of the thermocouples but are due to surface-tension-driven flow instability itself. Fourier spectral analysis showed a peak frequency at 0.14 Hz for the molten silicon bridge with a height of 10 mm and a temperature difference between the upper and lower interfaces of 100 K. This frequency agrees well with that of the temperature oscillation measured by the thermocouples (Fig. 7).

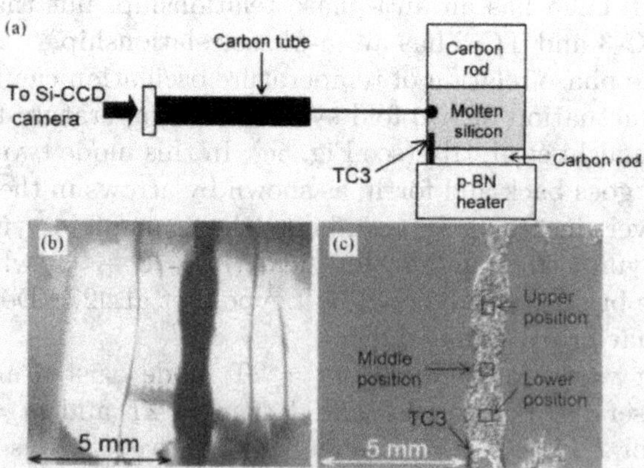

Fig. 6. Non-contact pyrometry for measuring temperature oscillation

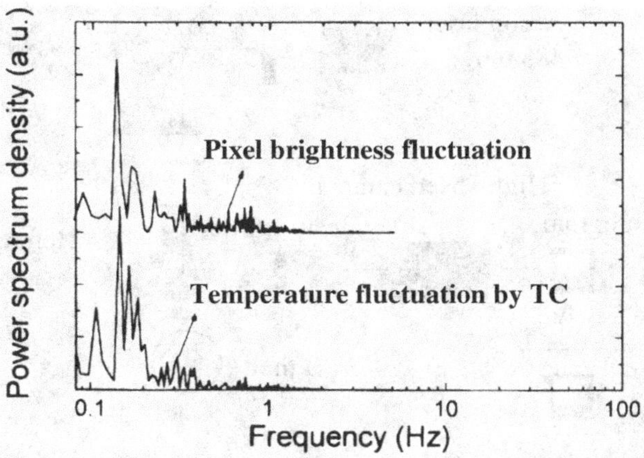

Fig. 7. Temperature oscillation measured by optical pyrometry (pixel brightness) and a thermocouple

Han et al. [12] and Cheng and Kou [6] also reported the oscillation frequency of a low-Pr-number liquid bridge, such as mercury and molten tin (5 and 1.5 Hz, respectively). However, these frequencies are attributed not to surface-tension-driven flow instability but to resonant oscillation. At a high frequency there is a single, independent peak due to resonant oscillation. This peak can be shifted to a lower frequency apparently by the aliasing effect when a conventional video sampling rate as low as 30 Hz is used.

3.3 Observation of Surface Oscillation by Phase-Shift Interferometry

Phase-shift interferometry [31,42], which is another non-contact diagnostic method, was used to observe oscillation of molten silicon surface under the earth's gravity, because it is difficult to reproduce surface oscillation by numerical modeling. Using phase-shift interferometry combined with a high-speed CCD camera (sampling rate: 500 Hz), we could detect not only surface oscillation due to flow instability but also surface oscillation with a resonant frequency. This was possible because we had overcome the frequency-shift problem due to the aliasing effect. Figure 8 shows the setup for the phase-shift interferometry; namely, an argon laser with a wavelength of 488 nm was used to distinguish reflection of the impinging light from actual emitted light from the sample at high temperature.

Analysis of the phase value showed that the amplitude of surface oscillation was as large as 1 - 2 μm, that is, larger than expected (Fig. 9). Figure 10 shows the Fourier spectrum of axial mode of surface oscillation. The peaks at frequency lower than 5 Hz are attributed to Marangoni flow instability, whereas the single peak at 52 Hz is attributed to the resonant frequency of the molten silicon bridge for the $(m, n) = (0, 2)$ mode. This frequency agrees with that estimated from the model proposed by Gañán and Barrero [11] that uses the

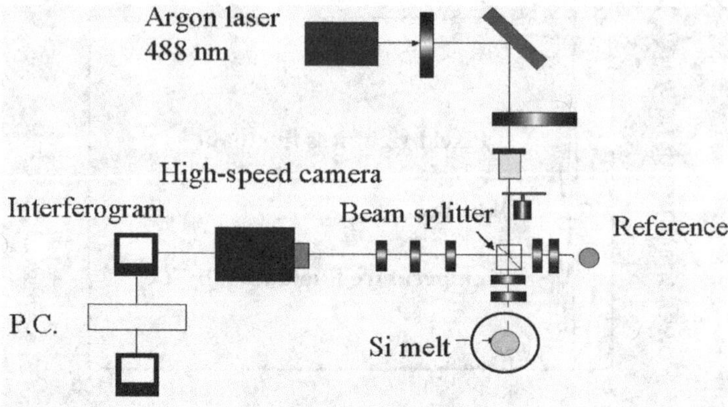

Fig. 8. Set-up for phase-shift interferometry

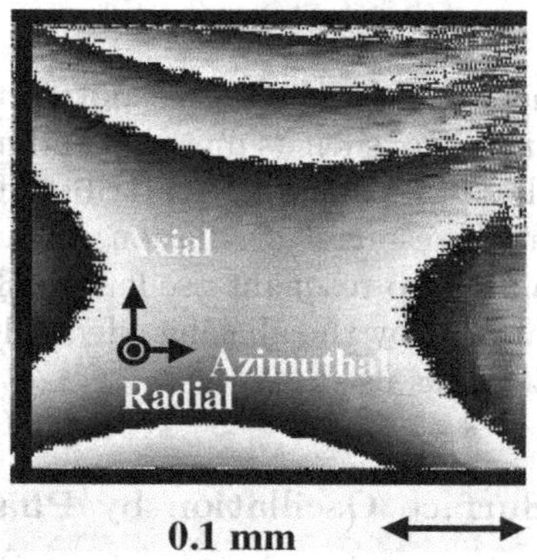

Fig. 9. Phase distribution profile of phase-shift interferometry

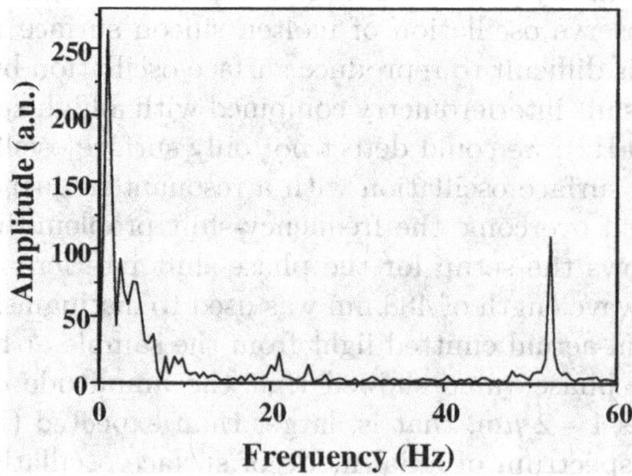

Fig. 10. Frequency analysis of axial mode of surface oscillation by phase-shift interferometry

surface tension, viscosity, and geometry of the bridge. Analysis of the oscillation at low frequencies due to the flow instability suggests that oscillation has standing-wave characteristics in the axial direction, and a traveling wave was observed in the azimuthal direction, as long as 10 s oscillation data was analyzed. Traveling characteristics appeared neither in the numerical modeling [15,22] nor the temperature-oscillation measurement (see Sect. 3.1); if a long-term observation using the phase-shift technique was available, azimuthal behavior could be clarified in more detail.

3.4 Oscillation of Melt/Crystal Interface Position

Oscillation of the melt/crystal interface position was observed by Schweizer et al. [40] in microgravity and by Sumiji et al. [43,44] on earth by using a laser microscope. This oscillation is caused by fluctuation of the non-axisymmetric temperature field within the molten silicon bridge. This was proved by simultaneous measurements of the melt/crystal interface position oscillation and temperature oscillation (Fig. 11). In these measurements, the thermocouple was set 45° -apart from the CCD camera observation point. When the temperature near the melt/crystal interface is increased, the interface moved to the crystal direction, i.e., "melt back" of the crystal occurred. When the temperature was lowered, abrupt crystal growth took place. In either case, growth striations must be formed in the grown crystals. When the melt/crystal interface oscillation was observed from two directions 135° apart, the phase relationship between these two observation points is in-phase and out-of phase. Null-correlation was also observed. The correlation was time-dependent. These observations suggest that there is non-axisymmetric temperature field within the molten silicon bridge and that it fluctuates at certain frequencies. The mode of fluctuation is deduced to be $m = 2 + 1$, as shown in Fig. 5c.

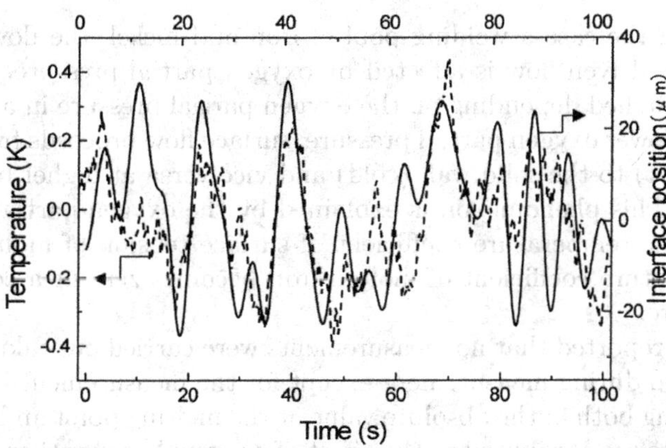

Fig. 11. Correlation between temperature oscillation and oscillation of the melt/crystal interface position

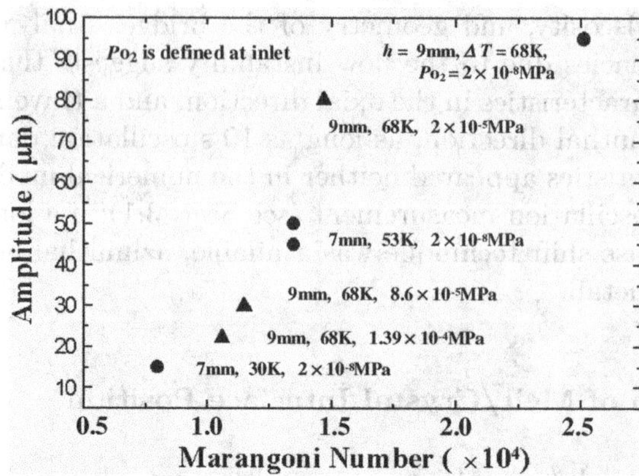

Fig. 12. Amplitude of melt/crystal position oscillation as a function of Marangoni number

Figure 12 shows that the amplitude of the melt/crystal interface oscillation is dependent on Marangoni number. This dependency is the same as that regarding the amplitude of temperature oscillation [29]. Note that the amplitude of the interface oscillation is also a function of oxygen partial pressure of an ambient atmosphere. This suggests that the surface-tension-driven flow is affected by oxygen partial pressure. Detailed discussion on the effect of oxygen partial pressure on surface-tension-driven flow is given in the following section.

4 Effect of Oxygen Partial Pressure on the Surface-Tension-Driven Flow of Molten Silicon

4.1 Dependence of Surface Tension and Temperature Coefficient on Oxygen Partial Pressure

As reported in the case a welding pool of iron and nickel, the flow direction of surface-tension-driven flow is affected by oxygen partial pressure. Namely, flow direction is switched depending on the oxygen partial pressure in an ambient atmosphere. At lower oxygen partial pressure, surface flow proceeds from the center of the pool (hot) to the solid wall (cold) and vice versa at higher oxygen partial pressure [24]. This phenomenon is explained by the oxygen-partial-pressure dependence of the temperature coefficient of surface tension of molten iron; that is, the temperature coefficient of molten iron becomes zero at a certain oxygen partial pressure.

Keene [18] reported that no measurements were carried out taking care of the effect of oxygen during measurement except for the measurement by Hardy [13]. Large scattering both in the absolute value at the melting point and the temperature coefficients are attributed to the effect of oxygen, because there is possibility that oxygen acts as a surfactant at the molten silicon surface, as same as in the case for molten iron and nickel. He recommended that measurement should be

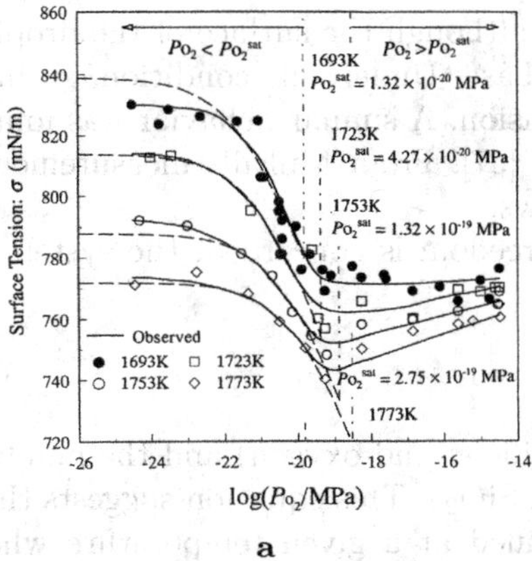

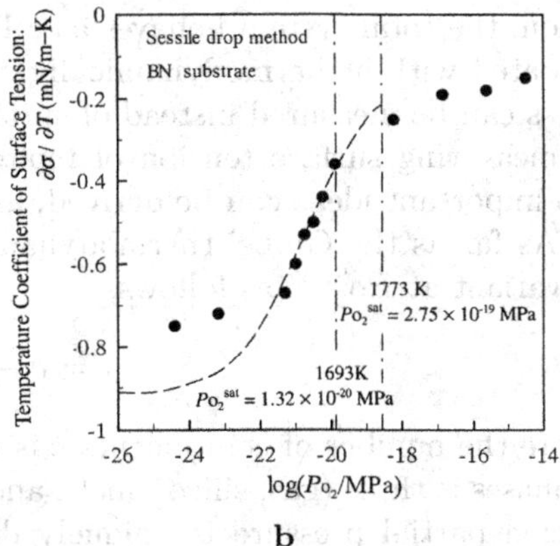

Fig. 13. **a** Surface tension of molten silicon and **b** its temperature coefficient as a function of oxygen partial pressure

carried out in a carefully controlled atmosphere with various oxygen partial pressures. He also recommended using levitation techniques for the measurement so that no contamination would be introduced from the crucible.

Mukai et al. used a sessile-drop method to measure the surface tension and its temperature coefficient of molten silicon in a carefully controlled atmosphere with various oxygen partial pressures (Figs. 13a and 13b) [25]. As shown in Fig. 13a, the surface tension value depends strongly on oxygen partial pressure, which was measured at the exit of the furnace, through which gas containing various amounts of oxygen flowed. Below the equilibrium oxygen partial pressure for a SiO_2 phase, Po_2^{SiO2}, surface tension decreases with increasing oxygen partial pressure. The equilibrium oxygen partial pressures for a SiO_2 phase are represented as dashed lines based on thermodynamical calculation. This oxygen partial pressure depends on temperatures, because the system is univariant. If we apply the Gibbs' adsorption isotherm to the experimentally obtained relationship between surface tension and oxygen partial pressure below Po_2^{SiO2}, the excess amounts of oxygen adsorbed at the melt surface Γ_o can be calculated from (4). This calculation suggests that there is one oxygen atom for ten silicon atoms at the melt surface, whereas oxygen concentration in the bulk melt is in the order of 10 ppm.

$$\Gamma_\mathrm{o} = -2(1/RT)(\partial\sigma/\partial \ln P_{\mathrm{O}_2}) = -(1/RT)(\partial\sigma/\partial \ln a) \qquad (4)$$

In this equation, R and T are gas constant and temperature, respectively; a is the activity of oxygen in the melt,which is assumed to be a consentration of $\underline{\mathrm{O}}_{\mathrm{Si}}$ in a solution system.

It should be noted that there is a kink in the surface-tension line in the vicinity of the equilibrium oxygen partial pressure for a solid SiO_2 phase, and above this point the surface tension seems to be constant or slightly increases with increasing oxygen partial pressure. This tendency suggests that above Po_2^{SiO2}, the surface of the silicon melt is covered with a thin SiO_2 film. Under this condition, the total system behaves as a liquid, although the surface of the droplet is coated with a thermodynamically solid phase. Under this condition, surface stress can be measured instead of surface tension. A similar behavior was found by measuring surface tension of molten tin [37]. From Mukai's measurements two important ideas can be derived, as follows.

As far as the Gibbs' thermodynamical freedom is concerned, the system is univariant at Po_2^{SiO2} as follows,

$$f = c - p + 2 = 1 \tag{5}$$

where the number of components c is two (silicon and oxygen) and the number of phases is three (gas, silicon melt, and solid SiO_2). This equation suggests that oxygen partial pressure is uniquely determined at a given temperature when three phases coexist; therefore, surface tension is uniquely determined at a given temperature when three phases coexist. Many workers have reported absolute values of temperature coefficient as small as $0.1\times10^{-3}\mathrm{N{\cdot}m^{-1}{\cdot}K^{-1}}$ (see for example [33]). From the above context, it is clear that these small values can be obtained without considering the effect of oxygen partial pressure and that they correspond to the temperature coefficient of surface stress. Thus, these small values are not realistic and should not be used for calculating the thermocapillary effect in the silicon melt.

4.2 Calibration of Oxygen Partial Pressure: Real Surface Tension at the Melt Surface

As mentioned in Sect. 4.1, surface tension and its temperature coefficient for molten silicon strongly depend on oxygen partial pressure. In Figs. 13a and 13b, oxygen partial pressure at the exit of the furnace was defined as Po_2^{exit}. Huan et al. [14] also measured surface tension of molten silicon as a function of oxygen partial pressure, which was defined at the entrance of the furnace as Po_2^{inlet}. The real oxygen partial pressure at the melt surface Po_2^{surface} is unknown because of experimental difficulty. Between Po_2^{inlet} and Po_2^{exit} there is a difference of over 15 orders of magnitude . Po_2^{exit} reflects the result of chemical reaction at the melt surface, whereas Po_2^{inlet} is a significant operating parameter. To explain this large deviation and to estimate the real value of surface tension at the melt surface, Ratto et al. [36] presented a new model, which takes account of the thermodynamics at the melt surface and oxygen transport to/from the melt surface, as follows.

$$P_{\mathrm{O_2}}^{\mathrm{inlet}} = P_{\mathrm{O_2}}^{\mathrm{surface}} + P_{\mathrm{Si}}^{\mathrm{saturation}} \sum_{j=1}^{2} \alpha_j K_j \frac{(1+Pe)}{(1+Pe/|\psi_j|)} P_{\mathrm{O_2}}^{\mathrm{surface}^{\alpha_j}} \tag{6}$$

$$P_{\mathrm{O_2}}^{\mathrm{exit}} = P_{\mathrm{O_2}}^{\mathrm{surface}} + P_{\mathrm{Si}}^{\mathrm{saturation}} \sum_{j=1}^{2} \alpha_j K_j \frac{Pe}{(1+Pe/|\psi_j|)} P_{\mathrm{O_2}}^{\mathrm{surface}^{\alpha_j}} \tag{7}$$

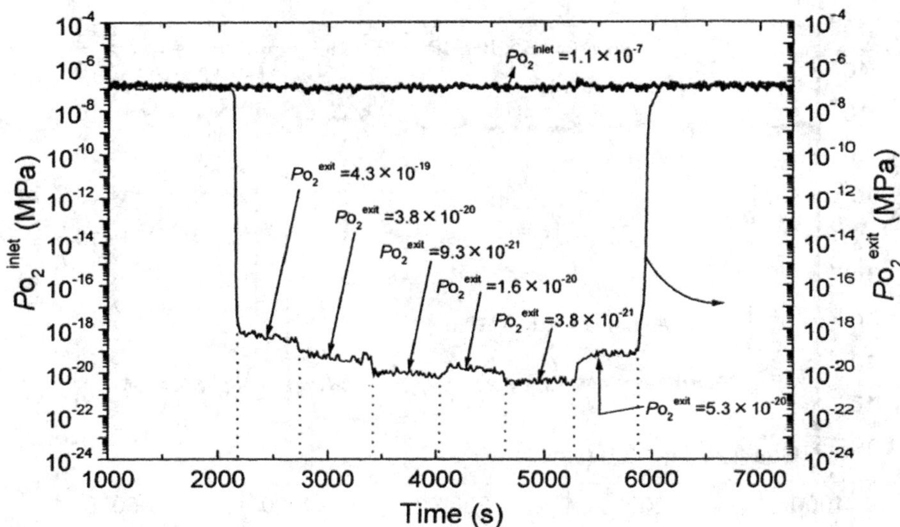

Fig. 14. The effect of the flow rate of inlet gas on the oxygen partial pressure measured at the exit

Here, α_j is the coefficient of oxygen for the jth reaction given as follows:

for j = 1, Si(l) + O_2(g) → SiO(g) and $\alpha_1 = 1/2$,

and

for j = 2, Si(l) + 1/2O_2(g) → SiO_2(s) and $\alpha_2 = 1$,

Where K_j is the dimensionless equilibrium constant for the jth reaction, ψ_j is the dimensionless diffusion constant of oxygen, and Pe is defined as the ratio of flow to diffusion of oxygen.

Azami and Hibiya [2] experimentally confirmed the above relationships, (6) and (7) by measuring Po_2^{inlet} and Po_2^{exit} simultaneously in a furnace containing a molten silicon droplet. A pure argon gas (6N) and a minute amount of argon gas containing 10% oxygen flowed through the furnace. As shown in Fig. 14, Po_2^{inlet} and Po_2^{exit} differ greatly, and Po_2^{exit} depends on the flow rate of mixed gas (0.5 to 5.0 l·min^{-1}), even though $Po_2^{inlet}$was fixed at 1.1 × 10^{-7}MPa. Po_2^{exit} was also measured as a function of Po_2^{inlet} at a constant flow rate of mixed gas (3.0 l·min^{-1}) (see Fig. 15). Although Po_2^{exit} depends on Po_2^{inlet}, Po_2^{exit} and Po_2^{inlet} are almost the same when $Po_2^{inlet}$was 7.5 × 10^{-4}MPa (and $Po_2^{exit}$was 4.8 × 10^{-4}MPa). This means that under this condition the silicon melt surface was coated with a SiO_2 film and that no more oxygen gas could react with the molten silicon. This $Po_2^{inlet}$of 7.5 × 10^{-4}MPa almost corresponds to the oxygen partial pressure for saturation, defined at the entrance, which corresponds to $Po_2^{surface}$ of 2.8 × 10^{-20}MPa as shown in Fig. 16. This figure was plotted by substituting the above experimental results into (6) and shows the relationships between $Po_2^{surface}$ and Po_2^{inlet}. It is clear that $Po_2^{surface}$ of 2.8 × 10^{-20}MPa is in the same order of magnitude as the theoretical oxygen partial pressure for saturation at 1688 K, i.e., $Po_2^{saturation}$of 1.1 × 10^{-20}MPa, and also Po_2 corresponding to the

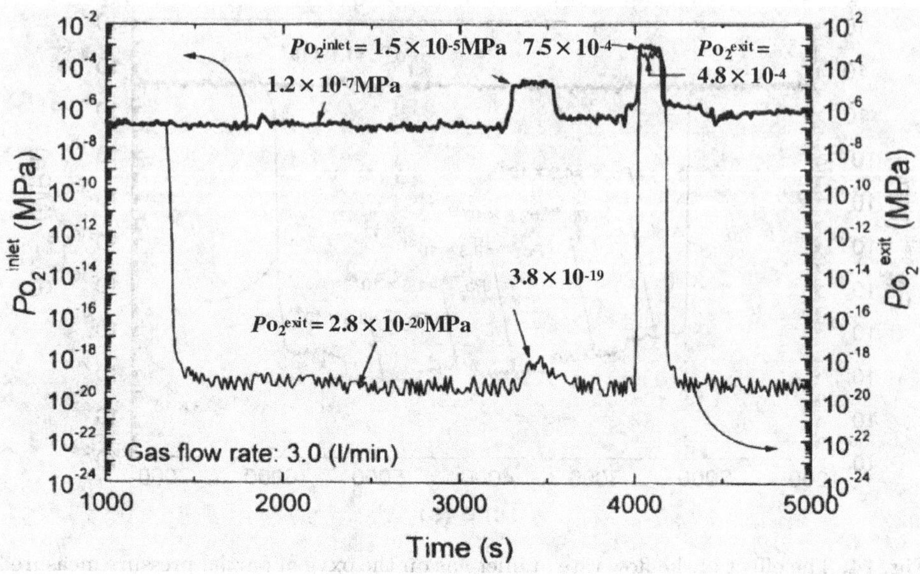

Fig. 15. Relationship between oxygen partial pressures measured at the inlet and that at the exit of the furnace

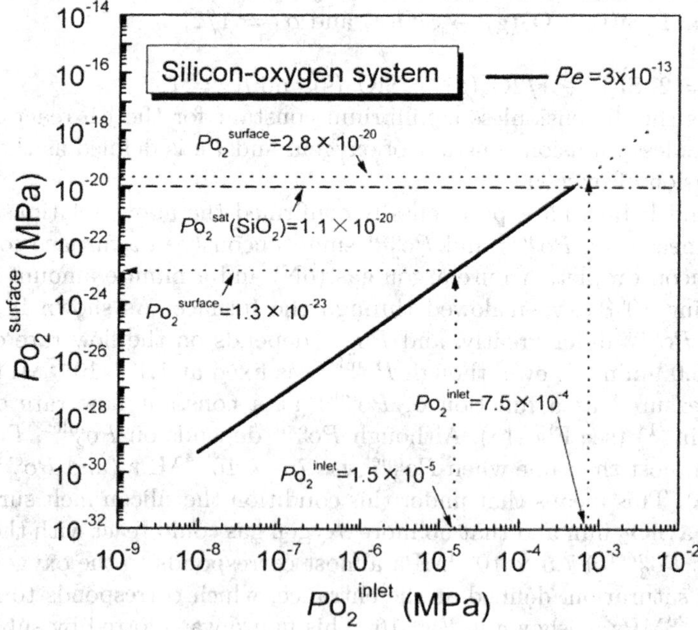

Fig. 16. Relationship between oxygen partial pressures at the melt surface and at the inlet of the furnace

kink in the measured-surface-tension line (see Fig. 13a). In the surface-tension measurement by Mukai et al., Po_2 was measured at the exit of the furnace, because gas flow rate was as low as 160 ml·m^{-1}; thus $Po_2^{surface}$can be estimated asPo_2^{exit}. Through the simultaneous measurements of Po_2^{inlet} and Po_2^{exit}, Ratto's model was shown to be valid for estimating $Po_2^{surface}$ either from Po_2^{inlet}or Po_2^{exit}.

4.3 Oxygen Partial Pressure Dependence of Surface-Tension-Driven Flow

As mentioned in the introduction, in Czochralski growth, a silicon melt is contained within a SiO_2 crucible, where oxygen dissolves into the melt

from the crucible. It is thus important to clarify the effect of oxygen on the surface-tension-driven flow of molten silicon. Accordingly, this oxygen effect has been investigated by observing flow velocity under microgravity [3] and temperature oscillation on earth [4].

Figure 17 shows the tracer motion in a molten silicon bridge visualized by X-ray radiography under microgravity during the parabolic flight of a jet plane (Gulf Stream II). Figure 18 shows tracer velocity within a liquid bridge with an aspect ratio of 1.0 as a function of oxygen partial pressure. In this experiment, $Po_2$was defined at the pressure the silica specimen ampoule was evacuated and sealed. This oxygen partial pressure corresponds to that defined as Po_2^{inlet}. Velocity was measured as a displacement of a tracer particle parallel to the axial direction of the liquid bridge in unit time. Tracer particle velocity would not reveal real flow velocity correctly, because the tracer-particle size (450 μm) is larger than the velocity boundary layer thickness of surface-tension-driven flow, i.e., less than 100 μm. Real surface velocity is expected to be one order of magnitude higher than the observed one, as predicted by Lan and Kou [20]. Nevertheless, it clearly depends on Po_2.

Figure 18 shows that flow velocity becomes zero above Po_2 of 10^{-4}MPa. At $Po_2^{inlet}$of 10^{-2}MPa flow was not observed, but precipitation of SiO_2 on the silicon melt bridge surface was observed. This result agrees well with experimental calibration of oxygen partial pressure mentioned in the previous section; at

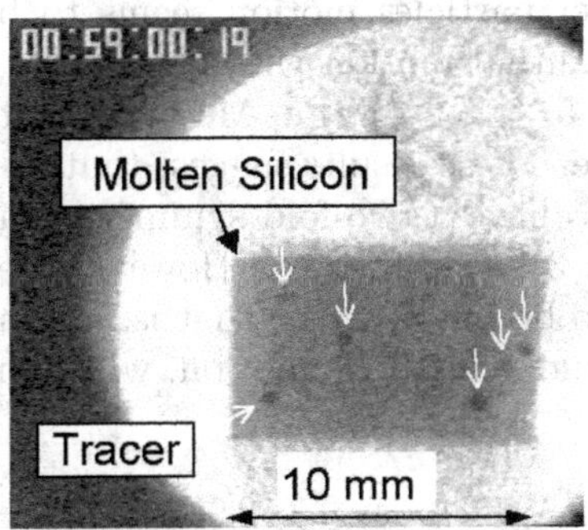

Fig. 17. Surface-tension-driven flow of molten silicon visualized by X-ray with trace particles under microgravity

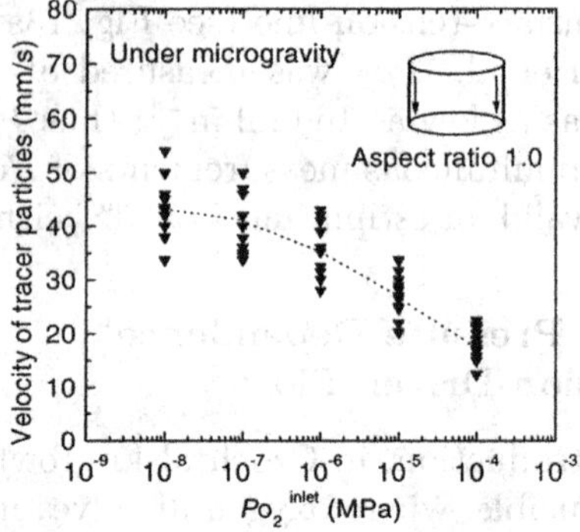

Fig. 18. Oxygen partial pressure (inlet) dependence of flow (tracer particle) velocity

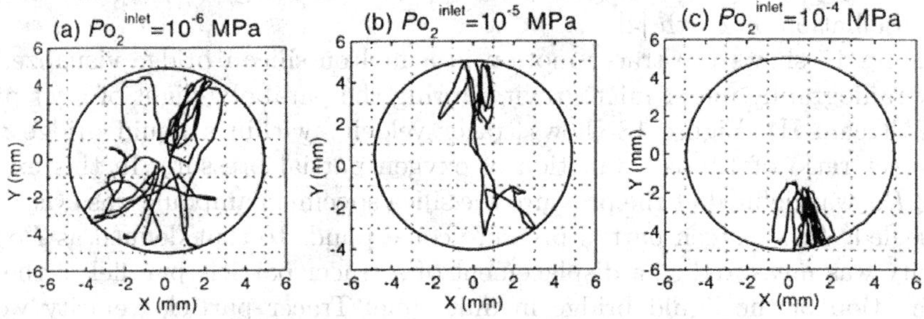

Fig. 19. Top view of a particle path as a function of oxygen partial pressure (inlet): **a** $P_{O_2}^{\rm inlet} = 10^{-6}$MPa, **b** $P_{O_2}^{\rm inlet} = 10^{-5}$MPa, and **c** $P_{O_2}^{\rm inlet} = 10^{-4}$MPa

$Po_2^{\rm inlet} = 7.5 \times 10^{-4}$ MPa ($Po_2^{\rm surface} = 2.8 \times 10^{-20}$ MPa) no more bare surface appeared.

Figures 19a, 19b, and 19c show the particle path from the top view of the 5-mm-high (aspect ratio Γ of 1) molten silicon bridge with a temperature difference between the upper and lower interfaces ΔT of 50 K [3]. It is clear that the flow mode depends on $Po_2^{\rm inlet}$. At $Po_2^{\rm inlet}$of 10^{-4}MPa ($Po_2^{\rm surface} \approx 7 \times 10^{-22}$ MPa; Ma $\approx$ 6000), tracer particles motion seems to be almost axisymmetric, but it showed small azimuthal motion within an angle of $\pi/6$. Below $Po_2^{\rm inlet}$of 10^{-5}MPa ($Po_2^{\rm surface} \approx 5 \times 10^{-24}$ MPa, Ma $\approx$ 9800) tracer particles cross the center of the liquid bridge. These results suggest that at $Po_2^{\rm inlet}$of 10^{-4}MPa, flow instability mode seems to have a two-fold symmetry with a twist motion, $m =$ 2T, as discussed in Sect. 3.1 (see Fig. 5c). However, below $Po_2^{\rm inlet}$of 10^{-5}MPa, instability mode seems to be $m = 2 + 1$; that is, a temperature field with two-fold symmetry has one-fold symmetric motion, which crosses the center part of the liquid bridge.

4.4 Effect of Oxygen Partial Pressure on Temperature Oscillation

Figure 20 shows how the oxygen partial pressure of an ambient atmosphere affects temperature oscillation mode for an oscillatory surface-tension-driven flow [4]. In this experiment, the melt height was 5 mm and temperature difference between the upper and lower interfaces ΔT was 26 K. When oxygen partial pressure Po_2^{inlet}increased from 3.5×10^{-7}MPa ($Po_2^{\mathrm{surface}} = 1.1 \times 10^{-26}$ MPa; $Ma = 6130$) to 1.8×10^{-5}MPa ($Po_2^{\mathrm{surface}} = 1.7 \times 10^{-23}$ MPa; $Ma = 4730$), temperature oscillation mode changed from oscillation with multiple frequencies to that with a single frequency (Figs. 20a and b). The time delay between the inlet of the oxygen-containing argon gas and the initiation of the change in temperature-oscillation mode is due to the lead time for substituting 6N-argon gas by oxygen-containing argon gas (Fig. 20c). For a 5-mm-high liquid bridge with ΔT of 15 K at $Po_2^{\mathrm{inlet}}$of 3.5×10^{-7}MPa (Ma=3530), temperature oscillation with a single peak was observed. Between the two above-mentioned Marangoni numbers (between 4730 and 6130), there must be a critical Marangoni number for transition from oscillatory flow with a single frequency to that with multiple frequencies, namely, Ma_{c3}. As flow velocity decreases with increasing Po_2(see Fig. 18), Marangoni flow is stabilized. Thus, flow with temperature oscillation

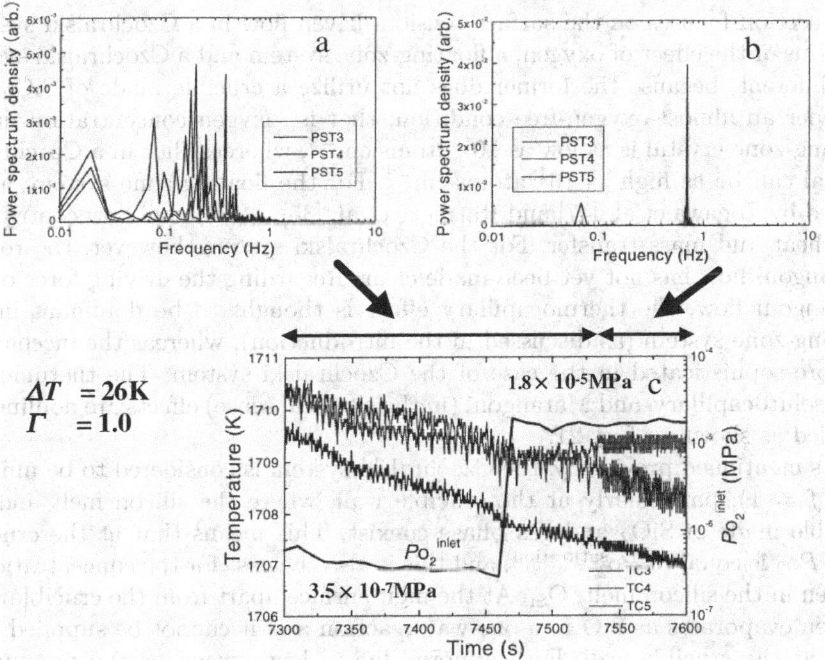

Fig. 20. Effect of oxygen partial pressure (inlet) on temperature fluctuation: **a** power spectrum density (PST3 - PST5) obtained through temperature oscillation measurement using thermocouples (TC3 - TC5) at $P_{O_2}^{\mathrm{inlet}} = 3.5 \times 10^{-7}$MPa, **b** $P_{O_2}^{\mathrm{inlet}} = 1.8 \times 10^{-5}$MPa, and (c) temperature oscillation

at multiple frequencies changes to that at a single frequency. Oscillatory flow with a single frequency has not yet been observed numerically.

In the course of the Marangoni-number calculation, we used rather high values of temperature coefficient, which correspond to low oxygen partial pressures. However, there is a possibility that we should use a rather low temperature coefficient, because one out of four thermocouple was damaged and the silicon melt touched the thermocouple sheath, made of SiO_2(we could use only three thermocouples, although four thermocouples were inserted in the silicon melt, as shown in Fig. 20). This implies that oxygen would have been dissolved into the silicon melt and lowered temperature coefficient of surface tension. On the other hand, in the case of a liquid bridge with a small Marangoni number, i.e., 1315, and a temperature coefficient $|\partial\sigma/\partial t|$ of $0.2 \times 10^{-3} N{\cdot}m^{-1}{\cdot}K^{-1}$, oscillation with a single frequency was also observed on earth [29].

A silicon single crystal was grown by a floating-zone method to observe the effect of oxygen partial pressure on the formation of dopant growth striations. Striation with a single frequency was observed, when grown at high oxygen partial pressure [45].

5 Flat Surface of a Czochralski Melt

This section focuses on the surface-tension-driven flow in a Czochralski system. In terms of the effect of oxygen, a floating-zone system and a Czochralski system are different, because the former does not utilize a crucible made of SiO_2 and is under an almost-oxygen-free condition; that is, oxygen concentration in the floating-zone crystal is as low as 10^{15}atoms$\cdot$cm^{-3}, whereas that in a Czochralski crystal can be as high as 10^{18}atoms$\cdot$cm^{-3}. For the floating-zone system, as reported by Togawa et al. [47] and Ratnieks et al. [35], Marangoni flow contributes to a heat and mass transfer. For the Czochralski system, however, the role of Marangoni flow has not yet been made clear. Regarding the driving force of the Marangoni flow, the thermocapillary effect is thought to be dominant in the floating-zone system (as discussed in the introduction), whereas the mechanism is more sophisticated in the case of the Czochralski system. The thermocapillary, solutocapillary, and Marangoni (in the classical sense) effects are nonlinearly coupled as shown in Fig. 21.

As mentioned previously, the Czochralski system is considered to be univariant ($f = 1$), particularly at the crucible wall, where the silicon melt and the crucible made of SiO_2 and gas phase coexist. This means that at the crucible wall, Po_2 is equal to $Po_2^{\text{saturation}}$, and this is also the case for the concentration of oxygen in the silicon melt, $\underline{O}_{Si}$. At the melt surface apart from the crucible wall, oxygen evaporates as SiO in a one-way reaction and it cannot be supplied as is done at the crucible wall. Furthermore, due to low oxygen partial pressure in the ambient atmosphere used for the Czochralski growth, oxygen is not supplied from a gas phase. This condition produces a gradient of oxygen concentration at the melt surface, so a thin SiO_2 film can not appear. This oxygen-concentration gradient from the crucible wall to the growing silicon crystal can be a driving

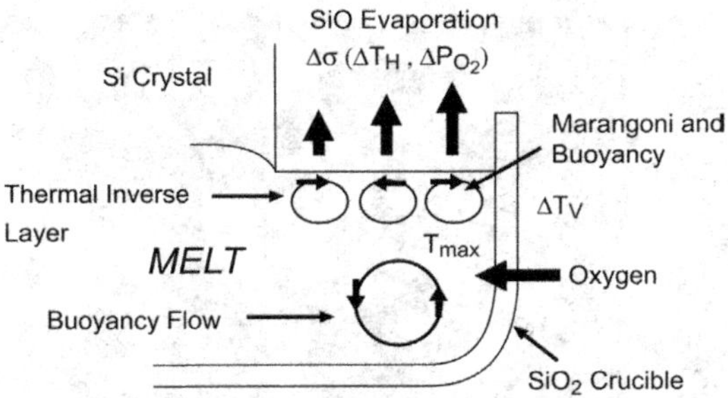

Fig. 21. Flow model within the Czochralski melt

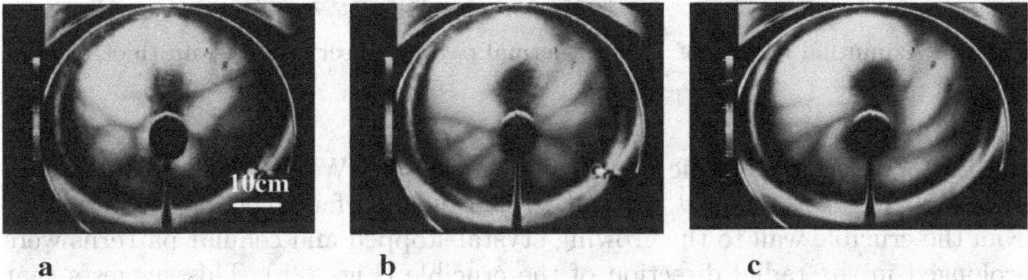

Fig. 22. Network patterns observed at the Czochralski melt: **a** without magnetic field and **b** with Magnetic field and **c** in the EMCZ (electromagnetic Czochralski) configuration

force for surface-tension-driven-flow, i.e., solutocapillary flow. Direction of the force is the same as that of thermocapillary effect induced by a temperature gradient between the crucible wall and the growing silicon crystal.

Due to intense heat radiation from the melt surface, temperature of the melt surface is lower than that of a bulk melt. Thus, Yi et al. [52] proposed that a thermal inverse layer and a cellular pattern are formed, and they estimated its contribution to the thermocapillary effect on the surface flow. Nakanishi et al. [30] confirmed the existence of the inverse layer experimentally; its depth was 6-7 cm. Yamagishi and Fusegawa [50] found a network pattern (cell structure) at the melt surface. Formation of a cell structure is attributed to the Rayleigh-Marangoni-Bénard mechanism in the thermal inverse layer, where both buoyancy and Marangoni (in a classical sense: see Fig. 1b) effects contribute to the formation of the pattern. For both Rayleigh and Marangoni mechanisms, the fact that temperature is higher at the bottom of the inverse layer than at the surface is essential.

Figure 22a shows a network (cellular) pattern observed by a CCD camera during crystal growth from a 700-mm-diameter crucible [5]. The network pattern moves slowly from the crucible wall to the center of the crucible. This motion is caused by a strong base flow, due to buoyancy beneath the inverse layer, whose direction is inbound just beneath the inverse layer (Fig. 21). We verified the correlation between the cellular pattern motion and the base flow by observing surface flow under two conditions: VMCZ (vertical magnetic Czochralski)

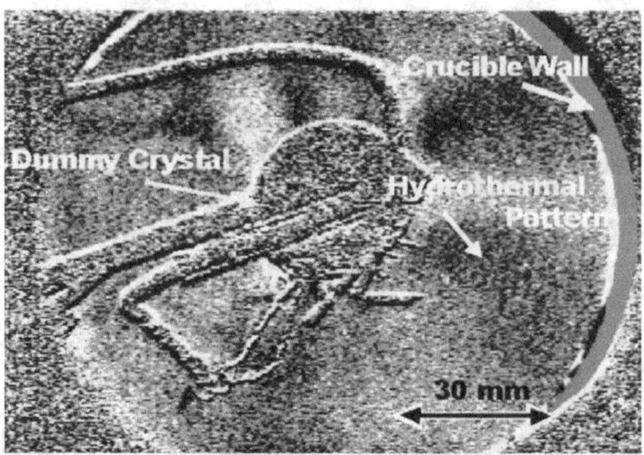

Fig. 23. Azimuthal motion of a hydrothermal pattern observed at 8-mm-thick shallow melt

and EMCZ (electromagnetic Czochralski) melts [49]. When a magnetic flux density of 50 mT was applied vertical to the melt surface, cellular pattern flow from the crucible wall to the growing crystal stopped and cellular patterns were prolonged in the radial direction of the crucible (Fig. 22b). This suggests that base flow is markedly damped and that cellular patterns were prolonged by the remaining thermocapillary and solutocapillary effects on the melt surface. Under the EMCZ condition, cellular patterns were twisted and rotated counter-clockwise (Fig. 22c). This is due to a strong azimuthal flow (counter-clock wise) driven by the Lorentz force present under the EMCZ condition. The dark lines in Figs. 22a–22c represent lower temperature than the other parts. The dark lines contact with the growing crystal. This suggests that the temperature field in the vicinity of the growing crystal is modulated at higher frequency than that of crystal rotation. Thus, network pattern would affect the quality of the crystal through introducing growth striation.

Figure 23 shows the surface of a shallow flat melt (8 mm thick). A strong flow from the crucible wall to the dummy crystal can be clearly observed, whereas a black stripe pattern can be seen to move azimuthally: on the right side of the melt the pattern moved counter-clockwise, and on the left side - clockwise. The same phenomenon was observed in a 3-mm-thick shallow melt. According to Smith and Davis [41], who theoretically studied hydrothermal instability, the deflection angle between flow direction and thermal-front-propagation direction is 80° for low-Pr-number fluids with a rectangular geometry. The observed black pattern can thus be attributed to a hydrothermal wave (even though our experiment was carried out in a circular geometry).

6 Perspective on Future Work

Surface-tension-driven flow in a liquid-bridge configuration of a low-Pr-number fluid, such as molten silicon, was found to undergo a transition from axisymmetric stationary flow to non-axisymmetric stationary flow at a threshold as

Ma_{c1}. This transition was found through numerical modeling. However, because this transition does not show any temperature oscillation, only flow visualization can detect it experimentally. X-ray flow visualization using a tracer particle is very difficult under a normal gravity condition, because the density of the tracer particle can be hardly adjusted to that of molten silicon, even though flow is surface-tension-dominant. On the other hand, a tracer particle can move in a molten silicon bridge under microgravity, even if a tracer particle has a large difference in density compared with that of molten silicon. We thus propose a flow-visualization experiment to confirm existence of the Ma_{c1} transition under microgravity aboard the International Space Station.

Czochralski growth is an important technique to produce silicon crystals for use in LSI. Understanding and controling surface flow in the Czochralski melt have not yet been sufficient. We need to collect fundamental data on physical chemistry of molten silicon at the melt surface. For example, two-dimensional *in-situ* measurement of surface tension in the Czochralski configuration is expected, so that surface-tension-driven flow and instability in the Czochralski melt can be more understood. Through this, the temperature field in the vicinity of the growing crystal, which affects crystal quality, can be clarified more in detail. Linear flow analysis for low-*Pr*-number fluid in the circular geometry has not been carried out much. Besides the experimental approach, numerical modeling [23] is also required.

Acknowledgment

Part of this study was supported by the Japan Space Utilization Promotion Center and the Japan Space Forum under the direction of the National Space Development Agency of Japan. We thank Prof. K. Mukai of Kyushu Institute of Technology and Prof. N. Imaishi for enlightening discussions. Thanks are also due to A. Hirao, K. Matsui, and Dr. G. Fujii of NEC Toshiba Space Systems for preparing the flight experiments.

References

1. T. Azami, S. Nakamura, T. Hibiya: J. Crystal Growth **231**, 82 (2001)
2. T. Azami, T. Hibiya: J. Crystal Growth **233**, 417 (2001)
3. T. Azami, S. Nakamura, T. Hibiya: J. Electrochem. Soc. **148**, G185 (2001)
4. T. Azami, S. Nakamura, T. Hibiya: J. Crystal Growth **223**, 116 (2001)
5. T. Azami, S. Nakamura, M. Eguchi, T. Hibiya: J. Crystal Growth **233**, 99 (2001)
6. M. Cheng, S. Kou: J. Crystal Growth **218**, 132 (2000)
7. A. Cröll, W. Müller, R. Nitsche: J. Crystal Growth **79**, 65 (1986)
8. A. Cröll, W. Müller-Sebert, R. Nitsche: Mater. Res. Bull. **24**, 995 (1989)
9. P. Dold, A. Cröll, F. Szofran, S. Nakamura, T. Hibiya, K. W. Benz: Space Station Utilization **1**, 31 (2000)
10. A. Eyer, H. Leiste, R. Nitsche: 'Crystal Growth of Silicon in Spacelab-1: Experiment ES-321 -'. In: *Proceedings of the 5th European Symposium on Materials Sciences under Microgravity - Schloß Elmau, November 5 - 7, 1984 (ESA SP-222)* pp.173- 182

11. A. Gañán, A. Barrero: Microgravity Sci. Technol. **3**, 70 (1990)
12. J.H. Han, Z. W. Sun, L.R. Dai, J.C. Xie, W.R. Hu: J. Crystal Growth **169**, 129 (1996)
13. S.C. Hardy: J. Crystal Growth **69**, 456 (1984)
14. X. Huan, S. Tokawa, S.-I. Chung, K. Terashima, S. Kimura: J. Crystal Growth **156**, 52 (1992)
15. N. Imaishi, S. Yasuhiro, Y. Akiyama, S. Yoda: J. Crystal Growth **230**, 164 (2001)
16. M. Jurish, W. Löser: J. Crystal Growth **102**, 214 (1990)
17. T. Kaiser, K.W. Benz: J. Crystal Growth **183**, 564 (1998)
18. B.J. Keene: surface Interface Anal. **10**, 367 (1987)
19. H.C. Kuhlmann: 'Thermocapillary Convection in Models of Crystal Growth', (Springer-Verlag, Berlin Heidelberg 1999)
20. C.W. Lan, S. Kou: J. Crystal Growth **108**, 351 (1991)
21. M. Levenstam, G. Amberg, T. Carlberg, M. Anderssson: J. Crystal Growth **158**, 224 (1996)
22. J. Leypoldt, H. C. Kuhlmann, H. J. Rath: J. Fluid Mech. **414**, 285 (2000)
23. Y.R. Li, N. Imaishi, T. Azami, T. Hibiya: 'Thermocapillary Flow in a Thin Annular Pool of Silicon Melt'. In: *Proceedings of the SPIE vol.4813, "Crystal Materials for Nonlinear Optical Devises and Microgravity Science"*, (2002) pp. 12-23
24. K.C. Mills, B.J. Keene, R.F. Brooks, A. Shirali: Philos. Trans. R. Soc. London, Ser. A **356**, 910 (1998)
25. K. Mukai, Z. Yuan, K. Nogi, T. Hibiya: ISIJ International **40**, S148 (2000)
26. S. Nakamura, K. Kakimoto, and T. Hibiya: 'Convection Visualization and Temperature Fluctuation Measurement in a Molten Silicon Column'. In: *Lecture Notes in Physics, Materials and Fluids under Low Gravity, Proceedings of the IXth European Symposium on Gravity-Dependent Phenomena in Physical Sciences Held at Berlin, Germany, 2-5 May 1995*, ed. by L. Ratke, H. Walter and B. Feuerbacher (Springer-Verlag, Berlin Heidelberg 1996) pp.343-349
27. S. Nakamura, T. Hibiya, K. Kakimoto, N. Imaishi, S. Nishizawa, A. Hirata, K. Mukai, S. Yoda, T. S. Morita: J. Crystal Growth **186**, 85 (1998)
28. S. Nakamura, T. Hibiya: Adv. Space Res. **24**, 1417 (1999)
29. S. Nakamura, T. Hibiya, N. Imaishi, S. Yoda, T. Nakamura, M. Koyama: Microgravity Sci. Technol. **12**, 56 (1999)
30. H. Nakanishi, M. Watanabe, K. Terashima: J. Crystal Growth **236**, 523 (2002)
31. K. Onuma, M. Sumiji, S. Nakamura, T. Hibiya: Appl. Phys. Letters. **74**, 3570 (1999)
32. J.R.A. Pearson: J. Fluid Mech. **4**, 489 (1958)
33. S.I. Popel, L.M. Shergin, B.V. Tsarevskii: Russ Metall. **2**, 72 (1976)
34. F. Preisser, D. Schwabe, A. Scharmann: J. Fluid Mech. **126**, 545 (1983)
35. G. Ratnieks, A. Muižnieks, L. Buligins, G. Raming, A. Muhlbauer: Magnitnaya Gidrodinamika **35**, 278 (1999)
36. M. Ratto, E. Ricci, E. Arato: J. Crystal Growth **217**, 233 (2000)
37. E. Ricci, L. Nanni, A. Passerone, Philos: Trans. R. Soc. London, Ser. A **356**, 857 (1998)
38. R. Rupp, G. Müller, G. Neumann: J. Crystal Growth **97**, 34 (1989)
39. D. Schwabe, R. Velten, A. Scharmann: J. Crystal Growth **99**, 1258 (1990)
40. M. Schweizer, A. Cröll, P. Dold, Th. Kaiser, M. Lichtensteiger, K. W. Benz: J. Crystal Growth **203**, 500 (1999)
41. M. Smith, S. Davis: J. Fluid Mech. **132**, 119 (1983)
42. M. Sumiji, S. Nakamura, K. Onuma, T. Hibiya: Jpn. J. Appl. Phys. **39**, 3688 (2000)

43. M. Sumiji, S. Nakamura, T. Hibiya: J. Crystal Growth **223**, 503 (2002)
44. M. Sumiji, S. Nakamura, T. Hibiya: J. Crystal Growth **235**, 55 (2002)
45. M. Sumiji, T. Azami, T. Hibiya: J. Crystal Growth, submitted
46. K. Takagi, M. Otaka, H. Natsui, T. Arai, S. Yoda, Z. Yuan, K. Mukai, S. Yasuhiro, N. Imaishi: J. Crystal Growth **233**, 399 (2001)
47. S. Togawa, Y. Nishi, M. Kobayashi: Electrochemical Society Proceedings **98-13**, 67 (1998)
48. M. Wanshura, V. Shevtsova, H.C. Kuhlmann, H.J. Rath: Phys. Fluids **7**, 912 (1995)
49. M. Watanabe, M. Eguchi, T. Hibiya: Jpn. J. Appl. Phys. **38**, L10 (1999)
50. H. Yamagishi, I. Fusegawa: J. Jpn. Assoc. Crystal Growth **17**, 304 (1990)
51. Y.K. Yang, S. Kou: J. Crystal Growth **222**, 135 (2001)
52. K.-W. Yi, K. Kakimoto, M. Eguchi, M. Watanabe, T. Shyo, T. Hibiya: J. Crystal Growth **144**, 20 (1994)

Low-Prandtl-Number Marangoni Convection Driven by Localized Heating on the Free Surface: Results of Three-Dimensional Direct Simulations

Thomas Boeck[1] and Christian Karcher[2]

[1] Laboratoire de Modélisation en Mécanique, Université Pierre et Marie Curie, 8 rue du Capitaine Scott, 75015 Paris, France
[2] Dept. of Mechanical Engineering, Ilmenau University of Technology, P.O. Box 100565, 98684 Ilmenau, Germany

Abstract. Marangoni convection in a square container with localized heating on the free liquid surface is simulated numerically using a pseudospectral Fourier-Chebyshev method. The extensive three-dimensional computations are performed for a low Prandtl number typical of liquid metals. Upon increasing the Marangoni number, the initially steady flow becomes oscillatory and eventually chaotic. The transitions reduce the spatial symmetries of the flow. Suitably defined integral velocity and temperature show scaling on the Marangoni number. The scaling exponents are compared with predictions from theoretical models for the heat transport in flow geometries relevant for the industrial process of electron beam evaporation.

1 Introduction

Convective heat transport is an ubiquitous phenomenon in nature as well as in industrial applications. Typically, convection is caused by buoyancy, but there are also cases where the Marangoni effect is dominant. Among them are certain metallurgical applications such as welding, crystal growth, and electron beam evaporation [2,7,16,17]. The numerical computations reported in this paper are motivated by the last of these three examples. We shall therefore briefly discuss electron beam evaporation and the convective heat transport associated with this important coating technology.

The electron beam evaporation process is realized inside a vacuum chamber with ultra high vaccum conditions. An electron beam gun with typically 10-100 kW beam power is targeted on the surface of a metal ingot placed inside a crucible. The beam heats up the ingot, which melts and eventually begins to evaporate. The metal vapor rising from the ingot deposits on a substrate placed above the crucible. The deposition rates of electron beam evaporation are significantly higher than those of other technologies. For the quality of the coating it is important to obtain sufficiently pure vapor, which clearly requires a clean melt. However, not only an appropriate ingot may be required. It may also be necessary to avoid contaminations, which can occur when the hot and chemically agressive liquid metal is in contact with a ceramic crucible wall. The remedy is to use water cooled crucibles of high thermal conductivity such as copper. This way, the melt cannot reach the wall since the wall temperature is

T. Boeck and C. Karcher, Low-Prandtl-Number Marangoni Convection Driven by Localized Heating on the Free Surface, Lect. Notes Phys. **628**, 157–175 (2003)
http://www.springerlink.com/

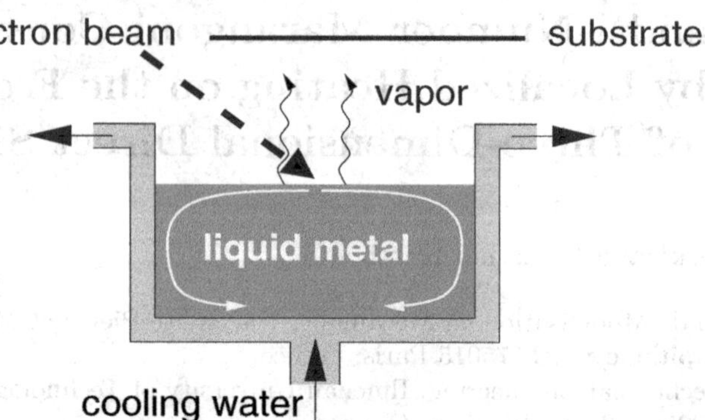

Fig. 1. Schematic diagram of electron beam evaporation from water-cooled crucibles

below melting temperature. As a downside, high temperature gradients have to be maintained in the melt. In particular, a large temperature difference exists on the free surface between the evaporating region and the liquid-solid interface. The temperature dependence of surface tension causes shear stresses directed towards the outer rim of the melt, which drive a flow indicated in Fig. 1. The strong convective heat transport associated with this Marangoni flow leads to poor energy efficiency. Typically, up to 5% of the beam power is used for evaporation. A significant fraction of the power is lost due to backscattering at the free surface, and more than half of it ends up in the cooling water. Reduction of the convective heat transport is the main option for an improvement of the energy efficiency since the loss due to backscattering depends for a given material only on the angle of incidence of the beam [1,8].

Progress on this complex engineering problem certainly requires detailed experimental work, in which the flow problem is only one aspect. Nevertheless, the understanding of the convective heat transport is an important issue. It can benefit from theoretical analyses and numerical simulations of simplified model systems such as the one studied in the present paper. Before we discuss the specific computational model and pseudospectral numerical method in Sect. 3 we shall first review theoretical estimates for the convective heat transport in Sect. 2. Section 4 describes the numerical results obtained from extensive three-dimensional direct numerical simulations, in particular flow patterns and scaling properties of integral quantities. We summarize our findings and present a few conclusions in the final Sect. 5.

2 Theoretical Estimates

The heat transport associated with Marangoni convection in a liquid with localized heating at the free surface has been estimated for laminar boundary layer flow and also for turbulent flow [12,15] using integral balances together with certain assumptions about the overall flow structure. The predictions in the laminar case also agree with more involved similarity solutions of a boundary layer model

[6]. The estimates given in [15] cover the limiting cases of large and small Prandtl numbers. Here we only present those for small Prandtl numbers typical of liquid metals.

2.1 Basic Assumptions and Equations

Basic assumptions in the analysis of [15] are a flat free upper surface with prescribed heat flux distribution and a purely thermocapillary flow. Free surface deformations and buoyancy are not taken into account. The forcing of the flow due to surface tension gradients is expressed by the free surface boundary condition

$$\rho\nu\frac{\partial \boldsymbol{v}_2}{\partial z} = \nabla_2\sigma, \tag{1}$$

where $\boldsymbol{v}$ is the velocity field, σ denotes the surface tension, ρ the fluid density and ν the kinematic viscosity. In this so-called Marangoni boundary condition, the free surface conicides with a coordinate plane $z = const.$ The subscript 2 indicates the projection of vectors onto this horizontal plane. For the surface tension it is assumed that it decreases linearly with the temperature T according to

$$\sigma = \sigma(T_r) - \gamma(T - T_r), \tag{2}$$

where T_r stands for a reference temperature.

Apart from the Marangoni boundary condition, the analysis in [15] makes use of the equations of motion for incompressible flow. For nondimensionalization of these equations we choose the depth h of the fluid layer as unit of length, ν/h as unit of velocity, and h^2/ν as unit of time, where ν denotes the kinematic viscosity of the fluid. Based on these units, the dimensionless evolution equations of the velocity field $\boldsymbol{v}$, the pressure p and the temperature T are

$$\frac{\partial \boldsymbol{v}}{\partial t} + (\boldsymbol{v}\cdot\nabla)\boldsymbol{v} = -\nabla p + \nabla^2\boldsymbol{v}, \quad \nabla\cdot\boldsymbol{v} = 0, \tag{3}$$

$$P\left\{\frac{\partial T}{\partial t} + (\boldsymbol{v}\cdot\nabla)T\right\} = \nabla^2 T, \tag{4}$$

where the Prandtl number $P = \nu/\kappa$ denotes the ratio of kinematic viscosity and thermal diffusivity κ. The unit θ for the temperature T is determined by the heat flux passing through the fluid layer from top to bottom. At the top surface (coinciding with the plane $z = 1$) the normal component of the heat flux density is

$$\frac{\lambda\theta}{h}\frac{\partial T}{\partial z} = q_{\max} f(x,y) \tag{5}$$

with T and z already being dimensionless. The other symbols in this expression are the thermal conductivity λ and the maxiumum value of the heat flux density $q_{\max}$. The dimensionless function f characterizes the spatial distribution of the heat flux. We fix θ by demanding that the prefactors in the previous equation become unity, which gives

$$\frac{\partial T}{\partial z} = f(x,y), \qquad \theta = \frac{h q_{\max}}{\lambda}. \tag{6}$$

The Marangoni boundary condition takes the dimensionless form

$$\frac{\partial \boldsymbol{v}_2}{\partial z} = Ma \nabla_2 T, \qquad Ma = \frac{\gamma \theta h}{\rho \nu^2}, \tag{7}$$

where the parameter Ma denotes the Marangoni number. To complete the formulation of the problem we assume that the no-slip sidewalls of the container are adiabatic, i.e. velocity and the heat flux vanish there. The no-slip bottom $z = 0$ of the liquid layer is assumed to be isothermal with the temperature $T = 0$.

Using the nondimensional equations and boundary conditions, one can derive two integral relations from (3) and (4). We multiply (3) by $\boldsymbol{v}$ and (4) by T and integrate over the volume Ω of the fluid. This gives

$$\int_\Omega (\nabla \boldsymbol{v})^2 \,\mathrm{d}x \,\mathrm{d}y \,\mathrm{d}z = -Ma \int_{z=1} \boldsymbol{v}_2 \cdot \nabla_2 T \,\mathrm{d}x \,\mathrm{d}y, \tag{8}$$

$$\int_\Omega (\nabla T)^2 \,\mathrm{d}x \,\mathrm{d}y \,\mathrm{d}z = \int_{z=1} f(x,y) T \,\mathrm{d}x \,\mathrm{d}y. \tag{9}$$

The first equation represents the kinetic energy budget with the left hand side denoting energy dissipation and the right hand side energy production due to the thermocapillary effect. Similarly, the second equation expresses a balance between thermal dissipation (left hand side) and production. Notice that terms containing time derivatives have been omitted in these equations because they vanish in a (statistically) stationary state.

2.2 Laminar Boundary Layer Flow

With these integral relations in hand we can now proceed with the laminar boundary layer case. The flow geometry in this situation should exhibit a velocity boundary layer of thickness Δ_U close to the free surface, which is shown schematically in Fig. 2. The characteristic velocity U in the boundary layer is much larger than the velocity in the recirculating region, which is supposed to occupy the entire bulk of the fluid. The temperature in the bulk is close to isothermal except for a region of size Δ_Θ, where it has the characteristic value Θ (cf. Fig. 2). This hot region is located where the heat flux is applied, i.e. where the function f is large. The goal of the analysis is to estimate the four unknowns U, Δ_U, Θ, Δ_Θ in terms of the parameters Ma and P. Two of the four necessary

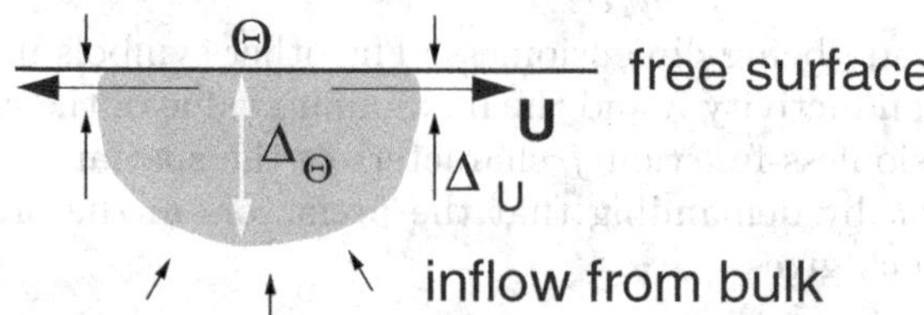

Fig. 2. Temperature and velocity scales in the hot region close to the free surface

relations are furnished by the integral balances. From the mechanical balance (8) we obtain

$$\Delta_U \frac{U^2}{\Delta_U^2} \sim Ma\Theta U. \tag{10}$$

The termal balance (9) gives

$$\Delta_\Theta \frac{\Theta^2}{\Delta_\Theta^2} \sim \Theta. \tag{11}$$

The other two relations are provided by the boundary layer scalings. From the boundary layer approximation to the momentum equation (3) we get

$$U^2 \sim \frac{U}{\Delta_U^2}. \tag{12}$$

The corresponding estimate for the temperature requires more care. Since the Prandtl number is small, the thermal diffusivity is large compared with the viscosity. The diffusive lengthscale of the temperature Δ_Θ is therefore large compared with the diffusive lengthscale of the velocity Δ_U. For this reason, the temperature is not sensitive to the small scale structure of the velocity field. The velocity in the boundary layer approximation of the heat equation is therefore not U but the characteristic velocity U_R of the fluid rising underneath the hot spot. We can estimate U_R from the mass flux of the rising fluid, which is approximately equal to $U_R \times$ (heated area of surface). This mass flux must be equal to the flux $U\Delta_U$ in the boundary layer. It follows that $U_R \sim U\Delta_U$, whereby the boundary layer approximation for the temperature becomes

$$P\Delta_U U \frac{\Theta}{\Delta_\Theta} \sim \frac{\Theta}{\Delta_\Theta^2}. \tag{13}$$

We can now compute the unknowns from the four relations (10-13). The results are

$$U \sim Ma^{1/2}P^{-1/2}, \tag{14}$$
$$\Delta_U \sim Ma^{-1/4}P^{1/4}, \tag{15}$$
$$\Theta \sim Ma^{-1/4}P^{-3/4}, \tag{16}$$
$$\Delta_\Theta \sim Ma^{-1/4}P^{-3/4}. \tag{17}$$

2.3 Turbulent Flow

The derivation of relations analogous to (14-17) for the turbulent case is also given in [15]. The overall structure of the flow is assumed to be the same as in the laminar case with surface jets directed from the hot spot towards the boundary of the container and a rising jet underneath the hot spot. In contrast to the laminar situation (where Δ_U characterizes the thickness of the surface jet) no additional lengthscale is introduced for the surface jets and the rising jet. The supporting argument from [15] is that turbulent jets have a finite opening angle [13]. We can therefore characterize the velocity field simply by the characteristic velocity U inside the jets.

For the temperature we assume that it can still be modelled as in the previous section, i.e. we retain the two parameters Θ and Δ_Θ. The thermal integral balance (11) is therefore unchanged, but the boundary layer scaling for the temperature in the hot region is different because we can now simply use U and not U_R as the typical velocity. The relation (13) thus becomes

$$PU\frac{\Theta}{\Delta_\Theta} \sim \frac{\Theta}{\Delta_\Theta^2}. \tag{18}$$

The final third relation is provided by the kinetic energy balance (8). In contrast to the laminar case we use the turbulent estimate U^3 for the left hand side of (8). Note that this is consistent with estimating the local energy dissipation rate $\varepsilon \sim u^3/h$ in dimensional units, where u is the integral velocity and h the integral lengthscale [9,18]. Using this estimate we get

$$U^3 \sim Ma\Theta U \tag{19}$$

from (8). Solving the three relations (11, 18, 19) for the unknowns we finally obtain the turbulent estimates

$$U \sim Ma^{1/3}P^{-1/3}, \tag{20}$$

$$\Theta \sim Ma^{-1/3}P^{-2/3}, \tag{21}$$

$$\Delta_\Theta \sim Ma^{-1/3}P^{-2/3}. \tag{22}$$

We see that the more efficient energy dissipation due to turbulence leads to a smaller exponent for the scaling of U on Ma. Moreover, Θ decreases faster with Ma (smaller exponent), i.e. the heat transport is also more efficient than in the laminar case.

3 Computational Model

3.1 Geometry and Boundary Conditions

In the numerical simulations to be reported in Sect. 4 we consider a square cavity, which is shown schematically in Fig. 3. The coordinate system is such that the bottom of the cavity is located in the plane $z = 0$ with $(x, y) = (0, 0)$ and $(x, y) = (L, L)$ as corner points. The governing equations of incompressible flow have already been listed in the previous section. The boundary conditions also carry over from there with one exception. At the sidewalls we have to replace the no-slip velocity boundary condition with the free-slip condition because we wish to use a Fourier-Chebyshev spectral method with Sine-/Cosine expansions in the lateral dimensions. The necessity of this change will become clear later when we discuss the numerical method in some detail. We remark that the lateral

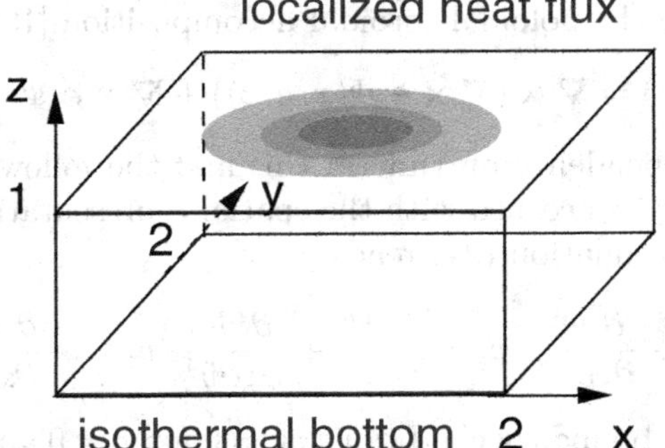

Fig. 3. Sketch of computational domain

free-slip conditions do not change the integral relations (14-17) and (20-22) from the previous section, i.e. the scaling relations in the laminar and turbulent case are not affected.

One important objective of the simulations is to realize a strong thermal forcing of the flow corresponding to high Reynolds and Peclet numbers. Boundary layers and turbulent flow will only occur at large values of these nondimensional quantities, which are therefore necessary if we wish to make any comparison with the theoretical estimates of the previous section. The additional parameters in the problem are chosen accordingly. For the aspect ratio of the box we select the rather low value of $L = 2$ for reasons of computational efficiency (large Reynolds numbers require high numerical resolution). For the Prandtl number we choose $P = 0.1$. This value is slightly larger than the Prandtl numbers of typical liquid metals. It helps us to achieve large Peclet and Reynolds numbers in the simulations since the ratio of these two quantities is just P. The final choice concerns the heat flux distribution f at the free surface. We assume a centered Gaussian profile of the form

$$f(x, y) = \exp\left[-\frac{(x-1)^2 + (y-1)^2}{2}\right], \tag{23}$$

which is compatible with the reflection and rotation symmetries of the square box. Narrower Gaussian profiles were also tried but not used any further because they required higher numerical resolution already at low values of Ma.

3.2 Numerical Method

The evolution equations (3,4) for the hydrodynamic variables are solved using a pseudospectral numerical method based on Fourier series in the horizontal directions x and y and a Chebyshev polynomial expansion in the vertical z–direction [5,10]. Notice that the essential steps in the construction of the numerical method parallel those of [4], where more details can be found.

Because of incompressibility, only two velocity components are independent. The velocity field can be represented in terms of two independent scalar quantities Ψ and Φ using the poloidal-toroidal decomposition [19]

$$\boldsymbol{v}(x,y,z) = \nabla \times (\nabla \times \boldsymbol{e}_z \Phi(x,y,z)) + \nabla \times \boldsymbol{e}_z \Psi(x,y,z). \tag{24}$$

We suppress the dependence on time in this and the following equations since we are currently only concerned with the spatial representation of the flow field. In component form, equation (24) reads

$$v_x = \frac{\partial \Psi}{\partial y} + \frac{\partial^2 \Phi}{\partial z \partial x}, \quad v_y = -\frac{\partial \Psi}{\partial x} + \frac{\partial^2 \Phi}{\partial z \partial y}, \quad v_z = -\frac{\partial^2 \Phi}{\partial x^2} - \frac{\partial^2 \Phi}{\partial y^2}. \tag{25}$$

The normal velocity boundary conditions $v_x = 0$ for $x = 0$ and $x = L$ and $v_y = 0$ for $y = 0$ and $y = L$ imply a Sine series expansion for v_x with respect to x and likewise a Sine series for v_y with respect to y:

$$v_x = \sum_{l>0} \sin(l\pi x/L)\hat{u}_l(y,z), \tag{26}$$

$$v_y = \sum_{m>0} \sin(m\pi y/L)\tilde{v}_m(x,z). \tag{27}$$

The same must be true for the each term in the representation of v_x and v_y through Φ and Ψ because of the mutual independence of the velocity potentials. We can therefore conclude that

$$\frac{\partial^2 \Phi}{\partial z \partial x} = \sum_{l>0} \sin(l\pi x/L)\hat{\phi}_l(y,z), \tag{28}$$

$$\frac{\partial \Psi}{\partial y} = \sum_{l>0} \sin(l\pi x/L)\hat{\psi}_l(y,z), \tag{29}$$

$$\frac{\partial^2 \Phi}{\partial z \partial y} = \sum_{m>0} \sin(m\pi y/L)\tilde{\phi}_m(x,z), \tag{30}$$

$$\frac{\partial \Psi}{\partial x} = \sum_{m>0} \sin(m\pi y/L)\tilde{\psi}_m(x,z). \tag{31}$$

To determine the expansion of Ψ we integrate $\partial\Psi/\partial y$ with respect to y and $\partial\Psi/\partial x$ with respect to x. A comparison between the resulting expressions shows that

$$\Psi(x,y,z) = \sum_{l>0}\sum_{m>0} \sin(l\pi x/L)\sin(m\pi y/L)\Psi_{lm}(z). \tag{32}$$

The same can be done for Φ. Here we find

$$\Phi(x,y,z) = \sum_{l>0}\sum_{m>0} \cos(l\pi x/L)\cos(m\pi y/L)\Phi_{lm}(z) \tag{33}$$

as Fourier representation. The temperature field is also represented by a double Cosine expansion

$$T(x,y,z) = \sum_{l\geq 0}\sum_{m\geq 0} \cos(l\pi x/L)\cos(m\pi y/L)T_{lm}(z) \tag{34}$$

because of the adiabatic sidewalls. The shear stresses on the sidewalls can be computed from Φ and Ψ, and turn out to be zero. For this reason our numerical method is restricted to lateral free-slip boundary conditions. We also note that in the numerical discretization the series (32-34) are replaced by finite sums with upper limits N_x and N_y for the indices l and m.

Evolution equations for the velocity potentials Φ and Ψ are derived by taking the curl and twice the curl of the momentum equation (3) and projection onto the vertical direction. We obtain two equations for the vertical velocity component $v_z = -\Delta_h \Phi$ and the vertical vorticity component $\omega_z = -\Delta_h \Psi$, where $\Delta_h = \partial_x^2 + \partial_y^2$ denotes the horizontal Laplace operator. In the following, we shall only use the quantities v_z and ω_z and do not explicitely compute Φ and Ψ. The complete system of evolution equations reads

$$\frac{\partial}{\partial t}\omega_z - \boldsymbol{e}_z \cdot \boldsymbol{\nabla} \times (\boldsymbol{v} \times \boldsymbol{\omega}) = \nabla^2 \omega_z, \tag{35}$$

$$\frac{\partial}{\partial t}\nabla^2 v_z + \frac{\partial}{\partial z}\boldsymbol{\nabla} \cdot (\boldsymbol{v} \times \boldsymbol{\omega}) - \boldsymbol{e}_z \cdot \nabla^2 (\boldsymbol{v} \times \boldsymbol{\omega}) = \nabla^4 v_z, \tag{36}$$

$$P\left(\frac{\partial}{\partial t}T + (\boldsymbol{v} \cdot \boldsymbol{\nabla})T\right) = \nabla^2 T. \tag{37}$$

The boundary conditions at the free surface are

$$\frac{\partial^2 v_z}{\partial z^2} - Ma\,\Delta_h T = v_z = \frac{\partial \omega_z}{\partial z} = \frac{\partial T}{\partial z} - f = 0 \qquad \text{at } z = 1, \tag{38}$$

where $\boldsymbol{\nabla} \cdot \boldsymbol{v} = 0$ has been used in the derivation of the first (Marangoni) boundary condition from (7). For the boundary conditions at the bottom we get

$$\frac{\partial v_z}{\partial z} = v_z = \omega_z = T = 0 \qquad \text{at } z = 0. \tag{39}$$

In order to derive the discrete representation of (35-37) we put $\zeta = \omega_z$, $\eta = \Delta v_z$, $\xi = v_z$ and introduce the wavevector

$$\boldsymbol{k} = \frac{l\pi}{L}\boldsymbol{e}_x + \frac{m\pi}{L}\boldsymbol{e}_y \tag{40}$$

to simplify notation. Fourier coefficients with indices l and m are denoted by, e.g. η_{lm}. For time differencing we use the implicit backward Euler method for the linear terms and the explicit second order Adams–Bashforth method for nonlinear terms. Advancing the solution from time level n to $n+1$ then requires the solution of four linear second order boundary value problems for each Fourier mode with the indices (l, m) according to the horizontal expansions (32-34). They read

$$\left(D^2 - \boldsymbol{k}^2 - \frac{1}{\Delta t}\right)\zeta_{lm}^{n+1} = -\frac{\zeta_{lm}^n}{\Delta t} - \mathrm{AB}\left\{[\boldsymbol{e}_z \cdot \boldsymbol{\nabla} \times (\boldsymbol{v} \times \boldsymbol{\omega})]_{lm}\right\}^n, \tag{41}$$

$$\left(D^2 - \boldsymbol{k}^2 - \frac{1}{\Delta t}\right)\eta_{lm}^{n+1} = -\frac{\eta_{lm}^n}{\Delta t}$$

$$+\mathrm{AB}\left\{\left[\frac{\partial}{\partial z}\boldsymbol{\nabla}\cdot(\boldsymbol{v}\times\boldsymbol{\omega})-\boldsymbol{e}_z\cdot\nabla^2(\boldsymbol{v}\times\boldsymbol{\omega})\right]_{lm}\right\}^n, \tag{42}$$

$$\left(D^2-\boldsymbol{k}^2\right)\xi_{lm}^{n+1}-\eta_{lm}^{n+1}=0, \tag{43}$$

$$\left(D^2-\boldsymbol{k}^2-\frac{P}{\Delta t}\right)T_{lm}^{n+1}=P\left(-\frac{T_{lm}^n}{\Delta t}+\mathrm{AB}\left\{[\boldsymbol{v}\cdot\boldsymbol{\nabla}T]_{lm}\right\}^n\right), \tag{44}$$

with $D=d/dz$ and $\mathrm{AB}\{f\}^n=(3f^n-f^{n-1})/2$ from the Adams–Bashforth formula. The boundary conditions for the Fourier coefficients are identical to (38) and (39) with exception of the Marangoni boundary condition and the thermal boundary condition at the free surface. The free surface boundary conditions become

$$\eta_{lm}+Ma\boldsymbol{k}^2T_{lm}=\xi_{lm}=\frac{\partial\zeta_{lm}}{\partial z}=\frac{\partial T_{lm}}{\partial z}-f_{lm}=0 \qquad \text{at } z=1, \tag{45}$$

with f_{lm} denoting the Fourier coefficients of the double Cosine expansion of $f(x,y)$ analogous to the temperature T (34). The boundary conditions at the bottom read

$$\frac{\partial\xi_{lm}}{\partial z}=\xi_{lm}=\zeta_{lm}=T_{lm}=0 \qquad \text{at } z=0. \tag{46}$$

An inspection of the system of equations and of the boundary conditions shows that the equations for ζ_{lm} and T_{lm} can be solved directly since they are independent of the other equations and suitable boundary conditions are available. The equations for η_{lm}, ξ_{lm} require a different approach because we cannot directly apply the no-slip condition $\partial\xi_{lm}/\partial z=0$ to the equation for η_{lm}. We solve these coupled equations by computing two linearly independent solutions of the homogeneous system for η_{lm}, i.e. we represent solutions as

$$\eta_{lm}=\eta_{lm}^{(0)}+\lambda\eta_{lm}^{(1)}+\mu\eta_{lm}^{(2)}, \tag{47}$$

$$\xi_{lm}=\xi_{lm}^{(0)}+\lambda\xi_{lm}^{(1)}+\mu\xi_{lm}^{(2)}, \tag{48}$$

where the functions with superscript 0 satisfy the inhomogeneous equations (42–43) and the boundary conditions $\eta_{lm}^{(0)}=0$ at $z=0$ and $z=1$. The solutions with superscripts 1 and 2 are obtained from the solution of the homogeneous equations (42–43 with zero right hand sides) with two linearly independent boundary conditions on η_{lm}. We take

$$\eta_{lm}^{(1)}(1)=\eta_{lm}^{(1)}(0)=1, \tag{49}$$

$$\eta_{lm}^{(2)}(1)=-\eta_{lm}^{(2)}(0)=1. \tag{50}$$

The unknown coefficients λ and μ are determined by inserting the above expressions into the Marangoni and the no-slip boundary condition at the bottom.

In the numerical simulations the size Δt of the time step is fixed. The auxiliary functions $\eta_{lm}^{(1,2)}$ and $\xi_{lm}^{(1,2)}$ only have to be computed once at the start for each wave vector $\boldsymbol{k}$. They are stored and reused at every time step.

In the discrete Chebyshev representation each of the boundary value problems (41–44) for the z–dependent Fourier coefficients at time level $n+1$ reduces to a tridiagonal system of linear algebraic equations. The boundary conditions are incorporated by means of the tau-method, which results in two filled rows. Nonlinear terms are computed pseudospectrally, i.e. in physical space using appropriate fast Sine, Cosine, and Chebyshev transforms with respect to the horizontal and vertical coordinates. The particular combination of transforms in the horizontal directions depends on the quanitity to be transformed. It is determined from the expansions (32-34) for the velocity potentials and the temperature, which are the fundamental quantities in our problem.

The algorithm is parallelized by distributing the Fourier coefficients across the processors such that the index m can run through its full range on each processor whereas the index l only runs through a subinterval. By that, only the Fourier transforms require inter-process communication. We compute them by the transpose method [11]. The program can be executed on a number of processors which is a power of two and which is smaller than N_x and N_y. Notice that N_x and N_y are powers of two because we only implement fast transforms to the base two as in [14]. For this reason, the total number of modes is $N_x + 1$ with respect to the x direction and $N_y + 1$ with respect to the y direction. These are odd numbers, which cannot be evenly distributed across the processors. The additional modes with index 0 are assigned to the first processor.

The implementation of the algorithm is based on an existing Fourier-Chebyshev code written in C for doubly periodic horizontal directions, which was developed for and used in previous work [3,4]. As before we use the Message Passing Interface (MPI) as communication library. The computational performance of our implementation should not have changed much by the adaptation to the lateral walls since the algorithmic differences are not substantial. The doubly periodic code achieved a speedup of 13.9 compared with sequential program execution when it was run on 16 processors of a SGI Origin 2000 computer with a single time step taking 0.2 seconds (numerical resolution $N_x = N_y = 64$, $N_z = 32$ collocation points). Details on the performance in the doubly periodic case are given in [4].

Compared with the doubly periodic code, the essential modifications concern the routines for the computation of the nonlinear terms, which include the Fourier transforms. The routines for solving the linear systems of equations (42–43) are unchanged. For validation of the code we therefore focus on the nonlinear terms and use Bénard-Maragoni convection with heating from below as a test case [3]. This is possible because even though the simulations in [3] were performed for periodic boundary conditions, certain flow patterns such as perfect hexagons (in a rectangular domain with aspect ratios $L_x = 4\pi/k$ and $L_y = 4\pi/\sqrt{3}k$, wavenumber $k = 1.99$) are compatible with lateral free-slip walls. This hexagonal flow is observed at Marangoni numbers slightly above the critical value for onset of convection. To adapt our code to the heating from below requires only minor changes in the heat equation. Using the same number of modes as in [3], we have successfully reproduced numerical data for the kinetic energy and the Nusselt number with the new code for lateral walls.

Table 1. Numerical parameters for selected runs performed at large values of Ma. The number of modes/collocation points in the spatial directions x, y and z are $N_x + 1$, $N_y + 1$ and $N_z + 1$. The integral velocity U is defined in (51), and the quantity T_c denotes the average value of the temperature at the center of the free surface

Ma	N_x	N_y	N_z	Δt	U	T_c
1.0×10^4	64	64	64	1.5×10^{-5}	1.03×10^2	0.345
2.0×10^4	128	128	64	1.0×10^{-5}	1.51×10^2	0.317
2.5×10^4	128	128	64	1.0×10^{-5}	1.69×10^2	0.307
3.0×10^4	128	128	64	9.0×10^{-6}	1.86×10^2	0.298
4.0×10^4	128	128	64	5.0×10^{-6}	2.15×10^2	0.285
5.0×10^4	128	128	64	5.0×10^{-6}	2.40×10^2	0.276
6.0×10^4	128	128	64	4.0×10^{-6}	2.62×10^2	0.268
7.0×10^4	128	128	64	3.5×10^{-6}	2.82×10^2	0.261
1.0×10^5	256	256	64	2.0×10^{-6}	3.35×10^2	0.247
1.4×10^5	256	256	64	1.5×10^{-6}	3.93×10^2	0.234
2.0×10^5	256	256	64	1.0×10^{-6}	4.40×10^2	0.220

4 Results

4.1 Flow Patterns

The numerical simulations with localized heating on the free surface were performed on the Cray T3E parallel computer of Dresden University of Technology using up to 64 processors. Table 1 lists the numerical parameters for the computationally most expensive runs at large Marangoni numbers. The simulations were typically stopped for times $t_{\max}$ of order unity. As initial data we mostly used the converged state obtained at the closest lower value of the Marangoni number.

We begin our discussion of the results with the temperature distribution in the case $Ma = 0$ of pure heat conduction. Figure 4d shows isotherms for stationary convection at $Ma = 1$ which, while not strictly conductive, are nevertheless indistinguishable from those for $Ma = 0$ because the Peclet number is of order 10^{-3}. We see that the free surface isotherms reflect the rotational symmetry of the imposed heat flux except near the corners, where the boundary distorts their circular shape. The isotherms in the vertical midplane (Fig. 4b) are densely spaced near the free surface and show some similartiy to the concentric pattern of a spherical or point source. At the bottom they are close to parallel with smaller gradients, which suggests that the heat flux density does not vary significantly across the bottom of the liquid.

General features of the velocity distributions (which are not restricted to low Marangoni numbers) are apparent in the contour plot of the vertical velocity

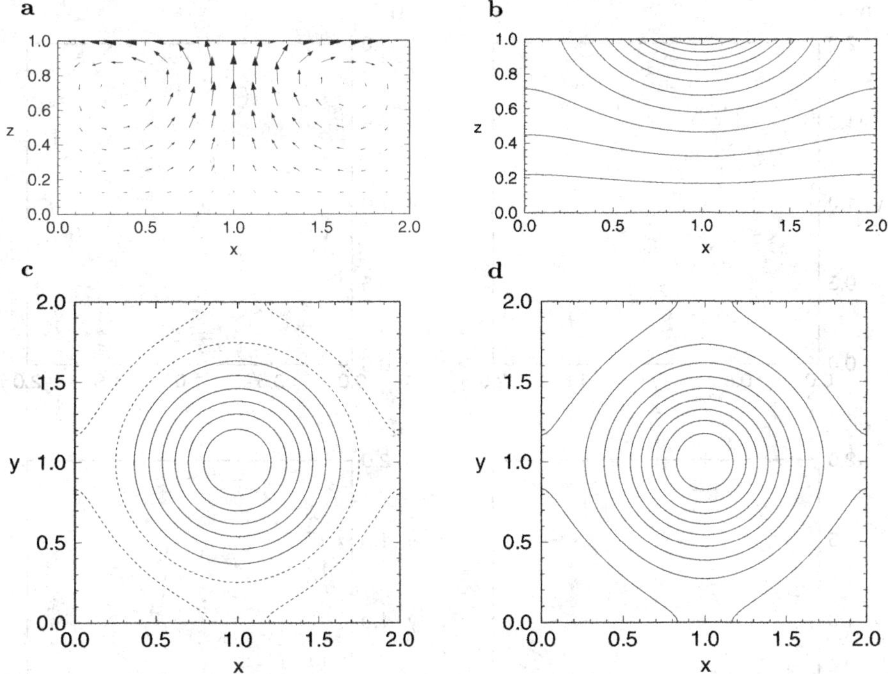

Fig. 4. Stationary convection at $Ma = 1$. **a** Vector plot of the velocity field in the vertical midplane $y = 1$. **b** Isotherms in the vertical midplane $y = 1$. **c** Contour lines of vertical velocity in the horizontal midplane $z = 0.5$, dashed lines indicate negative values. **d** Free surface isotherms

component in the horizontal midplane of the cell (Fig. 4c) and in the velocity vector plot for the vertical midplane (Fig. 4a). We see that the fluid rises in the center of the cell and descends at the lateral walls, i.e. the overall structure of the flow is that of a (deformed) toroid. The highest velocities in the vertical cross section occur at the free surface near the middle between the center and the outer wall. The stagnation points are closer to the free surface than to the bottom.

Besides these general properties, Fig. 4 demonstrates an important characteristic of the stationary flows at low Marangoni numbers, namely their spatial symmetries. The flows are invariant with respect to rotations by 90 degrees about the vertical axis and with respect to reflections about the vertical midplanes and diagonal planes. These invariances are maintained over a wide range of Marangoni numbers even though convection causes substantial changes in the isotherms and velocity distributions.

A qualitative change eventually occurs at $Ma \approx 2.4 \times 10^4$, where the flow becomes oscillatory. Figure 5 shows a series of snapshots for a single oscillation period at $Ma = 5 \times 10^4$. We have chosen this significantly larger Ma because at the threshold the oscillation amplitude is rather small. It is apparent that the Hopf bifurcation is accompanied by a loss of the reflection symmetries. Unfortu-

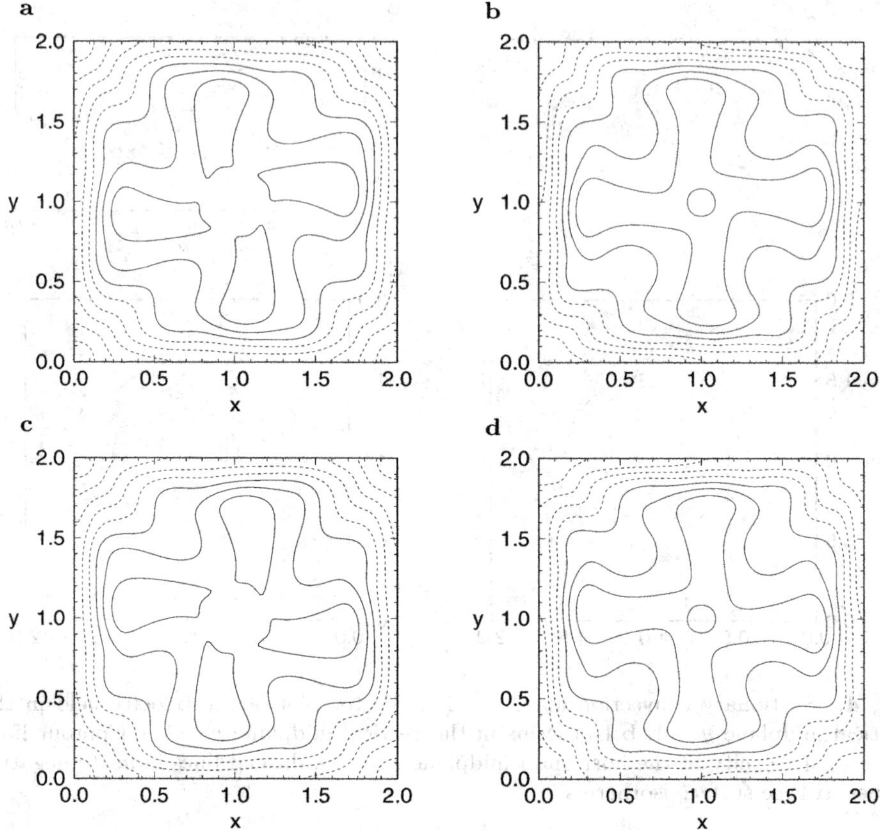

Fig. 5. Snapshots of a single oscillation period at $Ma = 5.0 \times 10^4$. The plots are equidistant in time and show contour lines of vertical velocity in the horizontal midplane $z = 0.5$. *Dashed lines* are used for negative values of v_z

nately, our numerical data are insufficient to determine whether the bifurcation is super- or subcritical. The threshold value $Ma \approx 2.4 \times 10^4$ was determined by tracing the the oscillatory branch downwards. We did not trace the stationary branch because of the slow convergence of such simulations.

Regular oscillations occur up to $Ma = 1.4 \times 10^5$. Chaotic motion was finally observed at $Ma = 2 \times 10^5$, which was the highest Marangoni number realized in the simulations. Figure 6a illustrates the regular or chaotic oscillations for these two Marangoni numbers using time series of the temperature in the center of the free surface. The spatial structure also becomes more and more complex upon increasing the Marangoni number. Figure 6 illustrates this with snapshots of isotherms in a vertical midplane (b,c) and snapshots of isolines of the vertical velocity in the horizontal midplane (d,e) for the two Marangoni numbers $Ma = 1.4 \times 10^5$ and $Ma = 2.0 \times 10^5$. As it has been assumed for the theory in Sect. 2, there is a tendency of the temperature field to concentrate in a small region beneath the upper surface, whereas the bulk becomes increasingly isothermal.

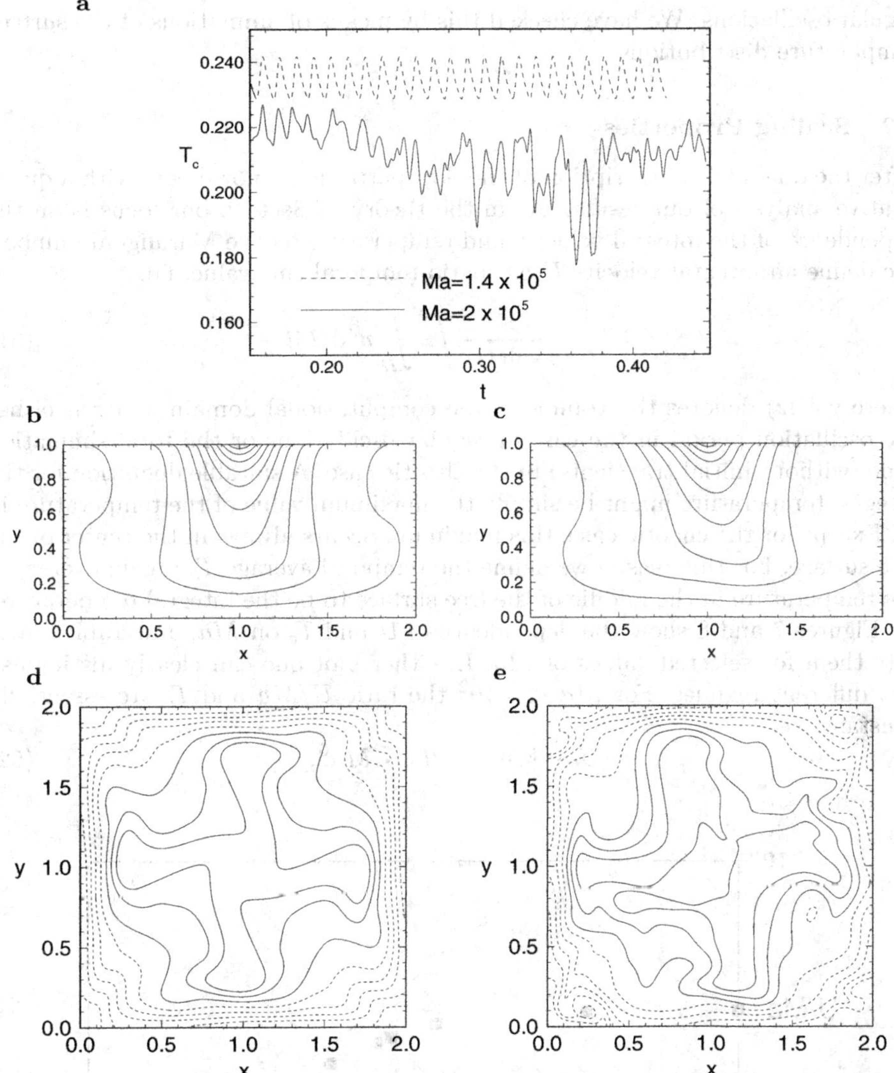

Fig. 6. **a** Time series of the temperature in the center of the free surface. **b,c** Isotherms in the vertical midplane. **d,e** Contour lines of vertical velocity in the horizontal midplane. Dashed lines are used for negative values of v_z. The snapshots **b,d** in the left column correspond to $Ma = 1.4 \times 10^5$ and (c,e) to $Ma = 2.0 \times 10^5$

We also note that the isotherms in the vertical midplane are asymmetric in either case, although this is not so obvious in the snapshot for $Ma = 1.4 \times 10^5$. The loss of the remaining spatial symmetry (rotation by 90 degrees about vertical axis) in the chaotic state is also apparent. It is partly responsible for the strong fluctuations of T_c shown earlier in Fig. 6a because the hottest spot on the surface moves about the center in the chaotic state whereas it stays in the center for the regular oscillations. We have checked this by means of animations of the surface temperature distributions.

4.2 Scaling Properties

After the qualitative description of the flow patterns we now begin with a quantitative analysis of our results. As in the theory of Sect. 2, our focus is on the dependence of the integral velocity and temperature on the Marangoni number. We define an integral velocity U as spatio-temporal rms value, i.e.

$$U^2 = \frac{1}{\tau \mathrm{vol}(\Omega)} \int_0^\tau \int_\Omega \boldsymbol{v}^2 \mathrm{d}\Omega \mathrm{d}t, \tag{51}$$

where $\mathrm{vol}(\Omega)$ denotes the volume of the computational domain and τ is either the oscillation period in the case of regular oscillations or the total simulation time (without initial transients) in the chaotic case. A suitable definition for the integral temperature might be simply the maximum value of the temperature in Ω. Except for the chaotic case, this maximum occurs always in the center of the free surface. For this reason we define the temporal average T_c (again over τ) of the temperature in the middle of the free surface to be the integral temperature.

Figures 7 and 8 show the dependence of U and T_c on Ma, and Table 1 also lists them for selected values of Ma. In either plot one can clearly distinguish two different regimes. For $Ma <\approx 10^2$ the ratio U/Ma and T_c are essentially constant, i.e.

$$U \sim Ma^1, \quad T_c \sim Ma^0, \tag{52}$$

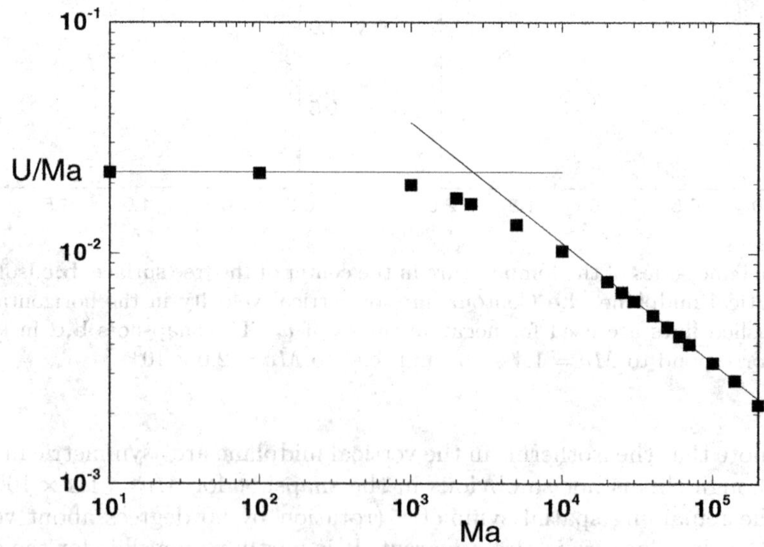

Fig. 7. Integral velocity U (defined by (51) as function of Ma. *Straight lines* correspond to the power laws (52) and (53)

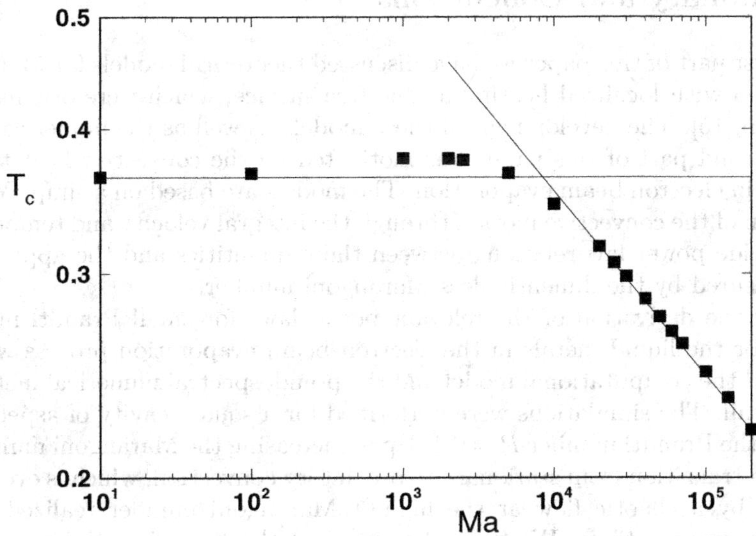

Fig. 8. Temporal average T_c of the temperature at the center of the free surface as function of Ma. *Straight lines* correspond to the power laws (52) and (53)

whereas for $Ma > 10^4$ both U/Ma and T_c are decreasing functions of Ma, which can be fairly well fitted by the power laws

$$U \sim Ma^{0.48}, \quad T_c \sim Ma^{-0.16}. \tag{53}$$

For the fits we have used the data between $Ma = 5 \times 10^4$ and $Ma = 1.4 \times 10^5$. The transition interval between the two regimes cannot be described by power laws.

The scaling behavior for small values of Ma can be easily explained. For small U the conductive temperature distribution is only weakly disturbed by the convective flow, i.e. T_c is independent of Ma. Furthermore, the momentum equation reduces to the linear Stokes equation in the limit of low Reynolds number, i.e. sufficiently small U. The magnitude of U is then proportional to the driving shear stresses at the free surface, which are proportional to Ma.

The scaling for large values of Ma suggests the existence of a laminar boundary layer flow as in the model of Sect. 2 since the exponent 0.48 for U is in remarkably good agreement with the predicted value 1/2. For T_c, the fitted exponent −0.16 has a larger deviation from the theoretical value −1/4. This is not particularly surprising since the isotherms shown in the vertical midplane in Fig. 6b,c are only beginning to justify the geometric assumptions of the model. A more pronounced thermal boundary layer should occur for larger velocities U. However, the flow need not necessarily be laminar for larger U since chaotic behavior appears for $Ma = 2 \times 10^5$. Perhaps a narrower prescribed heat flux distribution can help one to obtain better agreement.

5 Summary and Conclusions

In the first part of this paper we have discussed theoretical models for Marangoni convection with localized heating on the free surface, which were originally described in [15]. The development of these models as well as the numerical study in the second part of this paper was motivated by the convective heat transfer problem in electron beam evaporation. The models are based on a simple characterization of the convective motion through the integral velocity and temperature and provide power law relations between these quantities and the applied heat flux measured by the dimensionless Marangoni number.

After the derivation of the relevant power laws for small Prandtl numbers typical for the liquid metals in the electron beam evaporation process we have presented the computational model and the pseudospectral numerical method in some detail. The simulations were performed for a square cavity of aspect ratio two and the Prandtl number $P = 0.1$. Upon increasing the Marangoni number we observe a transition from stationary to oscillatory convection, which is eventually replaced by a chaotic flow at the highest Marangoni number realized in the extensive computations. We have also desribed the loss of spatial symmetries associated with the transition from one flow regime to the other.

The examination of suitably defined integral velocity and temperature revealed the existence of two distinct scaling regimes. For low values of the Marangoni number the temperature distribution remains close to the conductive state, whereas the velocity grows proportional to the Marangoni number. At the largest values of the Marangoni number the integral velocity grows proportional to the square root of the Marangoni number. This result is in good agreement with the prediction of the theoretical model of [15] for laminar boundary layer flow. Further analyses of the simulation data, in particular with respect to intrinsic lengthscales, are needed to ascertain the predominance of boundary layers in the flow. Verification of scaling on the Prandtl number (suggested by the theoretical model) is also desirable, but requires additional expensive simulations.

A fully turbulent regime could not be realized in the numerical simulations, although we have a first indication of its existence from the observed chaotic flow. We must also remark that our numerical method is not very efficient for the flow regimes at high Marangoni numbers. This is because the resolution requirements in the horizontal directions are dictated by the high velocity gradients near the sidewalls, whereas the Fourier expansion provides equidistant collocation points.

More realistic simulations are certainly desirable since numerous physical mechanisms and geometric details had to be omitted in the present work. E.g., the sidewalls should be no-slip and isothermal rather than free-slip and adiabatic since the heat loss in the actual electron beam evaporation occurs mainly through these sides. Surface deformation is also believed to be of importance [15] because of the high velocities.

Acknowledgements

We are grateful to A. Thess, who got us started with this problem, to A. Nepomnyashchy for interesting discussions and to the Zentrum für Hochleistungsrechnen at Dresden University of Technology for the extensive use of its parallel computers. Part of this work was supported by the by the Deutsche Forschungsgemeinschaft under grants Th497/9-2 and Th497/9-3.

References

1. G.D. Archard: J. Appl. Phys. **32**, 1505 (1961)
2. R. Bakish: *Introduction to Electron Beam Technology*, (Wiley, New York 1962)
3. T. Boeck, A. Thess: J. Fluid Mech. **399**, 251 (1999)
4. T. Boeck: *Bénard-Marangoni convection at low Prandtl numbers - Results of direct numerical simulations*, (Shaker Verlag, Aachen 2000)
5. C. Canuto, M.Y. Hussaini, A. Quarteroni, T. Zang: *Spectral Methods in Fluid Dynamics*, (Springer, Berlin, Heidelberg 1988)
6. C.L. Chan, M.M. Chen, J. Mazumder: Trans. ASME, J. Heat Transf. **110**, 140 (1988)
7. W.J. Cooper, R.D. Curry, K. O'Shea: *Environmental Applications of Ionizing Radiation*, (Wiley, New York 1998)
8. T.E. Everhardt: J. Appl. Phys. **31**, 1483 (1960)
9. U. Frisch: *Turbulence*, (Cambridge University Press, Cambridge 1995)
10. D. Gottlieb, S.A. Orszag: *Numerical Analysis of Spectral Methods: Theory and Applications*, CBMS-NSF Regional Conference Series in Applied Mathematics 26 (Society for Industrial and Applied Mathematics, Philadelphia 1977)
11. E. Jackson, Z.-S. She, S.A. Orszag: J. Sci. Comp. **6**, 27 (1991)
12. C. Karcher, R. Schaller, T. Boeck, C. Metzner, A. Thess: Int. Journal of Heat and Mass Transfer **43**(10), 1759 (2000)
13. L.D. Landau, E.M. Lifshitz: *Fluid Mechanics*, (Pergamon Press, Oxford 1987)
14. W. Press, B.P. Flannery, S.A. Teukolsky, W.T. Vetterling: *Numerical Recipes in C*, (Cambridge University Press, Cambridge 1993)
15. A. Pumir, L. Blumenfeld: Phys. Rev. E **54**, R4528 (1996)
16. S. Schiller, U. Heisig, S. Panzer: *Electron Beam Technology*, (Wiley, New York 1982)
17. J. Szekely: *Fluid Flow Phenomena in Metals Processing*, (Academic Press, New York 1979)
18. H. Tennekes, J.L. Lumley: *A First Course in Turbulence*, (MIT Press, Cambridge, London 1972)
19. O. Thual: J. Fluid Mech. **240**, 229 (1992)

Thermocapillary Flows and Vorticity Singularity

Eric Chénier[1], Claudine Delcarte[2], Guillaume Kasperski[2], and Gérard Labrosse[2]

[1] Université de Marne-la-Vallée, Laboratoire d'Etude des Transferts d'Energie et de Matière, 77454 Marne-la-Vallée Cedex 2, France
[2] Université de Paris-Sud XI, Laboratoire d'Informatique pour la Mécanique et les Sciences de l'Ingénieur, Bat. 508, B.P.133, 91403 Orsay Cedex, France

Abstract. The usual modelling of thermocapillarity introduces a vorticity singularity along the contact of free surfaces with solid boundaries. The liquid bridge hydrodynamics is adressed here, for its sensitivity to the size of a filtering length, δ, introduced by an explicit regularization of the singularity. Systematically following the convergence of the numerical results with δ shows that the stability properties of the axisymmetric flows, and their bifurcation maps, are correctly identified provided that this length scale is small enough. Although the singularity treatment is localized near the boundaries, the flow stability is controlled by an accumulation of mechanical power in the vicinity of the mid plane of the liquid bridge, that is *far* from the boundaries.

1 Introduction

Thermocapillary convection occurs in many industrial processes. The motivation of this study is to contribute to the understanding of the hydrodynamics of flows occuring in a floating zone crystal growth device, depicted in Fig. 1. The technique consists in applying an axisymmetric radiative heating to two aligned solid rods, a crystal seed and a feed rod. This creates an intersticial liquid region. Translating the material through the heat flux progressively melts the feed rod and allows the material to solidify on, as a single crystal. The resulting crystals are of great quality, thanks to the absence of any container pollution, but may however present some defects due to the unsteadiness of the flow in the molten part.

Numerous physical phenomena occur in this configuration. We will only retain the source of convection which is induced by temperature inhomogeneities along the free surface of the liquid zone. It is supposed to act in a fixed geometry. The proposed simplified configuration will then much differ from the actual crystal growth process, and nevertheless be very useful to point out an important aspect of its modelling. Furthermore, the flows will be restricted to be axisymmetric.

This communication is organized as follows. The next section makes a review of three definitions of surface tension and details the model of thermocapillary convection. In Sect. 3 are successively presented the liquid bridge configuration, the vorticity singularity generated by the classical model, a summary of the state of the problem of the moving contact line, an other singularity contained in the models of free surface isothermal flows, the way to treat the vorticity singularity,

E. Chénier, C. Delcarte, G. Kasperski, and G. Labrosse, Thermocapillary Flows and Vorticity Singularity, Lect. Notes Phys. **628**, 177–200 (2003)
http://www.springerlink.com/

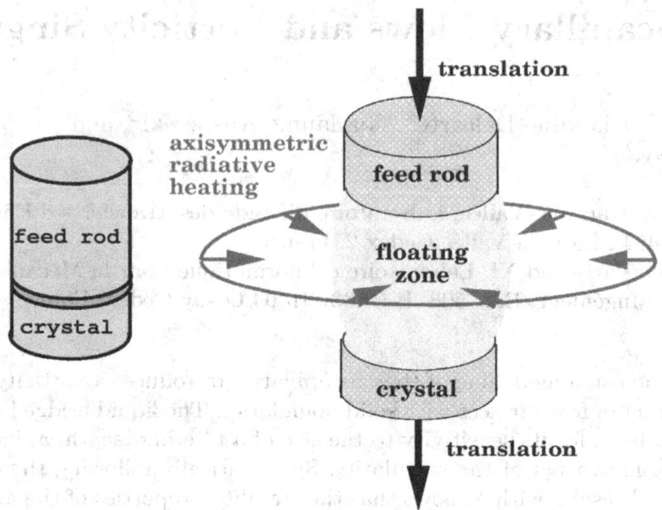

Fig. 1. Floating zone crystal growth device

and details on the adopted regularizations. In Sect. 4, the numerical approach is briefly exposed. Then the results are given and discussed in Sect. 5. The last section shows that a simple physical argument could be applied to modify the usual capillarity condition. But is it the only change one needs in order to recover a continuous physical configuration, as required at the macroscopic scale?

2 A Few Definitions

2.1 Surface Tension

The interface between two immiscible liquids, or between a liquid and a gas, possesses a potential energy. The surface density of this energy is called surface tension, σ. The minimization of the total surface energy explains why the shape of bubbles and droplets tends to become spherical (in the absence of any other force field). *Surface tension depends* on the nature of each fluid and also *on the local temperature and composition of the interface*. Its microscopic origin can be simply described as follows. Each molecule of a fluid deeply immersed in this fluid feels no resultant force, and can almost freely move. If it comes near a free surface the resultant of the molecular forces does not vanish anymore: the free surface is made of molecules subjected to a finite force field of molecular origin.

Different definitions of surface tension can be put forward. The microscopic approach, adopted by Adkins [1], is based on a causal analysis: this energy is due to the free bonds of the interface's atoms. In order to settle a predictive expression of σ, the author relates its value to the superficial density of atoms and to the latent heat of vaporization of the medium. The predicted value is in rather good agreement with experimental data. Landau and Lifchitz [21] give σ a thermodynamical definition which is non causal since it does not define the origin

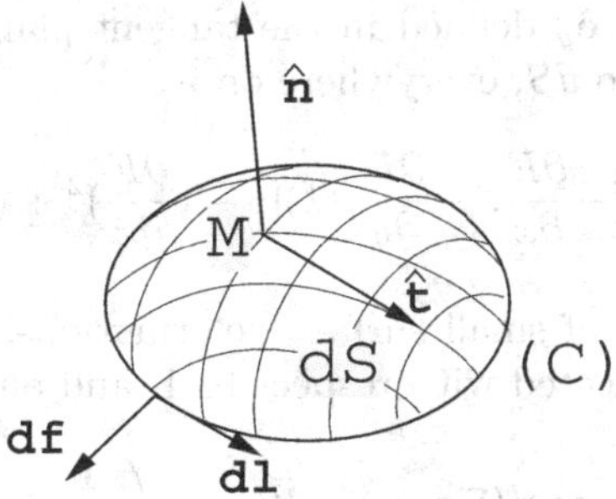

Fig. 2. Infinitesimal free surface element dS and related objects

of the energy. Surface tension is the energy required to create a unit interface area. A third definition is worth being given now for its subsequent use. The free surface is treated as an elastic membrane, on which a cut of infinitesimal length dl is made. Both sides of the cut experience then only a mechanical traction. To maintain them joined together an infinitesimal force of amplitude $df = \sigma\, dl$, replacing the action of surface tension, must be applied to them, normally to the cut and tangentially to the membrane. To proceed to superficial force balance one has to introduce oriented quantities. To this end, adopting the point of view of one fluid with respect to the other, let us define the outward unit normal $\hat{\mathbf{n}}$ to the free surface and a closed contour $\mathcal{C}$ containing an infinitesimal portion dS of the free surface about a given point M. The configuration is presented in Fig. 2. A unit tangential vector to the free surface at M, $\hat{\mathbf{t}}$, is also introduced. The resultant force acting on dS is

$$\mathbf{df} = -\oint_{\mathcal{C}} \sigma(x,y,z)\, \hat{\mathbf{n}}(x,y,z) \wedge \mathbf{dl},$$

x, y and z being the coordinates (defined in any system) along the path $\mathcal{C}$. This is a general expression since it takes into account the possible surface tension inhomogeneities, which is in fact essential for our configuration. This integral can be performed by adopting a local approximation, putting a reference system at M, with $\hat{\mathbf{n}}_M = \hat{\mathbf{dS}}$. At the first differential order of the local resultant, one gets:

- its component tangential to dS

$$\mathbf{df}.\hat{\mathbf{t}} = (\nabla_t \sigma)_{|M}\, dS; \tag{1}$$

- its component normal to dS

$$\mathbf{df}.\hat{\mathbf{n}} = -\sigma\, (\nabla_t \cdot \hat{\mathbf{n}})_{|M}\, dS.$$

These relations look difficult to apply with complex free surface shapes. However the normal component, for instance, is easy to evaluate if the local equation of the free surface is introduced, $z = F(x,y)$, the coordinate z being measured along $\hat{\mathbf{e}}_z \equiv \hat{\mathbf{n}}_M$, and $\hat{\mathbf{e}}_x$, $\hat{\mathbf{e}}_y$ defined in the tangent plane. This yields the local equation of the normal to dS, everywhere on it,

$$\hat{\mathbf{n}}(x, y, z) = (-\frac{\partial F}{\partial x}, -\frac{\partial F}{\partial y}, 1)[1 + (\frac{\partial F}{\partial x})^2 + (\frac{\partial F}{\partial y})^2]^{-1/2}.$$

With the approximation of small surface deformations, the term $(\nabla F)^2$ in the denominator can be neglected with respect to 1, and one gets

$$\mathbf{df}.\hat{\mathbf{n}} \equiv df_z \simeq \sigma\, (\nabla^2_{xy} F)_{|M}\, dS = \sigma \left(\frac{1}{R_1} + \frac{1}{R_2}\right) dS, \tag{2}$$

where R_1 et R_2 are the free surface curvature radii at M, respectively in the planes (x, z) and (y, z). It is recalled that indeed the curvature (> 0 or < 0) of a curve of equation $z = f(x)$ is $1/R = f\prime\prime/(1 + f\prime^2)^{3/2}$. The relation (2) is the well known Laplace law which predicts the pressure jump that occurs across the free surface. Applied to the axisymmetric liquid bridge of radius R, fixing its free surface to be a straight cylinder, it supplies the excess $\Delta p = \sigma/R$ of static pressure felt by the fluid inside the bridge.

The resulting tangential component corresponds to a stress. One will come back below to this component.

Those definitions hold for an interface defined as an ideal two-dimensional entity, that is a mathematical surface separating two media. In some cases, the interface can be given a finite thickness with internal properties and dynamics. Surface tension becomes an internal property of the interfacial medium, as density or temperature. Thick interfaces will be mentioned below and the reader is invited to refer to the review article of Anderson et al. [2]. Let us just note that those diffuse interface models are based on the existence of an internal capillarity (tension) obeying a Gibbs law for its dependence with the state variables. As mentioned in [24], in the static case, it has been proved that this description of internal surface tension was compatible with the classical model at the limit of zero interface thickness.

2.2 Thermocapillary Convection

Among the conditions required for a fluid layer to be in quiescent state with respect to its confinement, one is regarding the case with a free surface. No matter if there is or not an external acceleration field, such as gravity, the free surface must be isothermal and isocompositional, otherwise the fluid starts to flow. Indeed, as shown by (1), inhomogeneities of surface tension create a non zero resulting tangential force which the fluid cannot oppose to. The free surface acts then as source of momentum, which diffuses by viscosity to the bulk and generates the fluid motion, as illustrated in Fig. 3.

If surface tension only depends on temperature, its tangential gradient (see (1)) is

$$\nabla_t \sigma = \nabla_t T\, \frac{d\sigma}{dT}.$$

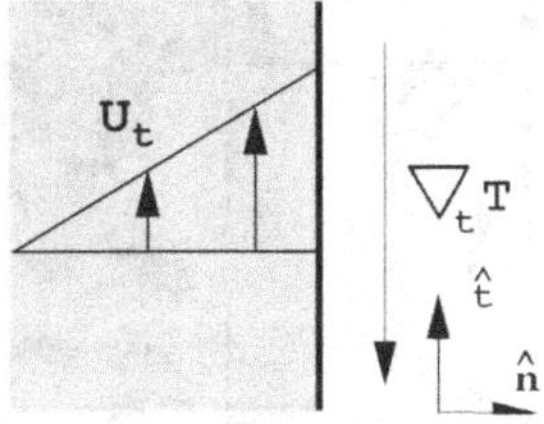

Fig. 3. Principle of thermocapillary convection

The last derivative is a material property of the free surface (in particular, it depends on the nature of each fluid) and it can be approximated by a constant if the temperature difference amplitude ΔT is small with respect to the mean temperature, noted T_0. In such a case, one has the simplified state equation,

$$\sigma = \sigma_0 + \gamma(T - T_0), \tag{3}$$

the constant γ being negative for most of the fluid-fluid interfaces. This leads to the following expression of the tangential force

$$\mathbf{df}_t = \gamma\,(\nabla_t T)\,dS. \tag{4}$$

The capillary stress is balanced by viscosity according to

$$\mu \nabla_n (\mathbf{v}.\hat{\mathbf{t}}) = \gamma \nabla_t T, \tag{5}$$

where μ is the momentum conductivity and $\mathbf{v}$ the fluid velocity. This relation defines a capillary velocity scale, V. Just expressing the order of magnitude of each term, and assuming that there is only one length scale, for instance R the radius of the bridge, one easily gets

$$V = \frac{\gamma}{\mu} \Delta T. \tag{6}$$

3 The Liquid Bridge Configuration

3.1 General Features

As announced from the very beginning, the simplest model is adopted. The liquid bridge is maintained by capillarity between two solid and isothermal planes where there is no phase change, neither melting nor solidification, and the free surface is straight and undeformable. A meridian cross section of the axisymmetric liquid zone is presented in Fig. 4. The zone radius and height are respectively denoted by R and $2R$: the length of the cylinder is equal to its diameter. The dynamics is supposed to be independent of the azimutal direction and the velocity is zero in this direction. This leads to a two-dimensional problem with $\hat{\mathbf{e}}_r$ and $\hat{\mathbf{e}}_z$ the radial and axial unit vectors respectively, r and z being the corresponding coordinates. The $z = 0$ origin is located on the mid-plane of the cavity.

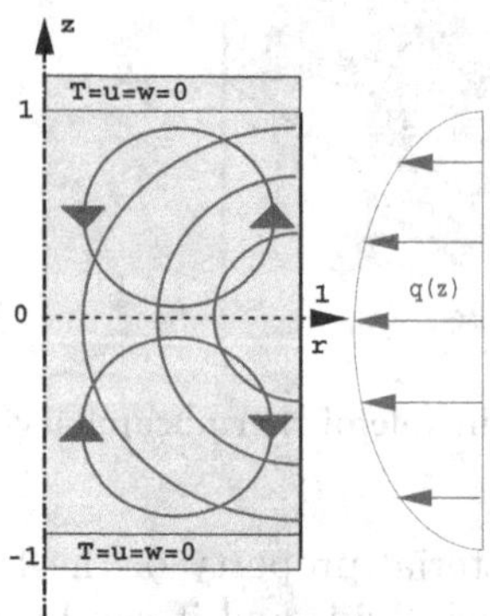

Fig. 4. Studied configuration

A zero gravity environment is taken. The free surface is submitted to an axisymmetric heat flux $Q(z) = Q_0.q(z)$, with $q(z)$ a non-dimensional function. The Newtonian Boussinesq liquid has constant density ρ, specific heat capacity c_p, thermal conductivity k, and a temperature dependent surface tension σ. The problem is made non-dimensional with the characteristic length scale R, velocity κ/R with κ the thermal diffusivity, time R^2/κ, and pressure $\rho\kappa^2/R^2$. The temperature difference magnitude, ΔT, is defined from the imposed heat flux and the fluid thermal conductivity, $\Delta T = Q_0 R/k$.

The liquid motion, with velocity $\mathbf{v} = u\hat{\mathbf{e}}_r + w\hat{\mathbf{e}}_z$, is governed by the following set of dimensionless balance equations:

$$\frac{\partial u}{\partial t} + (\mathbf{v}.\nabla)u = -\frac{\partial p}{\partial r} + Pr\,(\nabla^2 u - \frac{u}{r^2}), \tag{7}$$

$$\frac{\partial w}{\partial t} + (\mathbf{v}.\nabla)w = -\frac{\partial p}{\partial z} + Pr\,\nabla^2 w, \tag{8}$$

$$\frac{\partial T}{\partial t} + (\mathbf{v}.\nabla)T = \nabla^2 T, \tag{9}$$

$$\frac{1}{r}\frac{\partial (ru)}{\partial r} + \frac{\partial w}{\partial z} = 0, \tag{10}$$

with $\nabla = \hat{\mathbf{e}}_r\frac{\partial}{\partial r} + \hat{\mathbf{e}}_z\frac{\partial}{\partial z}$, $\nabla^2 = \frac{1}{r}\frac{\partial}{\partial r}(r\frac{\partial}{\partial r}) + \frac{\partial^2}{\partial z^2}$. Equations (7)–(10) respectively express radial and axial momentum, energy and mass transport, and are completed by the following boundary conditions :

- at $z = \pm 1$:

$$u = w = T = 0; \tag{11}$$

- at $r = 0$:

$$u = \frac{\partial w}{\partial r} = \frac{\partial T}{\partial r} = 0. \tag{12}$$

- at $r = 1$:

$$u = 0, \tag{13}$$

$$\frac{\partial w}{\partial r} = -Ma.\frac{\partial T}{\partial z}, \tag{14}$$

$$\frac{\partial T}{\partial r} = q(z). \tag{15}$$

This problem is thus characterized by two non-dimensional parameters:

- the Prandtl number, $Pr = \nu/\kappa$, is the ratio of the thermal to momentum characteristic diffusion times.
- the Marangoni number, $Ma = -\gamma Q_0 R^2/(k\mu\kappa)$ is the ratio of the characteristic thermocapillary (V) to thermal diffusion velocities.

It is worth mentioning that the Marangoni number can be identified as a Péclet number based on the thermocapillary velocity V. A Reynolds number can therefore be introduced via the Prandtl number. The stress boundary condition (14) amounts to impose the vorticity at the free surface, but indirectly, through a component of the thermal gradient which results from the flow it produces.

3.2 The Vorticity Singularities

Two boundary conditions imposed on the free surface are of flux type. The thermal condition is completely controlled by the experimentalist. And to comply with the requirement of isothermal solid boundaries, he has to design its experimental device in such a way that the lateral flux cancels at both extremities of the liquid bridge. We have thus chosen $q(z) = (1-z^2)^2$ to model the lateral heating distribution. The physical consistency we have just adopted for the thermal flux condition is also required for the flux of the momentum vertical component imposed by capillarity through (14) since it has to match the no-slip condition (11) at $z = +1$. But now this physical consistency is not easy to fulfill since the input $\frac{\partial T}{\partial z}$ to the flux of the momentum vertical component is controlled by the flow, while we would like it to vanish at $(r = 1, z = \pm 1)$. A simple argument leads to claim that, on the contrary, this thermal gradient will certainly never cancel there. Indeed, the heat supplied by the experimentalist to the fluid is advected, by the flow it generates, up to the solid fronts where it is conductively transferred to the solid with the dimensionless flux density $\frac{\partial T(r=1)}{\partial z}|_{z=\pm 1}$. And $\frac{\partial T}{\partial z}$ is expected to be a maximum at $(r = 1, z = \pm 1)$, and to increase as the flow velocity increases, i.e. with Ma. This analysis also holds when the free surface geometry is not compelled to be straight and undeformable [9].

A last comment deserves to be mentioned regarding these regions of high thermal gradients. Indeed the surface tension variation with temperature could depart somewhat from the adopted linear law, while keeping a continuous dependence. The viscosity, also, must depart from a given constant value, to reach an infinite value on the solid fronts. Obviously these considerations could just be taken as a first approach of reality in the presence of phase changes.

The usual thermocapillary convection modelling leads therefore to a vorticity singularity.

3.3 Singularities and Flows with Free Surface: A Shortcut through the Moving Contact Line Problem

The vorticity singularity we have just encountered in the naive modelling of thermocapillarity is not the first that Fluid Mechanics has to face when it deals with free surface flows. Indeed, the moving contact line problem has long been identified as presenting a major difficulty. It refers to the extremely common situation of the dynamic wetting of a solid surface by two Newtonian immiscible single-component isothermal fluids, one of them displacing its partner across a rigid solid surface: what are the kinematics, and dynamics, of the three-phase (liquid, liquid or gas, solid) contact line when it moves? Should one consider, for instance, that the fluid layer which wets the solid obeys a no-slip condition, though not adhering to the solid surface, acting like a ball bearings between the solid surface and the bulk flow? Is the vorticity singularity of thermocapillarity uniquely related to this difficulty, or is there something else to take into account?

The basic aspect of the wetting problem is that, beyond the molecular scales, several length (and time) scales are involved at the continuum level, and the macroscopic model that Hydrodynamics requires is not a trivial issue of the knowledge obtained at each scale level. As Dussan V. writes in [11] "One of the more conceptually difficult aspects associated with one viscous fluid displacing another ... is envisioning from a continuum point of view the mechanism by which this occurs". Are there relevant elements (experimental data or theoretical arguments) allowing us to claim that the local dynamic behaviors of the wetting are, or not, significantly linked to the bulk flow structure? It is still an open question.

This section is merely a brief survey of the litterature devoted to the moving contact line modelling. In two review articles, [11] and [13], a complete and comprehensive description is made of the state of knowledge reached in the eighties on the wetting physics.

Ingenious experiments (displacement of two immiscible fluids through a capillary tube, spreading of a drop of liquid on a flat solid surface, submerging of a tape into a bath of liquid, curtain coating) have supplied enlightening visual illustrations on (a) the rolling behavior of the wetting layer and (b) the complexity of the flow pattern in the interfacial region lying between the immiscible fluids (Fig. 5). They also led to nurter the controversy about the essential question of the existence of general boundary conditions that would allow one to solve the Navier-Stokes equations, though bypassing "the necessity of explicitely identifying the physics governing the dynamics of the fluids in the immediate vicinity of the moving contact line" [12]. The contact angle measurement is the central point of this question, subjected to the experimental difficulties of reaching its actual slope with the distance along the contact line and from the very beginning on the solid surface. For instance, it is claimed in [6] that "the dynamic contact angle is ... a functional of the flow field".

How to explore the physics of the fluid flows in the immediate vicinity of the contact line and of the interface between the immiscible fluids? Two different approaches might bring precious indications.

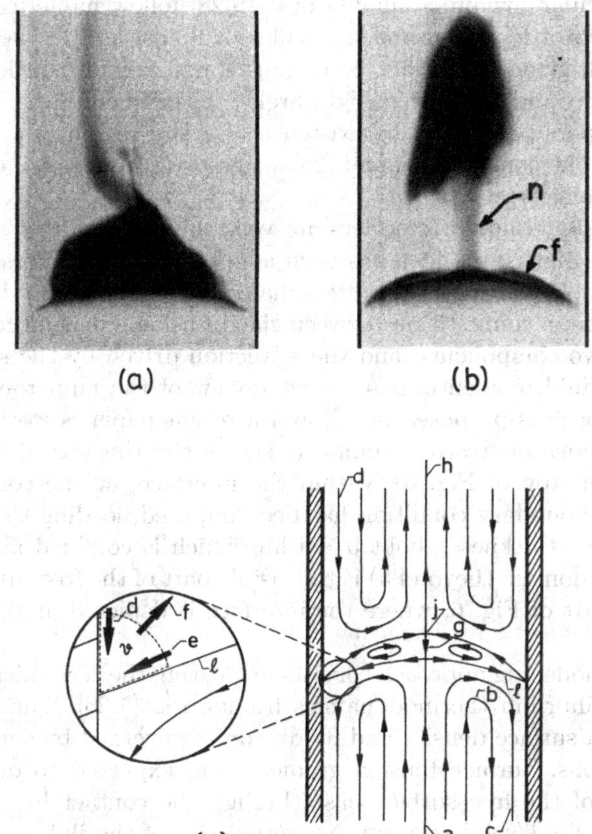

"Glycerine, the lower fluid, is displacing oil through a Plexiglas tube. Viewed from a frame of reference at rest with respect to the glycerine-oil interface, l, the walls appear to be moving with a constant downward velocity. For this system it is the glycerine that undergoes the rolling motion. Glycerine flows upward at the center, a, radially outward near the glycerine-oil interface, b, and downward close to the wall, c. The oil is the fluid with the more complex motion. Focusing attention on the motion of the oil near the contact line, we see that oil must be dragged into V at d and e due to the motion of the wall and the glycerine, respectively. Conservation of mass requires an outward flow at f. The motion of the oil far above the glycerine-oil interface is similar to that of the glycerine. Oil is dragged downward near the walls, d, and upward at the center, h. As a consequence of the axial symmetry, there is a bolus of oil "captured" within f and l near the glycerine-oil interface. This creates a stagnation-point flow within the oil at j, and a downward flow at g. The motion of the oil is visualized by injecting oil containing a black dye near the glycerine-oil interface (a). Note that the glycerine-oil interface billows into the oil. After a while, (b), dye outlines the bolus. The necking of the dye at n is evidence of the stagnation point within the oil (Dussan V. 1977)"

Fig. 5. Moving contact line experimental illustration, extracted from [10] with the original caption. (Reprinted with the permission of the American Institute of Chemical Engineers)

- The molecular dynamics simulations [19,28] follow numerically the space-time evolution of a few thousand molecules of fluids, and of the solid surface, interacting with given potentials. Numerous singular configurations can be explored in this way and phenomenological relations can be fitted or checked. Both just quoted papers point out the existence of a slip region of a few molecular diameters near the contact line, and [28] suggests "a breakdown of local hydrodynamics at atomic scales".

- Although liquid-liquid interfaces are very thin, about $10^{-7}\,cm$, or three to four molecular lengths, the idea has been adopted long ago (Poisson 1831) that its dynamics could be analyzed as if it had a finite thickness. Its equilibrium would result from a competition between the diffusion, creating countercurrents of the fluid's two components, and the advection driven by the surface tension forcing of the fluid momentum. A recent review of the numerous works made with this approach is proposed in [2], and a recent paper is worth being cited, [16]. Among many interesting points, it shows the theoretical pattern of the stream function (top of Fig. 6), within the interface, at the contact with the wall. A no-slip boundary condition has been imposed, leading to an inner layer (of dimensionless thickness about 5) within which is confined all the through-flow. The outer domain (beyond 5) is the visible part of the free surface, as shown by the lower part of Fig. 6, where the interface is depicted at the macroscopic scale.

A distinct modelling approach consists in treating the zero thickness interface in the non-equilibrium thermodynamics framework, [5,26]. The free surface is endowed with a surface density and its dynamics governed by surface state and balance equations. Surface tension gradients are expected to occur when the fluid particles of the free surface pass through the contact line, going from a fluid/fluid to a fluid/solid relation. No actual slip of the fluid over the solid is assumed. As claimed in [6], "The theory allows one to describe flows associated with moving contact lines ... as well-posed mathematical problems and interpret the results without invoking ad hoc auxiliary concepts".

Is it relevant to note that, in the thermocapillary case, surface tension gradients of thermal origin are explicitely imposed at the contact with the solid, merging those invoked in [6]?

3.4 How to Deal with the Singularity?

The study of the liquid bridge hydrodynamics by numerical experiments is valuable by the possibilities this means offers to isolate the behaviors that particular dynamical contributions may generate. Moreover the detailed insight that can be obtained of the liquid metal flows is very useful since these materials are opaque. Does the existence of a singularity require a specific methodological approach? Our answer is yes, and it does not refer to the choice of a numerical method against another. Indeed all schemes should supply equivalent results at the convergence with the mesh size and other working parameters. The goal is to proceed in such a way that the numerical results can be considered as physically interpretable, and, above all, observable in experiments performed with bound-

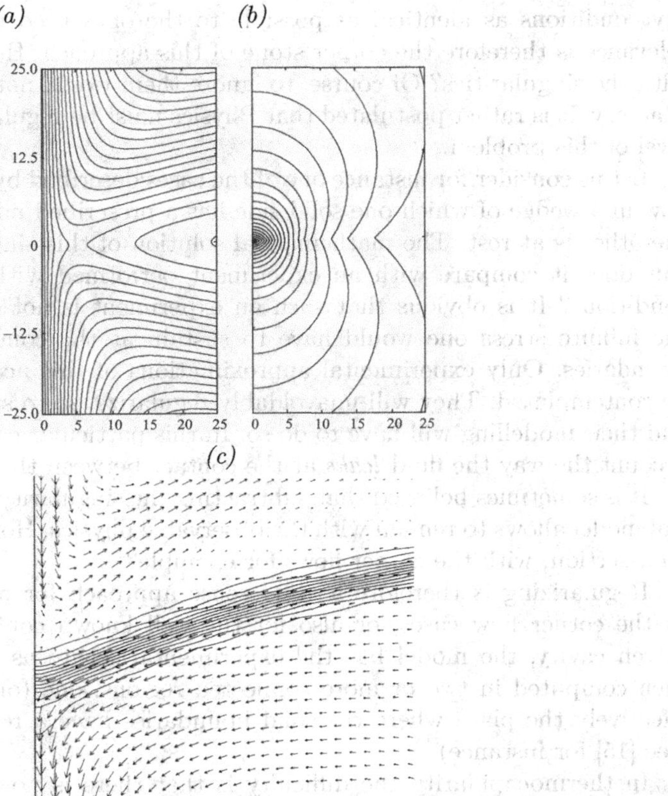

"(a) The stream function and (b) the chemical potential for the asymptotic flow at a contact line. The asymptotic interface thickness is 0 in these units, because $\epsilon \to 0$ faster than $\sqrt{\mu\kappa}$. The interface is located at the $y = 0$ axis. Both the stream function and the chemical potential are symmetric about the interface. The wall is located at $x = 0$ and is moving downward with unit speed. Because the boundary condition is no-slip, the velocity right next to the wall must be the same as that of the wall. An inner no-slip boundary layer is clearly shown. This inner layer, in which the velocity points downward through the interface, extends out to about $x = 5.1$. Past this inner no-slip boundary layer, the velocity through the interface is always upward. The chemical potential obeys the jump condition (...) at the interface, hence the kinks in the contour lines there. The direction in which a kink points depends on the direction of the velocity there, to the right for downward flow, to the left for upward.
Steady flow pattern (c) at the diffuse interface contact line. The upper fluid has one-hundredth the density and one thousandth the viscosity of the lower fluid."

Fig. 6. A moving contact line illustration from a diffuse interface model, extracted from [16] with original captions. (Reprinted with the permission of Cambridge University Press)

ary conditions as identical as possible to the ones used numerically. Physical relevance is therefore the corner stone of this approach. How, therefore, to deal with the singularities? Of course, to ignore them would not be scientifically satisfactory. It is rather postulated that Physics must be regular at the macroscopic level of this problem.

Let us consider for instance one of the cases described by [23] and [27], a local flow in a wedge of which one solid side has a prescribed non zero velocity while the other is at rest. The mathematical solution of this singular problem exists. But does it compare with an experiment performed with identical boundary conditions? It is obvious that such an experiment is not realizable because of the infinite stress one would have to sustain at the contact of the two solid boundaries. Only experimental approximations of this academic situation can be contemplated. They will unavoidably regularize, so to speak, the singularity, and their modelling will have to do so. In this particular case it has to take into account the way the fluid *leaks* at the contact between the solid boundaries.

It is sometimes believed that subtracting the singularity from the mathematical model allows to remain with the observable physics. How can one found such an assertion, with the corner flows for example?

Regularizing is therefore a good sense approach for modelling the reality. In the corner flow cases, or also for the well known configuration of the lid-driven cavity, the model has the experimental set-up as a guide. The flow is then computed in two or more connected regions, one (or two) of them being effectively the place where the solid boundaries have a relative sliding motion (see [15] for instance).

In thermocapillarity the difficulty is that there is no regularization model accepted as describing the physics of the contact free surface - solid boundaries. We are thus led

- to choose from scratch a function for regularizing the set of capillarity and no-slip conditions, (14) and (11), compatible with $\nabla \cdot \mathbf{v} = 0$ at the corners;
- to follow the convergence of the results with the filtering length δ introduced by this function;
- to check that the results do not depend upon the choice of the function made in the first step, but only on the filtering length scale size.

It may be useful to note here that the numerical schemes based on a local approximation of the derivatives do regularize the model, but implicitly. Their filtering length depends upon the meshing parameters and is thereby not straightforwardly interpretable physically. In this paper a pseudospectral method is used. It contains no implicit filtering.

3.5 The Regularized Problem

Boundary condition (14) is modified by introducing a function $f(z)$, such that $f(z = \pm 1) = 0$, which leads to

$$\frac{\partial w}{\partial r} = -Ma.\frac{\partial T}{\partial z}.f(z)\,. \tag{16}$$

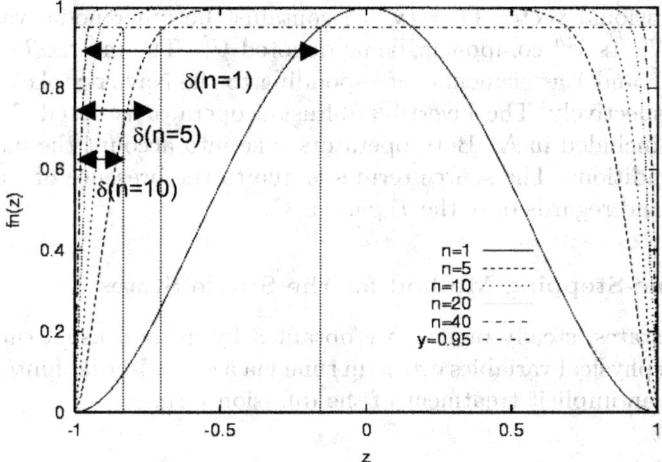

Fig. 7. Shapes of some $f_n^{(p)}(z)$ regularizing functions

Two very different regularizations have been used:

(a) A fixed or *passive* regularization in which

$$f(z) = f_n^{(p)}(z) = (1 - z^{2n})^2 \quad , \quad n \text{ positive integer} .$$

The filtering length $\delta(n)$ is defined as the distance over which the regularization acts according to $f_n^{(p)}(z = 1 - \delta(n)) = 0.95$, which leads to the length scale

$$\delta(n) = 1 - 0.0253^{1/2n},$$

of order $1/n$ for large n. Some $f_n^p(z)$ profiles are shown in Fig. 7.

(b) A flow dependent or *active* regularization suggested by the idea of a temperature dependence of the viscosity near the solid fronts. Here $f(z)$ reads

$$f(z) = f^{(a)}(T(r = 1, z)) \quad , \quad f^{(a)}(T) = exp[-\alpha\,(T_\infty - T)/T],$$

with α a positive constant and T_∞ the reduced temperature that the flow assigns to the free surface, *far* from the disks. The parameter α has been given the value 0.1. Some profiles of $f^{(a)}(T)$ are given in [9].

4 The Numerical Method

The system (7)-(16) is space-discretized with a Chebyshev collocation method having radial Gauss-Radau and axial Gauss-Lobatto grids [7]. A projection-diffusion algorithm is used to uncouple the velocity and pressure fields [4,20]. The resulting system is a set of ordinary temporal differential equations. It reads

$$\mathcal{T}\frac{d\,\mathbf{U}}{dt} = (\mathcal{N}(Ma) + \mathcal{L})\,\mathbf{U} + \mathbf{s}. \tag{17}$$

The N-dimensional vector $\mathbf{U} = (\mathbf{v}, T)$ contains the inner nodal values of the fields $\mathbf{v}$ and T, its kth component being denoted U^k. The matrix $\mathcal{T}$ is diagonal, with $(Pr)^{-0.5}$ and 1 as elements corresponding to the Navier-Stokes and energy equations respectively. The invertible diffusion operator is noted $\mathcal{L}$ and everything else is included in $\mathcal{N}$. Both operators take into account the nature of the boundary conditions. The source term $\mathbf{s}$ is due to the presence of the heat flux $q(z)$ in (15) and regards only the T part of $\mathbf{U}$.

4.1 A Time-Stepping Method for the Stable States

The stables states, steady or not, are obtained by using a usual time-stepping method. The physical variables evolve in time via a second-order finite differences scheme with an implicit treatment of the diffusion terms.

4.2 To Compute the Steady States, Stable or Not

For the understanding of Hydrodynamics, it is very instructive to determine, in the phase space, the trajectories of the steady states and whether they are stable or not. This requires to solve the steady version of (17), that is the fully non linear system

$$(\mathcal{N}(Ma) + \mathcal{L})\,\mathbf{U} + \mathbf{s} = 0. \tag{18}$$

An iterative Newton's method has to be implemented. Starting from a first guess $\mathbf{U}$ supplied by the time-stepping code, a correction $\delta\mathbf{U}$ is computed at each step from the linearization of (18),

$$(\mathcal{N}_{\mathbf{U}}(Ma) + \mathcal{L})\,\delta\mathbf{U} = -\,(\mathcal{N}(Ma) + \mathcal{L})\,\mathbf{U} - \mathbf{s}, \tag{19}$$

and used to update $\mathbf{U}$, according to $\mathbf{U} \leftarrow \mathbf{U} + \delta\mathbf{U}$, for the next step.

The Jacobian matrix $\mathcal{N}_{\mathbf{U}_0}(Ma) + \mathcal{L}$ is so large that it cannot be evaluated to a reasonable cost (despite its block structure) and (19) must itself be solved iteratively, preconditioned by a pseudospectral time-marching code, as proposed by [22].

4.3 The Continuation Curve

Rather than progressing in Ma with step by step sampled solutions, *continuous* paths of solutions are sought as function of Ma. From steady solutions known for successive values Ma_n, $(n = 1, ..., i)$, a first guess $\hat{\mathbf{U}}_{(i+1)}$ of the steady-state $\mathbf{U}_{(i+1)} = \mathbf{U}(Ma_{i+1})$ can be obtained by polynomial extrapolation, and inserted in (18) to compute $\mathbf{U}_{(i+1)}$. The Jacobian matrix may be singular at some value of the parameter, Ma. This is the signature of the presence of a bifurcation, and the continuation technique may fail there. At saddle-node bifurcations the continuation curve keeps continuous, but presents an infinite slope with Ma. The continuation is then carried out, but taking Ma as a dependent variable and one of the U^k's (defined at the beginning of this section) as a continuation variable.

4.4 Bifurcations

To know the linear stability of the steady solutions $\mathbf{U}$, the leading eigenvalues (those having the largest real parts) and the corresponding eigenvectors are calculated. Three successive steps are used.

(1) The leading eigenvalues are evaluated from the Arnoldi's method [3], implementing the technique described in [14] and [22]. The leading eigenvalues of a matrix A are also those of e^A, A standing for $\mathcal{N}_{\mathbf{U}}(Ma)+\mathcal{L}$. The exponential is estimated from the time-stepping code: a perturbation $\delta\mathbf{U} = (\delta\mathbf{v},\ \delta T)$ around $\mathbf{U}$ must indeed satisfy

$$\mathcal{T}\frac{d}{dt}\delta\mathbf{U} = (\mathcal{N}_{\mathbf{U}}(Ma) + \mathcal{L})\,\delta\mathbf{U}, \tag{20}$$

with homogeneous boundary conditions on the heat flux.

(2) Step (1) supplies an initial guess of the leading eigenvectors $\mathbf{X} + i\mathbf{Y}$ and corresponding eigenvalue $\lambda + i\omega$ for a given state $(\mathbf{U}, \mathbf{Ma})$. They are now improved by solving iteratively and accurately the system

$$\begin{aligned} (\mathcal{N}_{\mathbf{U}}(Ma) + \mathcal{L})\,\mathbf{X} &= \lambda\mathbf{X} - \omega\mathbf{Y}, \\ (\mathcal{N}_{\mathbf{U}}(Ma) + \mathcal{L})\,\mathbf{Y} &= \lambda\mathbf{Y} + \omega\mathbf{X}, \\ \mathbf{c}^T\mathbf{X} &= 1, \\ \mathbf{c}^T\mathbf{Y} &= 0. \end{aligned} \tag{21}$$

The vector $\mathbf{c}$ is chosen to impose normalization conditions, $\mathbf{c}^T$ being its transposed, such that $\mathbf{c}^T\mathbf{c} = \|\mathbf{c}\|^2$.

(3) The accurate localization of the bifurcation itself is performed by directly solving the fully nonlinear eigenvalue problem corresponding to $\lambda = 0$, via a Newton's method applied to the extended system of $3N + 2$ unknowns $(U^k, X^k, Y^k, k = 1, \ldots, N)$, ω and Ma:

$$\begin{aligned} (\mathcal{N}(\mathrm{Ma}) + \mathcal{L})\,\mathbf{U} + \mathbf{s} &= \mathbf{0}, \\ (\mathcal{N}_{\mathbf{U}}(\mathrm{Ma}) + \mathcal{L})\,\mathbf{X} &= -\omega\mathbf{Y}, \\ (\mathcal{N}_{\mathbf{U}}(\mathrm{Ma}) + \mathcal{L})\,\mathbf{Y} &= \omega\mathbf{X}, \\ \mathbf{c}^T\mathbf{X} &= 1, \\ \mathbf{c}^T\mathbf{Y} &= 0. \end{aligned} \tag{22}$$

If $\omega \neq 0$, a Hopf bifurcation is detected, otherwise it is either a saddle-node, pitchfork or transcritical bifurcation.

5 Results

5.1 Regularized Flows with $n = 1$

At any value of Pr, there exists a sufficiently small Ma value for which the flow is fully diffusive, dynamically and thermally, as shown on the left part of Fig. 8, where the streamlines are superimposed to the temperature field. From the temperature distribution, there is a minimum of surface tension at the center of the

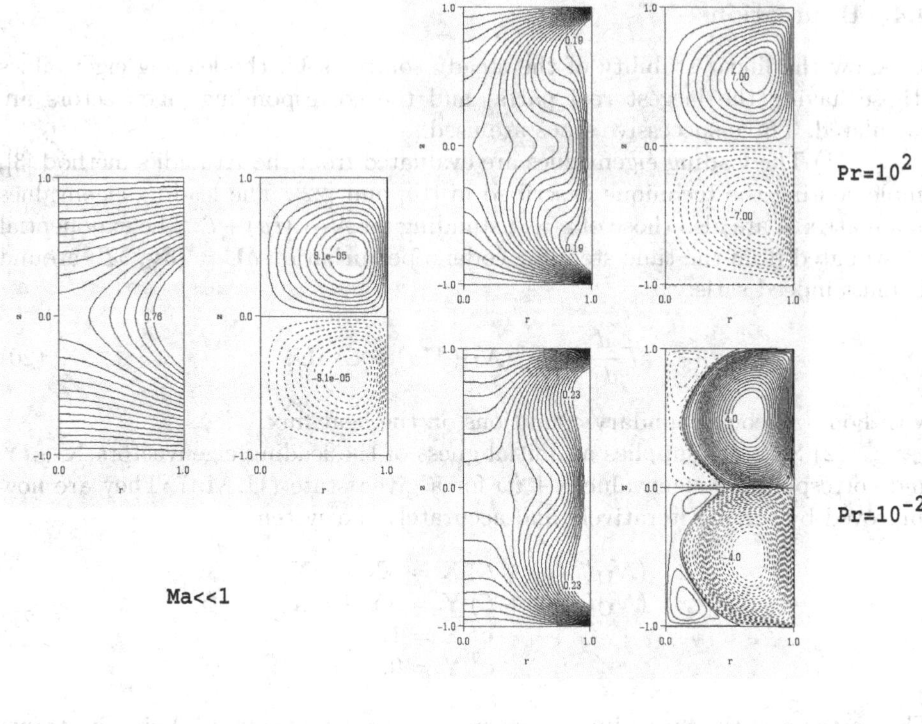

Fig. 8. Prandtl number effect on the solution

free surface and a maximum on each solid boundary. The fluid is thus driven from low to high surface tension regions, creating a two-cell flow configuration.

Increasing the value of the Marangoni number to $Ma = 10^5$ produces more convective thermal fields (it is recalled that Ma is a Péclet number based on the thermocapillary characteristic velocity). Heat is transported by conduction and convection. Convection globally flattens the temperature profile on the free surface but tightens its gradients close to the solid boundaries. If the thermal fields only slightly depend on Pr at fixed Ma, the flow field is in contrast very sensitive. Being diffusive at $Pr = 100$, it exhibits a pronounced convective structure at $Pr = 10^{-2}$. These flows are unstable with respect to oscillatory modes [17], with very different threshold laws depending on whether Pr is larger (constant Ma threshold) or smaller than 1 (constant Re threshold).

5.2 Quantitative Effects of Increasing n

Increasing the value of n directly increases the thermocapillary stress (see (16)). This increases the fluid velocity, which in turn tightens the thermal gradients near the solid boundaries, and consequently again increases the thermocapillary stress, and so on untill saturation. Increasing the parameter n of the regularizing

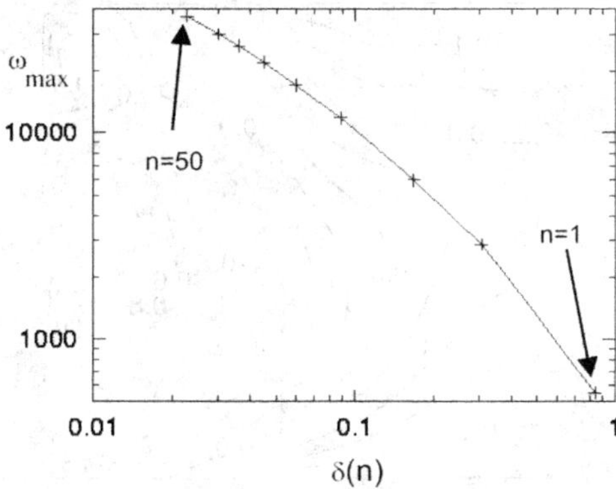

Fig. 9. Convergence with δ of the flow's vorticity maximum, $Pr = 1$, $Ma = 10^4$

function amounts to increasing the effective values of the Reynolds and Péclet numbers of the flow [18].

Figure 9 presents the evolution with $\delta(n)$ of the maximum vorticity of the steady flow for $Pr = 1$, $Ma = 10^4$. At lower Ma values, the convergence is reached rather quickly, for $n \approx 10$. Here it is not reached for $n = 50$, which corresponds to a regularizing function of polynomial degree 200. To get an accurate numerical solution in this case requires yet a (r, z) spectral decomposition of 200×320 modes.

5.3 Symmetry Breaking

Increasing the stiffness of the regularizing function does not only modify the convective regimes of the flows. It also gives an access to solutions with different symmetry properties. All the flows computed with $n = 1$ present a mirror-symmetry about the mid-plane, and remain stable until a Hopf bifurcation occurs at high Ma values [17]. Increasing n from $n = 2$ opens the access to non-symmetric steady state solutions. Figure 10 shows one of the non-symmetric solutions obtained for $n = 6$, $Ma = 600$ and $Pr = 10^{-2}$. Those solutions appear via successive undercritical pitchfork and saddle-node bifurcations [8] of the symmetric steady flows. The corresponding threshold values converge with n, as shown in Fig. 11 where Ma_i^c, $i = 0, 3$ respectively are the threshold values of the pitchfork and the three saddle nodes.

A non-symmetric solution has been found with the active thermal regularizing function with $\alpha = 0.1$, $Pr = 10^{-2}$, $Ma = 500$, of which the isolevels are displayed in Fig. 12. They look very much like those of Figure 10. This highlights the robustness of the symmetry breaking phenomenon. It must be emphasized, however, that non-symmetric solutions have only been reached with the continuation code. Due to the very low amplification rate of the destabilizing mode, to reach these states with a time-stepping code has been experienced to be of prohibitive computationnal cost, and a challenge for usual criteria for flow steadiness.

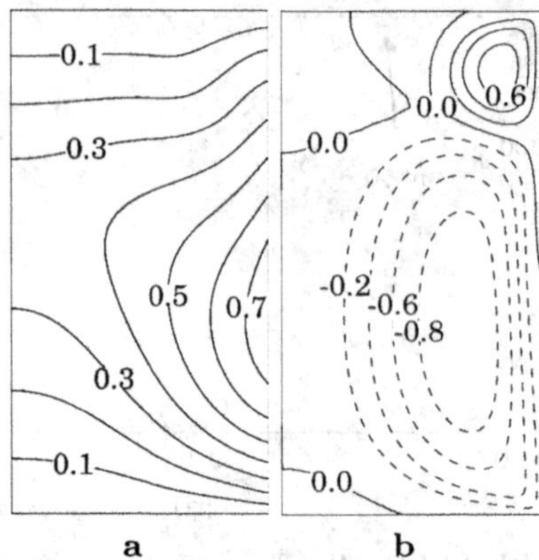

Fig. 10. Temperature **a** and streamfunction **b** isolines of the solution obtained with the passive regularizing function, $n = 6$, $Pr = 10^{-2}$, $Ma = 600$

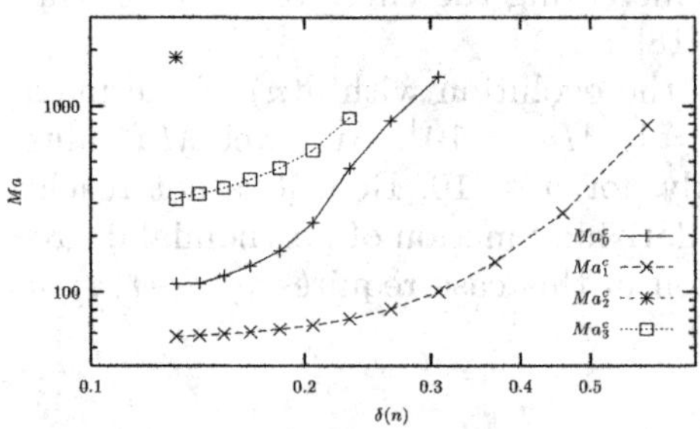

Fig. 11. Evolution of the Ma critical values with the filtering length $\delta(n)$, $Pr = 10^{-2}$, extracted from [9]

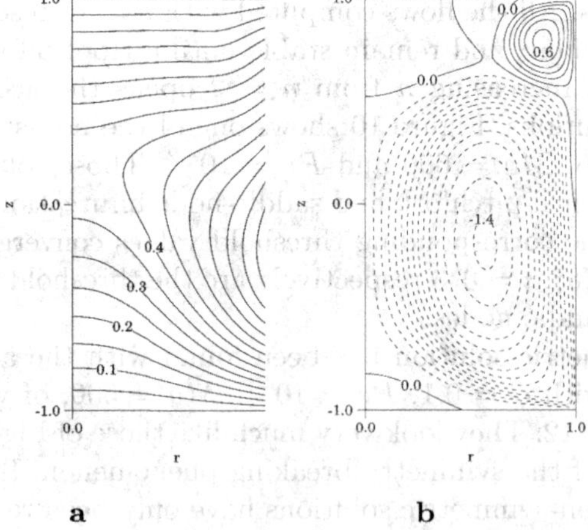

Fig. 12. Temperature **a** and streamfunction **b** isolines of the solution obtained with the active regularizing function, $\alpha = 0.1$, $Pr = 10^{-2}$, $Ma = 500$

5.4 Energetics of the Most Destabilizing Mode

The local growth rate of the kinetic energy of any destabilizing eigenmode can be decomposed as

$$\frac{\partial}{\partial t}\left(\frac{(\delta \mathbf{u})^2}{2}\right) = T_{I\delta v;v,r} + T_{I\delta v;v,z} + T_{I\delta w;w,r} + T_{I\delta w;w,z} - T_D + T_M + E,$$

where each term refers to a particular exchange of energy between the base state $\mathbf{v}$ and the perturbation $\delta\mathbf{v} = \delta u\ \hat{\mathbf{e}}_r + \delta w\ \hat{\mathbf{e}}_z$ velocities. The last term E contains all the contributions of which the volume integral cancels. The other terms are successively

$$T_{I\delta u;u,r} = -(\delta u)^2 \frac{\partial u}{\partial r} \quad , \quad T_{I\delta u;u,z} = -\delta u\,(\delta w\,\frac{\partial u}{\partial z}) \quad ,$$

$$T_{I\delta w;w,r} = -\delta w\,(\delta u\,\frac{\partial w}{\partial r}) \quad , \quad T_{I\delta w;w,z} = -(\delta w)^2 \frac{\partial w}{\partial z} \quad ,$$

$$T_D = Pr(\nabla \times \delta\mathbf{v})^2 \quad , \quad T_M = Pr\ \delta w \frac{\partial(\delta w)}{\partial r}\delta(r-1) \quad ,$$

where $\delta(.)$ is the Dirac function. The global rate reads then

$$\dot{E}_c = \frac{\partial}{\partial t}\left(\int_0^1\!\!\int_{-1}^1 \frac{(\delta\mathbf{v})^2}{2} r dr\, dz\right) = -D + M + I^{\delta u;u,r} + I^{\delta u;u,z} + I^{(\delta w;w,r)} + I^{(\delta w;w,z)}, \quad (23)$$

where all terms are volume integrals, such as $D = \iint T_D\ rdr\,dz$, and so on. All the data quoted in this section have been obtained with the passive regularization, $n = 13$ and $Pr = 10^{-2}$.

The surface of Figs. 13a,b and 14a–d gives the perturbation local power $\frac{\partial}{\partial t}\left(\frac{(\delta\mathbf{v})^2}{2}\right)$ as a function of $r \in [0,1]$ and $z \in [-1,0]$, i.e. in the lower half part

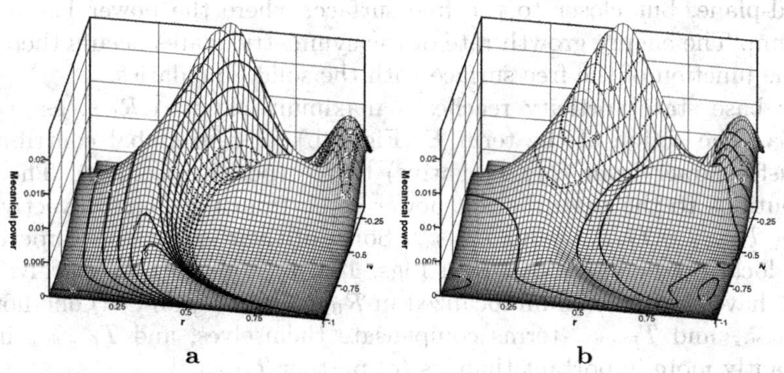

Fig. 13. The perturbation local power $\frac{\partial}{\partial t}\left(\frac{(\delta\mathbf{v})^2}{2}\right)$ of the most destabilizing mode, as function of $r \in [0,1]$ and $z \in [-1,0]$, for $Ma = 110 > Ma_c^0$, and isolines of **a** the steady state vorticity (extracted from [9]), **(b)** E

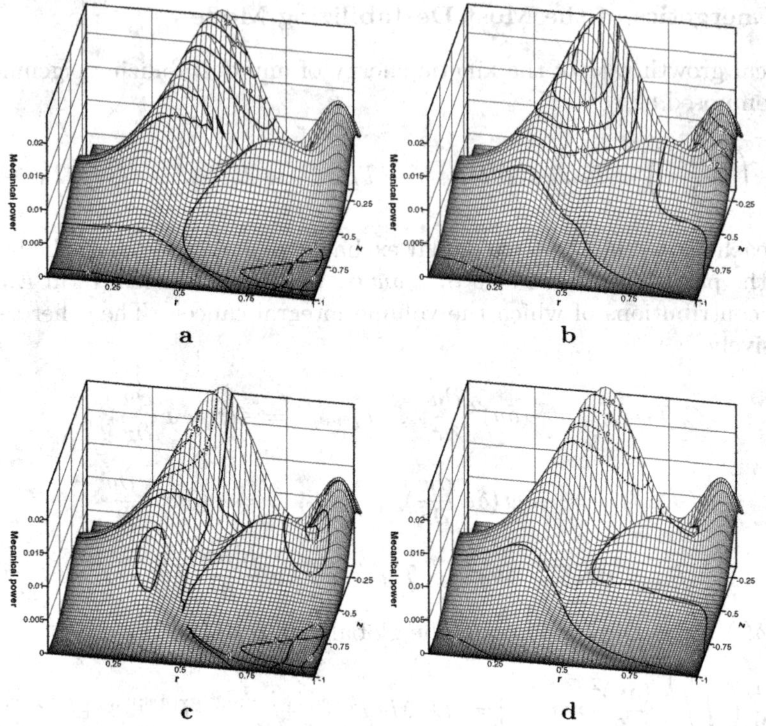

Fig. 14. The perturbation local power $\frac{\partial}{\partial t}\left(\frac{(\delta \mathbf{v})^2}{2}\right)$ of the most destabilizing mode, with some isolines of **a** $T_I^{\delta u;u,_z}$, **b** $T_I^{\delta w;w,_z}$, **c** $T_I^{\delta w;w,_r}$, **d** $T_I^{\delta u;u,_r}$

of the liquid bridge, for $Ma = 110$, slightly above Ma_c^0. Two regions are to be noticed, one (denoted $\mathcal{R}_{0.5}$) where the power has an absolute maximum, near the mid-plane Π, at about $r = 0.5$, and the second (denoted $\mathcal{R}_1$), still about the mid-plane, but closer to the free surface, where the power has a smaller maximum. The energy growth rate of the symmetric states occurs therefore *far* from the junction of the free surface with the solid boundaries.

The base state vorticity reaches a maximum close to $\mathcal{R}_{0.5}$, as shown by Fig. 13a. The action of the term E (Fig 13b), of zero global contribution, is to transfer power from $\mathcal{R}_{0.5}$ (negative) to $\mathcal{R}_1$ (positive isolines). The leading contributions to the perturbation power come from the four convective terms $T_{I\delta u;u,_z}$, $T_{I\delta w;w,_z}$, $T_{I\delta w;w,_r}$ and $T_{I\delta u;u,_r}$. Some of their isolines are superimposed on the local power distribution, in Figs. 14a, b, c and d respectively. All but $T_{I\delta w;w,_r}$ have their maximum localized in $\mathcal{R}_{0.5}$. From (a) and (d) one notes that the $T_{I\delta u;u,_z}$ and $T_{I\delta u;u,_r}$ terms compensate themselves, and $T_{I\delta w;w,_z}$ in (b) is significantly more important than its (c) partner $T_{I\delta w;w,_r}$.

The global quantities are now considered about $Ma_0^c(n = 13)$ as functions of Ma. In Fig. 15 are plotted all terms in (23) normalized by D. As already indicated, the growth of $\dot{E}_c$ with Ma is extremely small. The most contributing

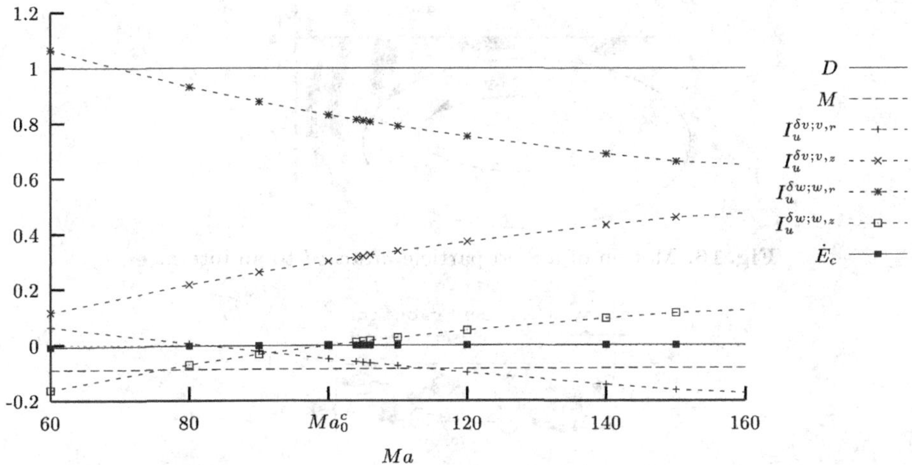

Fig. 15. Power balance of the most destabilizing mode of the symmetric steady flows as function of Ma about Ma_0^c

term, $I^{(\delta w;w,r)}$, decreases with Ma, while the next one, $I^{\delta u;u,z}$ increases. One notes also that the main term of the Figs. 14, namely $T_{I^{\delta w;w,z}}$ in (b), has an integrated contribution of modest importance, increasing however with Ma, from negative to positive supplies to $\dot{E}_c$, respectively before and after the transition.

6 Towards a Regular Model?

The previous results show that the flow regimes and thresholds converge with a vanishing length scale of regularization, explicitely introduced in the capillary condition. Are there physical arguments allowing to derive a physical regular model, instead of those artificial regularization functions? This section proposes an example of such an argument which could be used to add one term to the capillary condition and possibly remove the singularity.

Figure 16 presents an arbitrary flow in the vicinity of an interface between a pure liquid and a gas. Superimposed to this flow, the temporal evolution of a fluid particle *attached to the interface* is drawn. The fluid particle is progressively deformed. Its area on the interface decreases, and tends to zero for infinite times. The question which is here adressed is: what does the corresponding surface energy become? Indeed, the interface has an energy, its surface density being σ, the surface tension. It is recalled that this temperature-dependent quantity is related to the free bonds of atoms at the liquid interface.

From the principle of energy conservation, energy must be transfered to (or pumped from) the liquid volume. This is driven by the surface density of power $-\sigma\, div_s(\mathbf{V_s})$, div_s being the surface divergence operator acting on $\mathbf{V_s}$, the surface (or tangential) velocity of the liquid at the interface. It ensures a global energy

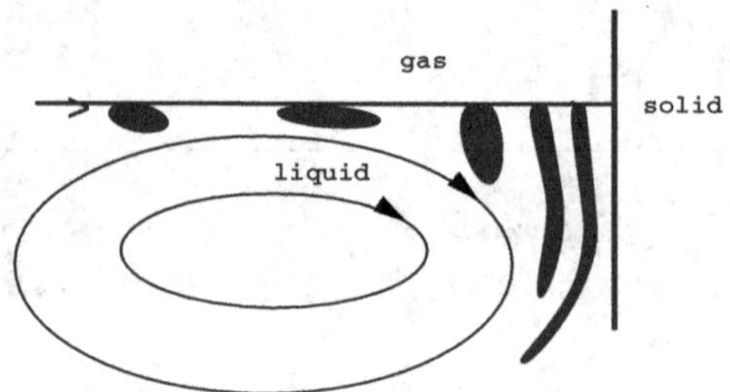

Fig. 16. Motion of a fluid particle *attached* to an interface

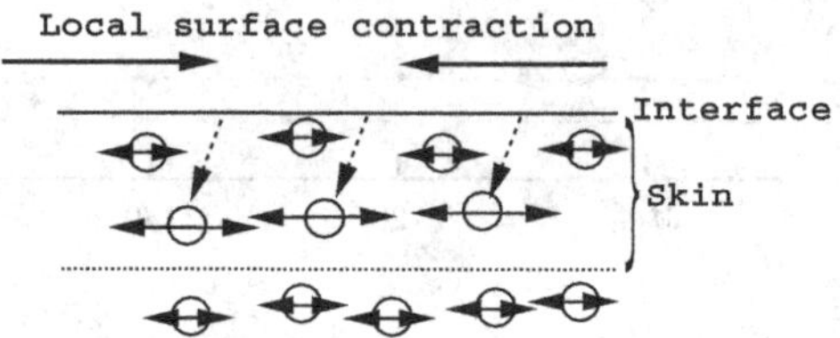

Fig. 17. Skin layer model, molecules leave the interface and transform their interfacial energy into kinetic energy, creating a skin pressure excess

conservation on a bounded steady interface since its surface integral is zero. The exchanges of power density with the fluid must modify the free surface internal energy.

Let us assume that such an alteration leads to an increase of the local pressure δP within an layer of mesoscopic *skin* length scale λ, as sketched in Fig. 17.

There is then a local increase of energy per unit surface $\lambda\delta P$. The time evolution of this quantity can be modelled by the conservation equation

$$\frac{\partial(\lambda\delta P)}{\partial t} = -\sigma div_s(\mathbf{V_s}) - \alpha^{-1}(\lambda\delta P), \tag{24}$$

which takes into account the surface power input and a linear law of dissipation introducing a time scale α. Instantaneous equilibrium leads to

$$\lambda\delta P = -\alpha\sigma div_s(\mathbf{V_s}). \tag{25}$$

The term $\lambda\delta P$ corresponds to repulsive forces: a surface stiffness acting against surface tension. It is straightforwardly inserted in the usual balance equation of tangential and normal surface stress

$$\nabla_{\mathbf{s}}\sigma - \sum_{k=1,2}(\underline{\mathbf{T}}_k.\mathbf{n}_k - (\mathbf{n}_k.\underline{\mathbf{T}}_k\mathbf{n}_k)\mathbf{n}_k) = \nabla_{\mathbf{s}}(-\alpha\sigma div_s(\mathbf{V_s})) \tag{26}$$

$$\sigma div_s(\mathbf{n})\mathbf{n} + \sum_{k=1,2}(-P_k\mathbf{n}_k + (\mathbf{n}_k.\underline{\mathbf{T}}_k.\mathbf{n}_k)\mathbf{n}_k) = (-\alpha\sigma div_s(\mathbf{V_s}))div_s(\mathbf{n})\mathbf{n} \tag{27}$$

In these equations, $k = 1, 2$ refers to the two fluids on either side of the interface, P_k is the pressure and $\underline{\mathbf{T}}_k$ is the anisotropic stress tensor of fluid k. $\mathbf{n}$ is a unit vector normal to the interface, $\mathbf{n}_k$ is directed outside of fluid k and $\nabla_\mathbf{s}$ is the interface gradient operator. The left hand sides of these equations correspond to classical theory, and the right hand sides to the inclusion of the surface stiffness effect.

Equations (26) and (27) show that the present approach leads to equations very similar to those developped in [25], without introducing interface intrinsic viscous property.

7 Conclusion

The usual modelling of thermocapillary convection contains a vorticity singularity along the contact line of the solid boundaries with the free surface. This singularity reveals the existence of a deep difficulty of understanding the underlying multiscale physics. If the numerical experiments can valuably help to improve the knowledge on the hydrodynamics of liquid metal they have to adopt a specific treatment of this singularity in order for their results to be able to interpret and predict the experimental observations. It has been shown that the stability thresholds of the steady axisymmetric flows are very sensitive to the small length scale size introduced as a new parameter by the regularization of the singularity. The base state which enjoys the property of mirror symmetry with respect to the mid-plane of the liquid bridge looses its stability, at a low value of the Marangoni number, to the benefit of one of the two nonsymmetric states that arise from a sequence of undercritical pitchfork and saddle-node bifurcations. To what extent is the physics correctly captured by an adhoc regularization? Is there a simple physical argument missing in the usual modelling of the capillary boundary conditions that could easily remove the singularity? The last section has presented an argument of this type which leads to add one more term in the boundary condition. Is it the missing term? Will it remove the singularity?

References

1. C.J. Adkins: *An introduction to thermal physics*, (Cambridge University Press 1987)
2. D.M. Anderson, G.B. Mc Fadden, A.A. Wheeler: Ann. Rev. Fluid Mech. **30**, 139 (1998)
3. W.E. Arnoldi: Q. Appl. Math. **9**, 17 (1951)
4. A. Batoul, H. Khallouf, G. Labrosse: C. R. Acad. Sci. (Paris) II **319**, 1455 (1994)
5. D. Bedeaux, A.M. Albano, P. Mazur: Physica A **82**, 438 (1976)
6. T.D. Blake, M. Brake, Y.D. Shikhmurzaev: Phys. Fluids **11**, 1995 (1999)
7. C. Canuto, M.Y. Hussaini, A. Quarteroni, T.A. Zang: *Spectral Methods in Fluid Dynamics*, (Springer Series in Computational Physics 1988).
8. E. Chénier, C. Delcarte, G. Labrosse: Eur. Phys. J. AP **2**, 93 (1998)
9. E. Chénier, C. Delcarte, G. Kasperski, G. Labrosse: Phys. Fluids **14**, 3109 (2002)

10. E.B. Dussan V.: AIChE J. **23**, 131 (1977)
11. E.B. Dussan V.: Ann. Rev. Fluid Mech. **11**, 371 (1979)
12. E.B. Dussan, V.E. Ramé, S. Garoff: J. Fluid Mech. **230**, 97 (1991)
13. P.G. de Gennes: Rev. Modern Phys. **57**, 827 (1985)
14. J. Goldhirsch, S. A. Orszag, B. K. Maulik: J. Sci. Comput. **2**, 33 (1987)
15. H.B. Hansen, M. A. Kelmanson: Computers Fluids **23**, 225 (1994)
16. D. Jacqmin: J. Fluid Mech. **402**, 57 (2000)
17. G. Kasperski, A. Batoul, G. Labrosse: Phys. Fluids **12**, 103 (2000)
18. G. Kasperski, G. Labrosse: Phys. Fluids **12**, 2695 (2000)
19. J. Koplik, J.R. Banavar, J.F. Willemsen: Phys. Rev. Letters **60**, 1282 (1988)
20. G. Labrosse, E. Tric, H. Khallouf, M. Betrouni: Num. Heat Transfert B **31**, 261 (1997)
21. L. Landau, E. Lifchitz: *Physique Théorique: Physique Statistique*, 4th ed. (Mir-Ellipse, 1994)
22. C.K. Mamun, L.S. Tuckerman: Phys. Fluids A **7**, 80 (1995)
23. H.K. Moffatt: J. Fluid Mech. **18**, 1 (1964)
24. P. Seppecher: Int. J. Engng. Sci., **34**, 977 (1996)
25. L.E. Scriven: Chemical Engineering Science **12**, 98 (1960)
26. Y.D. Shikhmurzaev: J. Fluid Mech. **334**, 211 (1997)
27. G.I. Taylor: *Aeronautics and Astronautics, p.12*, (Hoff and Vincenti, Pergamon Press 1960)
28. P.A. Thompson, M.O. Robbins: Phys. Rev. Letters **63**, 766(1989)

Unsteady Thermocapillary Flow and Free Surface Deformation in a Thin Liquid Layer

R. Balasubramaniam

National Center for Microgravity Research on Fluids and Combustion, Mail Stop 110-3, NASA Glenn Research Center, Cleveland, OH 44135, USA

Abstract. The thermocapillary flow in a thin liquid layer is analyzed when the free surface is permitted to deform. The thin liquid layer is differentially heated at the end walls. The rigid boundary at the bottom and the free surface are assumed to be thermally insulated. The liquid layer is assumed to be infinitely long in the third dimension. The analysis is performed both under normal gravity, where the hydrostatic force controls the deformation of the free surface, and in reduced gravity, where surface tension controls the deformation. Buoyant convection is assumed to be negligible in the liquid layer. A perturbation solution is constructed, with the height to length ratio (aspect ratio) of the liquid as the small parameter. The leading order solution at steady state corresponds to a one dimensional temperature distribution along the layer, a parabolic velocity field across the layer, with a vanishing flow rate, in a region away from the end walls, and a deformed free surface. The free surface deformation is such that mass accumulates near the cold end wall and is depleted near the hot end wall. The evolution of the leading order velocity and temperature fields, and the free surface deformation is considered. During the transient period when the shape of the free surface is evolving, the net flow rate across the liquid layer, as well as the effect of convective heat transfer, does not vanish. Convection tends to weaken the temperature gradient over most of the free surface, and thereby diminish the thermocapillary stress that drives the flow. One wonders whether the transient problem, in which convective heat transfer and free surface deformation are coupled, admits a solution that is oscillatory in time. To analyze this possibility, a linear bifurcation analysis of the steady state solution has been performed. A perturbation solution of the steady state fields has been utilized in the bifurcation analysis. The results show that neutrally stable disturbances that have an oscillatory behavior in time exist for various values of the system parameters.

1 Introduction

The flow generated by thermocapillarity, and its transition to oscillatory states, has been investigated theoretically and experimentally by many investigators. Such flows have been studied in liquid bridges, thin liquid layers and liquid pools that are not shallow (see reviews by Schwabe [6], Schatz and Neitzel [5]). Smith and Davis [8] analyzed the stability of thermocapillary flow in a thin liquid layer and identified an instability termed a "hydrothermal wave" that occurs at large Marangoni numbers due to the coupling of the velocity and temperature fields. In this mode, the disturbances propagate as traveling waves in directions that are parallel or perpendicular to the direction of the base flow, depending

R. Balasubramaniam, Unsteady Thermocapillary Flow and Free Surface Deformation in a Thin Liquid Layer, Lect. Notes Phys. **628**, 201–212 (2003)
http://www.springerlink.com/

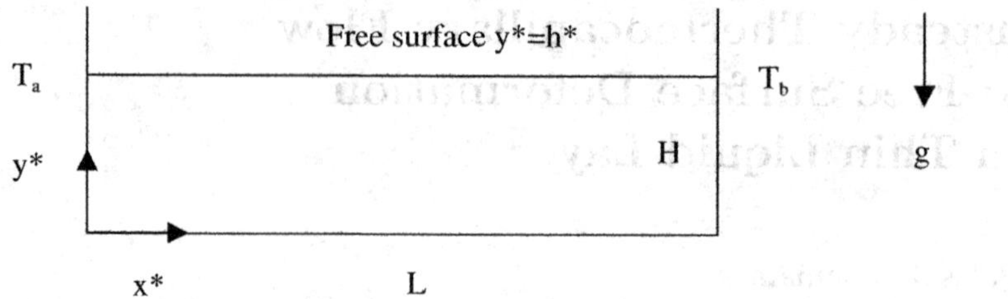

Fig. 1. Schematic of the thin liquid layer

on the Prandtl number. The deformation of the free surface is not important for the occurrence of hydrothermal waves, and the instability is predicted to occur in the absence of the deformation of the free surface.

The study by Smith and Davis [8] is briefly reviewed below. Consider the thermocapillary flow in a rectangular layer that is generated by differentially heated end walls, as depicted in Fig. 1. The ensuing temperature distribution on the free surface will be non-uniform, and the resulting thermocapillary stress generates a recirculating flow in the layer. The flow near the free surface is from the warm to the cold end wall, and the return flow near the bottom boundary is in the opposite direction. Smith and Davis assume that the free surface is located at $y^* = H$ and is undeformed by the flow. Let $x = \frac{x^*}{L}$, $y = \frac{y^*}{H}$. In the limit that the aspect ratio (*i.e.*, the depth to length ratio) of the layer tends to zero, Smith and Davis obtain a basic state in the layer in which

(i) the flow is a parallel flow in which the only non-zero component of the velocity is in the x direction, and is only a function of the y coordinate

(ii) the hydrodynamic pressure increases linearly with x

(iii) the temperature field is a superposition of a field linearly decreasing with x that is imposed by a similar field in the gas present above the free surface, and a flow induced field that is obtained by a balance between vertical conduction and horizontal convection. The flow induced field depends only on y and is such that the bottom boundary is cooler than the free surface. The flow induced field also has a vanishing heat flux at the bottom boundary and the free surface. The magnitude of the flow induced temperature field is proportional to the Marangoni number $M = \frac{(-\sigma_T)GH^2}{\mu\kappa}$ where σ_T is the temperature coefficient of surface tension, that is assumed to be a negative constant, G is the temperature gradient imposed on the free surface, H is the depth of the liquid and μ and κ are the viscosity and thermal diffusivity of the liquid. Smith and Davis also consider another basic state in which the liquid flows as in a plane Couette flow. We do not discuss this basic state and its stability.

The dynamics of the liquid layer is controlled by three dimensionless parameters, *viz.*, the Marangoni number, the Reynolds number (that equals M divided by the Prandtl number Pr) and a Biot number B that determines the heat transfer between the liquid and the gas. Of these, only the Marangoni number influences the basic state that has been described above. Smith and Davis investigate the linear stability of this basic state subject to infinitesimal disturbances expressed in normal modes. These are assumed to be traveling waves in the x and z directions, with an amplitude that depends on the y coordinate. This results in a differential eigenvalue problem for the growth rate of the disturbances, which is solved numerically for given values of the wavenumber in the x and z directions and the Marangoni, Prandtl and Biot numbers. The numerical calculations show that for Biot number equal to zero, the largest critical Marangoni number for neutral stability (*i.e,* vanishing real part of the growth rate) is $M = 398.5$, and occurs for $Pr \to \infty$. The imaginary part of the growth rate is non-vanishing. Therefore, at neutral stability, the basic state is unstable to disturbances that are oscillatory in time. Smith and Davis use the term "hydrothermal waves" for the unstable modes. The wave propagates in a direction close to the $\pm z$ direction for $Pr \to 0$, and propagates in directions close to the $(-x)$ direction for $Pr \to \infty$. The existence of hydrothermal waves in shallow liquid layers has been experimentally confirmed by Riley and Neitzel [4].

On the other hand, Kamotani *et al* [2] state that the Marangoni number alone does not correlate the onset of oscillatory flow in their experiments conducted in liquid layers, with an aspect ratio of unity, in a cylindrical container in microgravity. These authors observed the flow in a silicone oil with a kinematic viscosity of 2 cSt by visualizing the motion of tracer particles. They also observed the temperature distribution at the free surface by imaging it with an infrared camera.The experiments were performed in three different cylindrical containers, with varying diameters. The liquid depth was such that the aspect ratio was maintained to be unity in all the containers. The liquid was heated by a cylindrical resistance heater that was placed along the axis. The outer wall of the container was cooled and maintained at a constant temperature. Kamotani *et al* observed oscillatory flow when the temperature difference between the heater and the container wall exceeded a critical value ΔT_{cr}. However, they find that the critical Marangoni number $Ma_{cr} = \frac{(-\sigma_T)\Delta T_{cr} R}{\mu\kappa}$ (where R is the radius of the container) is dependent on the size of the container, and varies by about a factor of four between $R = 0.6\ cm$ and $R = 1.5\ cm$. They conclude that the Marangoni number alone does not characterize the onset of oscillatory flow, and that some other factor must be included. The authors state that the flow induced deformation of the free surface is the additional parameter. The deformation principally has two effects: (i) in a transient situation, there is a net transport of matter from the hot end wall to the cold end wall, and consequently the recirculation in the layer is not established until the transients decay (ii) the removal of matter near the hot end wall brings fresh fluid to the free surface whose temperature is different, and alters the surface temperature distribution. Thus the surface velocity, surface temperature variation and the deformation of the free surface are coupled. The authors derive a parameter that is a ratio of the estimated deformation of the free surface to an estimate of the thickness of the thermal boundary layer along the hot boundary, and conjecture that when this parameter is sufficiently large, the driving force for the flow (which is the free surface temperature variation) can be altered significantly, leading to oscillations. Kamotani *et al* find that this parameter correlates the onset conditions well for their space as well as terrestrial experiments. Therefore it appears that the free surface deformation is crucial in the transition to oscillatory flow states in their experiments.

The goal of the present study is to analyze the unsteady flow induced by thermocapillarity including the effect of the deformation of the free surface. To render the mathematical problem tractable, the analysis is performed for a thin liquid layer. The steady solution for the velocity and temperature fields in the layer, and the shape of the free surface are obtained first. To examine the possibility of other solutions, such as a periodic solution when these fields evolve from initial conditions, a bifurcation analysis of the steady state with respect to one-dimensional disturbances is performed.

2 Problem Formulation

Consider the thermocapillary flow in a thin liquid layer that is generated by differentially heated end walls (see Fig. 1). The liquid layer, with a free surface, is contained in a rectangular box and is of depth H, and length L. The container is assumed to be sufficiently long in the third dimension so that the flow in the layer is two-dimensional. We assume that $\epsilon = \frac{H}{L} \ll 1$. When the contact angle between the liquid and the material of the end wall is 90°, the equilibrium shape of the free surface is flat. We denote x and y to be the coordinates in the liquid parallel and perpendicular to the free surface, scaled by H and L respectively. The equilibrium free surface is located at $y = 1$. The left wall, located at $x = 0$, is maintained at a temperature T_a, and the right wall, located at $x = 1$, is maintained at a temperature T_b. We assume that $T_a > T_b$. Let $\Delta T = T_a - T_b$. The temperature coefficient of the surface tension $\frac{d\sigma}{dT}$ is denoted by σ_T and is assumed to be a negative constant. A reference velocity can be obtained by balancing the thermocapillary stress with a typical viscous stress at the free surface. This velocity scale is $U_R = \frac{(-\sigma_T)\Delta T\epsilon}{\mu}$, where μ is the viscosity of the liquid. The scaled temperature is defined as $T = \frac{T^* - T_b}{\Delta T}$. The scaled velocities in the x and y directions are denoted by u and v, and are normalized by U_R and ϵU_R respectively. The scaled hydrodynamic pressure p and the scaled time t are obtained by using the reference quantities $\mu U_R/(\epsilon H)$ and L^2/κ respectively, where κ is the thermal diffusivity of the liquid.

We assume that the flow in the liquid is incompressible and two-dimensional. All the physical properties with the exception of the surface tension are assumed constant. A formulation similar to that given below has also been developed by Smith and Vrane [9]. The governing equations are

$$\frac{\partial u}{\partial x} + \frac{\partial y}{\partial y} = 0 \tag{1}$$

$$\frac{\epsilon^2}{Pr}\frac{\partial u}{\partial t} + \epsilon^2 Re\left[u\frac{\partial u}{\partial x} + v\frac{\partial u}{\partial y}\right] = -\frac{\partial p}{\partial x} + \epsilon^2\frac{\partial^2 u}{\partial x^2} + \frac{\partial^2 u}{\partial y^2} \tag{2}$$

$$\frac{\epsilon^2}{Pr}\frac{\partial v}{\partial t}+\epsilon^2 Re\left[u\frac{\partial v}{\partial x}+v\frac{\partial v}{\partial y}\right]=-\frac{1}{\epsilon^2}\frac{\partial p}{\partial y}+\epsilon^2\frac{\partial^2 v}{\partial x^2}+\frac{\partial^2 v}{\partial y^2} \tag{3}$$

$$\frac{\partial T}{\partial t}+Ma\left[u\frac{\partial T}{\partial x}+v\frac{\partial T}{\partial y}\right]=\frac{\partial^2 T}{\partial x^2}+\frac{1}{\epsilon^2}\frac{\partial^2 T}{\partial y^2} \tag{4}$$

In these equations, $Re = \frac{U_R L}{\nu} = \frac{(-\sigma_T)\Delta T H}{\mu\nu}$ is the Reynolds number, $Ma = \frac{U_R L}{\kappa} = \frac{(-\sigma_T)\Delta T H}{\mu\kappa}$ is the Marangoni number, and $Pr = \frac{\nu}{\kappa}$ is the Prandtl number. If $\Delta T/L$ is used as the magnitude of the temperature gradient at the free surface, then the Marangoni number defined above is related to the Marangoni number M defined by Smith and Davis [8] by the relation $Ma = \frac{M}{\epsilon}$.

We assume the following boundary conditions. The liquid velocity vanishes at the rigid boundaries due to the no-slip and no penetration conditions.

$$u(x,0,t)=v(x,0,t)=0;\ u(0,y,t)=v(0,y,t)=0;\ u(1,y,t)=v(1,y,t)=0 \tag{5}$$

Let the free surface be represented as $y = h(x,t)$. We assume that $h, \frac{\partial h}{\partial x} \sim O(1)$ in ϵ. The physical slope of the free surface is $O(\epsilon)$, and at leading order the unit vectors normal and tangential to the free surface, $\hat{\eta}$ and $\hat{\xi}$ can be approximated to point in the y and x directions respectively. The kinematic, shear stress and normal stress balance at the free surface may be written as

$$\frac{1}{Ma}\frac{\partial h}{\partial t}(x,t)=v(x,h,t)-u(x,h,t)\frac{\partial h}{\partial x}=-\frac{\partial}{\partial x}\int_0^h u(x,y,t)dy \tag{6}$$

$$\frac{\partial u}{\partial \eta}(x,h,t)=-\frac{\partial T}{\partial x}(x,h,t) \tag{7}$$

$$Bo_s h(x,t)-Ca\, p(x,h,t)+2\epsilon^2 Ca\frac{\partial v}{\partial \eta}(x,h,t)=2\epsilon^2(1+Ca\, T(x,h,t))H_c \tag{8}$$

where H_c is the mean curvature of the free surface, $Bo_s = \frac{(\rho_l-\rho_g)gH^2}{\sigma_0}$ is the static Bond number, and $Ca = \frac{(-\sigma_T)\Delta T}{\sigma_0}$ is the Capillary number. In these expressions, ρ_l is the density of the liquid, ρ_g is the density of the ambient gas, g is the acceleration due to gravity, which points in the negative y direction, and σ_0 is a reference value for the surface tension.

The boundary conditions for the temperature field are that the end wall temperatures are prescribed, and the rigid boundary at $y = 0$ and the free surface are adiabatic.

$$T(0,y,t)=1;\ T(1,y,t)=0;\ \frac{\partial T}{\partial y}(x,0,t)=0;\ \frac{\partial T}{\partial \eta}(x,h,t)=0 \tag{9}$$

The contact angle between the liquid and the end wall will be assumed to be 90° where required. The total volume of liquid in the container is constrained to be equal to the volume initially present. The initial conditions for the velocity and temperature fields are assumed to be that at $t = 0$, the liquid layer is isothermal and quiescent.

For $\epsilon \ll 1$ we write the following asymptotic expansions.

$$u = u_0(x, y, t) + o(1), \qquad v = v_0(x, y, t) + \ldots, \qquad p = p_0(x, y, t) + \ldots,$$

$$T = T_0(x, t) + \ldots, h = h_0(x, t) + \ldots \tag{10}$$

Note that the leading order temperature field is one-dimensional, as the energy flow in the layer is essentially one-dimensional. Assuming that $Re \sim O(1)$, $Ma \sim O(1)$, the governing equations at leading order are

$$\frac{\partial p_0}{\partial x} = \frac{\partial^2 u_0}{\partial y^2} \tag{11}$$

$$\frac{\partial p_0}{\partial y} = 0 \tag{12}$$

$$\frac{1}{Ma}\frac{\partial h_0}{\partial t} = -\frac{\partial}{\partial x}\int_0^{h_0} u_0 dy \tag{13}$$

The energy equation at leading order can be obtained by a control volume approach to be

$$h_0\frac{\partial T_0}{\partial t} + Ma\frac{\partial T_0}{\partial x}\left(\int_0^{h_0} u_0 dy\right) = \frac{\partial}{\partial x}\left(h_0\frac{\partial T_0}{\partial x}\right) \tag{14}$$

Thus, at leading order, the velocity field in the liquid layer is essentially steady, and the hydrodynamic pressure is independent of vertical position in the layer.

At this order, the contribution of the viscous stress to the normal stress balance (8) can be neglected. When $Ca \ll 1$, the normal stress balance (8) can be simplified to

$$p_0(x, t) = Bo_e h_0(x, t) - \frac{\epsilon^2}{Ca}\frac{\partial^2 h_0}{\partial x^2} = Bo_e\left[h_0(x, t) - l\frac{\partial^2 h_0}{\partial x^2}\right] \tag{15}$$

where $Bo_e = \frac{Bo_s}{Ca} = \frac{\Delta\rho g H^2}{(-\sigma_T)\Delta T}$ and $l = \frac{\epsilon^2}{Bo_s} = \frac{\epsilon^2}{Bo_e Ca} = \frac{\sigma_0}{\Delta\rho g L^2}$. In the perturbation analysis, Bo_e and l (and hence $\frac{\epsilon^2}{Ca}$) are treated as constants that are independent of ϵ. In the limit $l \to 0$, the deformation of the free surface is controlled by gravity, and in the limit $l \to \infty$, surface tension controls the free surface deformation.

Using (15) in (11) and application of the boundary conditions yields the following solution for u_0.

$$u_0 = Bo_e\left(\frac{\partial h_0}{\partial x} - l\frac{\partial^3 h_0}{\partial x^3}\right)\left(\frac{y^2}{2} - yh_0\right) - y\frac{\partial T_0}{\partial x} \tag{16}$$

Thus the volumetric flow rate in the liquid layer at any x location is

$$\int_0^{h_0} u_0 dy = -\frac{Bo_e}{3} h_0^3 \left(\frac{\partial h_0}{\partial x} - l \frac{\partial^3 h_0}{\partial x^3} \right) - \frac{h_0^2}{2} \frac{\partial T_0}{\partial x} \tag{17}$$

Equation (17) may now be used in (13) and (14) to pose a coupled problem for the evolution of the temperature field and the deformation of the free surface. It should be mentioned that in the leading order analysis for the velocity field, the ability to impose boundary conditions at the end walls has been lost. However, we will do so indirectly in posing the problem for the surface deformation by requiring the volumetric flow rate (given in (17)) to vanish at the end walls.

The coupled problem at leading order for the temperature field and deformation of the free surface is

$$\frac{1}{Ma} \frac{\partial h_0}{\partial t} = \frac{\partial}{\partial x} \left[\frac{Bo_e}{3} h_0^3 \left(\frac{\partial h_0}{\partial x} - l \frac{\partial^3 h_0}{\partial x^3} \right) + \frac{1}{2} h_0^2 \frac{\partial T_0}{\partial x} \right] \tag{18}$$

$$h_0 \frac{\partial T_0}{\partial t} - Ma \frac{\partial T_0}{\partial x} \left\{ \left(\frac{Bo_e}{3} h_0^3 \left[\frac{\partial h_0}{\partial x} - l \frac{\partial^3 h_0}{\partial x^3} \right] + \frac{1}{2} h_0^2 \frac{\partial T_0}{\partial x} \right) \right\} = \frac{\partial}{\partial x} \left(h_0 \frac{\partial T_0}{\partial x} \right) \tag{19}$$

The boundary and initial conditions are that the end wall temperatures are specified; the contact angle between the liquid and each end wall is 90°; the volumetric flow rate vanishes at the end walls; the total volume of the liquid is constrained, and the liquid layer is isothermal and the free surface is initially undeformed.

$$T_0(0,t) = 1; \qquad T_0(1,t) = 0 \tag{20}$$

$$\frac{\partial h_0}{\partial x}(0,t) - \frac{\partial h_0}{\partial x}(1,t) - 0 \tag{21}$$

$$h_0 \frac{\partial^3 h_0}{\partial x^3} = \frac{3}{2 l Bo_e} \frac{\partial T_0}{\partial x} \qquad \text{at} \quad x = 0, 1 \tag{22}$$

$$\int_0^1 h_0 dx = 1 \tag{23}$$

$$T_0(x,0) = 0; \qquad h_0(x,0) = 1 \tag{24}$$

3 Steady State Solution

At steady state, the volumetric flow rate across any x location in the liquid layer must vanish. The velocity field can then be obtained to be

$$u_{0_{ss}}(y) = -\frac{dT_{0_{ss}}}{dx} \left(\frac{3}{4} \frac{1}{h_{0_{ss}}} y^2 - \frac{1}{2} y \right) \tag{25}$$

The steady state solution to the leading order temperature field and the shape of the free surface are governed by the following equations

$$T_{0_{ss}}(x) = \frac{1}{K}\int_x^1 \frac{dz}{h_{0_{ss}}(z)} \tag{26}$$

$$Bo_e\left[\frac{\partial h_0}{\partial x} - l\frac{\partial^3 h_0}{\partial x^3}\right] - \frac{3}{2}\frac{1}{K}\frac{1}{h_{0_{ss}}^2} = 0 \tag{27}$$

$$h'_{0_{ss}}(0) = h'_{0_{ss}}(1) = 0; \qquad \int_0^1 h_{0_{ss}}\, dx = 1 \tag{28}$$

where $K = \int_0^1 \frac{dz}{h_{0_{ss}}(z)}$ is an unknown constant. For a given value of Bo_e and l, $h_{0_{ss}}$ can be determined numerically. Assuming the deformation of the free surface to be small, a perturbation solution can be obtained to be

$$h_{0_{ss}}(x) = 1 + \frac{1}{Bo_e}\left[\frac{3}{2}x - \frac{3}{4} + \frac{3}{2}\frac{\sqrt{l}}{(1+e^{1/\sqrt{l}})}\left(e^{\frac{1-x}{\sqrt{l}}} - e^{\frac{x}{\sqrt{l}}}\right)\right] \tag{29}$$

From (29), it can be shown that the free surface deformation is proportional to $\frac{1}{Bo_e}$ for small values of l and is proportional to $\frac{1}{lBo_e}$ (which equals $\frac{Ca}{\epsilon^2}$) for large l.

Thus at steady state, the thermocapillary flow that is generated in the layer is a recirculating, nearly parallel flow in the region away from the end walls. The shape of the free surface is such that material is depleted near the end wall that is warm and accumulates near the cold wall. For any value of Ma, there is no effect of convection on the fields at steady state.

4 Bifurcation Analysis

That convection of energy has no influence on the fields at steady state for any value of the Marangoni number raises a question as to whether other solutions might exist that are different from the steady state solution determined above, or are time periodic, or lead to break up of the liquid layer. Experimental evidence (Schwabe *et al* [7]) shows that the first two possibilities do occur. The stability analysis performed by Smith and Davis [8] shows that time periodic solutions are possible for certain ranges of the parameters, even when the free surface is assumed not to deform.

We will address the question of the existence of other solutions to (18) to (24) by a bifurcation analysis of the steady state solution. The analysis is performed in a limited solution space. Rather than assume general three dimensional disturbances to the governing equations, as is typically done in stability analyses, we will investigate the existence of time periodic solutions to (18) to (23), when they are linearized about the steady state. The goal is to determine whether deformation of the free surface can lead to such time periodic solutions.

Such an analysis is motivated by the following heuristic argument. At steady state, there is no net volumetric flow rate across the depth of the liquid. The solution at steady state also shows that liquid is depleted in the vicinity of the warm wall and accumulates near the cool wall. Therefore, during the transient period while the flow was being established, the volumetric flow rate in the liquid layer must have been non-vanishing. To understand its consequences, let us assume that a disturbance is imposed to the liquid layer at steady state such that the net volumetric flow rate is from the warm wall to the cool wall. The net flow depletes material near the hot end wall, and this material accumulates near the cold end wall, since the boundaries are impenetrable. Therefore, the free surface is deformed. To maintain the same rate of conductive heat transfer as that at steady state, the temperature gradient must increase near the hot end wall where the material depletion reduces the area of cross-section, and decrease near the cold end wall. Now consider the effect of convection of energy. The net flow of liquid from the warm end wall to the cool end wall tends to homogenize the temperature along streamlines and hence weakens the temperature gradient over most of the free surface, compared to what prevails at steady state. Therefore, the driving force for the flow is altered by the disturbance. It is enhanced near the hot end wall where the material is depleted, and is reduced over much of the free surface, due to convective effects and material accumulation. The curvature gradient that results from the deformation of the free surface, together with surface tension, establishes a pressure gradient in the liquid layer that opposes the net volumetric flow caused by the disturbance and tends to generate a return flow. The variation in the hydrostatic head also makes an additional contribution to the pressure gradient. If the temperature gradient over most of the the free surface is sufficiently weakened, the opposing pressure gradient can establish a recirculation in the liquid layer overturning the forward motion, and a periodic, sloshing motion of the liquid and the deformation of the free surface can be envisioned.

Let

$$T_0 = T_{0_{ss}} + P(x)e^{\beta t}; \qquad h_0 = h_{0_{ss}} + Q(x)e^{\beta t} \tag{30}$$

Substituting (30) into (18) to (23) yields the following linearized equations

$$\frac{1}{Ma}\beta Q = \frac{d}{dx}\left[\frac{1}{2K}Q + \frac{1}{2}h_{0_{ss}}^2 P' + \frac{Bo_e}{3}h_{0_{ss}}^3 (Q' - lQ''')\right] \tag{31}$$

$$\beta h_{0_{ss}} P - MaT'_{0_{ss}}\left[\frac{1}{2K}Q + \frac{1}{2}h_{0_{ss}}^2 P' + \frac{Bo_e}{3}h_{0_{ss}}^3 (Q' - lQ''')\right] =$$

$$h_{0_{ss}} P'' + T'_{0_{ss}} Q + h'_{0_{ss}} P' + T'_{0_{ss}} Q' \tag{32}$$

$$P(0) = P(1) = Q'(0) = Q'(1) = \int_0^1 Q dx = 0;$$

$$lBo_e h_{0_{ss}}^3 Q''' - \frac{3}{2} h_{0_{ss}}^2 P' - \frac{3}{2K} Q = 0 \text{ at } x = 0, 1 \qquad (33)$$

For given values of Ma, Bo_e and l, (31)–(33) constitute a differential eigenvalue problem for β. The steady state solution must be determined from (26) and (27). In the results presented below, we use the perturbation solution for $h_{0_{ss}}$ given in (29). $T_{0_{ss}}$ is determined by substituting (29) in (26).

4.1 Numerical Results

The eigenvalue problem posed in (31)–(33) has been solved by a spectral-collocation method using Chebyshev polynomials (Orzag [3], Johnson [1]). An $N+1$ term approximation for $P(z)$ and $Q(z)$, where $z = 2x - 1$, was constructed by writing an expansion for P and Q in terms of Chebyshev polynomials $T_n(z)$, $n = 0, N$. Typically, a value $N = 25$ was used. A linear system of $2N+2$ equations was formed by satisfying (31) at $N-4$ collocation points $z_j = cos(\frac{\pi j}{N-5})$, $j, 0, N-5$, and (32) at $N-1$ collocation points $z_j = cos(\frac{\pi j}{N-2})$, $j, 0, N-2$, together with the boundary conditions given in (33). The generalized matrix eigenvalue problem was then solved using the solvers in Mathematica.

Figure 2a shows the neutrally stable curves (for which the real part of β is zero) for a case with $l = 1000$, for which surface tension controls the free surface deformation as can be expected in a reduced gravity environment. The imaginary part of β, which is proportional to the frequency of oscillation of the disturbances, is non-zero when Ma is approximately greater than 90. For Ma smaller than this value, the base state is unstable to a mode for which β_i is zero, and the disturbances can grow without exhibiting oscillations. Figure2b shows the dependence of β_i on Ma along the neutrally stable curve.

Figures 3 and 4 show similar curves for $l = 1$ and $l = 0.025$ respectively. These curves exhibit features that are qualitatively similar to that shown in Fig. 1 for $l = 1000$. The only difference is that for a case with $l = 0.025$, β_i is non-zero when Ma is non-zero.

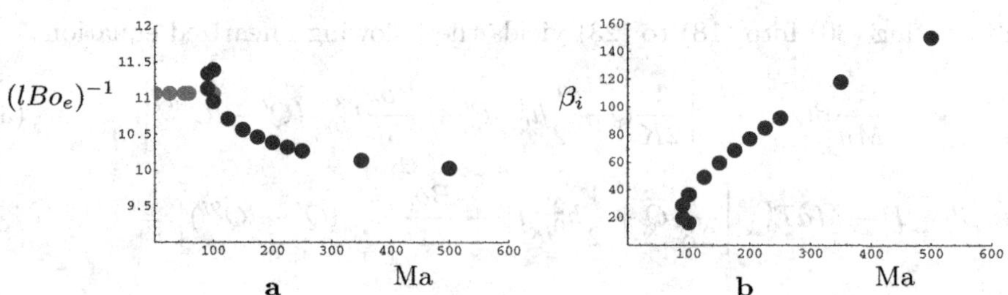

Fig. 2. Neutrally stable curves for $l = 1000$

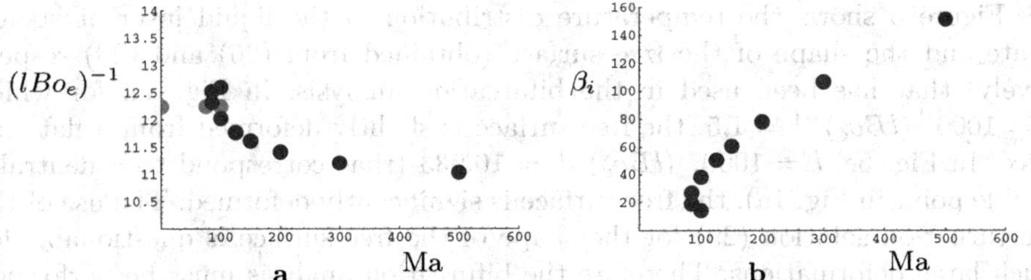

Fig. 3. Neutrally stable curves for $l = 1$

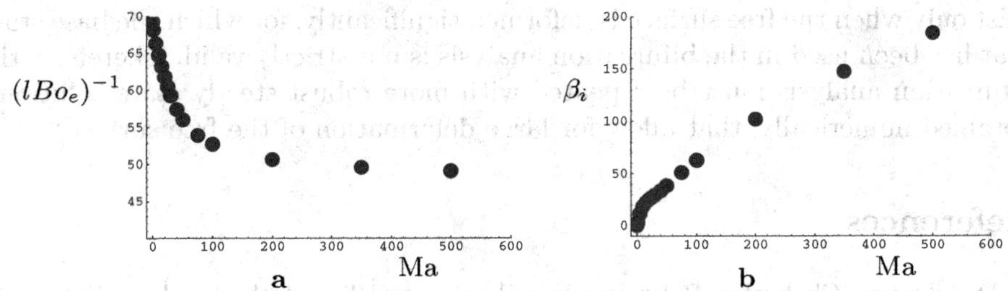

Fig. 4. Neutrally stable curves for $l = 0.025$

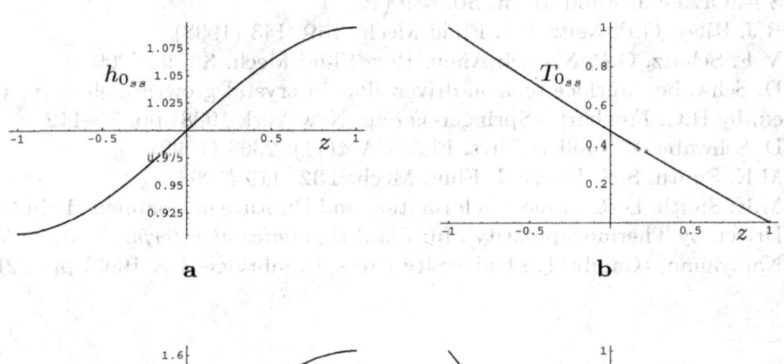

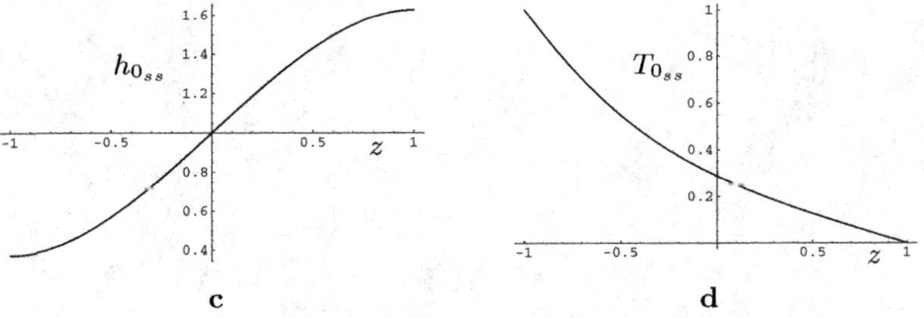

Fig. 5. Shape of the free surface and the temperature distribution in the liquid layer

Figure 5 shows the temperature distribution in the liquid layer at steady state and the shape of the free surface (obtained from (26) and (29) respectively) that has been used in the bifurcation analysis. In Fig. 5a, for which $l = 1000$, $(lBo_e)^{-1} = 1.5$, the free surface is slightly deformed from a flat surface. In Fig. 5c, $l = 1000$, $(lBo_e)^{-1} = 10.033$ (that correspond to a neutrally stable point in Fig. 1a), the free surface is significantly deformed. The use of the perturbation solution (29) for the shape of the free surface is questionable for such large deformations. Therefore the bifurcation analysis must be performed for a base state that is determined numerically from (26) to (28).

Thus, while neutrally stable oscillatory thermocapillary flow states are shown to exist in a thin liquid layer due to the coupling of convection of energy and the deformation of the free surface even in the absence of inertia, these flow states exist only when the free surface is deformed significantly, for which the base state that has been used in the bifurcation analysis is not strictly valid. Therefore, the bifurcation analysis must be repeated with more robust steady state solutions, obtained numerically, that allow for large deformation of the free surface.

References

1. D. Johnson: 'Chebyshev Polynomials in the Spectral Tau Method and Application to Eigenvalue Problems'. In: *NASA Contractor Report 198451*, (NASA Lewis Research Center, Cleveland, OH 2000)
2. Y. Kamotani, S. Ostrach, J. Masud: J. Fluid Mech. **410**, 211 (2000)
3. S.A. Orzag: J. Fluid Mech. **50**, 689 (1971)
4. R.J. Riley, G.P. Neitzel: J. Fluid Mech. **359**, 143 (1998)
5. M.F. Schatz, G.P. Neitzel: Annu. Rev. Fluid Mech. **33**, 93 (2001)
6. D. Schwabe: 'Surface-tension-driven flow in crystal growth melts'. In: *Crystals-11*, ed. by H.C. Freyhart, (Springer-Verlag, New York 1988) pp. 75–112
7. D. Schwabe, U. Möller: Phys. Fluids A **4**(11), 2368 (1992)
8. M.K. Smith, S.H. Davis: J. Fluid Mech. **132**, 119 (1993)
9. M.K. Smith, D.R. Vrane: 'Deformation and Rupture in Confined, Thin Liquid Films Driven by Thermocapillarity'. In: *Fluid Dynamics at Interfaces*, ed. by W. Shyy, R. Narayanan, (Cambridge University Press, Cambridge, UK 1999) pp. 221–233

The Influence of Static and Dynamic Free-Surface Deformations on the Three-Dimensional Thermocapillary Flow in Liquid Bridges

Hendrik C. Kuhlmann and Christian Nienhüser

ZARM – University of Bremen, Am Fallturm, 28359 Bremen, Germany

Abstract. The liquid-phase transport of heat and mass during the float-zone crystal-growth process is strongly affected by the surface-tension-driven flow caused by thermal gradients. In the half-zone model of this process the basic toroidal thermocapillary flow can become unstable to a three-dimensional and time-dependent state if the temperature gradients along the liquid–gas interface are increased. The corresponding two- and three-dimensional flows are analyzed here with emphasis on flow-induced dynamic deformations of the interface. The problem is treated using an asymptotic expansion of all relevant variables in the limit of high surface tension. The magnitude of the flow-induced surface deflection is quantified and the relative importance of different mechanisms is established. It is shown that the deformations are passive during the transition to oscillatory flow in high-Prandtl-number liquids as along as the capillary number is sufficiently small.

1 Introduction

Marangoni effects can play a significant role in the transport of momentum, heat, and mass in a number of physico-chemical systems. The effect occurs at the interface between a liquid and a gas, or between two immiscible liquids. The intermolecular attracting forces on a molecule of the liquid are balanced for molecules far away from the interface. Molecules near the interface, however, experience a net force directed into the liquid. Therefore, the potential energy of the molecules is higher at the interface than in the bulk and the liquid volume has a tendency to minimize its surface area. The layer at the interface within which the potential energy varies has a thickness of a few molecular diameters. From a macroscopic point of view the interface can be considered infinitely thin, and the forces arising in the layer can be considered to act on elements of the two-dimensional interface. The associated energy per surface area is called the surface tension. It can depend on a number of intensive variables. When it depends on the temperature, a variation of the temperature along the interface leads to a potential gradient. This gradient corresponds to a tangential force acting on the molecules at the interface and can drive a significant fluid motion: the thermocapillary flow. This effect is also called the thermal Marangoni effect.

Thermocapillary flows arise in the drying of paint [26], the spreading of liquid films [24], and in welding processes [5], to name only a few areas of application.

H.C. Kuhlmann and C. Nienhüser, The Influence of Static and Dynamic Free-Surface Deformations on the Three-Dimensional Thermocapillary Flow, Lect. Notes Phys. **628**, 213–239 (2003)
http://www.springerlink.com/

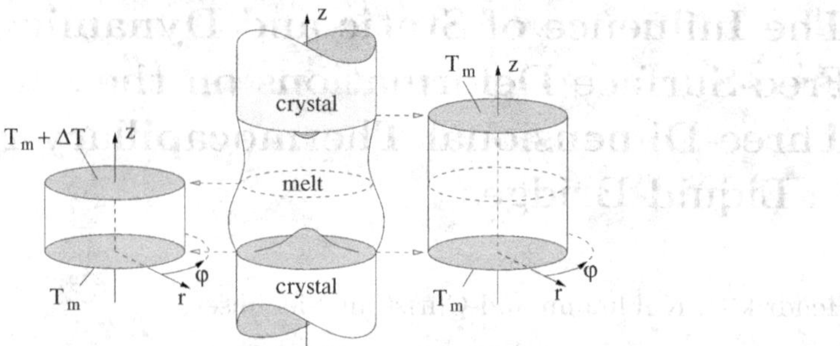

Fig. 1. Models for the float-zone process. The zones on the *left* and on the *right* side are called the half- and the full-zone model, respectively

Here we shall be concerned with the thermocapillary convection in liquid bridges. A liquid bridge consists of a liquid droplet captured between to solid adhesive surfaces. In the classical configuration the rigid surfaces are concentric parallel disks. The maximum length of a liquid bridge is limited and it depends on the magnitude and orientation of the gravity vector [18]. When the two bounding disks have different temperatures the temperature of the liquid must necessarily vary along the interface. This leads to a fluid motion via the thermocapillary effect.

The differentially heated liquid-bridge configuration is of fundamental interest in crystal growth. It is a simple model of the containerless float-zone crystal-growth process by which high-purity single crystals can be grown. In this process a cylindrical rod of a semiconductor material is slowly moved axially through a heated zone such that a liquid bridge is created. This is shown schematically in Fig. 1 (middle figure). Free-surface temperature gradients cannot be avoided and they drive a thermocapillary flow in the melt which transports heat, solutes, and impurities. The melt flow decisively influences the temperature and concentration fields just in front of the crystallizing solid–liquid interface, hence also the rate of crystallization and the chemical composition of the crystal. In particular, oscillatory flows are undesirable, because varying growth conditions cause concentration inhomogeneities in form of striations in the grown crystal. Therefore, different means for a suppression of flow oscillations have been applied like, e.g., crystal rotation, flow control by external heat sources and, most efficiently, by magnetic fields. This latter technique has recently received much attention (see e.g. [19]).

1.1 Thermocapillary Flow in Liquid Bridges

Models of the float-zone process have been considered since the early work of [2]. Since that time significant progress has been made in the numerical flow modeling. Also experimental models using transparent liquids have been investigated. Some of these have been carried out in space because buoyant convection, which is tightly coupled with the thermocapillary flow under gravity, can be eliminated under weightlessness conditions.

The numerical treatment of the float-zone process is complicated by the presence of a free surface and by the solidification boundaries. Therefore, numerical models have been devised which capture the essential features of the thermocapillary flow and which allow for a subsequent incorporation of more sophisticated effects. Figure 1 shows a floating zone (center) and the so-called half- (left) and full-zone model (right).

The full-zone is a liquid bridge heated symmetrically around its midplane. In this model the solidification interfaces are replaced by rigid disks being kept at equal and constant temperatures. Since the advancement of the external heater in the float-zone process is much slower than any other convective time scale, the axial through flow associated with the melting and solidifying interfaces can usually be neglected. The assumption of plane interfaces is a more serious restriction of the model. As long as the temperature difference is sufficiently small, the thermocapillary flow in this model consists of two toroidal vortices, mirror symmetric with respect to the midplane.

The most basic model is obtained, when the mirror-symmetric heating from the ambient is replaced by a rigid heated disk at the midplane of the full zone. Such a configuration is called a half zone. While the half-zone model cannot capture features which are associated with non-mirror-symmetrical flows which appear for higher temperature differences in the full zone, it still embodies the essential thermocapillary vortex flow. The half-zone configuration is much easier to use in model experiments with transparent fluids, because the temperature can be controlled by differentially heating the supporting disks. Also the numerical effort is less for the half zone, because a smaller number of grid points is required. For these reasons, the half zone has become one of the most popular model systems for thermocapillary flows. Before we consider this model in more detail, the conditions driving the thermocapillary flow will be introduced.

1.2 The Interfacial Boundary Conditions

At the free surface where the liquid and the ambient gas meet, a kinematic and a dynamic boundary condition must be satisfied. For an in-depth treatment of the boundary conditions, the reader is referred to [30].

Kinematic Condition

The kinematic boundary condition requires that a fluid element being originally at the free surface must remain at the free surface in the course of the fluid motion. If the vector $\boldsymbol{\xi}$ points at a fluid element on the interface, then the velocity field $\boldsymbol{U}$ on the interface is given by

$$\boldsymbol{U} = \frac{\mathrm{d}\boldsymbol{\xi}}{\mathrm{d}t}. \tag{1}$$

Let us use polar coordinates (r, φ, z) and describe the location of the free surface by the radial coordinate $r = h(z, \varphi, t) = \boldsymbol{\xi} \cdot \boldsymbol{e}_r$ as a function of the axial and azimuthal coordinates z and φ, and the time t. The radial component of (1) then gives us the kinematic boundary condition

$$U = \frac{\mathrm{d}h}{\mathrm{d}t} = \left(\frac{\partial}{\partial t} + \boldsymbol{U}\cdot\nabla\right) h(z,\varphi,t), \tag{2}$$

where U is the radial component of $\boldsymbol{U} = (U, V, W)$.

Dynamic Condition

For the dynamic boundary condition we must consider the force balance on the interface. Let $\boldsymbol{n}$ be the unit normal vector to the interface directed out of the liquid and into the gas. Then the force which the flow imposes on a surface element is $-\mathrm{S}\cdot\boldsymbol{n}$ where $\mathrm{S} = -P\mathrm{I} + \varrho\nu[\nabla\boldsymbol{U} + (\nabla\boldsymbol{U})^{\mathrm{T}}]$ is the stress tensor with I being the identity matrix and P the pressure in the liquid. Continuity of the momentum flux at the interface requires that the force exerted by the liquid on a surface element is exactly balanced by the force which the gas flow is exerting on it. In the absence of any further effects this leads to the dynamic condition $\mathrm{S}\cdot\boldsymbol{n} = \mathrm{S}_{\mathrm{gas}}\cdot\boldsymbol{n}$. If surface tension is associated with the interface, additional surface forces must be taken into account. The force balance then reads

$$\mathrm{S}\cdot\boldsymbol{n} = \mathrm{S}_{\mathrm{gas}}\cdot\boldsymbol{n} - \sigma(\nabla\cdot\boldsymbol{n})\boldsymbol{n} + (\mathrm{I} - \boldsymbol{n}\boldsymbol{n})\cdot\nabla\sigma. \tag{3}$$

The first additional term arises when the interface is not flat. For curved interfaces the well-known Laplace pressure in the liquid $\sigma(\nabla\cdot\boldsymbol{n})$ is acting on the interface. Note that $\nabla\cdot\boldsymbol{n} = R_1^{-1} + R_2^{-1}$ is just the sum of the two inverse radii of curvature R_1 and R_2. With our definition of $\boldsymbol{n}$ the Laplace pressure is positive when the liquid bulges outward. To minimize the surface energy, and hence the surface area, the bulge would have to recede. Therefore, the Laplace force is directed into the liquid, in vectorial form $-\sigma(\nabla\cdot\boldsymbol{n})\boldsymbol{n}$.

When the surface tension $\sigma(T)$, having units of energy per surface area, varies along the interface the tangential gradient of the surface tension represents a force per surface area acting in the direction of increasing surface tension. We can obtain the tangential gradient operator by taking the gradient and subtracting the component which is normal to the interface. The normal part of any vector $\boldsymbol{a}$ is given by $\boldsymbol{n}(\boldsymbol{n}\cdot\boldsymbol{a})$. Accordingly, we have the normal gradient operator $\boldsymbol{n}\boldsymbol{n}\cdot\nabla$ and the tangential gradient operator is $(\mathrm{I} - \boldsymbol{n}\boldsymbol{n})\cdot\nabla$. Therefore, the thermocapillary force per surface element is $(\mathrm{I} - \boldsymbol{n}\boldsymbol{n})\cdot\nabla\sigma$, which explains the last term in (3).

In the following we shall assume that the viscous stress exerted on the interface by the gas can be neglected due to the small dynamic viscosity of gases. If, furthermore, hydrostatic pressure variations in the gaseous atmosphere are negligibly small and if we measure the pressure in the liquid relative to the constant ambient pressure we obtain the simplified dynamic interface condition

$$\left[-P\mathrm{I} + \varrho\nu\left(\nabla\boldsymbol{U} + (\nabla\boldsymbol{U})^{\mathrm{T}}\right)\right]\cdot\boldsymbol{n} = -\sigma(\nabla\cdot\boldsymbol{n})\boldsymbol{n} + (\mathrm{I} - \boldsymbol{n}\boldsymbol{n})\cdot\nabla\sigma. \tag{4}$$

In case the density and viscosity of the ambient medium cannot be neglected the full interface condition (3) must be used.

1.3 The Half Zone

Owing to the adverse effect of flow oscillations on the crystals grown by the float-zone technique, much effort has been devoted to determine the conditions for the onset of oscillations in model systems and to understanding the physical oscillation mechanisms. To circumvent the complications associated with the unknown location of the interface, early models of thermocapillary flow have assumed, as a first approximation, that the interface is cylindrical and non-deformable, but can support tangential stresses. Laboratory experiments have confirmed the validity of this assumption when the bridges are small. For bridges of size 3 mm, e.g., the relative variation of the radius as a function of the height of the bridge is less than 5 % [34]. If the temperature difference is sufficiently small, the thermocapillary as well as the buoyant flow arise in the form of a toroidal vortex in the liquid bridge, provided the aspect ratio (= height/radius) is of the order of one. On an increase of the driving temperature gradient, the flow becomes three dimensional and the rotational symmetry is broken. The way in which this symmetry is broken differs fundamentally for low and for high Prandtl numbers [36]. For Pr $\lesssim$ 0.06, the rotational symmetry is lost to a stationary flow and the vortex becomes deformed in a saddle-like fashion [14,16]. For Pr $\gtrsim$ 1, on the other hand, the symmetry is lost to a time-dependent three-dimensional flow.[1] For these high Prandtl numbers and for a slightly supercritical temperature difference the flow can either be a pure traveling wave for which the non-axisymmetric part of the flow and temperature rotate like solid bodies in the clockwise or in the anti-clockwise direction, or it can be a pure standing wave with nodes and anti-nodes at periodic azimuthal locations. A recent overview on the flow modeling has been given by [9].

The above mentioned fundamental investigations did not take into account the deformability of the liquid–gas interface. However, the effect of the volume of fluid has been considered by [3,4] who assumed that the shape of the liquid is not influenced by the flow and is solely determined by the hydrostatic pressure. More recently, [22] quite systematically studied the influence of various factors, like, e.g, the volume of fluid and the gravity, on the linear stability of the thermocapillary flow in statically determined liquid bridges.

Dynamic flow-induced free-surface deformations in liquid bridges at higher Reynolds numbers have not yet been considered (but see [8,20,28]), because they are small under typical conditions. Nevertheless, some authors have assumed that the dynamic deformability of the liquid–gas interface plays a key role for the onset of oscillatory flows in high-Prandtl-number liquid bridges [6,7,25]. The good agreement, however, between experiments and numerical results obtained with a static interface for moderately large Prandtl numbers (Pr $\lesssim$ 15) indicate that dynamic deformations are only of minor importance regarding the first flow instability. Since experiments are underway [23] to measure the dynamic free-surface oscillations it seems be worthwhile to inquire more accurately into the

[1] The transition boundary for intermediate Prandtl numbers is much higher and has been calculated by [15].

structure and magnitude of the dynamic deformations to enable a comparison between experimental and numerical data.

2 Mathematical Description and Solution Technique

2.1 Formulation of the Problem

We consider a liquid bridge of volume $\mathcal{V}$ between two parallel, concentric disks of equal radii R, separated by a distance d (Fig. 2). Unless noted otherwise, gravity is assumed to act in the negative axial direction. The upper and the lower disks are kept at different but constant temperatures T_1 and T_2, respectively, with $T_1 = T_0 - \Delta T/2 < T_0 + \Delta T/2 = T_2$, where T_0 is the mean temperature $T_0 = (T_1+T_2)/2$. The fluid motion is driven by buoyancy forces and by thermocapillary forces which arise due to the temperature dependence of the density ϱ and the surface tension σ

$$\varrho(T) = \varrho_0 \left\{1 - \beta(T - T_0) + O\left[(T - T_0)^2\right]\right\}, \tag{5a}$$

$$\sigma(T) = \sigma_0 - \gamma(T - T_0) + O\left[(T - T_0)^2\right], \tag{5b}$$

where ϱ_0 and σ_0 are the density and the surface tension at the reference temperature T_0, β is the thermal expansion coefficient, and γ the coefficient of the surface tension.

In general, the viscosity and the thermal diffusivity will likewise depend on the temperature rendering the full Navier–Stokes equations quite complicated. However if we consider the limit of small temperature variations, the Navier–Stokes equations can be expanded in terms of the small buoyancy parameter $\delta = \beta\Delta T$. To keep the buoyancy forces in the equations, we must assume that the Grashof number (7) remains of order one, $\mathrm{Gr} = O(1)$. In the order $O(\delta^0)$ of this particular limit the Navier–Stokes, energy, and continuity equations reduce to the well-known Boussinesq equations [13]

$$\mathrm{Re}\left(\frac{\partial}{\partial t} + \boldsymbol{U}\cdot\nabla\right)\boldsymbol{U} = -\nabla P + \nabla^2\boldsymbol{U} + \frac{\mathrm{Gr}}{\mathrm{Re}}T\boldsymbol{e}_z, \tag{6a}$$

$$\mathrm{RePr}\left(\frac{\partial}{\partial t} + \boldsymbol{U}\cdot\nabla\right)T = \nabla^2 T, \tag{6b}$$

$$\nabla\cdot\boldsymbol{U} = 0, \tag{6c}$$

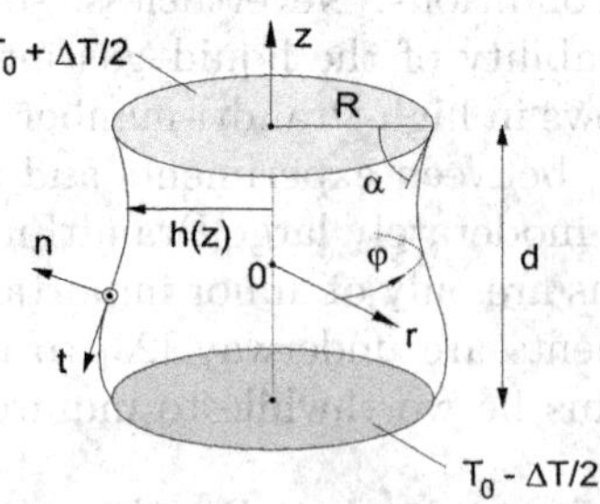

Fig. 2. Geometry of the liquid bridge

Table 1. Scaling of the variables. $U^\star$ is the characteristic thermocapillary velocity

Variable	t	z	r, h	U, V, W	P	T	$\mathcal{V}$
Scale	$d/U^\star$	d	R	$U^\star = \gamma \Delta T/\varrho_0 \nu$	$\gamma \Delta T/d$	ΔT	$\pi R^2 d$

where we use a cylindrical coordinate system (r, φ, z) and all variables including the velocity vector $\boldsymbol{U} = (U, V, W)$ and the pressure P have been made dimensionless using the scales given in Table 1. The scaled temperature field is defined as the deviation from the reference temperature: $(T - T_0)/\Delta T \to T$. The dimensionless groups governing the fluid motion are the Reynolds, Prandtl, and Grashof numbers

$$\mathrm{Re} = \frac{\gamma \Delta T d}{\varrho_0 \nu^2}, \ \mathrm{Pr} = \frac{\nu}{\kappa}, \ \mathrm{Gr} = \frac{g\beta \Delta T d^3}{\nu^2}, \ \mathrm{Ca} = \frac{\gamma \Delta T}{\sigma_0}, \ \mathrm{Bo} = \frac{\varrho_0 g d^2}{\sigma_0}, \ \mathrm{Bd} = \frac{\mathrm{Gr}}{\mathrm{Re}}, \tag{7}$$

where ν and κ denote the constant kinematic viscosity and thermal diffusivity, respectively. The capillary number (Ca) and the static Bond number (Bo) will appear in the dimensionless free-surface boundary conditions. Bd is the dynamic Bond number.

On the free surface the dynamic boundary condition (4) must be satisfied. In order to be consistent with the above Boussinesq approximation, we have to expand (4) for small temperature differences ΔT such that the Grashof *and* the Reynolds numbers remain of order $O(1)$. This latter condition is required to keep the thermocapillary driving force in the lowest order equations. In this limit both the relative density variation δ and the relative surface tension variation Ca are considered small. It is important to notice that both small parameters are related to each other by

$$\delta = \frac{\beta \sigma_0}{\gamma} \mathrm{Ca} = \frac{\mathrm{Bd}}{\mathrm{Bo}} \mathrm{Ca} \equiv \lambda \mathrm{Ca}. \tag{8}$$

Here $\lambda = \mathrm{Bd}/\mathrm{Bo}$ is a material parameter and independent of gravity. It is a measure for the relative variability of the density to that of the surface tension. Inserting the expansions (5a,5b), using relation (8), scaling the free surface boundary condition (4), and projecting it onto the normal ($\boldsymbol{n}$) and the tangential unit vectors ($\boldsymbol{t}$) yields

$$-\mathrm{Ca}P + \mathrm{Bo}\left(z - \frac{1}{2}\right) + \left[1 - \mathrm{Ca}T + O(\mathrm{Ca}^2)\right] (\nabla \cdot \boldsymbol{n})$$

$$+\mathrm{Ca}\left[1 - \lambda \mathrm{Ca}T + O(\mathrm{Ca}^2)\right] \boldsymbol{n} \cdot \left[\nabla \boldsymbol{U} + (\nabla \boldsymbol{U})^{\mathrm{T}}\right] \cdot \boldsymbol{n} = 0, \tag{9a}$$

$$\left[1 + O(\mathrm{Ca})\right] \boldsymbol{t} \cdot \nabla T + \left[1 - \lambda \mathrm{Ca}T + O(\mathrm{Ca}^2)\right] \boldsymbol{t} \cdot \left[\nabla \boldsymbol{U} + (\nabla \boldsymbol{U})^{\mathrm{T}}\right] \cdot \boldsymbol{n} = 0, \tag{9b}$$

where (9a) has been multiplied with Ca. The vector $\boldsymbol{t}$ denotes any one of the two linearly independent tangential vectors. Note that in addition to the dynamic pressure P relative to the ambient gas pressure the hydrostatic pressure

$P_{\text{hydrostatic}} = \text{Ca}^{-1}\text{Bo}(\frac{1}{2} - z)$ arises. This term results from the leading-order balance $O(\delta^{-1})$ of the Navier–Stokes equations the limit $\delta \to 0$.[2] In addition to (9a) and (9b) the kinematic condition (2) must be satisfied on the interface at $r = h(z, \varphi, t)$.

The contact lines where the liquid, solid, and gas meet are considered fixed. The conditions on the rigid isothermal walls at $z = \pm 1/2$ simply require

$$\boldsymbol{U} = 0 \qquad \text{and} \qquad T = \pm 1/2. \tag{10}$$

At the free surface, we shall always assume adiabatic boundary conditions ($\boldsymbol{n} \cdot \nabla T = 0$) for the temperature field.

In our formulation of the problem higher-order corrections to the bulk equations (6a–6c) are of the order $O(\text{Ca})$. In the boundary conditions (9a–9b) we have explicitly included the first-order ($O(\text{Ca})$) correction terms. While (6c) is the first approximation to the differential form of the mass conservation we shall also need, at a later stage, the exact integral form

$$M = \int_{\mathcal{V}} \varrho \, \mathrm{d}\mathcal{V}, \tag{11}$$

where the total mass M of the liquid bridge is constant. The form of (11) does not change when it is made dimensionless using the reference mass $\varrho_0 \pi R^2 d$. Then M is the mass relative to the mass of a cylindrical volume of fluid at the reference temperature T_0, and ϱ is the density relative to ϱ_0.

2.2 Expansion for Small Capillary Numbers

We are interested in the asymptotic behavior when the capillary number is small. This limit is of considerable interest, because the capillary number has been very small in most experiments on liquid bridges. Some representative data are given in Table 2.

For $\text{Ca} \to 0$ the normal component of the interfacial force balance (9a) is dominated by the Laplace pressure $\nabla \cdot \boldsymbol{n}$. This pressure can only be balanced

Table 2. Capillary numbers $\text{Ca} = \gamma \Delta T / \sigma_0$ at the critical temperature difference for the onset of oscillations ($\Delta T = \Delta T_c$) for some model fluids in typical geometries. Note that the critical temperature difference ΔT_c depends on the length scale

silicone oil (2 cSt)	$\text{Pr} \approx 25$	$\text{Ca} = O(10^{-2} \text{ to } 10^{-1})$
acetone	$\text{Pr} = 4.4$	$\text{Ca} = O(10^{-2})$
tin	$\text{Pr} = 0.02$	$\text{Ca} = O(10^{-3})$

[2] In a former publication of the authors [10] the term $-\lambda \text{BoCa} \int_{1/2}^{z} T \mathrm{d}z'$ erroneously appeared in the normal stress balance (9a). The present work corrects this mistake.

by a pressure jump between the liquid in the bridge and the ambient gas. The pressure jump P_s/Ca must be asymptotically large, i.e. $\sim \mathrm{Ca}^{-1}$. Thus we see that when expanding all fields in powers of Ca the leading-order terms just describe a static liquid bridge without any fluid motion.

To find the next higher corrections to the static liquid bridge we expand all unknowns as follows

$$\boldsymbol{U} = \boldsymbol{U}^{(0)} + \boldsymbol{U}^{(1)}\mathrm{Ca} + \boldsymbol{U}^{(2)}\mathrm{Ca}^2 + \ldots, \tag{12a}$$

$$T = T^{(0)} + T^{(1)}\mathrm{Ca} + T^{(2)}\mathrm{Ca}^2 + \ldots, \tag{12b}$$

$$P = \frac{P_s}{\mathrm{Ca}} + P^{(0)} + P^{(1)}\mathrm{Ca} + P^{(2)}\mathrm{Ca}^2 + \ldots, \tag{12c}$$

$$h = h^{(0)} + h^{(1)}\mathrm{Ca} + h^{(2)}\mathrm{Ca}^2 + \ldots, \tag{12d}$$

$$\mathcal{V} = \mathcal{V}^{(0)} + \mathcal{V}^{(1)}\mathrm{Ca} + \mathcal{V}^{(2)}\mathrm{Ca}^2 + \ldots. \tag{12e}$$

The volume $\mathcal{V}$ of the liquid bridge must also be expanded, since the mass is prescribed by (11) and the density is not constant.

The unit normal vector is given by

$$\boldsymbol{n} = \mathcal{N}^{-1}\left[1, -(\partial_\varphi h)/h, -\partial_z h/\Gamma\right]^{\mathrm{T}}, \tag{13}$$

where T denotes the transpose and $\mathcal{N}$ is the normalizing denominator. Corresponding expressions can be obtained for the tangential unit vectors. Since the normal and tangential unit vectors depend on the capillary number through the free surface location $h(\varphi, z, t)$, they must also be expanded

$$\boldsymbol{n} = \boldsymbol{n}^{(0)} + \boldsymbol{n}^{(1)}\mathrm{Ca} + \boldsymbol{n}^{(2)}\mathrm{Ca}^2 + \ldots, \tag{14a}$$

$$\boldsymbol{t} = \boldsymbol{t}^{(0)} + \boldsymbol{t}^{(1)}\mathrm{Ca} + \boldsymbol{t}^{(2)}\mathrm{Ca}^2 + \ldots. \tag{14b}$$

This expansion is the starting point for a more detailed analysis of the problem. A similar perturbation approach has been previously employed by [28,29] and [8] for the calculation of the Stokes flow and the dynamic surface deformations in cylindrical liquid bridges.

2.3 Thermocapillary Flow in a Fixed Domain

At leading order of Ca the normal stress condition (9a), and hence the interface shape, is independent of the flow and reduces to the Young–Laplace equation

$$\mathrm{Ca}P_s = \nabla \cdot \boldsymbol{n}^{(0)} + \mathrm{Bo}\left(z - \frac{1}{2}\right). \tag{15}$$

It is a second-order equation for the static interface location $h^{(0)}(z)$ which depends only on z, because the fixed contact lines which provide the boundary conditions for $h^{(0)}(z)$ are axisymmetric. In addition, $h^{(0)}(z)$ must satisfy the volume constraint $\mathcal{V} = \mathcal{V}^{(0)}$ determining the value of P_s.

To improve the leading-order static-state approximation without fluid motion, we consider the governing equations at $O(\mathrm{Ca}^0)$. The lowest-order flow

$[\boldsymbol{U}^{(0)}, T^{(0)}, P^{(0)}](r,z)$ is determined by the Boussinesq equations (6a–6c), subject to the rigid boundary conditions (10) and the tangential stress condition (9b) which must be considered at $O(\mathrm{Ca}^0)$

$$\boldsymbol{t}^{(0)} \cdot \nabla T^{(0)} + \boldsymbol{t}^{(0)} \cdot \left[\nabla \boldsymbol{U}^{(0)} + \left(\nabla \boldsymbol{U}^{(0)} \right)^{\mathrm{T}} \right] \cdot \boldsymbol{n}^{(0)} = 0, \qquad \text{on} \quad r = h^{(0)}(z). \quad (16)$$

For small Reynolds numbers the fluid motion will be axisymmetric. Hence, we only need to consider the tangential unit vector $\boldsymbol{t}^{(0)}(z)$ for which the azimuthal component vanishes.

From these equations we see that the flow at order $O(\mathrm{Ca}^0)$ is driven by buoyancy in the bulk and by thermocapillary shear stresses on the free surface. Equations (6a–6c), (10), (2), and (16) constitute the typical half-zone model with a static interface which is determined by (15) alone. This problem with a statically deformed interface has been treated numerically by a number of authors, e.g. [32], for gravity and zero gravity conditions.

2.4 Leading-Order Dynamic Surface Deformation

We have already seen that the static deformation $h^{(0)}$ at $O(\mathrm{Ca}^0)$ is determined by the *flow* at the lower order $O(\mathrm{Ca}^{-1})$, i.e. solely by the pressure P_s. Owing to the structure of the normal-stress equation (9a) a corresponding dependence also applies to the higher-order flow and surface-deformation fields.

Considering the normal-stress balance (9a) at $O(\mathrm{Ca}^1)$ we obtain the leading-order dynamic surface deformation $h^{(1)}(z,\varphi,t)$

$$\nabla \cdot \boldsymbol{n}^{(1)} = P^{(0)} + T^{(0)}\, \nabla \cdot \boldsymbol{n}^{(0)} - \boldsymbol{n}^{(0)} \cdot \left[\nabla \boldsymbol{U}^{(0)} + \left(\nabla \boldsymbol{U}^{(0)} \right)^{\mathrm{T}} \right] \cdot \boldsymbol{n}^{(0)}. \quad (17)$$

This is an equation for $h^{(1)}(z,\varphi,t)$, which is contained in $\boldsymbol{n}^{(1)}$, and it must be satisfied on $r = h^{(0)}(z)$ with $h^{(1)}(z = \pm 1/2) = 0$. It is clear that the flow-induced dynamic deformations of the interface $h^{(1)}$ depend on zeroth-order flow quantities only. Hence, the leading-order dynamic deformation $h^{(1)}$ does not couple back to the flow of order $O(\mathrm{Ca}^0)$ which creates it. The dynamic deformation $h^{(1)}$ is passive and completely determined by the flow field $[\boldsymbol{U}^{(0)}, T^{(0)}, P^{(0)}]$ of the classical half-zone model in the statically determined shape $h^{(0)}$.

The volume constraint for (17) is obtained by considering the dimensionless mass conservation (11)

$$M = \int_{\mathcal{V}} \varrho \mathrm{d}\mathcal{V} = \int_{\mathcal{V}^{(0)} + \mathrm{Ca}\mathcal{V}^{(1)} + \ldots} \left[1 - \lambda \mathrm{Ca} T + O(\mathrm{Ca}^2) \right] \mathrm{d}\mathcal{V}. \quad (18)$$

Since the leading order contribution is $M = \int_{\mathcal{V}^{(0)}} \mathrm{d}\mathcal{V}$, we obtain at $O(\mathrm{Ca})$ the condition

$$\mathcal{V}^{(1)} = \lambda \int_{\mathcal{V}^{(0)}} T^{(0)} \mathrm{d}\mathcal{V}. \quad (19)$$

For a solution of (17) explicit expressions for $\boldsymbol{n}^{(0)}$ and $\boldsymbol{n}^{(1)}$ are required. Using (13) we get

$$\nabla\cdot\boldsymbol{n}^{(0)} = \frac{\Gamma}{\mathcal{N}^{(0)}}\left(\frac{1}{h^{(0)}} - \frac{h''^{(0)}}{\Gamma^2 {\mathcal{N}^{(0)}}^2}\right), \tag{20a}$$

$$\nabla\cdot\boldsymbol{n}^{(1)} = \frac{1}{\Gamma {h^{(0)}}^3 {\mathcal{N}^{(0)}}^3}\Bigg[-\Gamma^2 h^{(0)} {\mathcal{N}^{(0)}}^2 h^{(1)} + {h^{(0)}}^2 h'^{(0)}\left(\frac{3h^{(0)}h''^{(0)}}{\Gamma^2{\mathcal{N}^{(0)}}^2} - 1\right) h^{(1)}_z$$
$$+ h^{(0)}h'^{(0)}\, h^{(1)}_{\varphi z} - \Gamma^2 h^{(0)}{\mathcal{N}^{(0)}}^2\, h^{(1)}_{\varphi\varphi} - {h^{(0)}}^3\, h^{(1)}_{zz}\Bigg]. \tag{20b}$$

Using (20a) we obtain the leading-order normal stress

$$\boldsymbol{n}^{(0)}\cdot\left[\nabla\boldsymbol{U}^{(0)} + (\nabla\boldsymbol{U}^{(0)})^{\mathrm{T}}\right]\cdot\boldsymbol{n}^{(0)}$$
$$= \frac{2}{{\mathcal{N}^{(0)}}^2}\left(\Gamma\partial_r U^{(0)} - h'^{(0)}\partial_r W^{(0)} - \frac{h'^{(0)}}{\Gamma}\partial_z U^{(0)} + \frac{{h'^{(0)}}^2}{\Gamma^2}\partial_z W^{(0)}\right). \tag{21}$$

Here the prime denotes the differentiation with respect to z. Inserting these expressions into (17) the resulting differential equation for $h^{(1)}$ takes the form

$$\left(C_1 + C_2\frac{\partial}{\partial z} + C_3\frac{\partial^2}{\partial\varphi\partial z} + C_4\frac{\partial^2}{\partial\varphi^2} + C_5\frac{\partial^2}{\partial z^2}\right) h^{(1)} = C_0. \tag{22}$$

It must be solved subject to the boundary conditions

$$h^{(1)}(\varphi, z = \pm 1/2) = 0, \tag{23}$$

and the volume constraint

$$\pi\mathcal{V}^{(1)} = \int_{-1/2}^{1/2}\int_0^{2\pi} h^{(0)}h^{(1)}\,\mathrm{d}\varphi\mathrm{d}z = \lambda\int_{-1/2}^{1/2}\int_0^{2\pi}\int_0^{h^{(0)}} T^{(0)}\, r\mathrm{d}r\mathrm{d}\varphi\mathrm{d}z, \tag{24}$$

which is obtained from (19) considering $\pi\mathcal{V} = \int_{\mathcal{V}} r\mathrm{d}r\mathrm{d}z\mathrm{d}\varphi$ at $O(\mathrm{Ca}^1)$. The factor π arises due to the scaling of the volume. The coefficients C_i in (22) depend on z. They are obtained as

$$C_0(z) = \Gamma {h^{(0)}}^3{\mathcal{N}^{(0)}}^3 P^{(0)} + \Gamma^2{h^{(0)}}^3{\mathcal{N}^{(0)}}^2\left(\frac{1}{h^{(0)}} - \frac{h''^{(0)}}{\Gamma^2{\mathcal{N}^{(0)}}^2}\right)T^{(0)} - \frac{2{h^{(0)}}^3\mathcal{N}^{(0)}}{\Gamma}$$
$$\times\left(\Gamma^3\frac{\partial U^{(0)}}{\partial r} - \Gamma^2 h'^{(0)}\frac{\partial W^{(0)}}{\partial r} - \Gamma h'^{(0)}\frac{\partial U^{(0)}}{\partial z} + {h'^{(0)}}^2\frac{\partial W^{(0)}}{\partial z}\right), \tag{25a}$$

$$C_1(z) = -\Gamma^2 h^{(0)}{\mathcal{N}^{(0)}}^2, \tag{25b}$$

$$C_2(z) = {h^{(0)}}^2 h'^{(0)}\left(\frac{3h^{(0)}h''^{(0)}}{\Gamma^2{\mathcal{N}^{(0)}}^2} - 1\right), \tag{25c}$$

$$C_3(z) = h^{(0)}h'^{(0)}, \tag{25d}$$

$$C_4(z) = C_1(z), \tag{25e}$$

$$C_5(z) = -{h^{(0)}}^3. \tag{25f}$$

C_0 represents the inhomogeneity of the linear system (22). It drives the deformation $h^{(1)}$ and depends on the zeroth-order velocity and pressure fields on the free surface at $h^{(0)}$. The bulk equations at $O(\mathrm{Ca}^1)$ will not be considered here, because they effect the interface deformation only at $O(\mathrm{Ca}^2)$.

3 Static Deformations

3.1 Methods of Solution

The equations governing the problem at $O(\mathrm{Ca}^0)$ and the interface problem at $O(\mathrm{Ca}^1)$ must be solved numerically. The solution strategy is the following. At first, the static shape $h^{(0)}$ is calculated by solving the Young–Laplace equation (15) together with the boundary conditions $h^{(0)}(z = \pm 1/2) = 1$ and $h'^{(0)}(z = 1/2) = -\Gamma \tan(\alpha - \pi/2)$, where α is the hot-wall contact angle which can be considered a parameterization of the mass M or, equivalently, the zeroth-order volume $\mathcal{V}^{(0)}$. The two-point boundary-value problem is solved for a prescribed angle α by a combined shooting and relaxation method [27].

Once the leading-order shape $h^{(0)}$ is known, the steady axisymmetric flow $(U_0^{(0)}, W_0^{(0)}, P_0^{(0)}, T_0^{(0)})$ is calculated by solving the zeroth-order flow problem (6a–6c). The subscript 0 indicates the axisymmetric basic flow. To that end the cross-section (r, z) of the liquid bridge bounded by $h^{(0)}(z)$ is mapped into a rectangular domain using body-fitted coordinates. On this rectangle, the system of partial differential equations is discretized using second-order finite differences and a non-equidistant grid, refined towards the rigid walls and the free surface. The resulting nonlinear difference equations are solved by Newton–Raphson iteration with damping.

The linear stability of the basic state $(U_0^{(0)}, W_0^{(0)}, P_0^{(0)}, T_0^{(0)})$ is calculated in the next step. Infinitesimal perturbations of the basic state are written as normal modes with integer azimuthal wave number m

$$(u, v, w, p, \theta)^{(0)}(r, \varphi, z, t) = \left(\hat{u}, \hat{v}, \hat{w}, \hat{p}, \hat{\theta}\right)^{(0)}(r, z) \times \exp(\mu t + im\varphi) + \text{c.c.}, \quad (26)$$

where the amplitudes $(\hat{\boldsymbol{u}}, \hat{p}, \hat{\theta})^{(0)} \in \mathbb{C}$ depend on r and z. By this ansatz and by linearization the three-dimensional partial differential equations for the perturbations are reduced to a two-dimensional linear problem. The discretization on the same grid as for the basic-state calculation finally leads to a generalized eigenvalue problem which is solved by inverse iteration. The linear stability boundary is then given by those parameters for which the maximum of the real parts of all possible temporal growth rates vanishes, i.e. by $\max[\Re(\mu)] = 0$.

More details on the solution procedure and the code validation can be found in [22]. They also calculated various stability boundaries and discussed the physical instability mechanisms. Some characteristic results are presented in the following.

3.2 Some Effects at Low Prandtl Numbers

The linear stability boundary of the two-dimensional basic flow is shown in Fig. 3a in form of a critical Reynolds number Re_c as a function of the contact angle α at the hot wall. Shown are data for three different small Prandtl numbers and for zero-gravity conditions $\mathrm{Bo} = \mathrm{Gr} = 0$. All modes are stationary with $\Im(\mu) = 0$. As a characteristic feature all stability curves show a minimum for $m = 2$ near a contact angle slightly less than 90°. For very small contact angles the liquid bridge becomes very slender and the critical mode changes to $m = 1$. Typical flow fields for a slender and a fat liquid bridge are shown in Fig. 4 for $m = 2$. In both cases the perturbation flow leads to a wavy deformation of the toroidal vortex as suggested by [14]. These authors pointed out the similarity of this instability with the instability of vortex rings [33]. The instability of ring vortices is caused by the elliptic instability mechanism. This mechanism relies on the presence of strain in the vortex core which is self-induced by the curvature of the core. The region where the elliptic mechanism is operative in the liquid-bridge vortex is indicated by the energy transfer near the center of the vortices in Fig. 4 (shaded). Here, the straining is not only self-induced, but rather a result of the boundaries of the domain and the thermocapillary shear stress. As shown by [22] there exists another contribution to the energy transfer which is due to a centrifugal effect. The corresponding energy transfer takes place near the outer streamlines of the vortex and can readily be identified in Fig. 4. Both elliptic and centrifugal effects work together to destabilize the two-dimensional basic flow. For slender liquid bridges the elliptic process dominates (Fig. 4a) while the instability is mainly due to centrifugal effects for fat liquid bridges (Fig. 4b). The presence of both destabilizing effects explains the minima of the critical curves in Fig. 3a. It is interesting to note that a similar instability arises near the upstream

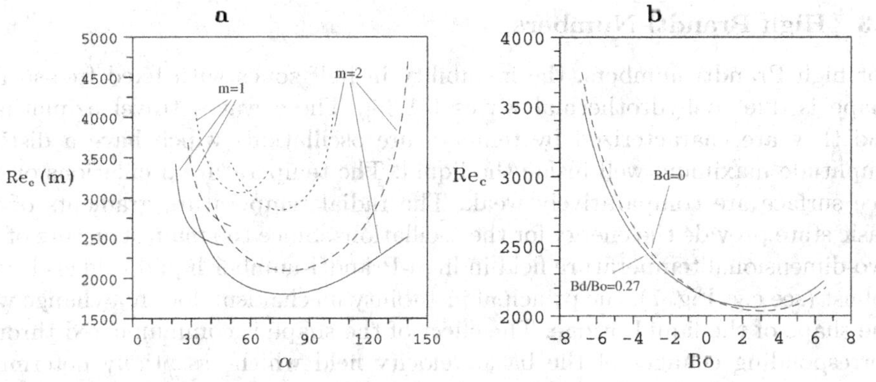

Fig. 3. **a** Neutral curves for low Prandtl numbers: $\mathrm{Pr} = 0$ (*full line*), $\mathrm{Pr} = 0.02$ (*long dashes*), and $\mathrm{Pr} = 0.04$ (*short dashes*). The other parameters are $\Gamma = 1$ and $\mathrm{Gr} = \mathrm{Bo} = 0$. **b** Neutral curves for $\mathrm{Pr} = 0.02$ as function of the static Bond number Bo for $\mathrm{Bd/Bo} = 0.27$ (*full line*) and for $\mathrm{Bd} = 0$ (*dashed line*). The other parameters are $\Gamma = 1$, $m = 2$, and $\mathcal{V}^{(0)} = 1$

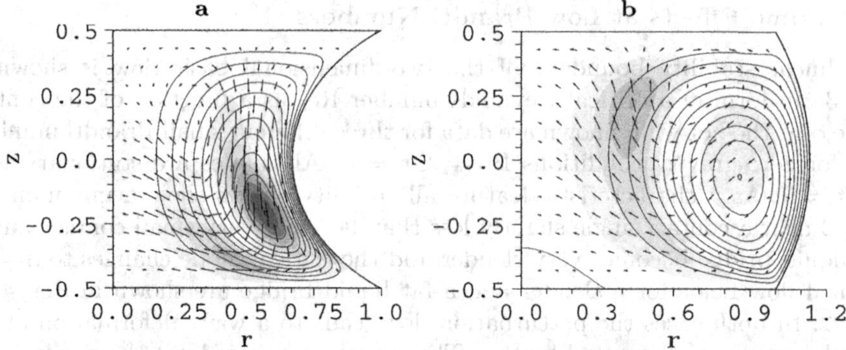

Fig. 4. Basic state streamlines, perturbation velocity vectors (*arrows*), and the azimuthal average of the local energy production rate $\int \boldsymbol{u} \cdot (\nabla \boldsymbol{U}_0^{(0)}) \cdot \boldsymbol{u} r \mathrm{d}\varphi$ (*shading*) at an angle φ at which the azimuthal perturbation velocity vanishes. The parameters are $\Gamma = 1$, Pr $= 0.02$, Re $= 2130$, Gr $=$ Bo $= 0$, $m = 2$, and $\alpha = 30°$ (**a**) and $\alpha = 120°$ (**b**)

boundary in finite-length cavities which model plane Poiseuille flow away from the end walls [1].

In a gravity field the linear stability of the low-Prandtl-number flow is mainly influence by the changes of the shape caused by the hydrostatic pressure. Buoyancy forces in the bulk play a minor role for typical parameters (Bd/Bo $= 0.27$). This can be seen in Fig. 3b which shows the linear stability boundary (full line) in comparison with the case in which buoyancy has been artificially neglected in the bulk (Bd $= 0$, dashed line), but not the gravity effect on the hydrostatic pressure.

3.3 High Prandtl Numbers

For high Prandtl numbers, the instability in half zones with fixed free-surface shape is due to hydrothermal waves [31,36]. These waves travel azimuthally and they are characterized by temperature oscillations which have a distinct amplitude maximum well inside the liquid. The temperature oscillations on the free surface are comparatively weak. The radial temperature gradients of the basic state provide the energy for the oscillations. Since the main structure of the two-dimensional temperature field in high-Prandtl-number liquid bridges is very robust (see e.g. Fig. 7), the principal instability mechanism does not change with the shape of the liquid bridge. The effect of the shape is communicated through corresponding changes of the basic velocity field which essentially determines the structure and location of the basic temperature gradient field [22].

Examples for the dependence of the linear stability boundaries on the shape (hot-wall contact angle α) are shown in Fig. 5a for zero gravity. As a characteristic feature, the critical Reynolds-number envelope of the neutral curves exhibits sharp maxima at which the critical mode changes. In the examples shown, this change of modes is associated with a change of the critical wave number m,

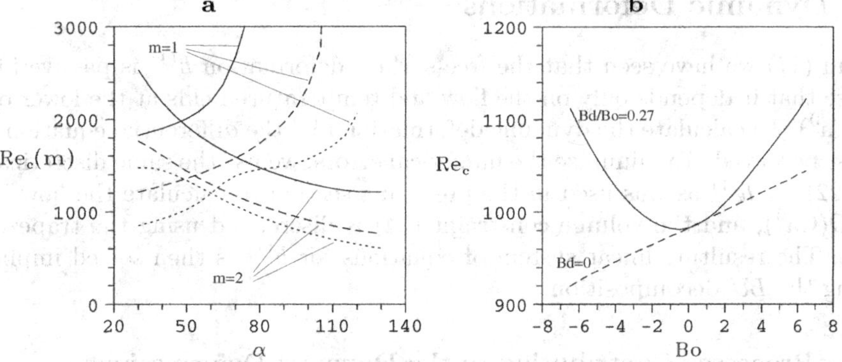

Fig. 5. **a** Neutral curves for high Prandtl numbers: Pr = 2 (*full line*), Pr = 4 (*long dashes*), and Pr = 7 (*short dashes*). The other parameters are $\Gamma = 1$, Bo = Bd = 0. **b** Neutral curves for Pr = 4 as function of the static Bond number Bo for Bd/Bo = 0.27 (*full line*) and for Bd = 0 (*dashed line*). The other parameters are $\Gamma = 1$, $m = 2$, and $\mathcal{V}^{(0)} = 1$

not of the instability mechanism. A similar trend has been observed also in the experiments of [17] and others.

Buoyancy forces are much more important for higher rather than for lower Prandtl numbers. The pure effect of hydrostatic shape changes would lead to a stabilization of the flow when heated from above (Bo > 0) and a destabilization when heated from below. From Fig. 5b it is seen that buoyancy bulk forces nearly always stabilize the flow for the present parameters (see also [35]) leading to a minimum of Re_c near Bo = 0. In particular, the counter-intuitive stabilization for heating from below (Bo < 0) is caused by a buoyancy-induced reduction of the radial temperature gradients in the bulk.

In this section we have discussed the most significant effects of the static shape of the liquid–gas interface on the flow stability in liquid bridges for an aspect ratio $\Gamma = 1$. The results are representative for many similar cases, and the mechanisms are the same. The details of the flow and its stability boundaries depend, of course, on the exact values of the parameters. For a more comprehensive analysis the reader is referred to [22].

It is important to note that we have found flow instabilities in the strict limit Ca → 0. In general, the numerical results are in good agreement with experimental measurements. The leading-order instability should not change qualitatively at higher orders of Ca as long as Ca is sufficiently small. However, small corrections should arise at $O(\mathrm{Ca})$. In the following we shall investigate the corrections for the location of the liquid–gas interface.

4 Dynamic Deformations

From (17) we have seen that the free-surface deformation $h^{(1)}$ is passive in the sense that it depends only on the flow and temperature fields at the lower order $O(\mathrm{Ca}^0)$. To calculate the dynamic deformation $h^{(1)}$ the differential equation (22) must be solved. To minimize the numerical errors, we use the same discretization of (22) for $h^{(1)}$ as was used in the preceding section to calculate the flow fields at $O(\mathrm{Ca}^0)$, and the volume constraint (24) is discretized using the trapezoidal rule. The resulting linear system of equations for $h^{(1)}$ is then solved implicitly using the LU-decomposition.

4.1 Processes Contributing to the Dynamic Deformations

The terms which arise in the normal stress balance (17) at $O(\mathrm{Ca})$ represent different physical processes. They are listed in Table 3. The first three terms appear additively in the inhomogeneity C_0 of the normal stress balance. They are due to the dynamic pressure, the viscous normal stress, and the variation of the surface tension with temperature. The fourth contribution describes the volume change of the liquid bridge due to a change of the mean temperature at $O(\mathrm{Ca}^0)$, hence the mean density, and enters the volume constraint (24).

Table 3. Terms in the normal-stress balance contributing to the dynamic interface deformation in $O(\mathrm{Ca}^1)$. The last column indicates the line types used in the figures

dynamic pressure:	$P^{(0)}$	(– – – – –)
viscous normal stress:	$-\boldsymbol{n}^{(0)} \cdot [\nabla \boldsymbol{U}^{(0)} + (\nabla \boldsymbol{U}^{(0)})^{\mathrm{T}}] \cdot \boldsymbol{n}^{(0)}$	(— - — -)
surface-tension variation:	$T^{(0)} \nabla \cdot \boldsymbol{n}^{(0)}$	(- - - - - -)
thermal expansion:	$\lambda \int_{-1/2}^{1/2} \int_0^{2\pi} \int_0^{h^{(0)}} T^{(0)} r \mathrm{d}r \mathrm{d}\varphi \mathrm{d}z$	(— - - - —)

Since the equation for the deformation (22) is linear, the total deformation $h^{(1)}$ is simply the sum of those interfacial deformations which would be caused by each individual term listed in Table 3.

To investigate the magnitude and structure of the surface-deformation field caused by each individual term only one term out of the first three contributions listed in Table 3 is taken into account in C_0 (25a) assuming a constant volume, $\mathcal{V}^{(1)} = 0$. The pure effect of the thermal expansion (fourth contribution in Table 3) is obtained by setting $C_0 = \Gamma {h^{(0)}}^3 {\mathcal{N}^{(0)}}^3 K$, where K represents a constant pressure, and by determining K such that the volume constraint (24) is satisfied. Summing up all terms then yields the total dynamic deformation $h^{(1)}$.

4.2 Two-Dimensional Steady Flow

The same code as in Sect. 3.1 is used to calculate the static shape of the liquid bridge $h^{(0)}(z)$, the flow and temperature fields $\boldsymbol{U}_0^{(0)}(r,z)$, $T_0^{(0)}(r,z)$, and the pressure gradient $\nabla P_0^{(0)}(r,z)$. From the latter, the zeroth-order pressure field on the free surface is obtained by integration along the free surface

$$P_0^{(0)}\left(r = h^{(0)}, z\right) = \int_{1/2}^{z} \boldsymbol{t}^{(0)} \cdot \nabla P_0^{(0)}\left(r' = h^{(0)}(z'), z'\right) \, \mathrm{d}s' + K, \tag{27}$$

where ds is the arc-length element and

$$\boldsymbol{t}^{(0)} = \frac{1}{\mathcal{N}^{(0)}} \begin{pmatrix} h'^{(0)}/\Gamma \\ 0 \\ 1 \end{pmatrix}. \tag{28}$$

Through the pressure $P_0^{(0)}$ the unknown integration constant $K = P_0^{(0)}(r = 1, z = 1/2)$ enters C_0 in (22) which, for axisymmetric flow, reduces to

$$\left(C_1 + C_2 \frac{\mathrm{d}}{\mathrm{d}z} + C_5 \frac{\mathrm{d}^2}{\mathrm{d}z^2}\right) h_0^{(1)} = C_0. \tag{29}$$

Here we have added the subscript 0 to $h^{(1)}$ to indicate that $h_0^{(1)}$ is the dynamic deformation caused by the axisymmetric basic flow. After discretization, using the boundary conditions $h_0^{(1)}(z = \pm 1/2) = 0$, the linear system of equations resulting from (29) is solved in the same way as (15) for $h^{(0)}$ (Sect. 3.1) and the integration constant K is determined by the volume constraint (24).

Low Prandtl Numbers

The only data available to date on dynamic deformations in liquid bridges are for Stokes flow under zero gravity in the limit Pr $\to 0$ [8,28]. These data have been used for a validation of the numerical code.

As an example, we consider Pr $= 0.02$. Figure 6a shows the decomposition of the total deformation $h_0^{(1)}$ into the different contributions for nearly creeping flow in a cylindrical liquid bridge under zero-gravity conditions. Due to the symmetry with respect to $z = 0$ of the $O(\mathrm{Ca}^0)$-flow problem for Re $\to 0$ (see, e.g. [9]), all deformation terms at $O(\mathrm{Ca})$ are anti-symmetric in z.

The volume change is negligible here, because the deviation $T^{(0)}$ from the linear conductive temperature profile is extremely small. The thermocapillary forces at lowest order $O(\mathrm{Ca}^0)$ drive a toroidal vortex (similar to those in Fig. 4) which is directed from the hot ($z = 1/2$, at the top in all figures) to the cold wall ($z = -1/2$) along the free surface. In the upper part of the half-zone, where the toroidal vortex flow is directed radially outward, the viscous normal stress is positive (the radial flow decelerates when approaching the free surface) and leads to an outward bulging of the interface. In the lower part the radial flow is directed away from the free surface leading to a constriction (dash-dotted line in Fig. 6a). Since the temperature in the upper part of the liquid bridge is higher than in the lower part, the surface tension is less in the upper than in the lower

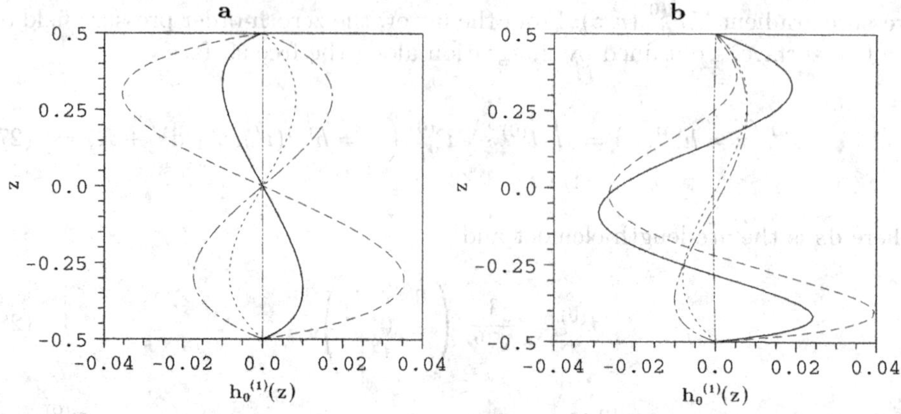

Fig. 6. Dynamic interface deformation $h_0^{(1)}(z)$ (*full line*) for Pr = 0.02, $\Gamma = 1$, $\mathcal{V}^{(0)} = 1$, and $\lambda = 0.27$ under zero gravity (Bo = Bd = 0) for Re = 10^{-4} (**a**) and at the critical point for the onset of three-dimensional flow at Re = Re$_c$ = 2130 (**b**). The different contributions to the dynamic deformation are indicated by the line types (see Table 3). The heating is from the top

half. Minimization of the surface energy then requires a receding of the free surface in the lower part of the zone (minimization of the free surface where the surface tension is high) and a bulging in the upper part. This effect reinforces the deformation caused by the viscous stress. The sum of both contributions, however, is more than compensated by the effect of the dynamic pressure field which has a dominating influence on the surface deformation. The logarithmic divergence of the pressure field at the hot and the cold corners ($P_0^{(0)} \to \mp\infty$ for $(r, z) \to (1, \pm 1/2)$, see [9,11]) leads to a constriction/bulging of the interface in the upper/lower half of the liquid bridge and explains the shape $h_0^{(1)}(z)$. Note that $h_0^{(1)}$ must be multiplied by Ca to obtain the non-dimensional surface deformation. For finite material and geometry parameters the condition for creeping flow, Re $\to 0$, requires $\Delta T \to 0$. Hence, also Ca $\to 0$ and the deformation tends to zero in this limit.

A corresponding decomposition of the dynamic deformation for a higher Reynolds number is shown in Fig. 6b. The surface deformation is markedly different from the case Re $\to 0$. As the most prominent inertia effect the high free-surface velocity causes a strong underpressure near $z = 0$ which leads to a constriction of the zone midway between both ends. In turn, the inertial-flow-induced pressure rises near both corners, where the flow is slower. This effect reinforces the bulging of the interface at the cold wall (caused by the cold-corner singularity), but it over-compensates the constriction of the interface at the hot-corner. As a result the free surface bulges near both ends. The contributions due to the variation of the surface tension and viscous normal stresses are of minor importance. Even the volume change is hardly visible on the scale of Fig. 6b. Since the dynamic deformations caused by inertia effects are nearly symmetric with respect to $z = 0$, while all other effects cause approximately antisymmetric deformations, the total dynamic deformation is lacking a definite symmetry when the Reynolds number is large.

High Prandtl Numbers

For high-Prandtl-number fluids, which are typically used in model experiments, the temperature and velocity fields are quite different. An example for the flow fields at order $O(\mathrm{Ca}^0)$ is shown in Fig. 7 for acetone ($\mathrm{Pr} = 4.38$).

When the Prandtl number is high the center of the vortex is located closer to the hot corner. Moreover, gravity usually has a significant influence, even if the length scale is in the mm range. The static interface shape is deformed and the vortex moves closer to the free surface, because the hot fluid transported downward along the free surface has a strong tendency to rise due to buoyancy in the return flow (cf. Sect. 3.3 and [22]).

Figure 8 shows the decomposition of the dynamic deformation $h_0^{(1)}$ caused by the two-dimensional flow from Fig. 7 at the critical Reynolds number [22,36]. The dynamic pressure and the change of the surface tension have a similar effect as in Stokes flow (Fig. 6a). However, the normal viscous stress promotes a bulging near both the hot and the cold corner. This can be explained in terms of the deceleration of the radial flow near the corners which can be seen from Fig. 7. Near the upper midplane the flow accelerates radially inward and the viscous normal stress leads to a negative contribution to $h_0^{(1)}$ near $z \approx 0.15$.

In a high-Prandtl-number flow, the surface tension is notably reduced from its isothermal value σ_0 over most parts of the free surface, except very close to the cold corner. Hence, there is a tendency to enlarge the free-surface area in the upper part of the liquid bridge at the expense of the free-surface area in the

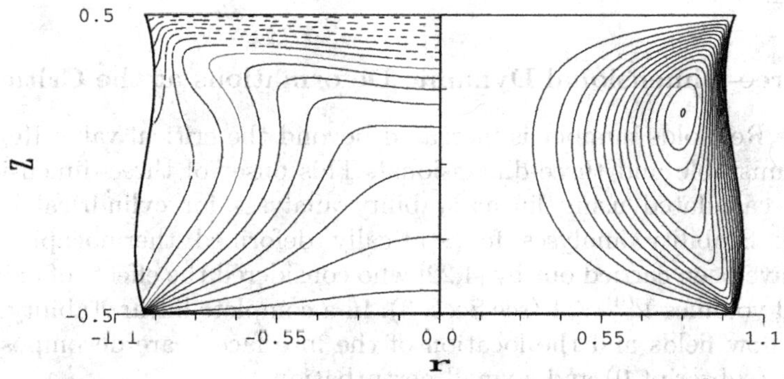

Fig. 7. Stream function (*right*) and temperature field (*left*) at $O(\mathrm{Ca}^0)$ for acetone with $\mathrm{Pr} = 4.38$, $\Gamma = 1$, and $\mathcal{V}^{(0)} = 1$ for heating from above. It is assumed that the radius of the liquid bridge is $R = 3.5\,\mathrm{mm}$ such that $\mathrm{Bd} = 1.10$ and $\mathrm{Bo} = 3.81$ ($\lambda = 0.288$). The Reynolds number corresponds to the critical value for the onset of three-dimensional flow, $\mathrm{Re} = \mathrm{Re}_c = 1011$. The mean temperature is $\langle T_0^{(0)} \rangle = -0.046$

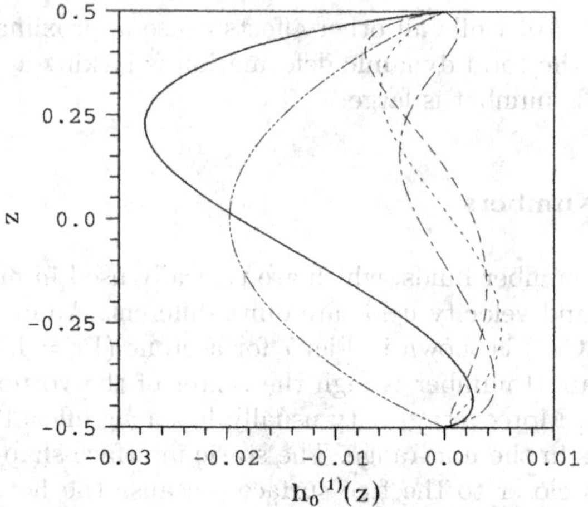

Fig. 8. Decomposition of the dynamic free-surface deformation $h_0^{(1)}(z)$ (*full line*) of an acetone liquid bridge heated from the upper side. All parameters as in Fig. 7 and line types as in Table 3

lower part (short dashes in Fig. 8). This amplifies the static deformation and is opposite to the dynamic deformation under zero gravity (see [10]).

The volume change, however, is the most significant effect. It results from a convective modification of the mean temperature from $\langle T_0^{(0)} \rangle = 0$ in the pure conductive state to $\langle T_0^{(0)} \rangle = -0.046$ for Re = 1011. For this reason the receding of the interface in the upper part of the bridge is relatively strong and the bulging in the lower part is restricted to a small region near the cold wall only.

For an acetone liquid bridge with a radius $R = 3.5\,\text{mm}$ the capillary number corresponding to Fig. 8 is $\text{Ca} = 1.5 \times 10^{-3}$. Hence, the total dynamic deformations due to the steady two-dimensional basic flow is very small and of the order of $0.17\,\mu\text{m}$.

4.3 Three-Dimensional Dynamic Deformations at the Critical Point

When the Reynolds number is increased beyond the critical value Re_c the flow becomes unstable and three-dimensional. This onset of three-dimensional flow has been calculated using linear stability analyses for cylindrical bridges by [15,21,36]. Stability analyses for statically deformed thermocapillary liquid bridges have been carried out by [4,22] who considered the effects of axial gravity forces and volumes $\mathcal{V}^{(0)} \neq 1$ (see Sect. 3). In a complete linear stability approach the total flow fields and the location of the interface h are decomposed into a basic flow (subscript 0) and a small perturbation

$$\begin{pmatrix} \boldsymbol{U}(r,\varphi,z,t) \\ P(r,\varphi,z,t) \\ T(r,\varphi,z,t) \\ h(\varphi,z,t) \end{pmatrix} = \begin{pmatrix} \boldsymbol{U}_0(r,z) \\ P_0(r,z) \\ T_0(r,z) \\ h_0(z) \end{pmatrix} + \varepsilon \begin{pmatrix} \boldsymbol{u}(r,z,\varphi,t) \\ p(r,z,\varphi,t) \\ \theta(r,z,\varphi,t) \\ h(z,\varphi,t) \end{pmatrix}, \tag{30}$$

where ε is a small parameter. For infinitely small perturbations ($\varepsilon \to 0$) the governing equations can be linearized. Due to the symmetries in t and φ the solutions of the linear stability problem at $O(\varepsilon^1\mathrm{Ca}^0)$ can thus be written in the form of normal modes

$$\begin{pmatrix} \boldsymbol{u}(r,z,\varphi,t) \\ p(r,z,\varphi,t) \\ \theta(r,z,\varphi,t) \\ h(z,\varphi,t) \end{pmatrix} = \begin{pmatrix} \hat{\boldsymbol{u}}(r,z) \\ \hat{p}(r,z) \\ \hat{\theta}(r,z) \\ \hat{h}(z) \end{pmatrix} e^{\mu t + im\varphi} + \text{c.c.}, \tag{31}$$

as in (26), where $\hat{h} \in \mathbb{C}$ depends on z.

If we want to calculate the dynamic free-surface deformations associated with the critical mode (31) we must consider three small parameters, ε, Ca, and δ, where δ is related to Ca through (8). We now insert (30) into the governing equations and consider the different orders of magnitude. At $O(\varepsilon)$ the usual linear stability problem (26) is recovered [15,21,22,36]. Evaluating the normal-stress balance (9a) at $O(\varepsilon\mathrm{Ca})$ we formally get the same equation (22) for $h^{(1)}$ as for the static deformation $h_0^{(1)}$. We only have to replace the flow and deformation fields of order $O(\mathrm{Ca}^0)$, which enter the coefficients C_0 through C_5 (25a–25f), by the normal mode $(\boldsymbol{u}^{(0)}, p^{(0)}, \theta^{(0)})$ of the statically deformed liquid bridge ($h^{(0)}$ from (15)) obtained at $O(\varepsilon)$.

The coefficients C_1 through C_5 are independent of φ. Since the critical mode $(\boldsymbol{u}^{(0)}, p^{(0)}, \theta^{(0)})$ and hence the inhomogeneity C_0 are harmonic in φ, the deformation at $O(\varepsilon\mathrm{Ca})$ must also be harmonic in φ. Therefore, we make the ansatz $h^{(1)}(z,\varphi,t) = \hat{h}^{(1)}(z)\mathrm{e}^{\mu t + im\varphi}$ leading to

$$\left[C_1 - m^2 C_4 + (C_2 + imC_3)\frac{\mathrm{d}}{\mathrm{d}z} + C_5 \frac{\mathrm{d}^2}{\mathrm{d}z^2}\right] \hat{h}^{(1)} = C_0 e^{-\mu t - im\varphi}. \tag{32}$$

This equation must be solved on $r = h^{(0)} = h_0^{(0)}$ with the boundary conditions $\hat{h}^{(1)}(z = \pm 1/2) = 0$. Since the temperature fluctuations $\theta^{(0)}$ are harmonic in the circumferential direction, the mean temperature is $\langle \theta^{(0)} \rangle = 0$ and the volume is preserved by the normal mode ($\mathcal{V}^{(1)} = 0$). The solution of (32) is obtained in the same way as for the steady axisymmetric dynamic deformations.

It is important to note that the dynamic deformations of the neutral mode at $O(\varepsilon\mathrm{Ca})$ are not coupled with the interfacial deformation of basic flow at $O(\mathrm{Ca}^2)$ as long as $m \neq 0$, because the base-flow deformation is axisymmetric. This is even true if we would take $\varepsilon \ll \mathrm{Ca}$ in which case the axisymmetric terms of $O(\mathrm{Ca}^2)$ are much larger than the non-axisymmetric terms of $O(\varepsilon\mathrm{Ca})$. Even for $\varepsilon \sim \mathrm{Ca}$ nonlinear terms generated by the perturbation flow at the order $O(\varepsilon^2)$ do not couple to the dynamic deformation at $O(\varepsilon\mathrm{Ca})$, because the terms of $O(\varepsilon^2)$ contain only spectral components with $m = 0$ (axisymmetric) and $2m$ (see e.g. [16]) which both differ from the wavenumber m of the neutral mode. Therefore, we need not further specify the type of limit in which ε and Ca tend to zero.

Since the neutral mode is only determined up to a constant factor, the absolute amplitude of the associated dynamic deformation cannot be calculated by the present approach. It is possible, however, to eliminate the unknown common factor by considering the ratio of the dynamic deformation to, e.g. the surface temperature of the neutral mode. Using the scalings from Table 1 we obtain

$$\frac{h^{(1)}_{\text{dim}}(z,\varphi)}{\theta^{(0)}_{\text{dim}}\left(h^{(0)},z,\varphi\right)} = \frac{h^{(1)}(z,\varphi)}{\theta^{(0)}\left(h^{(0)},z,\varphi\right)}\frac{\text{Ca}\,R}{\Delta T}, \tag{33}$$

for any point on the free surface, where the subscript 'dim' indicates dimensional quantities. For slightly supercritical driving ($\epsilon = (\text{Re} - \text{Re}_c)/\text{Re}_c \ll 1$) higher harmonics in circumferential direction remain sufficiently small and (33) should be comparable with experimental data. In the following we give typical examples for a low and for a moderately high Prandtl number.

Low Prandtl Numbers

Again we consider Pr = 0.02. The typical surface deformation caused by the stationary critical mode in a cylindrical bridge under zero gravity is shown in Fig. 9. The zeros of the temperature and the dynamic surface deformation are both at $\varphi = (n+1/2)\pi/2$. The extrema of the surface temperature of the neutral mode occur near the midplane $z \approx 0$, whereas the deformation extrema arise at $z \approx \pm 1/4$. While the extrema in the lower half of the zone are in phase with those of the temperature (outward bulging when the perturbation surface temperature is high), the deformation extrema in the upper half are out of phase by π relative to the temperature extrema. The ratio between the maximum absolute value of the surface temperature perturbation to the maximum absolute dynamic deformation is $\max h^{(1)}/\max\theta^{(1)}(r = h^{(0)}) = 1.56$.

The dynamic deformation is almost completely determined by the dynamic pressure field of the critical flow. The pressure, in turn, is governed by $\nabla^2 p^{(0)} = -2(\nabla \boldsymbol{U}_0) : (\nabla \boldsymbol{u}^{(0)})$ and depends the gradients of both the basic and the per-

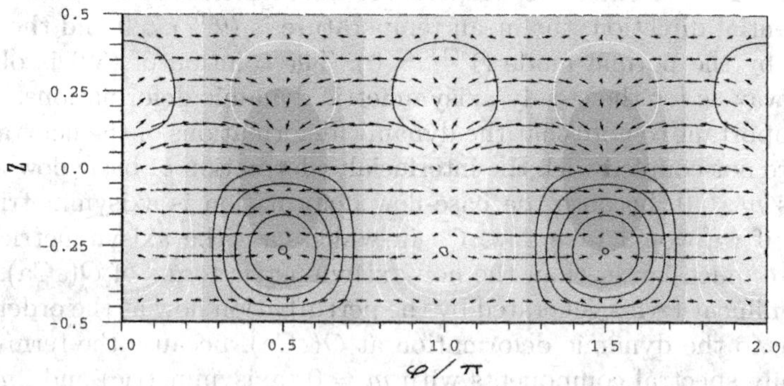

Fig. 9. Isotherms (*grey-shaded*; *dark*: $\theta^{(0)} > 0$, *bright*: $\theta^{(0)} < 0$), flow field (*arrows*), and *contours lines* of equal elevation of the interface (*black*: $h^{(1)} > 0$, *white*: $h^{(1)} \leq 0$) for the stationary critical mode for Pr = 0.02, Re = Re$_c$ = 2130, $\Gamma = 1$, $\mathcal{V}^{(0)} = 1$, $m_c = 2$, and Bd = Bo = 0

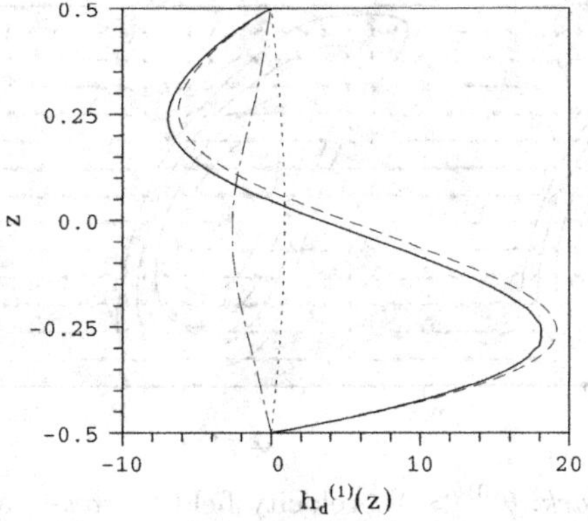

Fig. 10. Contributions to the deformation $h^{(1)}$ of the critical mode (*full line*) for $\mathrm{Pr} = 0.02$. Data are given at $\varphi = \pi/2$ where the perturbation temperature field takes a maximum. Parameters as in Fig. 9 and line types as in Table 3

turbation flow fields. Figure 10 shows the decomposition of the different contributions to the interfacial deformation at $\varphi = \pi/2$ where the temperature of the neutral mode takes its maximum value. The pressure distribution on the surface reflects the form of the neutral mode which, in a cut at $\varphi = \pi/2$, has the form of a vortex with center at the height $z \approx 1/4$ (see, e.g., Fig. 3a of [36]). The hot surface spot at $\varphi = \pi/2$ is associated with a reduction of the surface tension. This gives rise to a tiny bulging contribution. More important, but still much smaller than the pressure-induced deformation, is the receding of the interface caused by the viscous normal stress. It arises due to the free-surface flow towards the hot surface spot (Fig. 9) and the emerging radial inward flow due to continuity. Note that the azimuthal free-surface flow is in opposition to the Marangoni forces here, because the instability mechanism is essentially independent of the temperature field and inertia-induced [12,36] (see Sect. 3.2).

High Prandtl Numbers

For high Prandtl numbers the critical mode at $O(\varepsilon)$ arises as a pair of hydrothermal waves [36] which propagate in positive and negative φ directions. Here we consider the wave which propagates in the negative φ direction. Contrary to the stationary instability for low Prandtl numbers, the phase of the wave (31) depends on both r and z.

The flow, temperature, and dynamic-deformation fields for $\mathrm{Pr} = 4.38$ (corresponding to acetone) are shown in Fig. 11 under normal gravity conditions. For this moderately high Prandtl number the azimuthal phase of the surface-temperature wave is nearly independent of z. The phase of the dynamic deformation wave, however, depends notably on z. While the deformation wave is in phase with the temperature wave near the cold wall, it becomes approximately out of phase by π near the hot wall. The ratio between the maximum absolute value of the dynamic deformation to the maximum absolute value of the

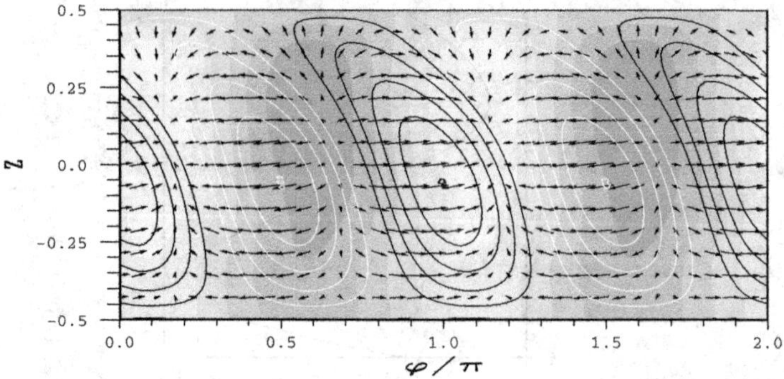

Fig. 11. Isotherms (*dark*: $\theta^{(0)} > 0$), velocity field (*arrows*), and contour lines of the dynamic deformations (*black*: $h^{(1)} > 0$, *white*: $h^{(1)} \leq 0$) of the critical mode for acetone with Pr = 4.38 for $\Gamma = 1$, $\mathcal{V}^{(0)} = 1$, and Re = Re$_c$ = 1011. The Bond numbers for normal gravity and $R = 3.5$ mm are Bd = 1.10 and Bo = 3.81 such that $\lambda = 0.288$

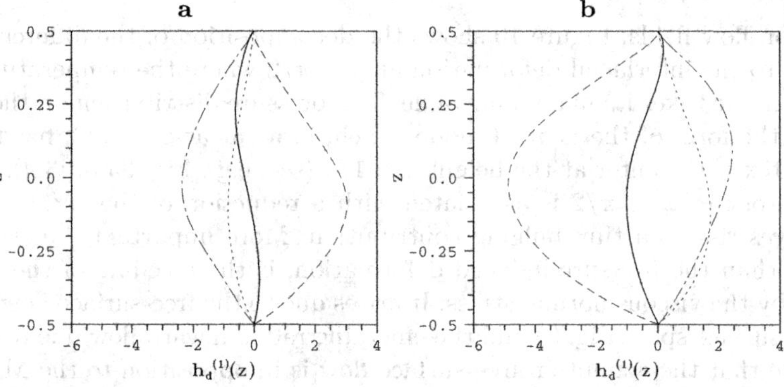

Fig. 12. Dynamic surface deformations (*full lines*) caused by the critical mode ($m_c = 2$) in acetone with Pr = 4.38, and its decomposition (for line types, cf. Table 3). Deformations are shown at azimuthal angles $\varphi_1 = 50°$ (**a**) and $\varphi_2 = 95°$ (**b**). Parameters as in Fig. 11

temperature perturbation is 0.05. For zero-gravity conditions the temperature and deformation waves have a similar structure as for gravity conditions. Also, the relative amplitude of the dynamic deformation to that of the temperature oscillation is of comparable magnitude [10].

The decomposition of the dynamic surface deformation into different contributions is shown in Fig. 12 for two azimuthal angles φ_1 and φ_2. As in [22] $\varphi_1 = 50°$ is defined by the angle at which the production of *thermal* energy by the radial transport (due to the basic flow) of basic-state temperature takes a local maximum in the plane $z = 0$ (Fig. 12a). The angle midway between two neighboring energy-transfer maxima defines φ_2. Since there are four maxima for $m = 2$ (see, e.g., Fig. 19 of [36]), we have $\varphi_2 = 95°$ (Fig. 12b).

The neutral flow is mainly driven by azimuthal Marangoni forces [36]. Therefore, the surface regions of high temperature from which the azimuthal surface flow originates experience an underpressure, whereas the cold regions are associated with a high pressure (no inertia effects). Since the pressure field is the largest contribution to the dynamic deformation, the interface tends to recede near the hot spots (near φ_2, Fig. 12b) whereas it tends to bulge near the somewhat colder surface areas (Fig. 12a). However, the receding near the hot spots is partly compensated by the viscous normal stress due to the bulk flow impinging on the free surface. Also the reduced surface tension associated with the hot spot tends to locally increase the surface area, hence adding to the bulging. Similar arguments apply to the cold-spot region.

The rapid change of the phase of the deformation wave near the hot wall can be explained by the viscous stress working against the pressure-induced deformation: only close to the hot wall the viscous stress in combination with the surface-tension effect dominates the pressure term, whereas the dynamic pressure contribution dominates in the lower half of the free surface. Further results on the dynamic surface deformation for other parameters can be found in [10].

5 Conclusions

The thermocapillary flow in liquid bridges has been considered with emphasis on the effects of static and dynamic free-surface deformations. The approach employed is based on an expansion of the governing equations in terms of a small capillary number Ca. The condition $\mathrm{Ca} \ll 1$ is not very restrictive, since Ca has been quite small in most model experiments, at least in those performed under normal gravity.

For the lowest-order expansion, $O(\mathrm{Ca}^0)$, the free-surface shape of the liquid bridge is solely determined by the static condition which is imposed by the Young–Laplace equation, independent of the flow. The effect of static deviations from a cylindrical shape, caused by the relative volume and/or the hydrostatic pressure difference, on the leading-order flow field were discussed. In addition, the linear stability of the axisymmetric flow with respect to general three-dimensional perturbations was analyzed and the most prominent causes effecting the critical Reynolds number were identified. For additional details, the reader is referred to [22].

At the next higher order, $O(\mathrm{Ca}^1)$, corrections to the location of the free surface arise. These deformations are flow-induced, hence dynamic. By decomposing the linear equations governing the dynamic deformation, the physical processes contributing to the surface corrections have been identified. In most situations, the dynamic pressure causes the dominant contribution to the dynamic deformation. Only for the steady high-Marangoni-number two-dimensional thermocapillary flow the shrinkage of the liquid bridge, caused by a reduced mean temperature, is the most important effect. The phase between of the time-dependent dynamic deformations associated with the onset of hydrothermal waves relative to the surface-temperature wave depends on the exact values of the parameters. This behavior is due to the viscous normal-stress-induced deformation which can locally compensate the dynamic deformation caused by the dynamic pressure.

The dynamic deformations at $O(\mathrm{Ca}^1)$ are caused by the flow fields at $O(\mathrm{Ca}^0)$. Therefore, the small dynamic deformations calculated here are passive in the sense that they cannot modify the flow at $O(\mathrm{Ca}^0)$, nor the three-dimensional flow instabilities which arise already at $O(\mathrm{Ca}^0)$. It can be expected that an active coupling between flow and dynamic free surface deformations, as suggested by [6,7] for large-Prandtl-number fluids, is only possible if the capillary number is $O(1)$. In this respect, a comparison of the present results with full-scale numerical simulations and with high-resolution surface-position measurements would be very interesting.

Acknowledgement

Part of this work has been supported by NASDA in the framework of the *Marangoni Modeling Research Project*.

References

1. S. Albensoeder, H.C. Kuhlmann: Eur. J. Mech. B/Fluids **21**, 307 (2002)
2. C.E. Chang, W.R. Wilcox: J. Crystal Growth **28**, 8 (1975)
3. Q.S. Chen, W.R. Hu: Intl. J. Heat Mass Transfer **41**, 825 (1998)
4. Q.S. Chen, W.R. Hu, V. Prasad: J. Crystal Growth **203**, 261 (1999)
5. T. DebRoy, S.A. David: Rev. Mod. Phys. **67**, 85 (1995)
6. Y. Kamotani, S. Ostrach: ASME J. Heat Transfer **120**, 758 (1998)
7. Y. Kamotani, S. Ostrach, M. Vargas: J. Crystal Growth **66**, 83 (1984)
8. H.C. Kuhlmann: Phys. Fluids A **1**, 672 (1989)
9. H.C. Kuhlmann: *Thermocapillary Convection in Models of Crystal Growth volume 152 of Springer Tracts in Modern Physics*, (Springer, Berlin, Heidelberg 1999)
10. H.C. Kuhlmann, C. Nienhüser: Fluid Dyn. Res. **31**, 103 (2002)
11. H.C. Kuhlmann, C. Nienhüser, H. J. Rath: J. Engng Math. **36**, 207 (1999)
12. H.C. Kuhlmann, H.J. Rath: J. Fluid Mech. **247**, 247 (1993)
13. L.D. Landau, E.M. Lifshitz: *Fluid mechanics*, (Pergamon Press, New York 1959)
14. M. Levenstam, G. Amberg: J. Fluid Mech. **297**, 357 (1995)
15. M. Levenstam, G. Amberg, C. Winkler: Phys. Fluids **13**, 807 (2001)
16. J. Leypoldt, H.C. Kuhlmann, H. J. Rath: J. Fluid Mech. **414**, 285 (2000)
17. J. Masud, Y. Kamotani, S. Ostrach: AIAA J. Thermophys. Heat Transfer **11**, 105 (1997)
18. J. Meseguer, L. A. Slobozhanin, J. M. Perales: Adv. Space Res. **16**, 5 (1995)
19. T.E. Morthland, J.S. Walker: J. Crystal Growth **158**, 471 (1995)
20. M. Mundrane, J. Xu, A. Zebib: Adv. Space Res. **16**(7), 41 (1995)
21. G.P. Neitzel, K.-T. Chang, D.F. Jankowski, H.D. Mittelmann: Phys. Fluids A **5**, 108 (1993)
22. C. Nienhüser, H.C. Kuhlmann: J. Fluid Mech. **458**, 35 (2002)
23. K. Nishino, S. Yoda: 'The role of dynamic surface deformation in oscillatory Marangoni convection in liquid bridge of high Prandtl number'. In: *Marangoni Convection Modeling Research*, ed. by H. Inokuchi, (NASDA Technical Memorandum TMR-000006E, 2000) pp. 43–71

24. A. Oron, S. H. Davis, S. G. Bankoff: Rev. Mod. Phys. **69**, 931 (1997)
25. S. Ostrach, Y. Kamotami, C. L. Lai: PhysicoChem. Hydrodyn. **6**, 585 (1985)
26. J.R.A. Pearson: J. Fluid Mech. **4**, 489 (1958)
27. W.H. Press, B.P. Flannery, S.A. Teukolsky, W.T. Vetterling: *Numerical Recipes FORTRAN*, (Cambridge University Press, 1989)
28. A. Rybicki, J.M. Floryan: Phys. Fluids **30**, 1973 (1987)
29. A. Rybicki, J.M. Floryan: Phys. Fluids **30**, 1956 (1987)
30. L.E. Scriven: Chem. Eng. Sci. **12**, 98 (1960)
31. M.K. Smith, S.H. Davis: J. Fluid Mech. **132**, 119 (1983)
32. L.B.S. Sumner, G.P. Neitzel, J.-P. Fontaine, P. D. Aversana: Phys. Fluids **13**, 107 (2001)
33. C.-Y. Tsai, S.E. Widnall: J. Fluid Mech. **73**, 721 (1976)
34. R. Velten, D. Schwabe, A. Scharmann: Phys. Fluids A **3**, 267 (1991)
35. M. Wanschura, H.C. Kuhlmann, H.J. Rath: Phys. Rev. E **55**, 7036 (1997)
36. M. Wanschura, V.S. Shevtsova, H.C. Kuhlmann, H. J. Rath: Phys. Fluids **7**, 912 (1995)

The Choice of the Critical Mode of Hydrothermal Instability in Liquid Bridge

Valentina M. Shevtsova, Mohamed Mojahed, Denis E. Melnikov, and Jean Claude Legros

Université Libre de Bruxelles, MRC, CP-165/62, 50, av. F.D.Roosevelt, B-1050 Brussels, Belgium

Abstract. The appearance and the development of thermoconvective oscillatory flows in a liquid bridge are explained. A theoretical and experimental study was conducted on a liquid bridge of aspect ratio $\Gamma = 1.2$ filled with a 10 cSt silicone oil. The experiments were carried out in terrestrial conditions for a wide range of volumes of liquid bridges. For the first time two branches of the stability diagram, (ΔT_{cr} vs. $Volume$), belonging to different azimuthal wave numbers were observed. The influence of the temperature of the cold rod on the onset of instability and on the critical azimuthal wave number was experimentally studied. The experiments show that the critical temperature difference is diminished with the increase of the temperature of the cold rod. 3-D numerical simulations have been done using the parameters as close as possible to the experimental conditions. In these calculations, both thermocapillary and buoyancy mechanisms of convection are taken into account and it is seen that the numerical results agree well with the experimental ones. A key reason for the good agreement between the model and experiments is that the numerical code takes into account the dependence of the viscosity upon the temperature.

1 Introduction

Due to their importance in crystal growth processes, convective flows and heat transfer in systems with free boundaries have been investigated intensively for different gravity conditions. These investigation have been conducted both numerically and experimentally. The often-used crystal-growth half-zone model consists of a finite cylindrical volume of fluid, confined between two concentric rigid disks, which are kept at different temperatures. In this model, usually called the liquid bridge, the temperature gradient along the free surface results in the convective flow due to buoyancy and thermocapillary effects.

To simplify the problem the majority of the theoretical studies assume the right circular free surface, which can be realized easily in microgravity conditions. The behaviour of 2-D flows and development of the instability is well understood for such a geometry [10], [12], and [19]. The flow is axisymmetric and steady if the temperature difference between the disks is small. When the temperature difference is increased and exceeds some critical value, ΔT_{cr}, the two-dimensional (2-D) toroidal flow undergoes a transition to time-dependent three-dimensional (3-D) flow. For small Prandtl numbers the first instability is stationary. Levenstamm and Amberg [6] showed that for higher Reynolds numbers this flow undergoes a second bifurcation, from a stationary to an oscillatory

V.M. Shevtsova, M. Mojahed, D.E. Melnikov, and J.C. Legros, The Choice of the Critical Mode of Hydrothermal Instability in Liquid Bridge, Lect. Notes Phys. **628**, 241–262 (2003)
http://www.springerlink.com/

flow. For liquids with larger Pr numbers, $Pr \geq 0.5$, two hydrothermal waves bifurcate directly from the steady state. In the hydrothermal wave (HTW) the thermal field is strongly coupled to the velocity field. These hydrothermal waves are characterized by an azimuthal wave number, which depends upon the Prandtl number and the aspect ratio among other factors. The azimuthal wave number m at the threshold of instability is called the critical mode.

The first empirical correlation for the determination of the azimuthal wave number, $m_{cr} \approx 2.2/\Gamma$, has been suggested by Preisser et al. [12] by analyzing the experimental data for a fluid of Pr=8.9. Here $\Gamma = d/R_0$ is the aspect ratio, d is the height of liquid bridge and R_0 is the radius. The slightly different correlation, $m_{cr} \approx 2.0/\Gamma$, has been obtained numerically for $Pr < 7$ assuming pure Marangoni convection [8,19].

By analogy to low Pr instability and that of a thin vortex ring [6] it follows that the azimuthal wave number may satisfy the relation $m = 2.5/\Gamma$. In reality the coefficient of proportionality for liquid bridges was found to be smaller than '2.5'. The numerical calculations by Wanschura et al. [19] showed that the relation $m_{cr} \approx 2.0/\Gamma$ remains also valid for small Pr number, although the mechanism of the instability has a different origin for relatively high ($0.5 \leq Pr \leq 7$) and low Prandtl liquids.

The above empirical relation $m_{cr} \approx 2.0/\Gamma$ does not hold for high Prandtl $Pr \geq 30$. Indeed, it has been theoretically shown by Xu and Davis [21] for high Prandtl numbers fluids that the critical azimuthal wave number m_{cr} is equal to one for the infinitely long cylindrical jet. Meanwhile experiments [1], [9], and [16] with different silicone oils demonstrate that the critical azimuthal mode corresponds to $m_{cr} = 1$ for $Pr \geq 30$ for the unit aspect ratio and aspect ratio close to one. In addition, the 3-D numerical calculations by Shevtsova et al. [17] have also confirmed that this empirical relation is not valid for $Pr = 35$.

In deformed liquid bridges, it turns out that the critical mode may depend upon the liquid volume. In some limits the empirical relation $m_{cr} \approx 2.0/\Gamma$ remains valid for non-cylindrical liquid bridges in the zero gravity condition. The radius of a liquid bridge changes with the height but for the zero-g condition the free surface shape is symmetrical with respect to the mid-plane. For small Prandtl numbers, $Pr = 0.01$, to take into account the shape effect, Lappa et al. [5] suggested the modified empirical relation $m_{cr} \approx 2.0/\tilde{\Gamma}$, where the radius of the liquid bridge at the mid-plane $\tilde{\Gamma} = h_{(z=d/2)}/d$ is used. It corresponds to the minimal radius of the zero-g configuration. Nienhüser et al. [11] have recently confirmed this formula by linear stability analysis and extended it up to $Pr = 4$.

For the case of high Prandtl numbers in terrestrial conditions the stability diagram (ΔT_{cr} *vs. Volume*) consists of two different oscillatory instability branches. For the case of 10 cSt silicone oil and aspect ratio $\Gamma = 4/3$ two branches corresponding to different azimuthal wave numbers have been experimentally obtained [16]. The branch on which ΔT_{cr} grows with increasing volume belongs to $m_{cr} = 1$, and the descending branch belongs to the azimuthal wave number $m_{cr} = 2$.

The question of the influence of the thermal conditions near the free surface of the liquid bridge on the onset of instability and the critical mode remains obscure. It is one of the points of present study. Usually the liquid bridge is assumed thermally insulated or the Biot number takes the heat exchange on the free surface into account. Wanschura et al. [20] have studied the role of the Biot number by using a linear stability analysis for the full zone in the case of radial heating. It was shown that the critical Reynolds number (Ma_{cr}) reaches a minimum near $Bi \approx 18$ and the type of instability depends on the Biot number, when $Pr = 4$, $\Gamma = 1$. Recently Schwabe et al. [14] have experimentally investigated heat loss and heating of the free surface in a hollow floating zone. They obtained that Ma_{cr} and the frequency increase significantly when the free surface is cooled.

The present experimental study is one of the first to investigate the influence of the different thermal conditions on the onset of instability of thermocapillary flow and the critical wave number in a half-zone model. The numerical calculations of the full 3-D non-steady Navier-Stokes equations have been done using the parameters as close as possible to the experimental conditions. An interpretation of some of the experimental results is proposed using the numerical data.

2 Experimental Set-Up

The current experiments were carried out to understand the influence of the interface deformation and of the thermal conditions around the liquid bridge on the threshold of instability and on the critical azimuthal wave number.

A wide spectrum of silicone oils is often used in terrestrial and microgravity conditions. The reasons for using high Prandtl number silicone oils are the reproducibility of the surface tension and their well-defined dependence upon temperature. Therefore the 10 cSt silicone oil ($Pr = 107$) was chosen as a working liquid. To increase the role of Marangoni convection in comparison with the buoyancy effect, the experiments were carried out in tiny liquid bridges. To study the thermal convection in 10 cSt silicone oil ($Pr = 107$) two different set-ups were developed. The scheme of the last one is shown in Fig. 1.

The upper rod was fixed in such a way, that three–dimensional movements are possible. A heating element (Resistor Minco $R \approx 100\,Ohms$) was mounted around the upper rod to heat the fluid from above. The lower rod was kept at a constant temperature using a thermoregulated water–cooling system. All experiments were done in air. The rods meant to hold the liquid zone were made from Aluminum alloy ($\lambda = 164W/mK$) and had the same diameter $2R = 6mm$. The present results correspond to a liquid zone of length $d = 3.6mm$ and as a result the aspect ratio was equal to $\Gamma = d/R = 1.2$.

To establish a floating zone the liquid was injected by a dedicated push syringe into a gap between rods. The push syringe allowed us to measure the volume of liquid with high accuracy. This is important, as the onset of time dependence is sensitive to the fluid volume. To prevent liquid creeping over the edge of the lower rod, the lateral surface was coated with anti-wetting fluid, FC723.

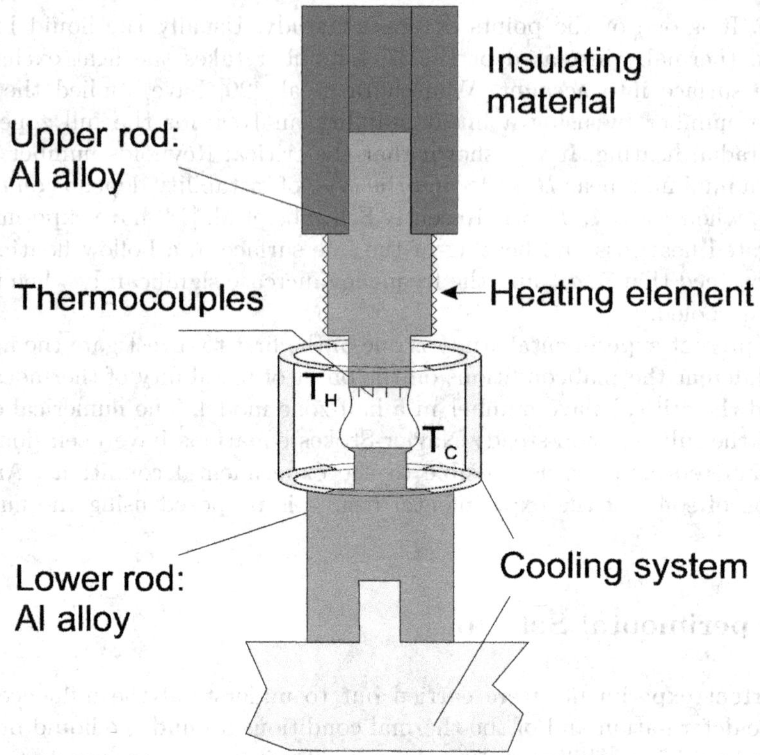

Fig. 1. Experimental set-up

The temperature oscillations due to time-dependent convection were measured by inserting five shielded thermocouples ($D = 0.25mm$) inside the liquid at the same radial and axial positions and different azimuthal angles. These thermocouples were embedded into the liquid through the upper rod, to prevent the disturbance of the free surface. Contrary to general opinion, we did not observe a strong influence of the amount of thermocouples on the critical ΔT_{cr}. A few experiments were repeated with 2 and 5 thermocouples and the comparison of the results were carefully made, see Sect. 5.3. It should be mentioned that the free surface, which is responsible for the driving force, was not disturbed. The choice of 5 thermocouples was made to determine without ambiguity the critical wave number and the type of hydrothermal wave. But 3 thermocouples were not enough to clearly determine the $m = 2$ mode.

All temperature signals given by thermocouples were amplified and band-pass filtered before being recorded by a computer. The signals were recorded with a time interval of 0.1 s. To protect the liquid bridge from the ambient air fluctuations the entire set-up was surrounded by a special glass box of large volume.

The first series of experiments was carried out to establish the stability diagram $(\Delta T_{cr}, V/V_0)$, for a liquid bridge. These experiments were conducted under the conditions described above, and they will be referred to as non-shielded liquid bridge experiments.

The second series was aimed to clarify the role of the temperature of the cold rod on the critical parameters. For each particular experiment the temperature of the cold rod was chosen and it was kept constant during the experiment. From one hand, such kind of study has been missed. Although Schwabe et al. [12,14] had carried out experiments when the working temperature in the liquid was about 300°C and Kawamura et al. [3] had used the temperature −20°C to avoid evaporation of silicone oil 1 cSt. From other hand, by changing successively the temperature of the cold rod from 9°C till 35°C the continuous Prandtl number can be experimentally reproduced. The definition of the Prandtl number is given through the temperature of the cold rod $Pr = \nu_{(T=T_{cold})}/k$.

For these experiments the liquid bridge was placed in a cylindrical pipe of larger internal diameter $2R = 12mm$. (See Fig. 1) The double walls of the pipe were filled by flowing water, the temperature of which could be easily controlled. The temperature of the water and the temperature of the cold rod were kept the same throughout this study, although they were not connected. This way the temperature of the ambient gas around the free surface could be controlled. The experiments under such conditions will be referred to as shielded liquid bridge experiments.

3 Numerical Model

The mathematical model for 3-D numerical simulation is chosen as close as possible to the experimental conditions. The characteristic parameters of the problem correspond to the physical properties of the working liquid which are listed in Table 1. Both thermocapillary and buoyancy mechanism of convection are taken into account. Basically the geometry of the theoretical model follows Fig. 1. Unlike the experiment the free surface is assumed cylindrical and non-deformable.

It is assumed in the model that two differentially heated horizontal flat concentric disks have radius R and are separated by a distance d. The temperatures T_h and T_0 ($T_h > T_c$) are prescribed at the upper and lower disks respectively, yielding a temperature difference of $\Delta T = T_h - T_c$. The temperature of the cold disk T_c is used as the reference, $T_0 = T_c$. The density, the surface tension and

Table 1. Physical properties of the silicone oil 10 cSt

ν	β	σ	k	$\gamma = d\sigma/dT$	ρ
m^2/s	$1/K$	N/m	m^2/s	N/mK	kg/m^3
10^{-5}	$1.08 \cdot 10^{-3}$	$20.1 \cdot 10^{-3}$	$0.95 \cdot 10^{-7}$	$-6.8 \cdot 10^{-5}$	934.

the kinematic viscosity are taken as linear functions of temperature. Therefore

$$\rho = \rho_o - \rho_o\beta(T - T_0), \qquad \text{where } \beta = -\rho_0^{-1}\frac{\partial\rho}{\partial T}$$

$$\sigma(T) = \sigma(T_0) - \sigma_T(T - T_0), \qquad \sigma_T = -\frac{\partial\sigma}{\partial T} = const.$$

$$\nu'(T) = \nu(T_0) + \nu_T(T - T_0), \qquad \nu_T = \frac{\partial\nu'}{\partial T} = const.$$

All other material properties are regarded as constant. The linear dependence of the kinematic viscosity upon temperature is widely used for theoretical and experimental studies and it is justified. For example, silicone oil 5cSt has the following temperature-dependence

$$\nu = a10^{bT}, \quad \text{where} \quad a = 7.091117 \cdot 10^{-6}\,\mathrm{m}^2/\mathrm{s} \quad b = -5.7259 \cdot 10^{-3}\,1/^\circ\mathrm{C}.$$

As **b** is a small parameter this expression can be written as

$$\nu = a10^{bT} = a(1 + b\,ln10\,(T - T_0) + \ldots) = a\left[1 - 1.318 \cdot 10^{-2}(T - T_0)\right],$$

which yields the linear law.

The governing Navier-Stokes, energy and continuity equations are written in non-dimensional primitive-variable formulation in a cylindrical co-ordinate system under the Boussinesq approximation:

$$\frac{\partial\mathbf{V}}{\partial t} + \mathbf{V}\cdot\nabla\mathbf{V} = -\nabla P + (1 + R_\nu(\Theta + z))\cdot\Delta\mathbf{V} + \mathbf{e}_z\,Gr\,(\Theta + z)$$
$$+R_\nu\cdot\mathbf{S}\times\nabla(\Theta + z), \tag{1}$$

$$\nabla\cdot\mathbf{V} = 0, \tag{2}$$

$$\frac{\partial\Theta}{\partial t} + \mathbf{V}\cdot\nabla\Theta = -V_z + \frac{1}{Pr}\cdot\Delta\Theta, \tag{3}$$

where $\mathbf{V} = (V_r, V_\varphi, V_z)$, $\Theta_0 = (T - T_0)/\Delta T$ is the dimensionless temperature with respect to the cold wall and Θ is the deviation from the linear temperature profile $\Theta = \Theta_0 - z$, and $\mathbf{S} = \partial V_i/\partial x_k + \partial V_k/\partial x_i$ is the strain rate tensor. The scales for time and pressure are $t_{ch} = d^2/\nu_0$ and $P_{ch} = \rho_0 {V_{ch}}^2$ respectively.

The following dimensionless parameters arise in the equations: the Prandtl number, the Grashof number, the Reynolds number and the aspect ratio i.e.,

$$Re = \frac{\sigma_T\Delta T d}{\rho_0\nu_0^2}, \qquad Gr = \frac{g\beta\Delta T d^3}{\nu_0^2}, \qquad Pr = \frac{\nu_0}{k}, \qquad \Gamma = \frac{d}{R_0}, \tag{4}$$

here k is the thermal diffusivity, g is the gravity acceleration.

A new parameter R_ν defines the relative variation of the kinematic viscosity.

$$R_\nu = \frac{\nu'(T_h) - \nu_0}{\nu_0} = \frac{1}{\nu_0}\frac{d\nu'}{dT}\Delta T, \qquad \nu = 1 + R_\nu(\Theta + z). \tag{5}$$

Here, ν - is dimensionless viscosity scaled by $\nu_0 = \nu(T_0)$.

At the rigid walls no slip and no penetration conditions are used and constant temperature is imposed:

$$\text{on the cold disk:} \quad \mathbf{V}(r,\varphi,z=0,t)=0, \quad \Theta(r,\varphi,z=0,t)=0, \tag{6}$$

$$\text{on the hot disk:} \quad \mathbf{V}(r,\varphi,z=1,t)=0, \quad \Theta(r,\varphi,z=1,t)=0. \tag{7}$$

On the cylindrical free surface ($r=1,\ 0\le\varphi\le 2\pi,\ 0\le z\le 1$) :

$$V_r=0, \quad (1+R_\nu(\Theta+z))\,\mathbf{S}\cdot\mathbf{e_r}+Re\left(\mathbf{e_z}\partial_z+\mathbf{e}_\phi\frac{1}{r}\partial_\phi\right)(\Theta+z) = 0. \tag{8}$$

Usually the free surface of the liquid bridge is assumed thermally insulated

$$\partial_r\Theta(r=1,\varphi,z,t)=0, \tag{9}$$

but in the present study it will not be always the case. The validity of this boundary condition will be discussed in the next section.

The time-dependent 3D equations, Eqs.(1)-(9) are solved by a finite volume method on a staggered grid. The physical domain $\bar{G}(0\le r\le 1,\ 0\le\varphi\le 2\pi,\ 0\le z\le 1)$ is discretized with a non-uniform cylindrical mesh in the radial and axial directions, and with an uniform mesh in the azimuthal direction. An implicit scheme for 2-nd order derivatives in the azimuthal direction and an explicit scheme for all the other spacial derivatives are applied. A combination of fast Fourier transform in the azimuthal direction and an implicit ADI method in the others directions is used for calculating the Poisson equation for the pressure.

The detailed description of the code validation and convergence on the grid can be found in [17].

4 Modelling of Heat Exchange on the Free Surface

Typically a numerical model has some limitations in comparison with experimental conditions. For liquid bridges one of the limitations can be dealt with by correct modelling of the heat transport on the free surface. The majority of the theoretical studies assume zero heat flux on the free surface, defined by (9).

To accommodate this boundary condition the temperature distribution in the ambient gas should correspond to the surface temperature distribution of the liquid. But the type of surface temperature distribution depends upon the Prandtl number. Now, a linear temperature profile along the free surface is observed for the small Prandtl numbers e.g., $Pr\approx 10^{-2}$. For the medium Prandtl numbers, $Pr\approx 1-7$, the temperature profile is of an **S** -shape. For large Prandtl numbers, ($Pr>30$) this **S**-shape is flattened a bit, and the temperature is almost constant in the central part of the free surface with strong variations near the hot and cold walls. The temperature profile, obtained by 3-D calculations for the experimental set of parameters and for $Pr=107$, is shown in Fig. 2. Probably, all of these cases cannot be described correctly with the frame wall of the same model.

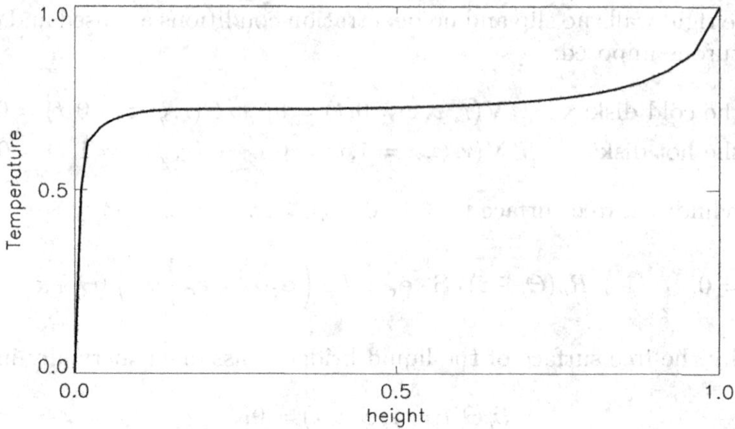

Fig. 2. Temperature distribution along the free surface. 3-D calculations when Pr=107, $Re = 110, R_\nu = -0.34969, Bi = 0$

On the boundary between the liquid surface and the gas Newton's law determines the heat transfer

$$\lambda \boldsymbol{n} \cdot \nabla T = -h(T - T_{amb}). \tag{10}$$

Here the dimensionless temperature is chosen as $T = T_0 + (\Theta + z)\Delta T$. Then Newton's law can be written as

$$\boldsymbol{n} \cdot \nabla \Theta = -\frac{hL}{\lambda}[(\Theta + z)\Delta T + \frac{T_0 - T_{amb}}{\Delta T}]$$

The heat transfer coefficient in dimentionless form is the Biot number i.e.,

$$Bi = \frac{hL}{\lambda_l},$$

where λ_l is the thermal diffusivity of the liquid and L is a length scale.

For comparison with experiments two points should be made clear: the value of the Biot number and the ambient temperature profile. The ambient temperature is usually assumed either constant or having a linear profile. The temperature of the gas far away from the surface (Bi=1) have been used by Neitzel [6] in the linear stability analysis. The study reveals an increase of the critical temperature difference (e.g. Marangoni number Ma_{cr}) when a non-zero Biot number is taken into consideration.

If T_{amb} is the linear temperature profile, e.g. $T_{amb} = T_0 + \frac{z}{L}\Delta T$ (10) will be written as

$$\boldsymbol{n} \cdot \nabla \Theta = -Bi\, \Theta \tag{11}$$

If T_{amb} is considered as a constant, Newton's law will be written as

$$\boldsymbol{n} \cdot \nabla \Theta = -Bi\, (\Theta + z + const), \tag{12}$$

where $const = (T_0 - T_{amb})/\Delta T$ is equal zero when $T_{amb} = T_0$. For the determination of the Biot number value, the main task is to evaluate the coefficient of heat transfer h with a good accuracy.

The Biot number is often obtained by the heat transfer calculated from the solid hot plate with a constant temperature to the liquid or to the gas. The values obtained such a way are used for the evaluation of heat transfer on the interface between liquid and gas, for example, from the surface of the falling liquid film to a gas in a technical applications. Here too the Biot number is obtained in a smarter way.

Following this the heat transfer from the gas side is described by the Nusselt number

$$Nu = \frac{Lh}{\lambda_{gas}} = f(Re_{gas}, Pr_{gas}).$$

From [4] it appears that heat flux, e.g. Nusselt number, is proportional to the square root of the Reynolds number

$$Nu = Re_{gas}^{1/2}\, F(Pr_{gas}). \tag{13}$$

where the Reynolds number is determined in a classical way $Re = LU_{gas}/\nu_{gas}$, following [4].

Then the heat transfer coefficient is determined as

$$h = \lambda_{gas}\, Re_{gas}^{1/2}\, F(Pr)/L$$

and Biot number $$Bi = \frac{\lambda_{gas}}{\lambda_l}\, Re_{gas}^{1/2}\, F(Pr) \tag{14}$$

One should notice that the Biot number does not contain the length scale. The dependence upon Prandtl number remains undetermined.

For high Pr numbers and laminar boundary layers this function can be fitted as

$$F(Pr) \approx 0.33\, Pr_{gas}^{1/3}, \quad \text{then} \quad Bi \approx 0.33 \frac{\lambda_{gas}}{\lambda_l}\, Re_{gas}^{1/2}\, Pr_{gas}^{1/3} \tag{15}$$

This formula is widely used in technical and scientific applications even for moderate Prandtl numbers.

For high Pr numbers and the turbulent boundary layer the dependence upon the Prandtl number can be written as $F(Pr) = const\, Pr^{3/4}$

For the liquids with Prandtl number close to one, $0.6 \le Pr \le 15$, Levich [7] has suggested the relation

$$F(Pr) = 0.5\, \alpha^{-1}(Pr), \tag{16}$$

where values $\alpha(Pr)$ are listed as the Table. For the present case of air, $Pr = k_{air}/\nu_{air} = 0.733$, it gives $\alpha(Pr) = 0.59$. The velocity of the gas U_{gas} is considered to be equal to the velocity of the liquid on the free surface. The 3-D calculations for $Bi = 0$ near the threshold of the instability, performed by the code described above, give the order of value $U_{surf} \approx 1.0\, cm/s$.

For air at the temperature 20°C the kinematic viscosity [4] is $\nu_{gas} = 0.15cm^2/s$, then $Re_{gas} = 3.84$. Also

$$\lambda_{gas} = 2.6 \cdot 10^3 erg\, s^{-1}\, cm^{-1} K$$

and

$$\lambda_{silicone\ oil} = 13.0 \cdot 10^3 erg\, s^{-1}\, cm^{-1} K$$

Substituting these physical values into (13) and using (16) we will get for the present experimental conditions

$$Bi = \frac{\lambda_{gas}}{\lambda_l} Re_{gas}^{1/2} F(Pr) \approx 0.48$$

The value of Biot number would be 3 times smaller that is, $Bi \approx 0.117$, if we would use (15).

5 Experimental Results

5.1 Stability Diagram

As the present experiments were conducted under terrestrial conditions the influence of gravity was unavoidable even for a tiny liquid bridge. The gravity effect results in the appearance of the bulk buoyancy force along with the existing Marangoni force and the deformation of the free surface. The relative importance of buoyancy and thermocapillary force is determined by the so called dynamic Bond number

$$Bo_{dyn} = \frac{Ra}{Ma} = \frac{Gr}{Re} = \frac{\rho g \beta d^2}{\partial\sigma/\partial T}$$

where Gr and Re are defined by (4) ($Ra = Gr * Pr$ and $Ma = Re * Pr$).

The volume of the liquid contained in the bridge and the deformation of the interface play an important role in the development of the instability. Previously it has been experimentally obtained by different scientific schools that for high Prandtl numbers the stability diagram, presented as a function of the critical Marangoni numbers versus variation of the liquid volume, consists of two different branches. Between these two oscillatory branches, there is a small range of volumes for which the steady flow is stable for very high values of the critical parameters, see Figs. 3–4. The same type of dependency was obtained numerically by linear stability analysis [2] when $Pr = 50$ and $\Gamma = 1.2$.

The present experimental study is one of the first which has determined that these two branches of the stability diagram, $(\Delta T_{cr}, V)$, correspond to different azimuthal wave numbers. The branch on which ΔT_{cr} grows with increasing volume belongs to $m_{cr} = 1$, and the descending branch belongs to the azimuthal wave number $m_{cr} = 2$. For the first time we have observed two different wave numbers along the stability diagram for the one of silicone oil 10 cSt, $Pr = 107$, when the aspect ratio was $\Gamma = 4/3$. These results were in contradiction with later numerical results in [2], where the both branches belong to the same wave number $m = 1$. To validate the comparison between numerical and experimental results, the experiments were repeated for the same aspect ratio as in [2], $\Gamma = 1.2$.

The experiments were conducted through a wide range of liquid bridge volumes, $(0.7 \leq V/V_0 \leq 1.2)$. Carrying out experiments in the slender liquid bridges is more simple than for rotund bridges, but there is a theoretical constraint for the small volumes. It is known [18] that for a given static Bond number $Bo_{st} = \rho g d^2/\sigma$ and slenderness $(L/2R)$ there is a minimal stable liquid bridge volume $(V/\pi d R^2)$. According to [18] for the present system, $Bo_{st} = 7.29$ and $\Gamma = 1.2$, this dimensionless minimal volume is a little bit less than $V \approx 0.7$. Therefore the minimal volume of liquid used in the experiments for slender liquid bridges was $V = 0.7$. The dynamic Bond number throughout this study was kept constant, $Bo_{dyn} = 2.3$.

The stability diagram $(\Delta T_{cr}, V)$ shown in Fig. 3 was obtained for 10 cSt silicone oil using the previous set up ($2R_0 = 6mm$, $d = 3.6mm$) not described above. The main points of the distinction are the following: the upper rod was made of brass; the central part of the lower rod, $2R = 4mm$, was made of glass and outer ring part was made of aluminum. This configuration permits visualization of the flow pattern from the bottom. The temperature of the cold rod was not strictly controlled but it was at room temperature, about $20-22°$C. Only 2 thermocouples of smaller diameter $D = 12\mu m$ were used. They were embedded from the upper rod at the same radial positions but at the different

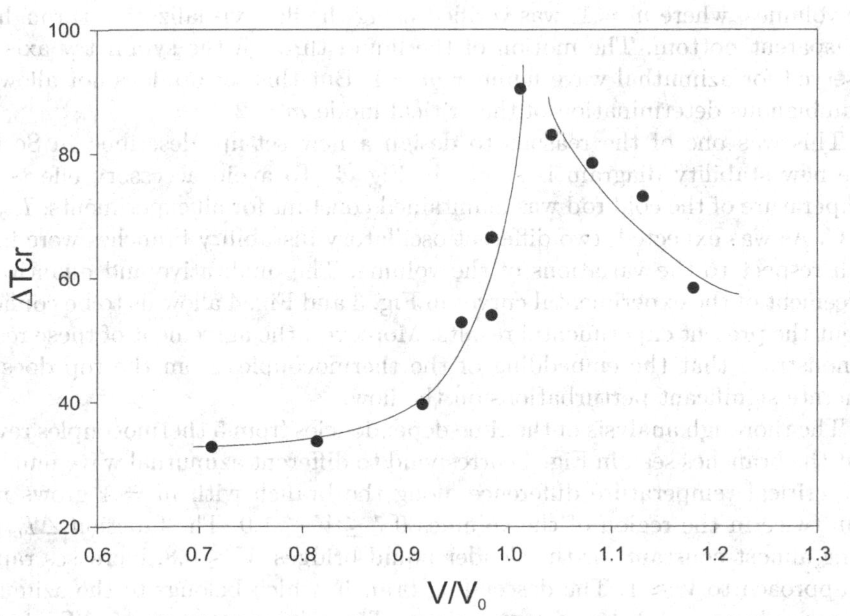

Fig. 3. Dependence of the critical temperature difference ΔT_{cr} upon a liquid bridge volume, $\Gamma = 1.2$, Pr=107, $Bo_{dyn} = 2.3$

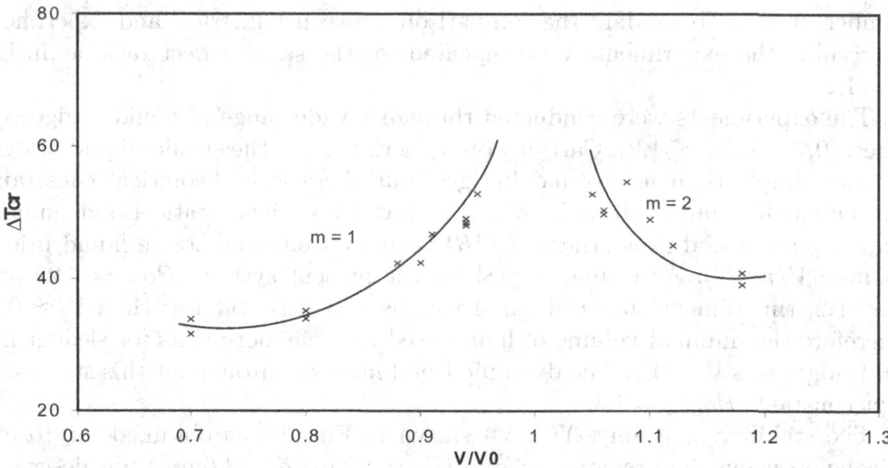

Fig. 4. Dependence of the critical temperature difference ΔT_{cr} upon a liquid bridge volume, $\Gamma = 1.2$, Pr=105, $Bo_{dyn} = 2.3$. The branches have different azimuthal wave numbers, $m = 1$ and $m = 2$

depths: 1.3mm and 0.7mm. In the azimuthal direction they were remotely placed by 180°. The stability diagram with good accuracy was obtained using this set up. Along with the analysis of the signals from thermocouples, the region of the volumes, where $m = 1$, was verified using the flow visualization through the transparent bottom. The motion of the liquid through the symmetry axis was observed for azimuthal wave number $m = 1$. But this set-up does not allow the unambiguous determination of the critical mode $m = 2$.

This was one of the reasons to design a new set-up, described in Sect. 3. The new stability diagram is shown in Fig. 4. To avoid accessory effects, the temperature of the cold rod was maintained constant for all experiments, $T_{cold} =$ 22°C. As was expected, two different oscillatory instability branches were found with respect to the variations of the volume. The qualitative and quantitative agreement of the experimental curves in Fig. 3 and Fig. 4 allow us to be confident about the present experimental results. Moreover, the agreement of these results demonstrate that the embedding of the thermocouples from the top does not generate significant perturbations on the flow.

The thorough analysis of the time dependencies from 5 thermocouples reveals that the branches seen in Fig. 4 correspond to different azimuthal wave numbers. The critical temperature difference along the branch with $m = 1$ grows more than twice in the region of the volumes $0.7 \leq V \leq 1.0$. The function $\Delta T_{cr}(V)$, being almost constant for the slender liquid bridges, $V \leq 0.8$, increases rapidly on approach to $V \approx 1$. The descending branch, which belongs to the azimuthal wave number m = 2, has a larger slope. The minimum value of ΔT_{cr} for the slender liquid bridges is smaller, then for rotund ones, compare $\Delta T_{cr} = 31$°C at $V \approx 0.72$ versus $\Delta T_{cr} = 40$°C at $V \approx 1.18$.

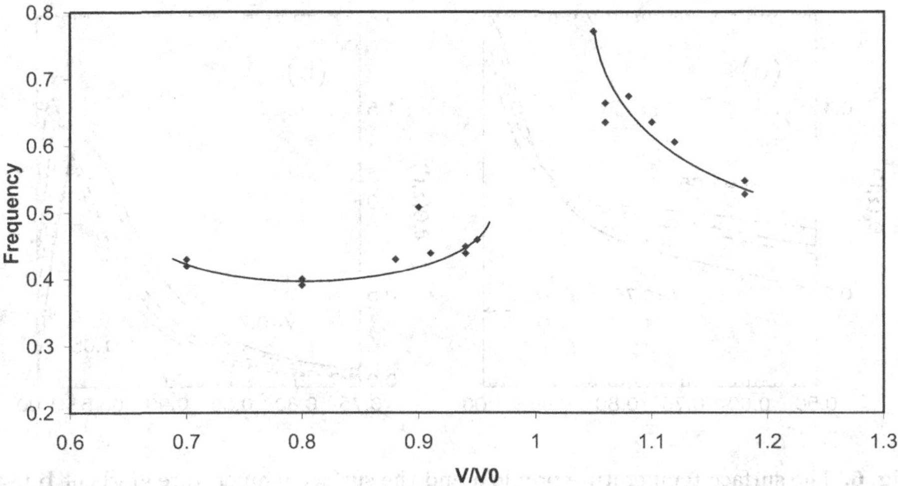

Fig. 5. Dependence of the critical frequency upon a liquid bridge volume, $\Gamma = 1.2$, $Bo_{dyn} = 2.3$

Our experiments do not confirm the numerical results [2], where both branches belong to the same wave number $m = 1$, although they correspond to the same aspect ratio $\Gamma = 1.2$. This may happen due to the different Prandtl number, Pr=50 versus Pr=107, or due to other reasons.

The value of the frequency at the threshold of instability is another important characteristic of the flow pattern. The critical frequency, f_{cr}, is also affected by a liquid bridge volume. Figure 5 shows that the dependency f_{cr} versus V consists of two branches similar to the stability diagram. The variation of the f_{cr} for the slender liquid bridges is weak, $0.39 \leq f_{cr} \leq 0.46$, and it has a pronounced minimum at $V = 0.8$. The branch for rotund liquid bridges has a steep descent. The critical frequency f_{cr} is smaller for the slender liquid bridges in comparison with a fat one. The range of the variation of f_{cr} is totally different for two branches. Moreover, two branches never have the same f_{cr}. The minimum f_{cr} for the fat liquid bridges is higher than any f_{cr} for the slender liquid bridges. This is one more strong point in favour of the fact that the second branch belongs to another wave number. The frequency skip (jump) is always observed when the system has changed the mode of the oscillatory regime.

The experimental result in deformed liquid bridges currently is ahead of the theory. The only explanation of the instability mechanism, presently existing, has been proposed by Shevtsova and Legros [15]. It pertains to the flow instability in deformable cylinders with wave number m = 0 and can be partially extended for arbitrary m.

For the no-deformable free surface no instability $m = 0$ was found in [15]. The main reason for the instability in the case of a deformable free surface is the existence of a cold finger near the hot disk. It appears that the stability of the system is very sensitive to the value of the upper contact angle. Surface

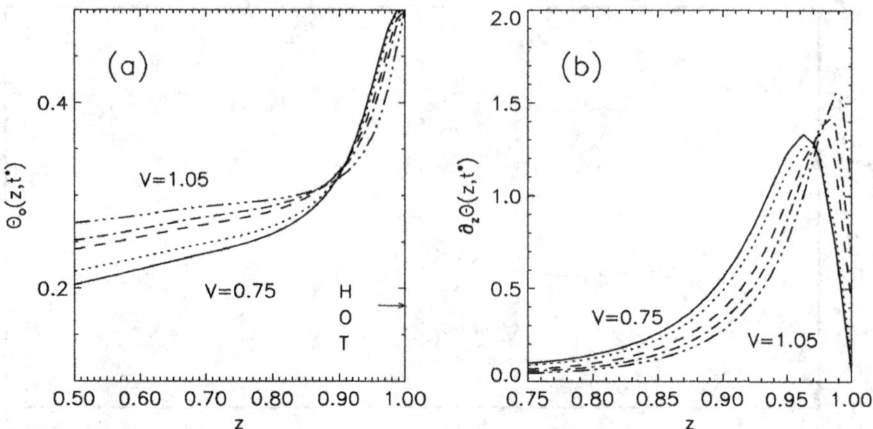

Fig. 6. The surface temperature profile **a** and the surface temperature gradient **b** near the hot rigid wall for different liquid bridge volumes. Numerical calculations have been done for the deformed free surface [15]; $Pr = 105$, $\Gamma = 4/3$, $Re/Re_{cr} = 1.03$

deformation changes the distribution of the temperature near the hot wall. In the case of a straight cylinder with an upper contact angle $\alpha = 90°$ the maximum of the temperature gradient is located on the hot wall. In the case of a deformed free surface the isotherms slide from the sharp corner into the interior (see Fig. 6a).

The peak of the local temperature gradient, driving the main flow, moves away from the hot wall when the upper contact. For instance, for a volume V = 0.8, the maximum of the temperature gradient is located at 5 – 7% of the distance from the hot wall, see Fig. 6b. The positions of this maximum and of the steep branch of the temperature gradient, with a positive slope, coincide with the area of influence of the cold finger on the free surface. In this case, a small disturbance of the surface temperature has a very strong impact on the accelerations or decelerations of the main flow. It means that the smaller ΔT should be applied to destabilize the flow. On the other hand, the deformation of the liquid bridge produces a narrow neck near the hot wall. This is particularly true under gravitational conditions, when the contact angle is below 90°, even for liquid bridges with $V > 1$. As a result, the distance between the cold finger and the free surface is shorter and the sensitivity of the free surface is increased. With increasing volume, at a fixed aspect ratio, the neck widens and it is necessary to apply a higher ΔT to initiate oscillations.

From the above explanations we can draw the conclusion that the position of the gap between branches should depends upon aspect ratio. This is in agreement with what was observed in different experiments, [13].

5.2 Existence of Mixed Modes

A change of flow structure with a increase of temperature difference study beginning from the threshold of instability up to the second transition is studied. The

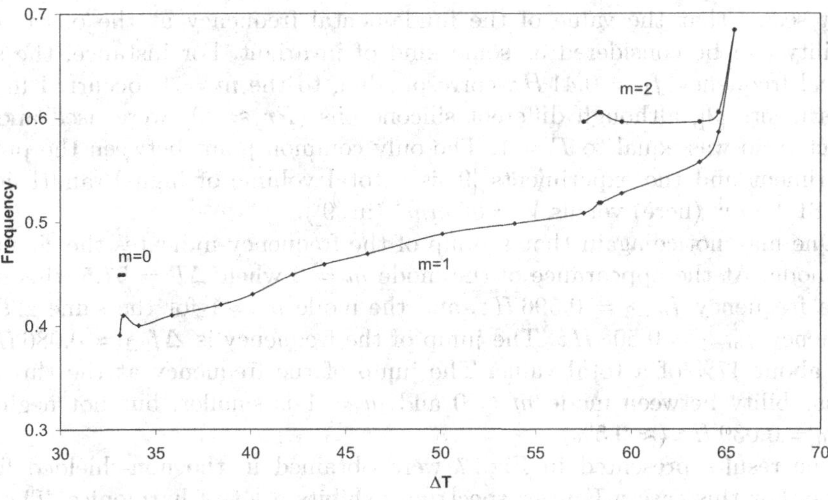

Fig. 7. The dependence of the frequency upon the temperature difference in the supercritical area, $(\Delta T \geq \Delta T_{cr})$. Experimental results, $V = 0.8, \Gamma = 1.2$

experiments were conducted using the set-up described in Sect. 3 in shielded and non-shielded liquid bridges for $V = 0.8$. To describe the flow above the onset of instability a new parameter

$$\epsilon = (\Delta T - \Delta T_{cr})/\Delta T_{cr}$$

describing the distance from the critical point is introduced.

The variation of the temperature with time at different azimuthal positions was recorded by five thermocouples. The signals from different thermocouples were compared, and the decision about the critical azimuthal mode was taken analyzing the phase shift between signals. Moreover, the type of hydrothermal wave can be also determined: either standing or travelling one. The analysis of the time series and of the corresponding phase shift between thermocouples tells us that the instability began nearly as a standing wave with a wave number m = 1 which then switched to a m = 1 travelling wave at $\epsilon = 0.2$. At the next step the Fourier analysis is applied to the time-dependent signals. The detailed analysis of the power spectrum reveals a mixed mode m = 0 and m = 1 at the threshold of instability. The dependence of the frequency upon the temperature difference in the supercritical area $(\Delta T > \Delta T_{cr})$ is shown in Fig. 7. The critical ΔT is found to be $33.2 \pm 0.1°$C. Two independent fundamental frequencies appear just above the onset of time-dependent convection $f_0 = 0.449$Hz $(m = 0)$ and $f_1 = 0.41$Hz $(m = 1)$. The mode $m = 0$ exists only in the narrow region $0 \leq \epsilon < 0.05$. The frequency of the second one, $m = 1$, slowly increases by moving further into the supercritical area $(\Delta T > \Delta T_{cr})$. Somewhat far from the onset of the oscillatory instability $(\epsilon \geq 0.7)$ an additional frequency appears, which corresponds to the wave number $m = 2$. This mode coexists with mode m = 1 within some range of ϵ $(0.7 < \epsilon < 1.0)$ and then becomes dominant.

It seems that the value of the fundamental frequency at the onset of instability can be considered as some kind of invariant. For instance, the same critical frequency $f_1 = 0.41\,Hz$ corresponding to the $m = 1$, occurred in past experiments [9], although different silicone oils ($Pr \approx 40$) were used and the aspect ratio was equal to $\Gamma = 1$. The only common point between the present experiment and the experiments [9] is a total volume of high Prandtl liquid: $V = 81.4\,mm^3$ (here) versus $V = 85\,mm^3$ (in [9]).

One may notice again that a jump of the frequency indicates the change of the mode. At the appearance of the mode $m = 2$ when $\Delta T = 57.5$, this mode has a frequency $f_{m=2} = 0.596\,Hz$, and the mode $m = 1$ for the same ΔT has frequency $f_{m=1} = 0.508\,Hz$. The jump of the frequency is $\Delta f_{2,1} = 0.086\,Hz$ or it is about 17% of a total value. The jump of the frequency at the threshold of instability between mode $m = 0$ and $m = 1$ is smaller, but not negligible $\Delta f_{1,0} = 0.039\,Hz$ ($\approx 9.5\%$).

The results presented in Fig. 7 were obtained in the non-shielded liquid bridge. For this case a Fourier spectrum exhibits a lot of harmonics. The first super harmonic $f = 2f_1$ with a distinguishable tiny amplitude appears in the spectrum practically after vanishing of the mode $m = 0$. Similar experiments were conducted in the shielded liquid bridge. Unlike the previous case, the power spectrum contains only the fundamental frequency $f = 0.459\,Hz$ which slowly grew with increasing ΔT. The level of the noise in the Fourier spectrum was very low and the first harmonics with distinguishable amplitude appear far from the critical point, $\epsilon \approx 0.29$

5.3 Variation of the Temperature of the Cold Rod

One of the goals of the present experimental study is to investigate the influence of the temperature of the cold endwall on the onset of instability. These experiments were carried out in the shielded liquid bridge, when the temperature of the water in the surrounding pipe was equal to the temperature of the cold rod. An experimental study has been done for three different volumes of liquid contained in the bridge while the temperature of the lower rod was varying in a large range, $10°\mathrm{C} < T_{cold} < 32°\mathrm{C}$.

The experimental results for the smallest volume, $V/V_0 = 0.7$, are presented in Fig. 8. The critical temperature difference, ΔT_{cr}, is shown as a function of the temperature of the cold rod, T_{cold}. The crosses correspond to the experimental points. They can be fitted by the linear law

$$\Delta T_{cr} = 39.03 - 0.347 \cdot T_{cold}$$

and the solid line in Fig. 8 draws this curve fit. The critical temperature difference strongly diminishes with the increase of T_{cold}. Indeed, the ΔT_{cr} reduces by 25%, when the temperature of the cold rod is changed by 21°C. The variations of the temperature of the cold rod alter the mean temperature in the system, and, of course, the characteristic value of the viscosity, which is somewhat sensitive to the temperature. Hence, the graph also captures the dependence of ΔT_{cr} upon the Prandtl number. For this particular volume of liquid, $V/V_0 = 0.7$, the azimuthal wave number remains unchanged with the variation of the temperature of the cold rod, $m = 1$.

For a larger volume of the liquid, $V = 0.8$, the dependence ΔT_{cr} versus T_{cold} remains a decreasing function. But, it was found that not only the critical ΔT_{cr} (or Marangoni number) strongly depends upon the mean temperature in the system, but the critical wave number is switched from m = 1 to $m = 2$ at some $T_{cold} = T^*_{cold}$. The dependence of the critical temperature difference upon the temperature of the cold rod for this case is shown in Fig. 9, where the crosses indicate the experimental points. The careful analysis of the temperature time series, $T(t)$, exhibits, that the two various groups of the experimental points belong to different wave numbers. The results can be fitted by straight lines with similar slopes:

$$\Delta T_{cr} = 45.89 - 0.556 \cdot T_{cold} \qquad (m = 1)$$
$$\Delta T_{cr} = 47.09 - 0.542 \cdot T_{cold} \qquad (m = 2)$$

As it follows from Fig. 9 the mode $m = 1$ is always observed for the lower mean temperatures, e.g. when the temperature of the cold rod varies in the range $10°\mathrm{C} < T_{cold} < 20.7°\mathrm{C}$. The mode $m = 2$ exists for $T_{cold} > 20.7°\mathrm{C}$. Thus, the transition between the modes takes place at $T^*_{cold} = 20.7°\mathrm{C}$. A lot of experiments were conducted in the vicinity of this point, $T^*_{cold} = 20.7°\mathrm{C}$, which have confirmed a strict separation of the domains of existence of the different

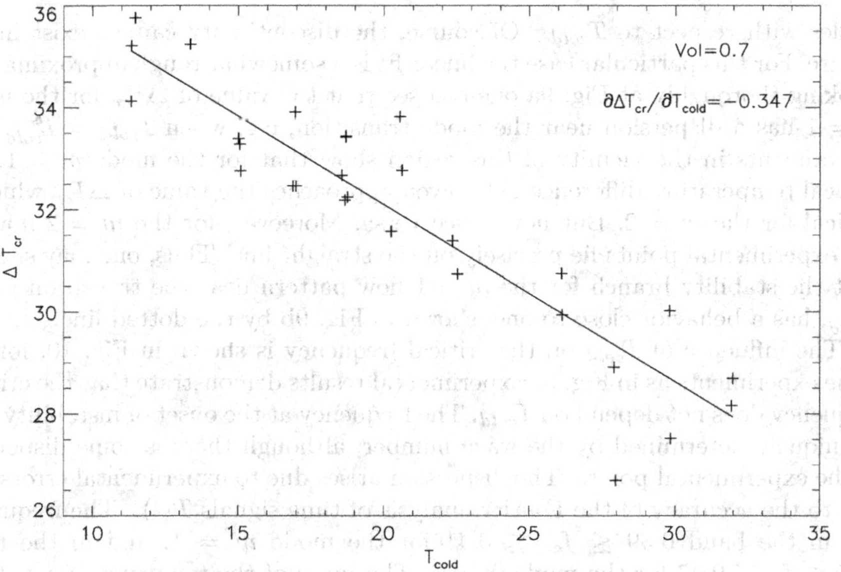

Fig. 8. The dependence of the critical temperature difference upon the temperature of the cold endwall, $V = 0.7$, $\Gamma = 1.2$. The crosses correspond to the experimental points and solid line draws the linear interpolation of the results

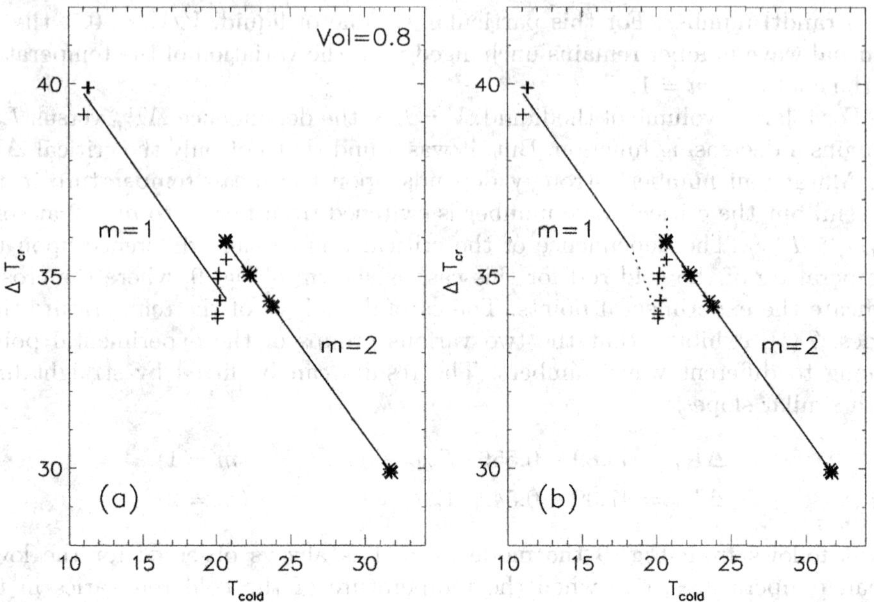

Fig. 9. The dependence of the critical temperature difference upon the temperature of the cold rod, $V = 0.8$, $\Gamma = 1.2$. The crosses correspond to the experimental points and solid lines draw the linear approximations; (b) the judicious fitting near the mode transition is shown by the dotted line

modes with respect to T_{cold}. Of course, the discontinuity cannot exist in the nature. For this particular case the linear fit is a somewhat rough approximation. Looking thoroughly at Fig. 9a one can see that the value of ΔT_{cr} for the mode $m = 1$ has a dispersion near the mode transition, e.g. when $T_{cold} = T^*_{cold} \pm \delta$. Experiments in the vicinity of this region show that for the mode $m = 1$, the critical temperature difference ΔT_{cr} even approaches the value of ΔT_{cr} which is typical for the $m = 2$. But never vice versa. Moreover, for the $m = 2$ flow all the experimental points lie precisely on the straight line. Thus, one may suggest that the stability branch for the $m = 1$ flow pattern near the transition point, T^*_{cold}, has a behavior close to one, shown in Fig. 9b by the dotted line.

The influence of T_{cold} on the critical frequency is shown in Fig. 10, for the same experiments as in Fig. 9. Experimental results demonstrate that the critical frequency does not depend on T_{cold}. The frequency at the onset of instability, f_{cr}, is uniquely determined by the wave number, although there is some dispersion of the experimental points. The dispersion arises due to experimental errors and due to the accuracy of the Fourier analysis of time signals $T(t)$. The frequency lies in the band $0.39 \leq f_{cr} \leq 0.41$ for the mode $m = 1$, and in the band $0.45 \leq f_{cr} \leq 0.47$ for the mode $m = 2$. The jump of the frequency, $\Delta f \approx 0.06$, also indicates a distinction of the mode.

With further increasing of the liquid volume, $V/V_0 = 0.9$, the dependence of ΔT_{cr} versus T_{cold} turns out to be similar to that of $V/V_0 = 0.7$. Not only are

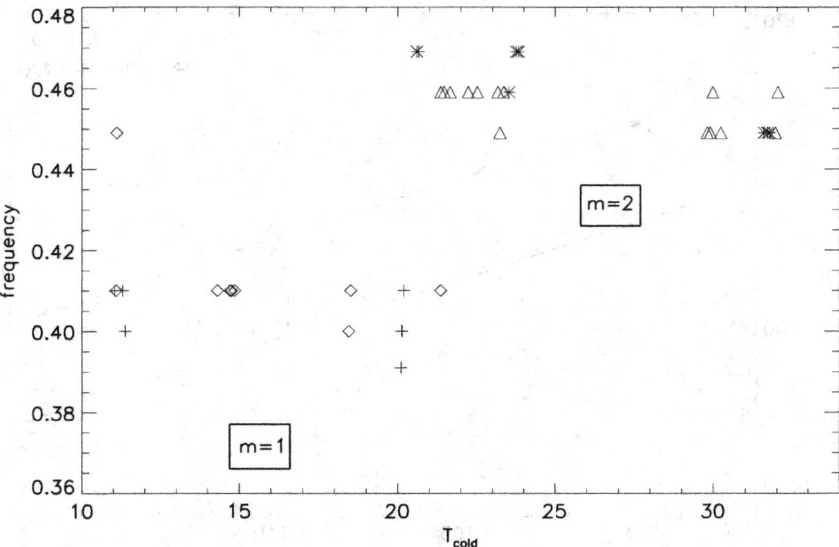

Fig. 10. The dependence of the critical frequency upon the temperature of the cold rod. The f_{cr} does not depend upon the temperature of the cold endwall. The different modes of flow organization have various critical frequencies

Table 2. The evolution of the dependence ΔT_{cr} vs. T_{cold} with the volume of liquid

V/V_0	0.7	0.8	0.9
$\partial\Delta T_{cr}/\partial T_{cold}$ $(m=1)$	−0.347	−0.556	−0.363
$\partial\Delta T_{cr}/\partial T_{cold}$ $(m=2)$		−0.542	

the slopes of the linear fitting comparable, but the critical wave numbers are also equal, i.e., $m = 1$. The evolution of the slopes of the linear fittings for the various volumes are presented in Table 2. The comparison of the data in Table 2 manifests a distinctive behaviour of the flow structure for the $V/V_0 = 0.8$

Analyzing the shapes of the interfaces for the different volumes, we have found that in the case of $V/V_0 = 0.9$ the shape of the experimental liquid bridge approaches the right-circular cylinder more than that correspond to other smaller or larger volumes. This allows us to make a comparison with 3-D numerical simulations for the right-circular cylinder. Therefore the experimental results for the $V/V_0 = 0.9$ are shown in Fig. 11 as a function Re_{cr} versus Pr instead (ΔT_{cr} vs. T_{cold}).

We remind, that for re-calculation of the experimental points via Re and Pr

$$Re_{cr} = \frac{\sigma_T \Delta T_{cr} d}{\rho_0 \nu_0^2}, \qquad Pr = \frac{\nu_0}{k}$$

the value of the viscosity at the temperature of the cold endwall is used, $\nu_0 = \nu(T_{cold})$. Usually the value of the viscosity is given in handbooks at room temper-

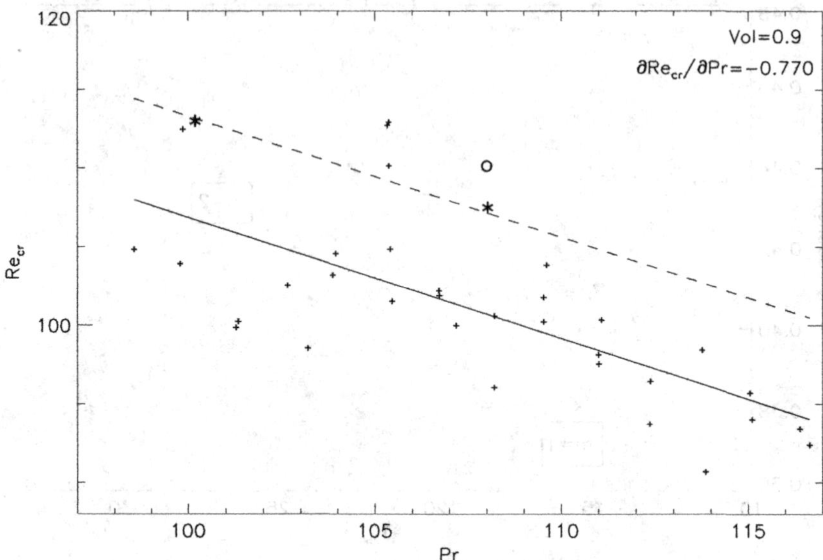

Fig. 11. The dependence of the critical Reynolds number on the Prandtl number, $V = 0.9$, $\Gamma = 1.2$. The crosses correspond to the experimental points and solid line draws the linear interpolation of them; the stars and the dashed line correspond to the numerical results for $Bi = 0.48$; the circle indicates the numerical result for $Bi = 5.0$

ature (T_r) or for some short range of temperatures. Therefore to make the plot, the values of viscosity at T_{cold} have been calculated using the linear dependence

$$\nu(T_{cold}) = \nu(T_r) + \nu_T(T_r - T_{cold}), \qquad \nu_T = \partial_T \nu = const. \tag{17}$$

The 3-D numerical calculations have been done for the same dependence of viscosity upon temperature as in (17), which in the present case corresponds to $R_\nu = -0.38148$. The calculations have been performed for different Biot numbers using the linear temperature profile in the ambient gas, i.e., using (11).

The solid line in Fig. 11 corresponds to the linear fit of the experimental points, which are shown by small crosses. The dashed line displays the linear fit of the numerical points obtained for $Bi = 0.5$, which are shown by stars at $Pr = 100$ and at $Pr = 107.83$. Numerical and experimental results are in excellent qualitative and quantitative agreement. The slopes of the linear fits are equal, and the calculated values of Re_{cr} are only 5% higher than the experimental ones. This divergence can be explained by the assumptions about the shape of the interface. Note, that the key point for this agreement is that the numerical code takes into account the dependence of the viscosity upon the temperature throughout the bulk. To emphasize this, the calculations with a constant bulk viscosity, $R_\nu = 0$, and $\nu_0 = \nu(T_{cold})$ have been done for $Pr = 107.83$. The determined value of the critical Reynolds number, $Re_{cr} = 140$, exceeds the experimental results by 40%. The divergency is so large, that it is impossible to put this point on Fig. 11.

The numerical values of the Re_{cr} nearly coincide for different thermal conditions: in the absence of the heat transfer through the interface $Bi = 0$ and for the value $Bi = 0.48$ estimated in Sect. 4. Although one may find that the Re_{cr} for $Bi = 0.48$ is a little bit higher than for $Bi = 0$. In the scale of Fig. 11 they are not distinguishable. To clarify the role of the Biot number the simulations have been done with an intentionally magnified value of Biot number, $Bi = 5.0$. The effect of Biot number on the Re_{cr} is much weaker than variability of the viscosity. The numerical value for this case, $Bi = 5.0$, $Re_{cr} = 110$ is shown by circle in Fig. 11. The presence of the Biot number tends to stabilize the thermocapillary flow, although the difference is not significant, it is only about 3%, e.g. $Re_{cr} = 107$ for $Bi = 0$ versus $Re_{cr} = 110$ for $Bi = 5.0$

Note, that the way of the presentation of the experimental results is also important. For instance, for the same experimental results $V = 0.9$ the value of the slope $\partial Re_{cr}/\partial Pr$ is almost twice as large as $\partial \Delta T_{cr}/\partial T_{cold}$, e.g. 0.770 versus 0.363.

6 Conclusions

The appearance and the development of the thermoconvective oscillatory flows are investigated experimentally and numerically in a liquid bridge of aspect ratio $\Gamma = 1.2$ formed by 10 cSt silicone oil. The experiments were carried out under terrestrial conditions in a tiny liquid bridge for a wide range of liquid bridge volumes, $0.7 \leq V/V_0 \leq 1.2$.

Two different set-ups were built to study the dependence of critical temperature difference upon volume of a liquid. It is seen that the embedding of the thermocouples from the top does not generate significant perturbations on the flow, e.g. the values of the critical parameters remain independent of the number of thermocouples. The tests have been done with 2 and 5 thermocouples.

For the first time two branches of the stability diagram, (ΔT_{cr} vs. V/V_0), belonging to different azimuthal wave numbers were experimentally obtained. The branch on which ΔT_{cr} grows with increasing volume belongs to $m_{cr} = 1$, and the descending branch belongs to the azimuthal wave number $m_{cr} = 2$. The jump of the critical frequency between the different branches confirms the distinction of the wave numbers.

An experimental study has been done when the temperature of the lower rod is varied in a large range $10°\text{C} < T_{cold} < 32°\text{C}$. The variation of T_{cold} results in the change of the mean temperature in the system and some parameters, e.g. Prandtl number. It is seen that ΔT_{cr} decreass with increasing the temperature of the cold endwall. For one particular volume of liqud, $V/V_0 = 0.8$, the azimuthal wave number switches from $m = 1$ to $m = 2$ when $T_{cold} = 20.7°\text{C}$, although for the volumes $V/V_0 = 0.7$ and $V/V_0 = 0.9$ the azimuthal wave number remains constant, $m = 1$.

The full 3-D non-steady Navier-Stokes equations have been solved numerically in cylindrical columns. Both thermocapillary and buoyancy mechanisms of convection are included. The characteristic parameters were chosen as close as possible to the experimental ones. Numerical models have some limitations in comparison with experimental conditions. Some of the limitations can be reduced by using variable viscosity throughout the bulk and by correct modelling of the heat transport on the free surface. As such the dependence of the viscosity upon the temperature is taken into account. A method of the determination of the Biot number is suggested, according to which $Bi \approx 0.5$ for the present physical system. The numerical and experimental results are then in good agreement.

References

1. L. Carotenuto, D. Castagnolo, C. Albanese, R. Monti: Phys. Fluids **10**, 555 (1997)
2. Q.S. Chen, W.R. Hu: Int.J.Heat Mass Transfer **41**, 825 (1998)
3. H. Kawamura, I. Ueno, S. Tanaka, D. Nagano: 'Oscillatory, chaotic and turbulent thermocapillary convections in a half-zone liquid bridge'. In: *Proceedings of 2nd Int. Symp. Turbulence& Shear Flow Phenomena (TSFP2)*, (Stockholm, Sweden 2001)
4. L.D. Landau, E.M. Lifshitz: *Hydrodynamics*, (Nauka, Moscow, Russia 1976)
5. M. Lappa, R. Savino, R. Monti: Int. J. Heat Mass Transfer **44**, 1983 (2001)
6. M. Levenstamm, G. Amberg: J. Fluid Mech. **297**, 357 (1995)
7. V.G. Levich: *Physiochemical Hydrodynamics*, (Prentice-Hall, Inc., Englewood Cliffs, N.J. 1962)
8. J. Leypoldt, H.C. Kuhlmann, H.J. Rath: J. Fluid Mech. **414**, 285 (2000)
9. K.A. Muehlner, M. Schatz, V. Petrov, D. McCormic, J.B. Swift, H.L. Swinney: Phys. Fluids **9**, 1850 (1997)
10. G.P. Neitzel, K.T. Chang, D.F. Jankowski, H.D. Mittelman: Phys. Fluids A **5**(1), 108 (1993)
11. C. Nienhüser, H.C. Kuhlmann, H.J. Rath: J. Fluid Mech. **458**, 35 (2002)
12. F. Preisser, D. Schwabe, A. Scharmann: J. Fluid Mech. **126**, 545 (1983)
13. M. Sakurai, N. Ohishi, A. Hirata: Adv. Space Res. **24**, 1379 (1999)
14. D. Schwabe, S. Frank: J. Jpn. Soc. Microgravity Appl. **15**, Supplement II, 431 (1998)
15. V.M. Shevtsova, J.C. Legros: Phys. Fluids **10**, 1621 (1998)
16. V.M. Shevtsova, M. Mojahed, J.C. Legros: Acta Astronautica **44**, 625 (1999)
17. V.M. Shevtsova, D.E. Melnikov, J C. Legros: Phys. Fluids **13**, 2851 (2001)
18. L.A. Slobozhanin, J.M. Perales: Phys. Fluids A **5**, 1305 (1993)
19. M. Wanschura, V.M. Shevtsova, H.C. Kuhlmann, H.J. Rath: Phys. Fluids **5**, 912 (1995)
20. M. Wanschura, H.C. Kuhlmann, H.J. Rath: 'Instability of the thermocapillary flow in symmetrically heated full liquid zones'. In: *Proceedings of the Joint Xth European and VIth Russian Symposium on Physical Science in Microgravity,Russia, St.Petersburg*, ed. by V.S. Avduevsky, V.I. Polezhaev, **1**, p. 172 (1997)
21. J.-J. Xu, S.H. Davis: Phys. Fluids **27**, 1102 (1984)

Thermocapillary Droplet Migration on an Inclined Solid Surface

Marc K. Smith[1], Steven W. Benintendi[2], and Cavelle P. Benjamin IV[1]

[1] The George W. Woodruff School of Mechanical Engineering, Georgia Institute of Technology, Atlanta, Georgia 30332-0405, USA
[2] Department of Mechanical and Aerospace Engineering, University of Dayton, Dayton, Ohio 45469, USA

Abstract. Active control of the position of a liquid droplet on a solid surface is a crucial part in the design of discrete fluid management technology for microfluidic applications. One way to accomplish this control is to impose specially shaped thermal fields upon the droplet and/or the solid surface. The imposed temperature gradient produces a surface-tension-driven flow inside the droplet that forces the motion of the contact line. When the imposed temperature gradient is large enough, this motion causes the droplet to migrate in the direction of decreasing temperature. In this paper, a detailed lubrication theory is presented that describes this internal flow and the subsequent contact-line motion in a thin droplet. Results are presented to show that this technique can be used to drive a droplet up an inclined solid surface against the force of gravity.

1 Introduction

Free surface flows driven by a surface-tension gradient have been studied for over a hundred years. One of the first published accounts of this effect was by Marangoni [30], who used it to explain the presence of wine tears on the side of a wine glass. Thermally induced, surface-tension gradients (thermocapillarity) are important in those applications where heat transfer and free surface flows are present. Thermocapillarity has been used to explain some of the instabilities seen in heated liquid films [34], and various other phenomena seen in diverse engineering applications such as materials processing, laser welding, liquid film cooling, and film coating. Over the last decade or so, attention has been focused on using surface-tension gradients to control the motion and behavior of liquid films or droplets. Such control can be used to suppress free-surface instabilities [31,36], drive the motion of liquid films over surfaces [5], control the motion of liquid droplets on solid surfaces [41], and enhance heat transfer in engineering systems [33]. In the relatively new area of microfluidics, surface-tension control can be used to provide fluid management in microdevices for biomedical applications, such as genetic assays, for chemical processing and sensors, and for environmental testing.

In the present paper, thermocapillary forces will be studied as a means to control the motion of a liquid droplet on a solid surface. The model problem used in this study is a two-dimensional, liquid droplet on a non-uniformly heated or cooled, inclined solid surface, bounded above by a passive gas. The liquid droplet

M.K. Smith, S.W. Benintendi, and C.P. Benjamin IV, Thermocapillary Droplet Migration on an Inclined Solid Surface, Lect. Notes Phys. **628**, 263–289 (2003)
http://www.springerlink.com/

is small (appropriate for microsystems) and so the parameter regime is one in which inertial forces are negligible, but the forces associated with surface tension, surface-tension gradients, gravity, viscous effects, and contact-line physics are all the same order of magnitude. It will be shown that for a given solid-liquid-gas system, thermocapillarity can be used to externally control the flow inside the droplet, the motion of the contact lines, and thus the bulk motion of the droplet on the solid surface.

A contact line is formed when three immiscible material phases are in mutual contact, such as a water droplet on a glass substrate in an air environment. The theoretical analysis of droplet migration faces the formidable problem of a moving contact line. Huh and Scriven [27] and Dussan and Davis [11] showed that using the no-slip boundary condition in this problem gives rise to a nonintegrable shear-stress singularity at the contact line. Dussan and Davis [11] showed that the singularity was the result of a dynamic incompatibility in the mathematical modeling assumptions. Contact-line motion gives rise to a multi-valued velocity field at the contact line that cannot be obtained using the continuous field equations that describe fluid motion.

A second difficulty in modeling this problem is to describe the true nature of the contact angle or the slope of the free surface at the contact line. Under static conditions, the solid-liquid-gas interactions at the contact line are accounted for by prescribing the value of the contact angle, thus providing a boundary condition for the shape of the free surface. When the contact line moves, it is not clear if the contact angle should be specified, or whether it is a function of other variables, such as the contact-line speed.

Both of these modeling issues can be isolated from the study of the bulk motion of the droplet because the length scale associated with the physics of the contact line is so much smaller that the length scale of the droplet. Thus, the moving contact-line problem should be studied using asymptotic techniques, as done by Hocking [22,23], Hocking and Rivers [24], Dussan [12,13], Huh and Mason [28], Cox [8], and many others. The difficulty with this is that at a microscopic length scale very near to the actual contact line the governing physics is still not known. Many modeling assumptions have been explored, such as slip of the fluid along the solid surface [8,11–13,19,22–24,28], a fixed microscopic contact angle [8,22–24], precursor films [9,44], non-Newtonian fluids [11], disjoining potential forces [25,35], dynamic contact-angle conditions [2,21], and non-equilibrium thermodynamics [39,40], for example. Given a particular model for the contact-line physics and its solution for the local flow field, the result is then matched to the flow field far from the contact line using asymptotic techniques. Thus, no matter what the physics in the contact-line region, the effect on the flow of the bulk droplet is reduced to an effective boundary condition for the bulk flow. The purpose of the asymptotic method is to determine the correct form of this boundary condition.

A more pragmatic approach to determining the effective boundary conditions at the contact line is to use an empirical correlation. One popular correlation is a relation between the apparent contact-line speed and the apparent contact angle. The apparent contact angle is the angle that the free surface makes with the solid surface when viewed on a macroscopic length scale. The experiments of Schwartz and Tejeda [37], Hoffman [26], Tanner [43], and Chen [7] provide such a relation. Many authors, [2,16,19–21,41] for example, have used this approach to successfully compute contact-line motion.

One of the first papers studying the motion and control of a liquid droplet on a solid surface was Greenspan [19]. He considered the spreading or retraction of an isothermal axisymmetric droplet due to a nonequilibrium initial shape. The model consisted of a Maxwell slip law and a linear relationship between the apparent contact angle and the apparent contact-line speed. By making the equilibrium contact angle a decreasing function of position on the solid surface, he showed that the droplet would migrate in the direction of decreasing contact angle. Experimentally, this non-uniform wettability can be realized by coating the solid surface in a specified manner. Chaudhury and Whitesides [6] experimentally confirmed this type of behavior by placing a water droplet on a specially prepared silicon wafer. When they inclined the solid surface from the horizontal, the droplet migrated uphill against the force of gravity.

The sliding of a liquid droplet down an isothermal, inclined solid surface due to gravity was examined by Hocking [22] using a thin, two-dimensional droplet model and asymptotic techniques. Dussan and Chow [14] extended this model to a thin, three-dimensional droplet and Dussan [15] removed the small-slope limitation. Dimitrakopoulos and Higdon [10] considered a three-dimensional droplet with arbitrary contact angles governed by a Stokes-flow approximation. They used a spectral boundary element method to compute the droplet motion.

Ehrhard and Davis [16] considered a thin droplet spreading on a uniformly heated or cooled solid surface. Their results showed that when the solid surface is uniformly heated, thermocapillarity flows are established that retard the spreading process, and when the solid surface is uniformly cooled, spreading is enhanced. Ehrhard [17] performed experiments on isothermal and uniformly heated or cooled droplets that compared favorably with the earlier theoretical predictions of Ehrhard and Davis [16]. Other work on heated surfaces include the spreading of a volatile droplet by Anderson and Davis [1] and the spreading of a film by Lopez et al. [29].

The work of Ehrhard and Davis [16] on a thin, two-dimensional droplet heated or cooled from below was extended by Smith [41] to include a temperature gradient imposed along the solid surface. The thermocapillary flow induced by this temperature gradient caused the droplet to migrate down the temperature gradient. Benintendi [3] demonstrated this same thermocapillary migration for a thin, three-dimensional droplet.

In the current work, the model of Smith [41] is extended to include thermocapillarity droplet migration on an *inclined* solid surface. The thermocapillary stress, gravitational body force, and capillary pressure interact with each other such that the droplet moves either up or down the solid surface depending on the imposed temperature gradient. Representative results for the quasi-steady droplet velocity and width will be presented. The parameter ranges where steady droplet migration is possible will also be shown.

An additional purpose of this work is to thoroughly discus the theoretical and numerical analyses that are employed. In particular, the use of lubrication theory to describe droplet migration is highlighted. This presentation should help new researchers in this field to easily understand these powerful asymptotic techniques and how they are used in this and similar problems.

2 Problem Formulation and Analysis

In this section, the model problem used to describe the forced thermocapillary migration of a liquid droplet on a non-uniformly heated solid surface is posed and analyzed. In the following subsections, the model geometry is defined, the governing equations and boundary conditions are presented and scaled, and the equations are simplified using an asymptotic analysis (lubrication theory) to produce the evolution equation that describes the shape of the droplet. Lastly, a force balance for the droplet motion is presented and discussed.

2.1 Model Definition

The model problem for this study is sketched in Fig. 1. The two-dimensional liquid droplet is composed of an incompressible Newtonian liquid with constant density ϱ, dynamic viscosity μ, thermal conductivity k, and specific heat c. The solid surface is inclined at the angle α with respect to the horizontal. A Cartesian coordinate system is used with the x-axis directed along the solid surface and the z-axis directed normal to the solid surface. The heating of the solid surface is modeled by imposing a linear temperature profile on the surface $T_s = T_0 - bx$, where T_0 is the average temperature of the solid surface and $b = -\mathrm{d}T_s/\mathrm{d}x > 0$ is the imposed temperature gradient in the $-x$-direction. The surrounding gas is passive with a constant pressure p_∞ and an imposed temperature profile $T_g = T_\infty - bx$, where T_∞ is constant.

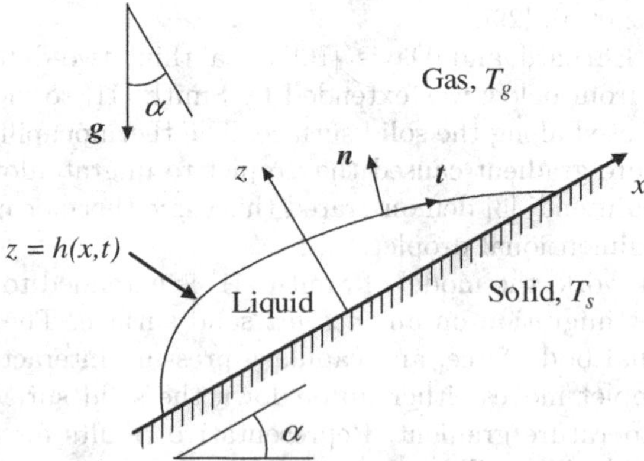

Fig. 1. A sketch of a two-dimensional liquid droplet on a non-uniformly heated or cooled, inclined solid surface

The velocity $\boldsymbol{v} = (u, w)$, pressure p, and temperature T of the liquid droplet are governed by the continuity, momentum, and energy equations:

$$\nabla \cdot \boldsymbol{v} = 0 , \tag{1a}$$

$$\varrho \{\boldsymbol{v}_t + (\boldsymbol{v} \cdot \nabla)\boldsymbol{v}\} = -\nabla p + \varrho \boldsymbol{g} + \mu \nabla^2 \boldsymbol{v} , \tag{1b}$$

$$\varrho c \{T_t + \boldsymbol{v} \cdot \nabla T\} = k \nabla^2 T , \tag{1c}$$

where t is time, the gravity vector $\boldsymbol{g} = -g(\sin\alpha, \cos\alpha)$, g is the acceleration of gravity, and letter subscripts denote partial differentiation with respect to that variable.

The boundary conditions on the solid surface $z = 0$ are the no-slip conditions and the imposed thermal forcing:

$$u = w = 0 , \quad T = T_s = T_0 - bx . \tag{2}$$

The free surface of the liquid droplet is defined by the surface $z = h(x, t)$. The unit normal vector $\boldsymbol{n}$, unit tangent vector $\boldsymbol{t}$, and curvature κ of the free surface are:

$$\boldsymbol{n} = (-h_x, 1)/(1 + h_x^2)^{1/2} , \tag{3a}$$

$$\boldsymbol{t} = (1, h_x)/(1 + h_x^2)^{1/2} , \tag{3b}$$

$$\kappa = -h_{xx}/(1 + h_x^2)^{3/2} . \tag{3c}$$

A linear equation of state for the surface tension of the free surface σ is used:

$$\sigma = \sigma_0 - \gamma(T - T_0) , \tag{4}$$

where σ_0 is the surface tension at the temperature T_0 and γ is the negative rate of the change of surface tension with respect to temperature.

The free-surface boundary conditions are the kinematic condition, the normal- and tangential-stress balances, and the energy balance:

$$h_t + u h_x - w = 0 , \tag{5a}$$

$$\boldsymbol{n} \cdot \mathbf{S} \boldsymbol{n} = -\sigma\kappa , \tag{5b}$$

$$\boldsymbol{t} \cdot \mathbf{S} \boldsymbol{n} = \boldsymbol{t} \cdot \nabla\sigma , \tag{5c}$$

$$-k \nabla T \cdot \boldsymbol{n} = h_g (T - T_g) . \tag{5d}$$

Here, $\mathbf{S}$ is the viscous stress tensor for the liquid:

$$\mathbf{S} = -p\mathbf{I} + \mu \begin{bmatrix} 2u_x & u_z + w_x \\ u_z + w_x & 2w_z \end{bmatrix} , \tag{5e}$$

in which $\mathbf{I}$ is the identity tensor. The constant h_g is the convective heat transfer coefficient for heat transfer from the free surface of the liquid droplet to the surrounding passive gas. The imposed temperature of the gas phase is:

$$T_g = T_\infty - bx . \tag{5f}$$

In this model, evaporation of the liquid is neglected and so the volume of the droplet V_0 remains constant:

$$\int_{\tilde{c}_l(t)}^{\tilde{c}_u(t)} h(x,\, t)\, \mathrm{d}x = V_0 \,. \tag{6}$$

Here, $x = \tilde{c}_l(t)$ is the location of the contact line on the lower or left side of the droplet and $x = \tilde{c}_u(t)$ is the location of the contact line on the upper or right side of the droplet.

To completely pose this problem, further modeling conditions must be specified near the contact lines to describe the appropriate physics active in these regions. It is not clear at this time what these conditions should be, as described in the introduction. However, the purpose of this work is to use lubrication theory to describe the *bulk motion* of the droplet as forced by an imposed temperature gradient. For this, only a constitutive relation for the speed of the apparent contact line U_{CL} is needed. One popular relation is:

$$U_{\mathrm{CL}} = \begin{cases} K(\phi - \phi_a)^m, & \phi > \phi_a \\ 0, & \phi_r < \phi < \phi_a \\ -K(\phi_r - \phi)^m, & \phi < \phi_r \end{cases} , \tag{7}$$

where ϕ is the apparent contact angle and the remaining parameters are material constants: the apparent advancing and receding contact angles, ϕ_a and ϕ_r, the mobility parameter, K, and the mobility exponent, m. This relationship is supported by experimental evidence with $m = 3$ for isothermal spreading [43] and for thermocapillary forced spreading [16]. The relation is introduced now for scaling purposes.

2.2 Lubrication Scaling

The above equations and boundary conditions—not counting any additional contact-line modeling conditions—describe the general problem of a two-dimensional liquid droplet with a temperature-dependent surface tension forced by an imposed temperature gradient in the solid surface and by gravity. In many interesting physical situations the apparent advancing contact angle is very small, i.e., $\phi_a << 1$. This implies that the height of the droplet is much smaller than its width and that the slope of the free surface is very small. Under these conditions, lubrication theory works well to describe the bulk behavior of the droplet.

To set up this problem for a lubrication analysis, the conventional symbol for a small parameter ϵ is introduced:

$$\epsilon = \phi_a \,. \tag{8}$$

A length scale based on the droplet volume is defined as:

$$L_s = \sqrt{V_0/\epsilon} \,. \tag{9}$$

A velocity scale is chosen based on a measure of the mobility or speed of the apparent contact line:

$$U_s = K\phi_a^m \, . \tag{10}$$

Lastly, the appropriate temperature scale is defined as:

$$\Delta T = T_0 - T_\infty \, . \tag{11}$$

The model equations (1–7) are scaled as follows:

$$x^* = x/L_s \, , \quad z^* = z/(\epsilon L_s) \, , \quad h^* = h/(\epsilon L_s) \, , \tag{12a}$$

$$u^* = u/U_s \, , \quad w^* = w/(\epsilon U_s) \, , \quad t^* = t/(L_s/U_s) \, , \tag{12b}$$

$$p^* = p/\left\{\mu U_s/(\epsilon^2 L_s)\right\} \, , \quad T^* = (T - T_\infty)/\Delta T \, , \tag{12c}$$

$$\sigma^* = \sigma/\sigma_0 \, , \quad \phi^* = \phi/\phi_a \, . \tag{12d}$$

Here, the length scale L_s is a measure of the width of the droplet and ϵL_s is a measure of the thickness or height of the droplet. Lengths in the x- and z-directions use these respective scales. The x-velocity scale is the contact-line mobility U_s and the z-velocity scale is chosen to balance both terms in the continuity equation. Time is scaled on a convective time scale, i.e., the time it takes for the droplet to move its own width. A viscous pressure scale is used that balances the pressure gradient along the droplet with the viscous shear stress normal to the droplet. The difference between the average temperature of the solid surface and the average gas temperature is used to scale the temperature of the liquid relative to the average gas temperature. Surface tension is scaled on its reference value and all apparent contact angles are scaled on the apparent advancing contact angle.

The scaled equations are written as follows, where the "star"-notation is dropped for brevity and clarity. The governing equations are:

$$u_x + w_z = 0 \, , \tag{13a}$$

$$\epsilon^2 R\left\{u_t + uu_x + wu_z\right\} = -p_x - G\sin(\alpha)/C + \epsilon^2 u_{xx} + u_{zz} \, , \tag{13b}$$

$$\epsilon^4 R\left\{w_t + uw_x + ww_z\right\} = -p_z - \epsilon G\cos(\alpha)/C + \epsilon^4 w_{xx} + \epsilon^2 w_{zz} \, , \tag{13c}$$

$$\epsilon^2 RPr\left\{T_t + uT_x + wT_z\right\} = \epsilon^2 T_{xx} + T_{zz} \, . \tag{13d}$$

The boundary conditions on the solid surface $z = 0$ are:

$$u = w = 0 \, , \quad T = 1 - Nx \, . \tag{14}$$

The boundary conditions on the free surface $z = h(x, t)$ are:

$$h_t + uh_x - w = 0 \, , \tag{15a}$$

$$\boldsymbol{n} \cdot \mathbf{S}\boldsymbol{n} = -\sigma\kappa/C \, , \tag{15b}$$

$$\boldsymbol{t} \cdot \mathbf{S}\boldsymbol{n} = -\boldsymbol{t} \cdot \nabla T/\Delta C \, , \tag{15c}$$

$$-\nabla T \cdot \boldsymbol{n} = B(T + Nx) , \tag{15d}$$

with

$$\nabla = \left(\epsilon \frac{\partial}{\partial x}, \frac{\partial}{\partial z} \right) , \tag{15e}$$

$$\boldsymbol{n} = (-\epsilon h_x, 1)/(1 + \epsilon^2 h_x^2)^{1/2} , \quad \boldsymbol{t} = (1, \epsilon h_x)/(1 + \epsilon^2 h_x^2)^{1/2} , \tag{15f}$$

$$\kappa = -h_{xx}/(1 + \epsilon^2 h_x^2)^{3/2} , \quad \sigma = 1 - \epsilon^2 \frac{C}{\Delta C} T , \tag{15g}$$

$$\mathbf{S} = -p\mathbf{I} + \epsilon \begin{bmatrix} 2\epsilon u_x & u_z + \epsilon^2 w_x \\ u_z + \epsilon^2 w_x & 2\epsilon w_z \end{bmatrix} . \tag{15h}$$

The constant droplet volume condition is:

$$\int_{\tilde{c}_l(t)}^{\tilde{c}_u(t)} h(x,t) \, \mathrm{d}x = 1 . \tag{16}$$

The scaled constitutive relation for the speed of the apparent contact line is:

$$U_{\mathrm{CL}} = f(\phi) = \begin{cases} (\phi - \phi_A)^m, & \phi > \phi_A \\ 0, & \phi_R < \phi < \phi_A \\ -(\phi_R - \phi)^m, & \phi < \phi_R \end{cases} . \tag{17}$$

The dimensionless groups in these equations are defined as follows:

$$\text{Reynolds number,} \quad R = \frac{\varrho U_s L_s}{\mu} , \tag{18a}$$

$$\text{Bond number,} \quad G = \frac{\varrho g L_s^2}{\epsilon \sigma_0} , \tag{18b}$$

$$\text{Prandtl number,} \quad Pr = \frac{\mu c}{k} , \tag{18c}$$

$$\text{imposed temperature gradient,} \quad N = \frac{b L_s}{\Delta T} , \tag{18d}$$

$$\text{Biot number,} \quad B = \frac{\epsilon h_g L_s}{k} , \tag{18e}$$

$$\text{capillary number,} \quad C = \frac{\mu U_s}{\epsilon^3 \sigma_0} , \tag{18f}$$

$$\text{thermocapillary number,} \quad \Delta C = \frac{\mu U_s}{\epsilon \gamma \Delta T} , \tag{18g}$$

$$\text{advancing contact angle,} \quad \phi_A = 1 , \tag{18h}$$

$$\text{receding contact angle,} \quad \phi_R = \frac{\phi_r}{\phi_a} . \tag{18i}$$

2.3 An Analysis Using Lubrication Theory

The set of equations and boundary conditions (13–17) with additional contact-line conditions can be solved numerically using a variety of different methods, such as front tracking, volume of fluid, level sets, or finite elements. These techniques can provide a good numerical solution for arbitrary values of the parameter set if sufficient resolution is used in the computation. One drawback of such a purely numerical solution is that the dominant physical forces in the problem can be identified only by a careful examination of the results of the simulation together with (possibly) some numerical experimentation with the values of the parameter set.

In the present work an alternative methodology is used, one based on asymptotic techniques. The idea is to exploit the fact that the droplet is very thin so that an approximate solution can be attained. This technique has three main advantages over the full numerical methods mentioned above. First, taking the "thin droplet" limit reduces the number of relevant parameters in the problem. Second, the asymptotic differential equations and boundary conditions are expressed only in terms of the relevant (and dominant) physical processes at work in this limit. This allows for a simple interpretation and explanation of the results in terms of the dominant physics. Lastly, the numerical solution of the asymptotic problem is simpler and faster than a full numerical simulation.

The thinness of the droplet is measured by the small parameter ϵ. With the scalings defined in (12), all of the dependent variables in the problem are $\mathcal{O}(1)$ quantities with respect to ϵ. In addition, all of the dimensionless groups defined in (18) should be considered $\mathcal{O}(1)$ quantities in the limit of small ϵ.

The asymptotic approximation of this problem continues by writing all of the dependent variables, except the free-surface shape h, as asymptotic expansions in the small parameter ϵ as follows:

$$(u,\, w,\, p,\, T) = (u_0,\, w_0,\, p_0,\, T_0) + \epsilon(u_1,\, w_1,\, p_1,\, T_1) + \dots \,. \tag{19}$$

These expansions are substituted into the governing equations and boundary conditions (13–15) and the resulting terms to each order in ϵ are grouped together. The leading-order or $\mathcal{O}(1)$ problem is written next, but the zero-subscript is dropped for simplicity since the analysis will not be taken any farther in this work. The governing differential equations are:

$$0 = u_x + w_z \,, \tag{20a}$$

$$0 = -p_x - G\sin(\alpha)/C + u_{zz} \,, \tag{20b}$$

$$0 = -p_z \,, \tag{20c}$$

$$0 = T_{zz} \,. \tag{20d}$$

The solid-surface boundary conditions at $z = 0$ are:

$$u = w = 0 \,, \quad T = 1 - Nx \,. \tag{21}$$

The free-surface boundary conditions at $z = h(x, t)$ are:

$$0 = h_t + u h_x - w \; , \tag{22a}$$

$$p = -h_{xx}/C \; , \tag{22b}$$

$$u_z = -\{T_x + h_x T_z\}/\Delta C \; , \tag{22c}$$

$$-T_z = B(T + Nx) \; . \tag{22d}$$

These $\mathcal{O}(1)$ equations are known as the core-flow or outer problem. They are strictly valid away from small end regions or boundary layers of length $\mathcal{O}(\epsilon)$ near each contact line of the droplet. This is a typical characteristic of a singular perturbation problem, signaled in this case by the disappearance of the second partial derivatives with respect to the x-variable in the momentum and energy balances in the limit of small ϵ. The core-flow problem is easily integrated. The result is:

$$T = 1 - Nx - \frac{Bz}{1 + Bh} \; , \tag{23a}$$

$$p = -h_{xx}/C \; , \tag{23b}$$

$$u = F\left\{hz - z^2/2\right\} + Sz \; , \tag{23c}$$

$$w = -F_x\left\{hz^2/2 - z^3/6\right\} - F h_x z^2/2 - S_x z^2/2 \; . \tag{23d}$$

Here, F is the longitudinal body force acting on the droplet and S is the thermocapillary shear stress acting on the free surface. These quantities are given by:

$$F = \frac{h_{xxx}}{C} - \frac{G\sin(\alpha)}{C} \; , \tag{23e}$$

$$S = \frac{N}{\Delta C} + \frac{B}{\Delta C}\frac{h_x}{(1 + Bh)^2} \; . \tag{23f}$$

These results show that in this limit the pressure across the droplet thickness is constant with a capillary-pressure jump across the free surface. This pressure has a parametric variation in the x-direction due to the variation in the droplet thickness h with x. The temperature across the droplet is governed by pure conduction from the imposed temperature on the solid surface to a convective heat flux to the passive gas. The velocity in the x-direction is driven by three forces. Two are the body forces in F characterized in (23e). One of these is the longitudinal gradient of the capillary pressure (the first term in F) and the other is the gravitational body force along the solid surface (the second term in F). The third force is a thermocapillary shear stress S characterized in (23f). This stress arises from the imposed temperature gradient on the solid surface (the first term in S) and from the nonuniform temperature of the free surface due to variations in the droplet thickness (the second term in S). The z-velocity is whatever is required to satisfy continuity and no-penetration on the solid surface. It is shown here for completeness, but it will not be required in what follows.

The shape of the droplet is determined by solving an evolution equation for the free surface that is based on the continuity equation. The simplest way to derive this equation is to first define q, the volume flow rate of the droplet in the x-direction:

$$q = \int_0^h u \, dz \; . \tag{24}$$

The derivative of q with respect to x is:

$$q_x = \int_0^h u_x \, dz + u(z=h) \, h_x \; .$$

Now, integrate the continuity (13a) across the droplet thickness, and use the kinematic condition (15a) on the free surface and q_x above as follows:

$$\int_0^h \{u_x + w_z\} \, dz = 0 \; ,$$

$$\int_0^h w_z \, dz + \int_0^h u_x \, dz = 0 ,$$

$$w(z=h) - w(z=0) + q_x - u(z=h) \, h_x = 0 \; ,$$

$$w(z=h) - u(z=h) \, h_x + q_x = 0 \; ,$$

$$h_t + q_x = 0 \; . \tag{25}$$

This evolution equation is valid for arbitrary values of ϵ. To use it at leading order, integrate u from (23c) to obtain the volume flow rate q in terms of the droplet shape h:

$$q = Fh^3/3 + Sh^2/2 \; . \tag{26}$$

The evolution equation (25) for the droplet shape h is a partial differential equation that is first-order in time and fourth-order in space (the F-term in q has three x-derivatives). To integrate it, a single initial condition is needed, which can be whatever droplet shape one would like to start from. In addition, four end conditions are required; two at each contact line of the droplet. To determine these end conditions, a series of boundary-layer or scaling transformations on the governing equations and boundary conditions (13–16) are needed in order to examine the flow solution locally near each contact line. As an example of this analysis, consider the lower contact-line region at $x = \tilde{c}_l$.

The first scaling transformation isolates the end region of the droplet. It is:

$$\xi = (x - \tilde{c}_l)/\epsilon \; . \tag{27}$$

The only part of the transformed problem set that is needed to complete the specification of the leading-order core-flow problem is the normal-stress free-surface boundary condition (15b). To leading-order, this transformed boundary condition is:

$$\bar{\kappa} = \bar{h}_{\xi\xi} = 0 \; , \tag{28}$$

where the overbar indicates an end-region variable. This equation shows that surface tension completely dominates the normal-stress balance and so the free surface has zero curvature. In two dimensions, this shape is a straight line:

$$\bar{h} = A_1 + A_2 \xi \,, \tag{29}$$

where A_1 and A_2 are unknown constants determined from matching to adjacent flow regions.

Inside the end region there are two more nested boundary layers very close to the actual contact line where different forces dominate the flow physics. The first one is a region where viscous forces balance surface tension resulting in a slight bending of the free surface. Even closer to the actual contact line another set of physical balances comes into play, although the precise nature of the physics here has not yet been determined, as discussed in the introduction. Whatever the nature of this physics, the resultant shape of the free surface in these two contact-line regions $\tilde{h}$ will be described by some function of an inner variable η. This shape must match to the shape of the droplet in the end region given by (29). A leading-order match is:

$$\lim_{\xi \to 0} \bar{h} = \lim_{\eta \to \infty} \tilde{h} = \beta \,,$$

where β is a constant that reflects the relevant physics from the inner two contact-line regions. This parameter could be either positive or negative depending on whether the free surface very near to the actual contact line has positive or negative curvature respectively. This matching condition determines the first constant from the end region:

$$A_1 = \beta \,. \tag{30}$$

The remaining constant A_2 in (29) is determined by matching to the core-flow free-surface shape h. To do this, express the end-region shape $\bar{h}$ in terms of the core-flow variable x:

$$\bar{h} = \frac{A_2}{\epsilon} \left(x - \tilde{c}_l + \frac{\beta \epsilon}{A_2} \right) \,. \tag{31}$$

The core-flow free-surface shape h will be determined from a numerical solution of the evolution (25). To match to the end-region shape $\bar{h}$, h is written as a Taylor series expansion about the position of the *apparent contact line* c_l:

$$h(x,\, t) = h(c_l,\, t) + h_x(c_l,\, t)\,(x - c_l) + \ldots \,.$$

The droplet thickness is defined to be zero at the apparent contact line and the slope of the droplet is the apparent contact angle:

$$h(c_l,\, t) = 0 \,, \quad h_x(c_l,\, t) = \phi_l \,.$$

Thus,

$$h(x,\, t) = \phi_l\,(x - c_l) + \ldots \,. \tag{32}$$

Matching (31) and (32) gives:

$$A_2 = \epsilon\phi_l \ , \quad c_l = \tilde{c}_l - \beta/\phi_l \ . \tag{33}$$

This matching process shows that the core-flow shape is determined in terms of the position of the apparent contact line. This position is expressed in terms of the physics of the inner contact-line regions using the single constant β, which is expected to be very small, i.e., $\beta << \epsilon$. At the apparent contact line, the apparent contact angle ϕ_l is determined by the force balance maintained in the core flow near the end region. Another outcome of matching to this inner contact-line region would be a relationship between the contact-line speed and the apparent contact angle. The derivation of such a relation, like the one shown in (17) will not be done here.

The second boundary condition for the core-flow free-surface shape at the lower contact line is determined by considering the total volume flux from the end region. The free-surface shape in the end region written in core variables is:

$$\bar{h} = \phi_l \ (x - c_l) \ ,$$

where $(x - c_l) = \mathcal{O}(\epsilon)$. This shape is that of a triangle or wedge and its volume is:

$$V_l = \phi_l(x - c_l)^2/2 = \mathcal{O}(\epsilon^2) \ .$$

The control volume mass balance in the end region is:

$$\frac{\mathrm{d}V_l}{\mathrm{d}t} = q_{\mathrm{CL}} - q(c_l, t) \ ,$$

where q_{CL} is the volume flux from the inner contact-line region into the end region (assumed to be small), and $q(c_l, t)$ is the volume flux from the end region into the core region. Since $V_l = \mathcal{O}(\epsilon^2)$, the leading-order volume flux into the core region is:

$$q(c_l, t) = q_{\mathrm{CL}} \approx 0 \ . \tag{34}$$

A similar analysis is applied to the upper end region of the droplet to produce two more boundary conditions for the core-flow problem. All together, the evolution equation and the proper boundary and initial conditions for the core flow are:

$$h_t + q_x = 0 \ , \tag{35a}$$

$$h(c_l, t) = 0 \ , \quad q(c_l, t) = 0 \ , \tag{35b}$$

$$h(c_u, t) = 0 \ , \quad q(c_u, t) = 0 \ , \tag{35c}$$

$$\int_{c_l(t)}^{c_u(t)} h(x, t) \, \mathrm{d}x = 1 \ , \tag{35d}$$

$$h(x, 0) = H(x) \ , \tag{35e}$$

where $H(x)$ is any initial shape satisfying the constant droplet volume condition (35d).

Note that the equation for a constant droplet volume (35d) is automatically satisfied by the other equations in system (35). To see this, consider the volume of the droplet:

$$V(t) = \int_{c_l(t)}^{c_u(t)} h(x, t)\,\mathrm{d}x\,.$$

This is accurate to leading order because the volume of the liquid in the end regions is $\mathcal{O}(\epsilon^2)$ and the difference between the actual contact-line positions and the apparent contact-line positions is assumed to be much smaller than ϵ. Take the time derivative of this droplet volume and use the evolution equation and boundary conditions of system (35) as follows:

$$\begin{aligned}
\frac{\mathrm{d}V}{\mathrm{d}t} &= \frac{\mathrm{d}}{\mathrm{d}t}\int_{c_l(t)}^{c_u(t)} h(x, t)\,\mathrm{d}x\,, \\
&= \int_{c_l(t)}^{c_u(t)} h_t\,\mathrm{d}x + h(c_u, t)\frac{\mathrm{d}c_u}{\mathrm{d}t} - h(c_l, t)\frac{\mathrm{d}c_u}{\mathrm{d}t}\,, \\
&= \int_{c_l(t)}^{c_u(t)} h_t\,\mathrm{d}x\,, \\
&= -\int_{c_l(t)}^{c_u(t)} q_x\,\mathrm{d}x\,, \\
&= q(c_l, t) - q(c_u, t)\,, \\
&= 0\,.
\end{aligned}$$

Thus, the volume of the droplet is constant in time. The constant-volume condition (35d) is therefore a dependent condition and it is implicitly satisfied by the droplet evolution equation and its boundary conditions. When the evolution equation is integrated in time, this condition is used only on the initial shape.

The evolution equation and its supporting conditions in system (35) do not model the motion of the contact lines. When this evolution equation is solved, it determines the lower and upper apparent contact angles as functions of time. These apparent contact angles are the forcing conditions seen by the flow in the contact-line region. The response of the contact line to this forcing is determined by a detailed local analysis of the contact-line region. Since the purpose of this work is to understand the bulk motion of the droplet, this local analysis is not done here. In its place, the constitutive relation for the contact-line speed given in (17) together with the appropriate initial conditions for the positions of the apparent contact lines are used to model the contact-line motion.

In its final form, the following system of equations describes the motion of a thin, two-dimensional liquid droplet on an inclined, non-uniformly heated solid surface:

$$h_t + q_x = 0\,, \quad q = Fh^3/3 + Sh^2/2\,, \tag{36a}$$

$$F = \frac{h_{xxx}}{C} - \frac{G\sin(\alpha)}{C}\,, \quad S = \frac{N}{\Delta C} + \frac{B}{\Delta C}\frac{h_x}{(1+Bh)^2}\,, \tag{36b}$$

$$h(c_l, t) = 0\ , \quad h(c_u, t) = 0\ , \quad q(c_l, t) = 0\ , \quad q(c_u, t) = 0\ , \tag{36c}$$

$$\int_{c_l(t)}^{c_u(t)} h(x, t)\, \mathrm{d}x = 1\ , \quad h(x, 0) = H(x)\ , \tag{36d}$$

$$\phi_l(t) = h_x(c_l, t)\ , \quad \phi_u(t) = -h_x(c_u, t)\ , \tag{36e}$$

$$\frac{\mathrm{d}c_l}{\mathrm{d}t} = -f(\phi_l)\ , \quad \frac{\mathrm{d}c_u}{\mathrm{d}t} = f(\phi_u)\ , \tag{36f}$$

$$c_l(0) = -a_0\ , \quad c_u(0) = a_0\ , \tag{36g}$$

$$f(\phi) = \begin{cases} (\phi - \phi_A)^m, & \phi > \phi_A \\ 0, & \phi_R < \phi < \phi_A \\ -(\phi_R - \phi)^m, & \phi < \phi_R \end{cases} . \tag{36h}$$

Here, a_0 is the initial half-width of the droplet.

When this set of equations for the core flow is integrated for a moving contact line problem, the solution has a logarithmic singularity at the contact line. One popular way to avoid this singularity is to use a slip model in the computation, as done by [2,19,21,22] and many others. This procedure will not be discussed further since this singularity is not a problem in the small capillary number limit used in the remainder of this work.

2.4 Small Capillary Number Limit

The system of (36) can be integrated numerically for $\mathcal{O}(1)$ values of the given parameters. A backward Euler method in time and a finite difference or a psuedo-spectral method in space are good choices for this task. However, further simplifications can be made by exploiting the typically small values of two parameters. The value of the capillary number is about 10^{-2} for a water droplet of radius 1 mm moving at about 1 mm/s. In microfluidic applications, the capillary number may be even smaller. To use this small parameter, define $\hat{q} = Cq$ and rewrite the evolution equation (36a) as:

$$Ch_t + \hat{q}_x = 0\ .$$

Taking the limit as $C \to 0$, with $\hat{q} = \mathcal{O}(1)$ gives the steady form of the evolution equation $\hat{q}_x = 0$. This is immediately integrated and the no-flux contact-line conditions are used to yield $\hat{q} = 0$ all along the droplet. The physical interpretation of this limit is that the flow inside the droplet comes to equilibrium (steady state) much faster than the contact lines can move. In doing so the free-surface position becomes steady and the apparent contact angles are determined. These angles then force the motion of the contact lines as given by the contact-line speed conditions (36f). Note that in this limit, the initial condition for the droplet shape must be dropped, since the h_t-term in the evolution equation is lost. The initial positions of the apparent contact lines are still prescribed. This is appropriate since the contact-line-motion conditions (36f) are first order in time.

Another small parameter is the Biot number, which measures the heat transfer from the liquid droplet to the passive gas. For air flowing past a small water droplet at about 1 mm/s, this parameter is less that 10^{-3}. Before taking the limit of small Biot number in the droplet flux equation, $\hat{q} = 0$, the following parameters are defined:

$$\hat{N} = CN/\Delta C \; , \quad \hat{M} = CB/\Delta C \; . \tag{37}$$

These parameters are assumed to remain $\mathcal{O}(1)$ in the limit of small capillary number and small Biot number. The parameter $\hat{N}$ is a measure of the thermocapillary stress arising from the imposed temperature gradient along the solid surface and $\hat{M}$ measures the thermocapillary stress arising from the temperature difference between the solid surface and the passive gas interacting by conduction through the non-uniform thickness of the droplet.

The droplet flux equation $\hat{q} = 0$ is a third-order differential equation and so three boundary conditions are needed: zero thickness at both apparent contact lines and the constant droplet volume condition. The volume condition is used here because the two volume-flux end conditions are automatically satisfied by the droplet flux equation.

With these parameter simplifications, the following non-linear differential equation describes the shape of the droplet $h(x)$:

$$\frac{1}{3}h^3 \left\{ h_{xxx} - G\sin(\alpha) \right\} + \frac{1}{2}h^2 \left\{ \hat{N} + \hat{M}h_x \right\} = 0 \; , \tag{38a}$$

$$h(c_l) = h(c_u) = 0 \; , \quad \int_{c_l}^{c_u} h(x) \, \mathrm{d}x = 1 \; . \tag{38b}$$

Note that this equation shows that the droplet shape h has a singularity in its third derivative at both contact lines since h is zero at these two points.

2.5 Droplet Momentum Balance

In the previous two sections, lubrication theory was used to derive the equations (38) that describes the shape of the droplet. Equation (38a) is a mass balance for the flow inside the droplet. Another physical balance of interest is the global momentum balance for the droplet. This balance is easily written using the free-body diagram shown in Fig. 2. Here, the thin liquid droplet is migrating uniformly to the right (up the inclined surface) with the velocity $U(t)$. It is subject to the following forces: surface tension at the upper and lower contact lines (σ_u and σ_l), the gravitational body force, and the shear force from the solid surface F_τ. The force balance in the x-direction in *dimensional* variables is:

$$\varrho V_0 \frac{\mathrm{d}U}{\mathrm{d}t} = \sigma_u \cos(\phi_u) - \sigma_l \cos(\phi_l) - \varrho g V_0 \sin(\alpha) + F_\tau \; . \tag{39}$$

This balance is simplified for this lubrication problem using the fact that both of the contact angles are small. Thus, expanding the cosine terms with a two-term Taylor series gives the result:

$$\varrho V_0 \frac{\mathrm{d}U}{\mathrm{d}t} = \sigma_u - \sigma_l - \frac{1}{2}\sigma_u \phi_u^2 + \frac{1}{2}\sigma_l \phi_l^2 - \varrho g V_0 \sin(\alpha) + F_\tau \; .$$

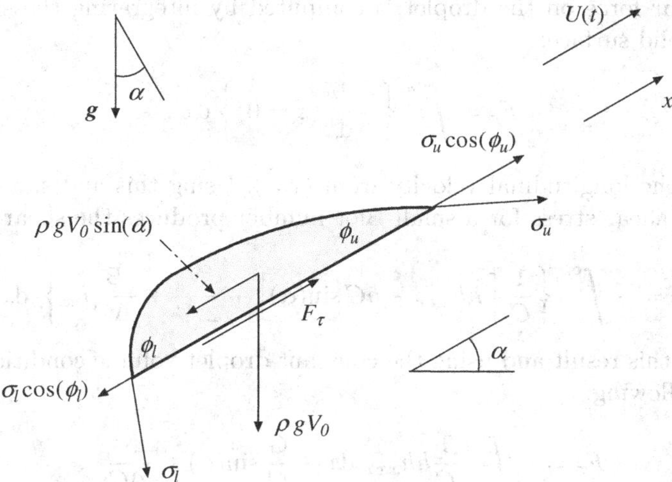

Fig. 2. A free-body diagram for the two-dimensional, thin liquid droplet on a non-uniformly heated or cooled, inclined solid surface

The surface tension in this balance is evaluated at each contact line. To determine this, combine the linear equation of state for the surface tension (4) and the prescribed linear temperature profile on the solid surface from (2) to obtain:

$$\sigma = \sigma_0 + \gamma b x \ .$$

With this, the momentum balance becomes:

$$\varrho V_0 \frac{\mathrm{d}U}{\mathrm{d}t} = \gamma b \ell + \frac{1}{2}\sigma_0 \left\{\phi_l^2 - \phi_u^2\right\} + \frac{1}{2}\gamma b \left\{\phi_l^2 c_l - \phi_u^2 c_u\right\} - \varrho g V_0 \sin(\alpha) + F_\tau \ ,$$

where $\ell = c_u - c_l$ is the length of the droplet. Scaling this balance using the lubrication scales and parameters defined in Sect. 2.2 produces the following result (with the "star"-notation dropped):

$$\epsilon^2 R \frac{\mathrm{d}U}{\mathrm{d}t} = \epsilon^2 \frac{N}{2\Delta C}\left\{\phi_l^2 c_l - \phi_u^2 c_u\right\} + \frac{1}{2C}\left\{\phi_l^2 - \phi_u^2\right\} + \frac{N}{\Delta C}\ell - \frac{G}{C}\sin(\alpha) + F_\tau \ .$$

To leading-order in ϵ the droplet force balance is:

$$0 = \frac{1}{2C}\left\{\phi_l^2 - \phi_u^2\right\} + \frac{N}{\Delta C}\ell - \frac{G}{C}\sin(\alpha) + F_\tau \ . \tag{40}$$

Thus, thermocapillary migration of a thin liquid droplet exhibits a static, i.e., no acceleration, balance of forces along the solid surface. The shear force from the solid surface due to fluid motion inside the droplet balances the gravitational body force, the differential surface tension force at the contact lines caused by the imposed temperature profile, and the differential surface-tension force at the contact lines caused by the difference in the two contact angles.

The shear force on the droplet is computed by integrating the shear stress along the solid surface:

$$F_\tau = \int_{c_l}^{c_u} \left\{ -\frac{\mathrm{d}u}{\mathrm{d}z}(z=0) \right\} \mathrm{d}x \,,$$

where u is the longitudinal velocity from (23c). Using this and simplifying the free-surface shear stress for a small Biot number produces the shear-force integral:

$$F_\tau = -\int_{c_l}^{c_u} \left\{ \frac{1}{C} \Big[h h_{xxx} - hG\sin(\alpha) \Big] + \frac{N}{\Delta C} + \frac{B}{\Delta C} h_x \right\} \mathrm{d}x \,.$$

Integrating this result and using the constant droplet volume condition of (36d) gives the following:

$$F_\tau = -\int_{c_l}^{c_u} \frac{1}{C} h h_{xxx} \,\mathrm{d}x + \frac{G}{C}\sin(\alpha) - \frac{N}{\Delta C}\ell \,. \tag{41}$$

This result shows that the shear force on the droplet has only three components: a capillary-pressure-gradient force (the h_{xxx}-term), a gravitational body force (the G-term), and a surface-tension force due to the imposed temperature gradient (the N-term). Note that the thermocapillary shear stress on the droplet due to the uniform heating (the B-term) integrates to zero.

Substituting the shear force from (41) into the force balance equation (40) yields the following result:

$$\frac{1}{2}\left\{\phi_l^2 - \phi_u^2\right\} - \int_{c_l}^{c_u} h h_{xxx} \,\mathrm{d}x = 0 \,. \tag{42}$$

This interesting relation for the droplet shape is *independent* of any of the physical parameters of this problem. It provides a check on the computed shape of the free surface, especially since h_{xxx} is singular at both contact lines as noted at the end of the previous section. The physical interpretation of this relation is that it is the x-component of the normal-force balance on the free surface. The first term is the net surface tension force at the contact lines in the x-direction due to a difference in the apparent contact angles. The second term is the net capillary force on the free surface of the droplet in the x-direction.

Equation (42) can be derived independently by considering a force balance on a differential element of the free surface as shown in Fig. 3. The balance of surface tension and pressure forces in the x-direction, as shown on the free-body diagram in the figure, yields the dimensional relation:

$$-\sigma\cos(\theta)\big|_x - p(z=h)h_x\big|_x \Delta x + \sigma\cos(\theta)\big|_{x+\Delta x} = 0 \,,$$

where θ is the local slope of the free surface. Using $p(z=h) = \sigma\kappa$, dividing by Δx, and taking the limit as $\Delta x \to 0$ gives:

$$\frac{\mathrm{d}\cos(\theta)}{\mathrm{d}x} = \kappa h_x \,.$$

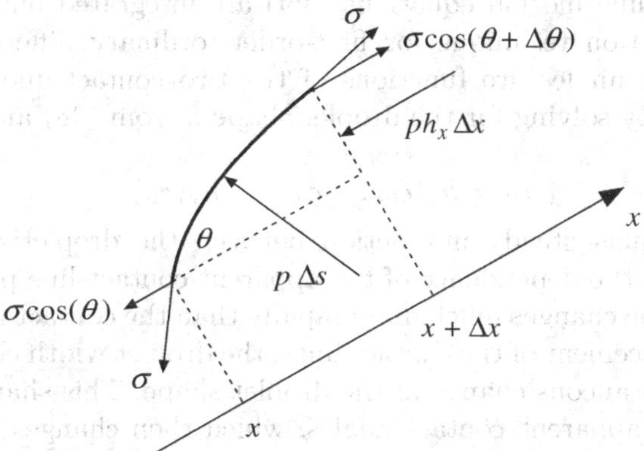

Fig. 3. A free-body diagram for the surface tension and pressure forces on an elemental piece of the free surface. The arclength of this portion of the free surface is given by Δs

Integrating this along the length of the droplet, integrating the curvature term by parts, and setting $h\kappa = 0$ at each contact line produces:

$$\cos(-\phi_u) - \cos(\phi_l) = -\int_{c_l}^{c_u} h\kappa_x \, \mathrm{d}x \; . \tag{43}$$

This general relation for the droplet shape can be written for a thin droplet by letting the contact angles be small and setting $\kappa = -h_{xx}$. The result is (42).

3 Numerical Method

The motion of a thin droplet in the small-capillary-number limit is a quasi-steady problem that can be integrated numerically. The motion of the contact lines is determined by integrating the pair of coupled, first-order ordinary differential equations (36f) with the initial conditions (36g). In this work, the initial shape is chosen to be that of a motionless, fully spread, isothermal droplet on a horizontal surface centered at the origin. The initial contact-line positions are $\pm a_0$, where a_0 is the half-width of this initial droplet shape. The initial shape is found by integrating the equation $H_{xxx} = 0$, obtained from (36a) under the given conditions, together with the contact conditions and the constant droplet volume condition. The result is:

$$H(x) = \frac{3}{4a_0^3}\left\{a_0^2 - x^2\right\} \; . \tag{44}$$

The initial half-width a_0 is found from the fully spread condition that the contact angle equals the advancing contact angle. For the lower or left contact angle, this condition is $H_x(-a_0) = \phi_A = 1$. This yields the half width:

$$a_0 = \sqrt{3/2} \; . \tag{45}$$

The contact-line motion equations (36f) are integrated numerically using a standard integration technique for first-order, ordinary differential equations. The two contact angles are functions of the two contact-line positions. They are determined by solving for the droplet shape h from (38) and then using the relations:

$$\phi_l = h_x(c_l) \ , \quad \phi_u = -h_x(c_u) \ .$$

Note that the quasi-steady interaction between the droplet's motion and its shape is through the dependence of the apparent contact-line positions on time. The droplet shape changes much more rapidly than the contact lines move. Thus, for a given displacement of the contact lines, the droplet width changes leading to an almost instantaneous change in the droplet shape. This shape change causes a change in the apparent contact angles, which then changes the speed of the contact lines and so their resulting displacement.

The final numerical task is to solve (38) for the droplet shape. This can be done using a number of different methods, including shooting, finite differences, or collocation. A simple and accurate technique combining Newton iteration and Chebyshev collocation is described next.

First, the nonlinear differential equation (38a) ($\hat{q} = 0$) is linearized using the Newton-Kantorovich method described by Boyd [4] and used by [2,32,41,42]. This method applies Newton's method directly to the nonlinear differential equation by using Frechet or functional derivatives in h. To demonstrate, let h_i be the droplet shape at the i-th iteration and define the increment function Δ as:

$$\Delta = h_{i+1} - h_i \ .$$

The function Δ satisfies a linear differential equation obtained from Newton's method on the equation $\hat{q}(h) = 0$. In functional form, this is:

$$\frac{\partial \hat{q}}{\partial h}(h_i)\,\Delta = -\hat{q}(h_i) \ .$$

Evaluating this yields the linear equation for the increment function:

$$\frac{1}{3}h_i^3 \Delta_{xxx} + \frac{1}{2}h_i^2 \hat{M} \Delta_x + \{h_{ixxx} - G\sin(\alpha)\}\, h_i^2 \Delta + \left\{\hat{N} + \hat{M} h_{ix}\right\} h_i \Delta = -\hat{q}(h_i) \ , \tag{46a}$$

$$\Delta(c_l) = \Delta(c_u) = \int_{c_l}^{c_u} \Delta \, \mathrm{d}x = 0 \ . \tag{46b}$$

To start the iteration, the initial shape H from (44) is used. This shape has a volume of one and so Δ has a volume of zero as shown in (46b).

The linear equations (46) are solved at each iteration step using a Chebyshev collocation method. The first step is to transform the problem domain from $c_l \le x \le c_u$ to $-1 \le \xi \le 1$. Then, the transformed increment function $\bar{\Delta}$ is written as a series of n Chebyshev functions $T_k(\xi)$ as follows:

$$\bar{\Delta} = \sum_{k=0}^{n-1} \delta_k T_k(\xi) \ , \tag{47}$$

where $\{\delta_k, k = 0 : n - 1\}$ are n constants. This series is substituted into a transformed version of the differential equation (46a) and the result is evaluated on $m = n - 3$ interior collocation points defined as:

$$\xi_j = \cos\left\{\frac{\pi(j - 1/2)}{m}\right\}, \quad j = 1 : m \,. \tag{48}$$

The series for $\bar{\Delta}$ is also substituted into the three boundary conditions (46b). All together, the result is a system of n linear equations for the n unknown constants δ_k. After arranging these constants into the column vector $\underline{\delta}$, the system of equations can be written in standard form as:

$$\underline{\underline{A}}\,\underline{\delta} = \underline{b}\,,$$

where $\underline{\underline{A}}$ is a known n by n matrix and $\underline{b}$ is a known column vector of length n. This system of equations is easily solved using standard techniques for linear systems.

Once the constants δ_k are found, the increment function Δ is determined by properly evaluating and transforming the series (47) and the droplet shape is updated. The iteration is halted when Δ falls below some error criterion. Either the maximum of the absolute value of Δ or the root-mean-square of Δ can be used for this error test.

4 Results

The results presented in this section show how the heating of the droplet interacts with the gravity force directed along the solid surface. The following parameters are held fixed in all that follows: the advancing contact angle $\phi_A = 1$, by the scaling used; the receding contact angle $\phi_R = 0.8$; the mobility exponent $m = 3$, as determined by experimental results on spreading; and the Bond number $G = 5$. The three parameters that are varied are the strength of the imposed temperature gradient $\hat{N}$, the strength of the uniform temperature difference between the solid surface and the passive gas $\hat{M}$, where $\hat{M} > 0$ represents uniform heating from below and $\hat{M} < 0$ represents uniform cooling, and the solid surface inclination angle α.

For an inclination angle of $\alpha = 20°$, no uniform heating or cooling ($\hat{M} = 0$), and an applied temperature gradient of $\hat{N} = 0.1$ the droplet behaves as shown in Fig. 4. Figure 4a shows the transient development of the droplet half-width and the droplet velocity, and Fig. 4b shows the final steady-state droplet shape. In this case, the imposed thermocapillary stress is not large enough to overcome the pull of gravity and so the droplet ends up sliding down the solid surface at a constant velocity. The final steady-state shape has a typical capillary bump on the lower (downhill) side that has been reported by many authors, see [5,18,22,38,44] for example.

For an inclination angle of $\alpha = 20°$, a uniform cooling of $\hat{M} = -1$, and an applied temperature gradient of $\hat{N} = 0.6$ the thermocapillary stress is increased

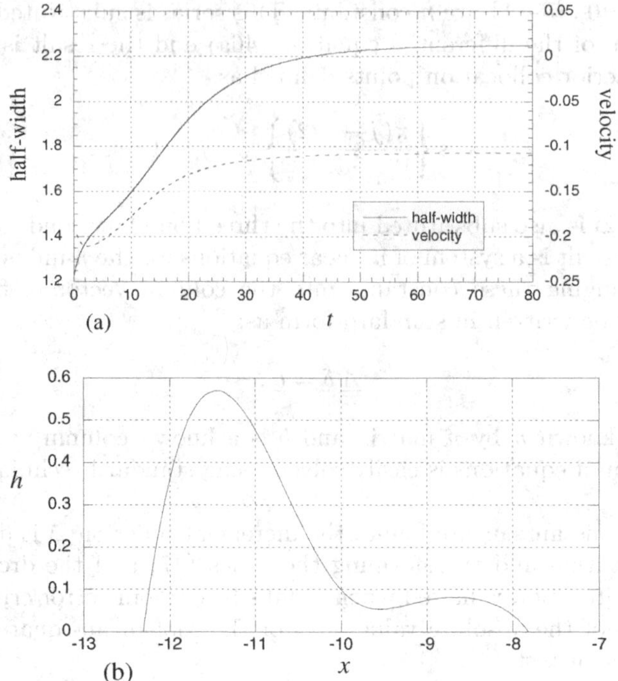

Fig. 4. a The transient behavior of the droplet half-width and velocity with the parameters $\alpha = 20°$, $\hat{M} = 0$, $\hat{N} = 0.1$, $G = 5$, $\phi_R = 0.8$, and $m = 3$. **b** The final steady-state droplet shape for $t > 80$

significantly. Figure 5a shows the transient development of the droplet half-width and the droplet velocity, and Fig. 5b shows the final steady-state droplet shape. Now, the thermocapillary stress is enough to overcome the pull of gravity and the droplet attains a constant-velocity migration *up* the solid surface. The final steady-state shape is very similar to that seen by Smith [41] for thermocapillary droplet migration on a horizontal surface.

In Fig. 6, the final state of the droplet is shown using domain plots in the parameter space ($\hat{N}$, $\hat{M}$) for three different values of the inclination angle α. The results for $\alpha = 0$, shown in Fig. 6a, are identical to those of Smith [41] for a horizontal solid surface. Here, two different steady-state configurations of the droplet are possible. Region I is no migration. In this region, the droplet remains fixed in position on the solid surface because contact-angle hysteresis is present ($\phi_A = 1$ and $\phi_R = 0.8$). Even though the contact lines do not move, there is an internal thermocapillary driven flow caused by the applied temperature gradient on the solid surface. In Region II, constant-velocity migration to the right occurs because the imposed temperature gradient is strong enough to overcome contact-angle hysteresis. Region III is where the imposed temperature gradient is so large that the model breaks down completely. The boundary of this region is marked

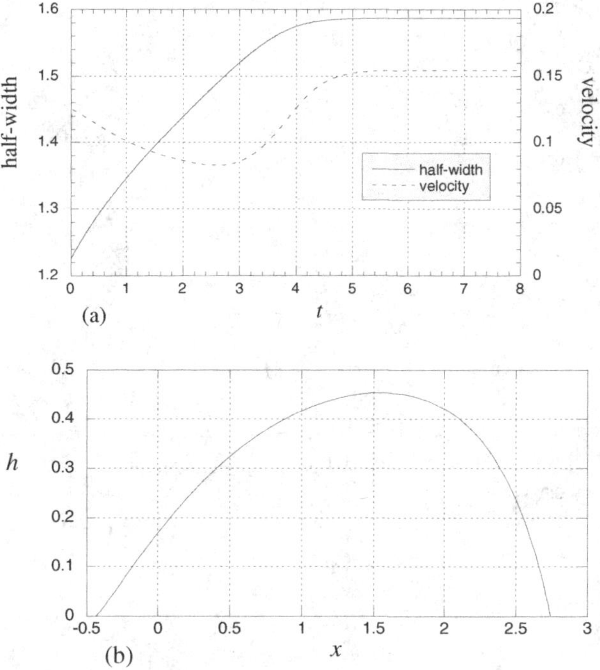

Fig. 5. **a** The transient behavior of the droplet half-width and velocity with the parameters $\alpha = 20°$, $\hat{M} = -1$, $\hat{N} = 0.6$, $G = 5$, $\phi_R = 0.8$, and $m = 3$. **b** The final steady-state droplet shape for $t > 10$

by the curve defined when the lower contact angle of the droplet becomes zero. At this point, the contact line speed condition (7) is no longer valid. A better model for the receding contact-line speed is needed to accurately predict the droplet behavior for these large values of the imposed temperature gradient.

When the inclination angle is increased to $\alpha = 10°$, the three regions seen in Fig. 6a are still present, but their boundaries have shifted upward in the direction of increasing $\hat{N}$ as shown in Fig. 6b. Here, gravity tends to pull the droplet down the solid surface, and so a larger temperature gradient is needed to counter this force at steady state. For this value of the inclination angle, Region IV appears for small values of the imposed temperature gradient $\hat{N}$. In this region, gravity is large enough to overcome contact-angle hysteresis and so the droplet is pulled down the solid surface at a constant velocity. When the temperature gradient imposed on the solid surface is large enough, the thermocapillary shear stress on the droplet balances contact-angle hysteresis and gravity and the droplet will remain fixed in place (Region I). Note, however, that the droplet still has an internal thermocapillary driven flow. For larger values of the imposed temperature gradient the droplet moves up the solid surface at a constant velocity.

Lastly, Fig. 6c shows the results when the inclination angle is increased to $\alpha = 60°$. The previous four regions remain (although shifted upwards) and

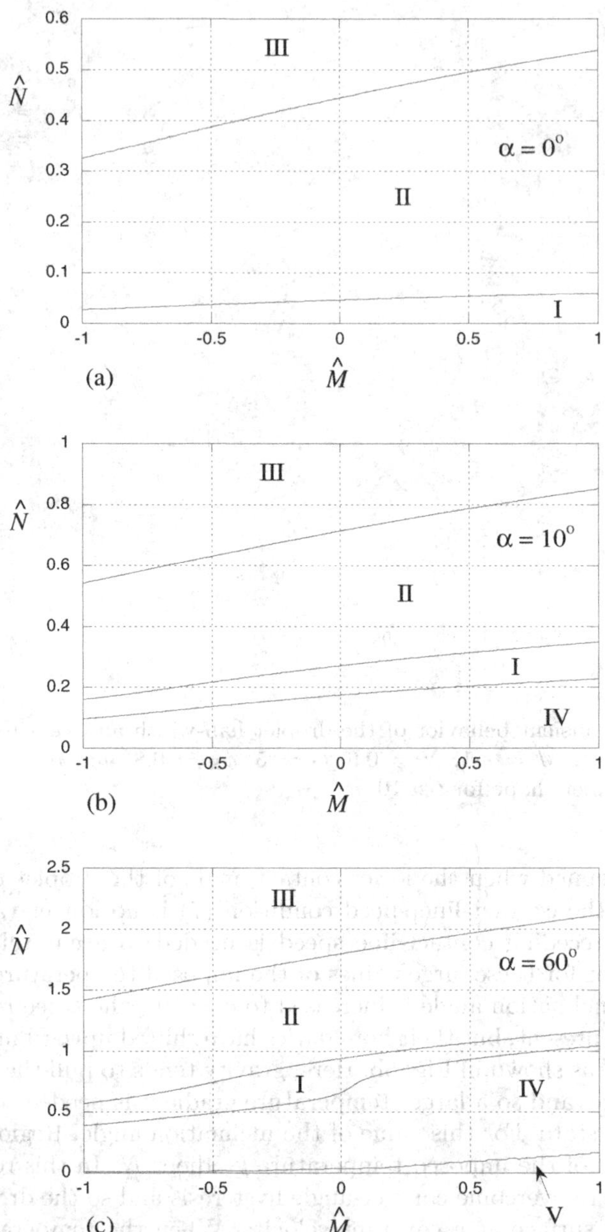

Fig. 6. Domain plots showing the various regions of steady-state droplet behavior in the parameter space of $\hat{N}$ versus $\hat{M}$ with $G = 5$, $\phi_R = 0.8$, and $m = 3$. (a) $\alpha = 0°$, (b) $\alpha = 10°$, and (c) $\alpha = 60°$. The marked regions are: I, no contact-line motion; II, constant-velocity, thermocapillary migration up the inclined surface; III, model not valid; IV, constant-velocity, gravity-driven migration down the inclined surface; and V, transient behavior. The boundary between regions II and III is where $\phi_l \approx 0$

Region V appears for small values of $\hat{N}$. In this region, the droplet behaves in a transient manner. The lower contact line is driven by gravity to the left while the upper contact line is driven by thermocapillarity to the right. This motion stretches the droplet and will ultimately lead to droplet breakup. Thus, in Region V, no steady-state behavior is possible.

5 Discussion

The behavior of the droplet, as shown in the domain plots of Fig. 6, can be explained by examining the dominant physical processes expressed in the steady droplet-shape equation (38a). When the droplet is isothermal and the angle of inclination is not too large, the droplet just slides down the solid surface pulled by gravity. When $\hat{M} = 0$ and $\hat{N} \neq 0$, the imposed temperature gradient on the solid surface causes a thermocapillary stress that drives a flow in the droplet to the right. This reduces the flow caused by gravity and decreases the lower apparent contact angle. Thus, the droplet slows. When the imposed temperature gradient is large enough, the droplet stops and is held in place on the solid surface by contact-angle hysteresis (Region I in Figs. 6b,c). For a larger imposed temperature gradient, the thermocapillary stress causes a flow in the droplet that increases the upper contact angle and this, in turn, causes the droplet to move up the solid surface against gravity (Region II in Figs. 6b,c).

When the solid surface is horizontal and uniformly heated or cooled, the only forces on the droplet are the body force caused by the capillary pressure gradient and the thermocapillary stress caused by temperature variations on the free surface resulting from thickness variations along the length of the droplet. When the droplet is cooled from below, the free surface is hotter at the center than at the contact lines. This causes a thermocapillary stress that forces a flow toward each contact line. This flow causes the droplet to spread out and to become thinner. When the droplet is heated from below, the opposite occurs and the droplet becomes fatter. This effect was originally explained by Ehrhard and Davis [16].

When the solid surface is inclined and uniformly cooled, the droplet moves down the surface, but it is thinner because of the thermocapillary stress caused by the uniform cooling. Imposing a temperature gradient along the solid surface adds another thermocapillary stress that causes a flow to the right in the droplet. This slows the droplet down, as before, but in this case the thinner droplet makes the imposed temperature gradient more effective. The reason for this is seen in the droplet-shape equation (38a). Here, the volume flux due to the gravitational body force is proportional to h^3 while that due to the thermocapillary shear stress is proportional to h^2. As the droplet thickness decreases, the thermocapillary shear stresses dominates. The result is that the magnitude of the imposed temperature gradient needed to stop the droplet and then to make it move up the solid surface is smaller when the droplet is uniformly cooled. This effect is reflected in the positive slope of all of the boundaries between the different regions in Figs. 6a,b,c.

6 Conclusions

Microfluidic devices for diverse applications such as genetic assays, chemical sensors, and environmental testing need some form of fluid management to control and manipulate the liquid samples used during their operation. One way to do this is to apply specially shaped thermal fields on the free surface of a sample droplet and/or the underlying solid surface so that surface-tension-driven flows can be created inside the droplet. These flows change the apparent contact angle of the droplet and thus force the contact line to move in a desired direction. This kind of control of the contact line means that the droplet can be moved to any given position on the solid surface when required.

Thermocapillary droplet control has been verified analytically using a simple fluid model of the process. The model was analyzed using lubrication theory and results were presented that show the required temperature gradient for constant-velocity migration as a function of the angle of inclination of the solid surface and the amount of contact-angle hysteresis. It was shown that the forcing technique is most effective when the droplet is thin. This is also where the approximations used in lubrication theory become more accurate.

The development of the model equations using lubrication theory was described in detail. The theory is a singular perturbation theory in which end regions near the contact lines are analyzed as boundary layers. The flow away from the contact lines (the core region) is described as a locally parallel flow driven by gravity, capillary pressure, and thermocapillary stress. The analysis of the end regions produces effective boundary conditions that are imposed on the flow and droplet shape in the core region. These effective boundary conditions are applied at the positions of the apparent contact lines. The position of the apparent contact line is related to the position of the actual contact line through a constant that reflects the inner physics of the contact-line region.

The most important of the effective boundary conditions from the end-region analysis is a relation describing the speed of the apparent contact line in terms of the apparent contact angle. Since there is much controversy on the exact form of this relation, an empirical relation based on experimental evidence is used in this work. When the proper contact-line speed and apparent contact-line position relations are derived using the appropriate physics for the contact-line region, these results can be applied directly to this lubrication model by modifying the relevant boundary conditions and recomputing the results.

References

1. D.M. Anderson, S.H. Davis: Phys. Fluids **7**, 248 (1995)
2. S.W. Benintendi, M.K. Smith: Phys. Fluids **11**, 982 (1999)
3. S.W. Benintendi: Thermocapillary Migration of a Three-Dimensional Liquid Droplet on a Solid Surface. Ph.D. Thesis, Georgia Institute of Technology, Atlanta, GA (1999)
4. J.P. Boyd: *Chebyshev & Fourier Spectral Methods*, Lecture Notes in Engineering 49 (Springer-Verlag, New York 1989)

5. A.M. Cazabat, F. Heslot, S.M. Troian, P. Carles: Nature **346**, 824 (1990)
6. M.K. Chaudhury, G.M. Whitesides: Science **256**, 1539 (1992)
7. J.-D. Chen: J. Colloid Interface Sci. **122**, 60 (1988)
8. R.G. Cox: J. Fluid Mech. **168**, 169 (1986)
9. P.G. de Gennes: Rev. Modern Physics **57**, 827 (1985)
10. P. Dimitrakopoulos, J.J.L. Higdon: J. Fluid Mech. **395**, 181 (1999)
11. E.B. Dussan, S. H. Davis: J. Fluid Mech. **65**, 71 (1974)
12. E.B. Dussan V: J. Fluid Mech. **77**, 665 (1976)
13. E.B. Dussan V: Ann. Rev. Fluid Mech. **11**, 371 (1979)
14. E.B. Dussan, R.T.-P. Chow: J. Fluid Mech. **137**, 1 (1983)
15. E.B. Dussan: J. Fluid Mech. **151**, 1 (1985)
16. P. Ehrhard, S.H. Davis: J. Fluid Mech. **229**, 365 (1991)
17. P. Ehrhard: J. Fluid Mech. **257**, 463 (1993)
18. R. Goodwin, G. M. Homsy: Phys. Fluids **3**, 515 (1991)
19. H.P. Greenspan: J. Fluid Mech. **84**, 125 (1978)
20. H.P. Greenspan, B.M. McCay: Stud. Appl. Math. **64**, 95 (1981)
21. P.J. Haley, M.J. Miksis: J. Fluid Mech. **223**, 57 (1991)
22. L.M. Hocking: Q. J. Mech. Appl. Maths. **34**, 37 (1981)
23. L.M. Hocking: Q. J. Mech. Appl. Maths. **36**, 55 (1983)
24. L.M. Hocking, A.D. Rivers: J. Fluid Mech. **121**, 425 (1982)
25. L.M. Hocking: Phys. Fluids **7**, 1214 (1995)
26. R.L. Hoffman: J. Colloid Interface Sci. **50**, 228 (1975)
27. C. Huh, L.E. Scriven: J. Colloid Interface Sci. **35**, 85 (1971)
28. C. Huh, S.G. Mason: J. Fluid Mech. **81**, 401 (1977)
29. P.G. Lopez, S.G. Bankoff, M.J. Miksis: J. Fluid Mech. **324**, 261 (1996)
30. C. Marangoni: Nuova Cimento **5**, 239 (1871)
31. A.C. Or, R.E. Kelly, L. Cortelezzi, J. L. Speyer: J. Fluid Mech. **387**, 321 (1999)
32. A. Oron: Phys. Fluid **12**, 1633 (2000)
33. A. Oron, S.G. Bankoff: Physics Fluid **13**, 1107 (2001)
34. J.R.A. Pearson: J. Fluid Mech. **4**, 489 (1958)
35. L.M. Pismen, Y. Pomeau: Phys. Rev. E **62**, 2480 (2000)
36. R.J. Riley, G.P. Neitzel: J. Fluid Mech. **359**, 143 (1998)
37. S.W. Schwartz, S.B. Tejeda: J. Colloid Interface Sci. **38**, 359 (1972)
38. L.M. Schwartz: Phys. Fluids **1**, 443 (1989)
39. Y.D. Shikhmurzaev: J. Fluid Mech. **334**, 211 (1997)
40. Y.D. Shikhmurzaev: Phys. Fluids **9**, 266 (1997)
41. M.K. Smith: J. Fluid Mech. **294**, 209 (1995)
42. M.K. Smith, D.R. Vrane: 'Deformation and Rupture in Confined, Thin Liquid Films Driven by Thermocapillarity'. In: *Fluid Dynamics at Interfaces*, eds. W. Shyy, R. Narayanan, (Cambridge University Press, United Kingdom 1999) pp. 221–233
43. L.H. Tanner: J. Phys. D: Appl. Phys. **12**, 1473 (1979)
44. S.M. Troian, E. Herbolzheimer, S.A. Safran, J.F. Joanny: Europhys. Lett. **10**, 25 (1989)

Electrocapillary Flows

Duane Johnson

Department of Chemical Engineering, University of Alabama, Tuscaloosa, AL 35487-0203, USA

Abstract. Two common capillary phenomena that have been investigated are the thermocapillary and the solutocapillary phenomena where the interfacial tension is a function of the temperature and surface concentrations, respectively. This article will focus on a lesser-known capillary phenomena: electrocapillarity. Electrocapillarity occurs when an electric field is applied tangentially to an electrically charged interface, and is a natural extension of the two better known capillary flows. The first section of this paper will compare the thermodynamic derivation of electrocapillarity and the electric Maxwell stress tensor. It will be shown that these two seemingly different phenomena are actually the same. The remaining sections will include several reasonably simple experiments that demonstrate electrocapillary flows under different situations.

1 Introduction

In the past several decades, a reasonable understanding of particular capillary flows and their importance in industrial applications has been developed. Among the capillary phenomena that have been mainly investigated are the thermocapillary and the solutocapillary phenomena where the interfacial tension is a function of the temperature and surface concentrations, respectively. Thermocapillary flows exist when a temperature gradient is applied tangentially to a liquid–fluid interface. The interfacial tension decreases as the temperature increases and therefore the interfacial tension is lower in regions of higher temperature. A variation in the interfacial tension causes the interface to move and subsequently drags with it the two bulk phases in contact with the interface. Solutocapillary flows occur when there exists a concentration of an absorbed species on the interface. This occurs, for example, when a drop of surfactant is placed on a liquid–fluid interface. For most surfactants, the interfacial tension decreases as more surfactants are absorbed. The decrease in the interfacial tension near the region of the added surfactant causes the interface to move away from the drop.

This article will focus on a lesser known capillary phenomena: electrocapillarity. Electrocapillarity occurs when an electric field is applied tangentially to an electrically charged interface, and is a natural extension of the two better known capillary flows. The next section will draw a comparison between the thermodynamic derivation of electrocapillarity and the electric Maxwell stress tensor. It will be shown that these two seemingly different phenomena are actually the same. The remaining sections will include several reasonably simple experiments that demonstrate electrocapillary flows under different situations.

D. Johnson, Electrocapillary Flows, Lect. Notes Phys. **628**, 291–304 (2003)
http://www.springerlink.com/

2 Electrocapillarity

Electrocapillarity is a surface phenomenon that exists when an electric field is applied tangentially to a fluid–fluid interface that has a non-zero surface charge [15,22]. This charge can be present even in the absence of an electric field, for example when an ionic surfactant is absorbed. The strong effect of an ionic surfactant on the surface tension has been known for sometime in the thermodynamics and chemistry communities [12,15], but has seemed to receive little attention – with the exception of the Taylor–Melcher leaky dielectric model [22] – in the fluid dynamics communities [19]. This is surprising considering the number of interesting phenomena that are present in electrocapillary flows.

This novel flow occurs in several important applications such as the control of fluid motion in microfluidic systems [8], electrically enhanced heat transfer [7,16,17], aluminum production from cryolite (Hall-Héroult cell) [9,18], electric emulsification [20], and electrospraying in inkjet printers. Another exciting potential application is the production of supersaturated dissolved oxygen (SSDO) using electrolysis. SSDO can be used in biofermentation reactions, paper pulp bleaching, or other "green" replacements for oxidation reactions.

Experimental electrocapillary curves for mercury in contact with HCl acid are shown in Fig. 1 [15]. There are also many aqueous–organic systems with ionic surfactants that exhibit strong electrocapillarity. The curves are parabolic with

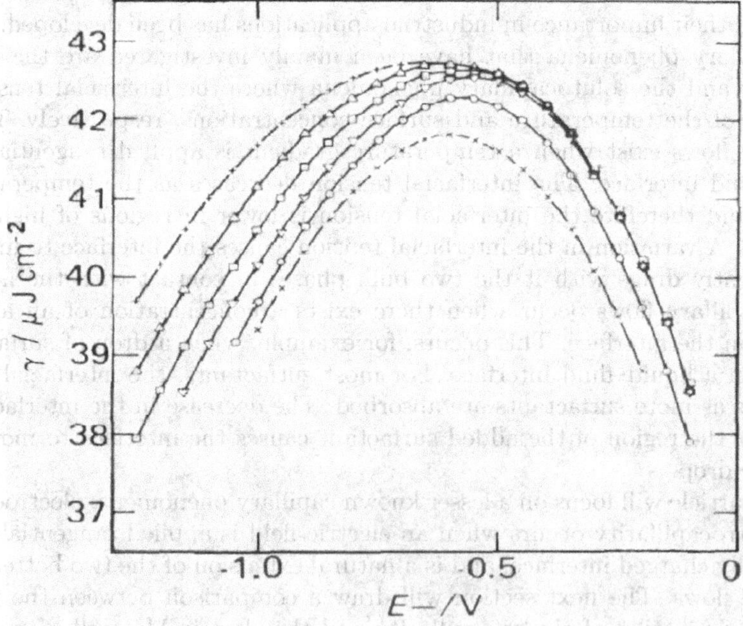

Fig. 1. Plot of the surface tension versus applied electric field [15]. The slope of the curve gives the excess surface charge, and the maximum of the curve gives the electric field value of zero surface charge

the slope equal to the surface excess charge. The point of maximum surface tension, which corresponds to zero surface charge, is found at the maximum of these curves. Notice that this does not occur at zero voltage. This indicates that even when no electric field is applied, a non-zero surface charge exists. For a small voltage drop of only 0.6 V, a 10% drop in the interfacial tension can be seen, demonstrating the effect is appreciable. Therefore, in applications where electric fields are applied across immiscible fluid interfaces, electrocapillary-driven flows will be significant. To compare the electrocapillary effect with other well-known capillary phenomena, such as thermocapillarity, one can look at the two Marangoni numbers.

$$Ma_T = \frac{\partial\gamma/\partial T\ (\Delta T)\, d}{\mu\kappa}, \quad \text{and} \quad Ma_E = \frac{\partial\gamma/\partial\varphi\ (\Delta\varphi)\, d}{\mu\eta}$$

where Ma_T is the thermal Marangoni number, and Ma_E is the electrical Marangoni number. T is the temperature, γ is the surface tension, φ is the electric potential, d is the depth of the fluid layer, μ is the fluid viscosity, $\eta = d^2\,\sigma/\epsilon$ is the electrical diffusivity, and κ is the thermal diffusivity. A well-studied fluid in thermocapillary phenomenon is silicone oil. It is known that small temperature gradients (say 1 °C), with Marangoni numbers around 50, can cause substantial flows on the order of 1 cm/sec [10]. Using the curve in Fig. 1, we can calculate a Marangoni number on the order of 1,500 for mercury; 30 times larger than the equivalent thermal Marangoni number. Therefore, some fluid–fluid systems will show very large flows in the presence of very small potential differences. This is in contrast to leaky dielectric experiments, where large electric fields on the order of kV are commonly used.

There appear to be two different ways of looking at the electrocapillary effect. The first involves the electric Maxwell stresses acting on the interface (the Taylor–Melcher leaky dielectric model), and the second involves the Gibbs free energy of the surface, as is typically done in the thermocapillary literature. The following is a simple derivation that demonstrates that these two ways of looking at the problem are actually the same. However, looking at the problem from two different aspects can lead to different and useful insight into novel flows.

We start by giving the stress balance across an arbitrary, deflecting interface. The symbol $[A] = A_1 - A_2$, where A_1 is some property of one fluid, and A_2 is the same property of the other fluid. The stress balance we will consider here is the sum of the viscous and Maxwell stresses, the pressure contribution, and the surface deformation.

$$[\boldsymbol{T}^\mu + \boldsymbol{T}^e] \cdot \boldsymbol{n} = \boldsymbol{n}\,[p] + \gamma\boldsymbol{n}\nabla\cdot\boldsymbol{n} \tag{1}$$

where $\boldsymbol{n}$ is the unit outward normal, p is the pressure, $\boldsymbol{T}^\mu$ is the viscous stress tensor, $\boldsymbol{T}^e$ is the electric Maxwell stress tensor, and γ is the interfacial tension. We will ignore the magnetic field contributions. The viscous and electric stress tensors are given by:

$$\boldsymbol{T}^\mu = \mu\left(\nabla\boldsymbol{v} + (\nabla\boldsymbol{v})^T\right) = \mu\left(\frac{\partial v_i}{\partial x_j} + \frac{\partial v_j}{\partial x_i}\right) \tag{2}$$

$$\boldsymbol{T}^e = \epsilon\epsilon_0\left(\boldsymbol{E}\boldsymbol{E} - \frac{1}{2}|E^2|\,\boldsymbol{I}\right) = T^e_{ij} = \epsilon\epsilon_0\left(E_iE_j - \frac{1}{2}\delta_{ij}E^2\right) \tag{3}$$

where $\boldsymbol{v}$ is the velocity vector, ϵ is the electric permittivity, and $\boldsymbol{E}$ is the electric vector field.

In the thermocapillary flow literature [14], the stress is usually balanced by the variation of the surface tension.

$$[\boldsymbol{n}\cdot(-p\boldsymbol{I} + \boldsymbol{T}^{\mu})] = \gamma\boldsymbol{n}\nabla\cdot\boldsymbol{n} - (\boldsymbol{I} - \boldsymbol{n}\boldsymbol{n})\cdot\nabla\gamma = \gamma\boldsymbol{n}\nabla\cdot\boldsymbol{n} - \nabla_s\gamma \tag{4}$$

where $\boldsymbol{I}$ is the identity tensor, and $\nabla_s = (\boldsymbol{I} - \boldsymbol{n}\boldsymbol{n})\cdot\nabla$ is the surface gradient operator. The first term on the right hand side represents the stress caused by surface deformation, and the second term represents the stress caused by variation in the surface tension. The interfacial tension is found from a thermodynamic calculation of the Gibbs' free energy of the interface [11].

$$-\mathrm{d}\gamma = s\,\mathrm{d}T + \sum_i \Gamma_i\,\mathrm{d}\tilde{\mu}_i + \sum_i z_i\,\Gamma_i\,F\,\mathrm{d}\varphi \tag{5}$$

Here, s is the surface entropy, T is the temperature, Γ_i is the concentration of absorbed species i, $\tilde{\mu}_i$ is the chemical potential, z_i is the charge of species i, F is Faraday's constant, and φ is the electric potential. The first term represents the thermocapillary forces, the second term represents the solutocapillary forces, and the third term represents the electrocapillary forces. The Lippman equation is derived by taking the derivative with respect to the electric potential and holding the temperature and chemical potential constant [3].

$$\left(\frac{\partial\gamma}{\partial\varphi}\right)_{T,\tilde{\mu}_i} = -\sum_i z_i\,\Gamma_i\,F = -\sigma_s \tag{6}$$

The symbol σ_s represents the free surface charge on the interface, and is the slope of the graph in Fig. 1.

If we balance the different stress terms in (1) and (4), we get.

$$[\boldsymbol{T}^e]\cdot\boldsymbol{n} = \nabla_s\gamma \tag{7}$$

Equation (7) shows that the projection of the Maxwell stress tensor on the interface is balanced by the interfacial tension gradient. Next, we find the tangential component of (7), by taking the dot product with some vector tangential to the interface, $\boldsymbol{t}$.

$$\boldsymbol{t}\cdot[\boldsymbol{T}^e]\cdot\boldsymbol{n} = \boldsymbol{t}\cdot\nabla_s\gamma = -\sigma_s\,\boldsymbol{t}\cdot\nabla\varphi \tag{8}$$

$$\boldsymbol{t}\cdot[\boldsymbol{T}^e]\cdot\boldsymbol{n} = \boldsymbol{t}\cdot\left[\epsilon\epsilon_0\left(\boldsymbol{E}\,(\boldsymbol{E}\cdot\boldsymbol{n}) - \frac{1}{2}|E^2|\,\boldsymbol{n}\right)\right] = \boldsymbol{t}\cdot[\epsilon\,\epsilon_o\boldsymbol{E}\,(\boldsymbol{E}\cdot\boldsymbol{n})] \tag{9}$$

Equating the right hand side of (8) and (9), and using the identity $-\nabla\varphi = \boldsymbol{E}$, we have:

$$[\epsilon\epsilon_0\,(\boldsymbol{E}\cdot\boldsymbol{n})] = \sigma_s \tag{10}$$

We can interpret this result along two lines of thought. The first is where an electric field, $\boldsymbol{E}$, applied normal to an interface induces a surface charge, σ_s. The second way we can look at this is if a surface charge exists in the absence of a normal applied electric field, the surface charge creates a stress along the interface in the presence of a tangential electric field via the Maxwell stress tensor. In the following sections, several experiments will be used to demonstrate both interpretations. The first scenario, where an applied electric field induces a surface charge, is the basis of the leaky dielectric experiments. The latter experiments demonstrate how considering the surface tension gradients can lead to flows that are similar to known experiments in the thermocapillary literature.

3 Leaky Dielectric Experiments

One of the classical experiments to demonstrate the electrocapillary phenomena was performed by Taylor and Melcher [24]. In their experiment, two electrodes were placed vertically on the sides of a container and another electrode was placed on the top of the container at an angle to the horizontal direction (Fig. 2). The container was filled partially with corn oil so that the fluid surface touched the bottom of the lowest part of the top electrode. A voltage difference of 20 kV was applied between the two electrodes and the corn oil began to flow. In this experiment, both a tangential and a normal electric field are needed. The normal component of the electric field induces a positive surface charge at the interface via the Lippman equation (6). The tangential component of the electric field then acts on this induced surface charge to produce a coulombic force on the interface that pushes the interface from right to left.

We can also analyze this problem by looking at the thermodynamics of the interfacial tension. As the induced surface charge is positive, the interfacial tension decreases in the direction of increasing potential. The higher electric potential on the right side of the container produces a lower interfacial tension on the right side of the container. The interfacial tension pulls on the interface from right to left causing the counter clockwise flow. Note that because the interfacial charge is zero in the absence of an electric field, a rather large electric potential is necessary to first create an interfacial charge. Using (10) and the surface charge

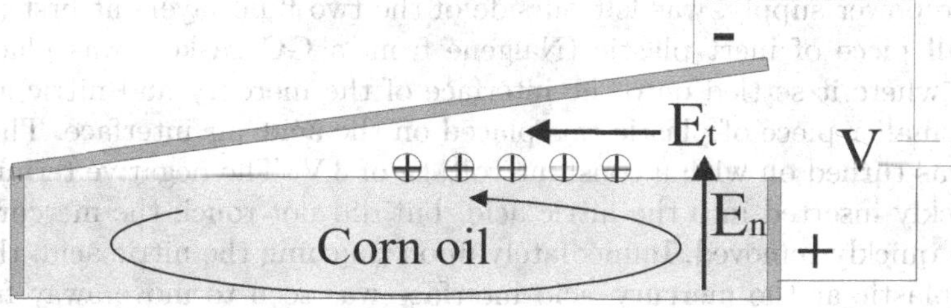

Fig. 2. Schematic of the leaky dielectric experiment by Melcher and Taylor. A large normal component of the electric field induces a positive surface charge and a tangential electric field acts on the surface charge to cause capillary flow

values from Fig. 1, we can estimate the electric field needed to generate the same amount of surface stress present in the mercury–acid system (see below). From the slopes of the curves in Fig. 1, the surface charge, σ_s, is estimated to be on the order of $10^{-2}\,\mathrm{J\,V^{-1}\,m^{-2}}$. Assuming all of the voltage drop in the leaky dielectric system occurs across the interface, we can estimate the electric field necessary to induce the surface charge.

$$E = \frac{\sigma_s}{\epsilon\epsilon_0} \tag{11}$$

Using the value of the permittivity of free space, $\epsilon_0 = 8.85{*}10^{-12}\,\mathrm{C^2\,N^{-1}\,m^{-2}}$ and assuming the permittivity of the oil is of order one, the order of the electric field is $10^9\,\mathrm{V/m}$. This is indeed a rather large electric field and demonstrates that systems with even a small surface charge are equivalent to leaky dielectric systems with huge electric fields.

4 Experiments with Non-zero Surface Charge

We next discuss a series of three experiments that demonstrate electrocapillary flows when the interface is charged. The non-zero surface charge in the absence of an electric field may be present due to thermodynamically favorable conditions – as in the mercury–acid system (Fig. 1) – or when an ionic surfactant is added intentionally to the interface. The first experiments will consist of mercury and nitric acid in a petri dish. The second and third experiments will show electrocapillary flows for a potassium chloride aqueous solution and an organic methyl isobutyl ketone (MIBK) solution with an insoluble cationic surfactant, cetyl pyridinium chloride (CPC) [6]. The last two experiments are directly analogies to two well-known thermocapillary experiments where the driving force for the capillary flows is applied either tangentially or perpendicular to the charged interface.

4.1 Mercury–Nitric Acid Experiment

In the first experiment, a loop of copper wire was placed in the bottom of a layer of mercury and the wire was connected to the positive terminal of a small, constant voltage, power supply. A second wire, connected to the negative terminal of the power supply, was left outside of the two fluid layers at first (Fig. 3). One small piece of inert plastic (Nalgene from a GC gasket) was placed into the fluid where it settled onto the interface of the mercury and nitric acid and another smaller piece of plastic was placed on the acid–air interface. The power supply was turned on with a constant voltage of 3 V. The negative terminal was then quickly inserted into the nitric acid, but did not touch the mercury layer, and then quickly removed. Immediately upon touching the nitric acid, the small piece of plastic at the mercury–acid interface was seen to move away from the wire (Fig. 4). Once the wire was removed, the gasket moved back to its original location, but at a much slower speed. The experiment was then repeated with no applied voltage, and the gasket was not seen to move at all. It is interesting to

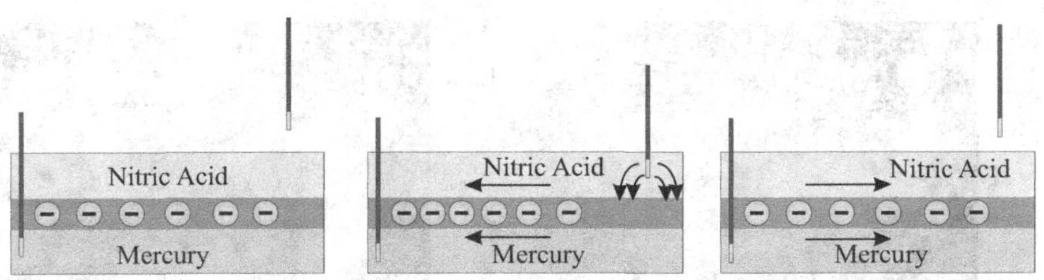

Fig. 3. Schematic of the elastic electrocapillary effect. **a** An insulated copper wire with an exposed end is placed inside a layer of mercury that has a layer of nitric acid above it. **b** Another copper wire is quickly inserted into the nitric acid layer and removed. The application of the voltage causes the interface to move away from the wire. **c** Once the wire is removed, the interface returns more slowly back to its original location

note the elastic behavior of the interface. This is in contrast to, say, the addition of a drop of soap to a water–air interface. When the soap is added to the interface, the interface immediately shoots out radially. However, the interface never recoils from the spreading, and eventually stops moving. The electrocapillary effect appears to be caused by the interfacial charges. When the wire is placed in the acid, the electric field pushes the negatively charged interface away from the inserted wire (Fig. 3). When the wire is removed, the compressed charges on the interface repel each other and push the interface back. This can also be explained in terms of the electrocapillary curves. The electric field near the inserted wire decreases the charge density near the wire. As the surface charge is negative, a smaller charge decreases the surface tension near the inserted wire causing the interface to move. When the wire is removed, the interfacial charge is reestablished, possibly by the diffusion of ions from solution. This would account for the relatively longer times needed to push the gasket back to its original position.

4.2 Tangential Electric Fields with Ionic Surfactants

The second experiment is analogous to the well-known thermocapillary experiment where a layer of fluid is placed in a rectangular container with a hot wall (say the right wall) and a cold wall (say the left wall) [10]. In this experiment, the interface moves from regions of low surface tension (the hot wall) to regions of high surface tension (the cold wall). The fluid then returns along the bottom of the container in a circular fashion.

In the second experiment, a 1 M KCl solution was made with DI water. A small amount of 50 μm, hollow glass spheres was added to the water for PIV flow imaging [1]. A small amount of an organic salt, tetra methyl ammonium iodide (TMAI) and a cationic surfactant, CPC, were added to ACS grade MIBK, which was then carefully added on top of the aqueous layer. The two solutions were placed in a rectangular Pyrex container with two stainless steel electrodes (Fig. 5a). The positive electrode did not touch the water to prevent electrolysis. The system was allowed to sit for several hours to allow the fluid motion to come to rest, and to allow the surfactant to come in chemical equilibrium with the

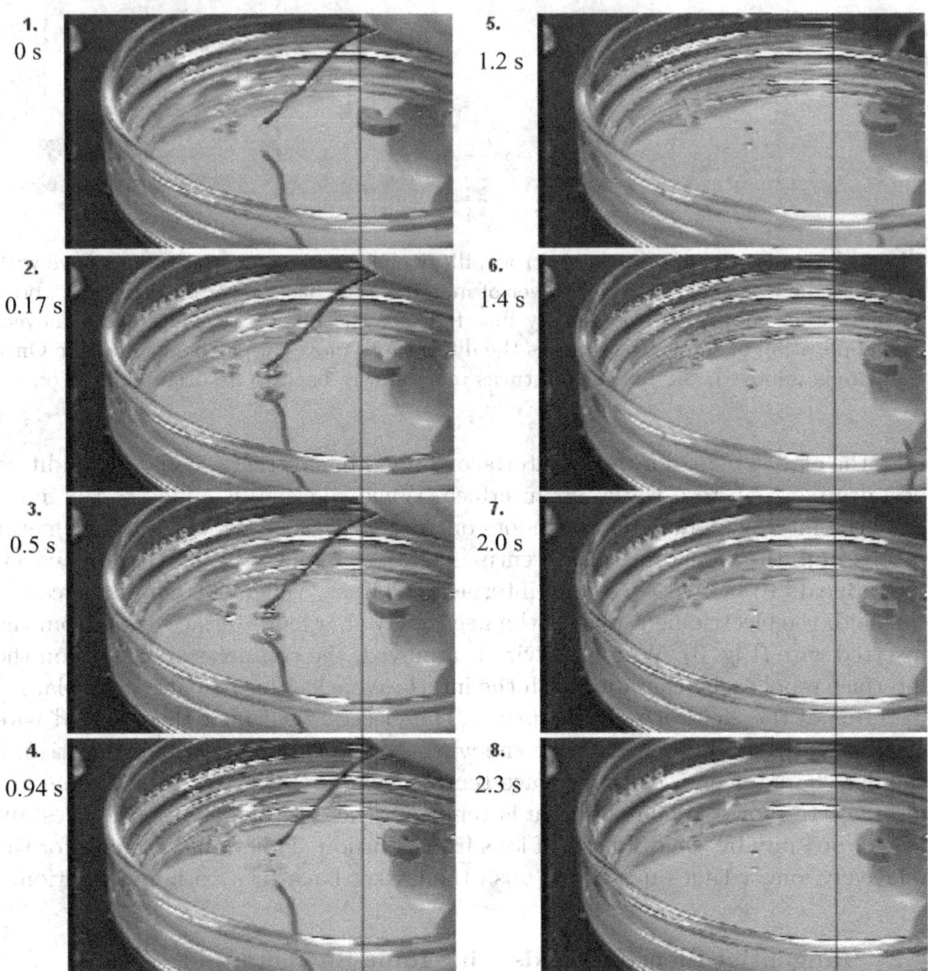

Fig. 4. Sequence of eight images for the experiments with mercury and nitric acid. A piece of plastic is inserted at the mercury-nitric acid interface and at the nitric acid-air interface. The time each image was acquired is given to the left of the picture. The line drawn through each image references the original location of the piece of plastic

aqueous solution. MIBK is barely soluble in water and no fluid motion was seen due to solutal Marangoni stresses [23].

A constant 18 V was then applied across the two electrodes and the fluid immediately began to flow. If the fluid motion was due to the electrocapillary effect, then it should move from the positive electrode to the negative electrode along the interface. This is indeed what happens (Fig. 6). Because the surfactant is a positive ion, the surface charge is positive ($\sigma_s > 0$). The potential drop from the positive electrode to the negative electrode is negative ($\Delta\varphi < 0$), and therefore, according to the Lippman equation, the surface tension should increase

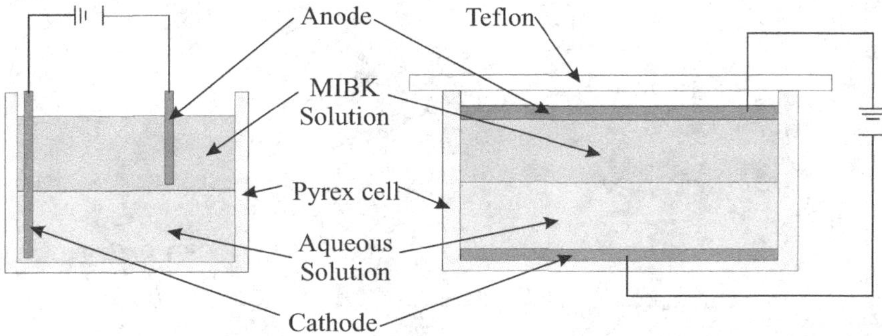

Fig. 5. Schematic of the two experimental cells: a) horizontal electric potential gradient and b) vertical potential gradient.

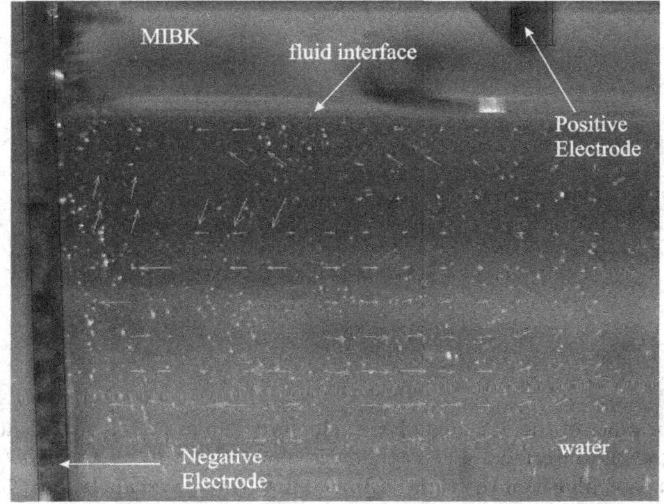

Fig. 6. Grayscale image of the electrocapillary-driven flow. The top fluid layer is MIBK, the bottom fluid layer is an aqueous 1 M KCl solution, and a small amount of CPC cationic surfactant has been added to the interface. The fluid flows from the positive electrode to the negative electrode with only a small applied voltage. The positive (right) electrode is kept out of the water to prevent electrolysis. The velocity vectors superimposed on the water layer were produced by PIV software

from the positive electrode to the negative electrode. As the surface tension is larger at the negative electrode, the interface will be pulled towards the negative electrode, as seen in Fig. 6.

To convince ourselves, the experiment was repeated without the surfactant. In each case, whether an aqueous salt, an organic salt, or neither salt was added, there was no fluid motion when the surfactant was absent. This ruled out the

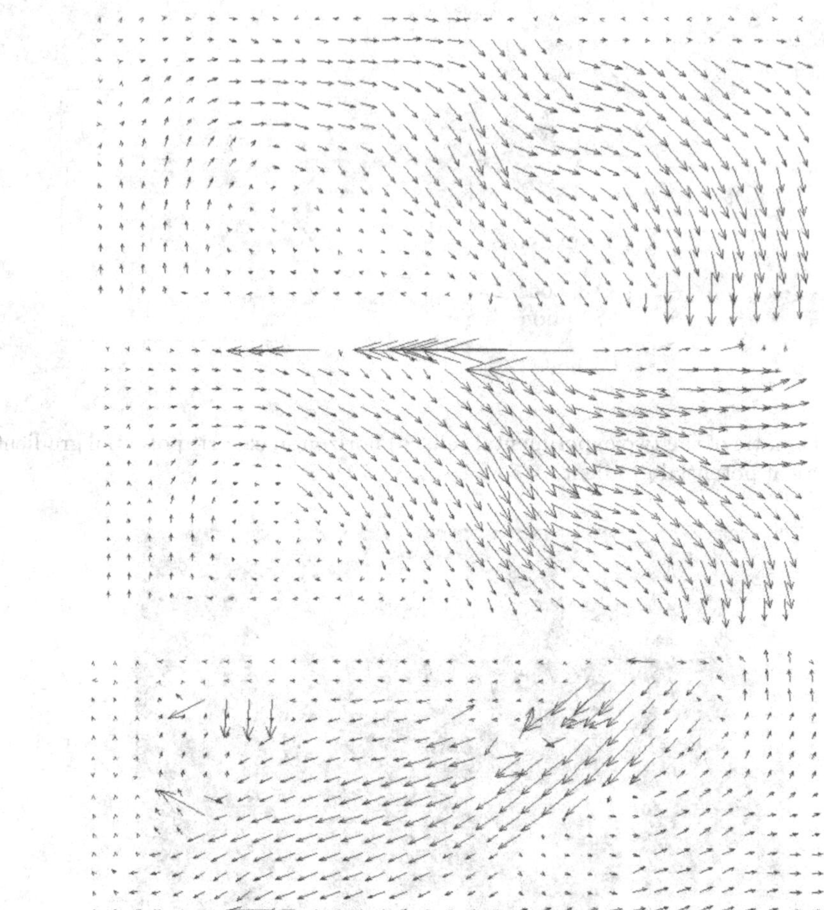

Fig. 7. Vector plots of the aqueous phase velocity when applying a tangential potential gradient across the interface. The cathode is along the wall on the left, the anode is above the aqueous phase on the right. **a** Initial flow three seconds before 25V is applied. **b** One second after the field is applied. **c** Fifty seconds after the voltage is applied

possibility that electromigration and subsequent adsorption of the salt ions was driving the fluid motion. This demonstrates an important point that not all ions are equally surface active. For example, it is quite possible that a number of potassium or chloride ions are present on the interface. However, the thermodynamics favors the ions in solution and the net surface charge due to the salt ions is negligible.

Another interesting aspect of this experiment was the dynamic behavior of the fluid flow (Fig. 7). Initially, a steady flow was seen due to colloidal-capillary flows [5]. When the voltage was applied, the surface quickly shot from the positive electrode to the negative electrode (right to left). After a few seconds, the surface flow only extended partially to the negative electrode and a small vortex was seen below the positive electrode. Even for these small voltages, the flow appeared to be nearly turbulent and complicated three-dimensional flows developed.

4.3 Perpendicular Electric Fields with Ionic Surfactants

In the third experiment, a voltage potential was applied normal to the interface (vertical potential field). This is very similar to the Marangoni–Bénard instability experiments [2], [13,14,21] where a vertical temperature gradient is applied across two immiscible fluids. In the thermocapillary experiments, a temperature gradient does not cause any flow until it is increased beyond a critical value (critical Marangoni number). Imagine a tank filled with two fluids with a hot top plate and a cold bottom plate (this eliminates the possibility of buoyancy-driven convection). If some part of the interface deflects temporarily upwards, the peak of the deflection will become hotter than the lower positions of the interface (Fig. 8). This newly established temperature gradient along the interface causes a variation in the surface tension. If the temperature gradient is large enough, the surface tension gradient will overcome the viscosities and thermal diffusivities of the two fluids and steady flow will be observed.

A completely similar scenario occurs when an electric field is applied perpendicular to a charged interface. During the application of the vertical potential difference, a perturbation of the interface will experience different electrical potentials. The variation in the electric potential along the interface causes a variation in the interfacial tension. If the interfacial tension gradient is large enough to overcome the viscosities and electrical diffusivities of the two fluids, then persistent flow can be seen.

The electrocapillary experiments were set up as shown in Fig. 5b using the same fluids as in the previous tangential electric field experiment. When a small voltage was applied, no flow was seen. When the voltage was increased to 50 V, however, small circular vortices were seen near the interface (Fig. 9). When the surfactant concentration was reduced, the flow was not seen for voltages of up to 100 V. At this point, oxygen bubbles were forming due to electrolysis of the aqueous phase and the experiment was stopped.

Even more interesting was the case when more surfactant was added. When a voltage potential of 40 V was applied, no fluid motion was initially detected. Suddenly, the interface rippled like a sheet in the wind. This sudden onset of in-

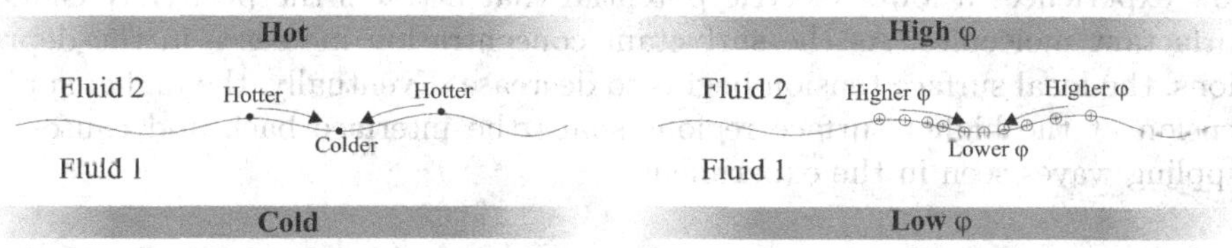

Fig. 8. Mechanism of the thermocapillary and electrocapillary flows when the gradient (either temperature or electric) is imposed perpendicular to the interface. Perturbations to the interface cause a variation of the applied field on the interface. These variations give rise to interfacial tension gradients that can lead to continuous surface flow

Fig. 9. Velocity vector plots of the flow near the interface during a perpendicularly applied electric field. The top of the picture is at the aqueous–organic interface and the bottom of the picture is about 1 cm below the interface. Small vortices are seen near the interface, but the flow does not extend much below the interface

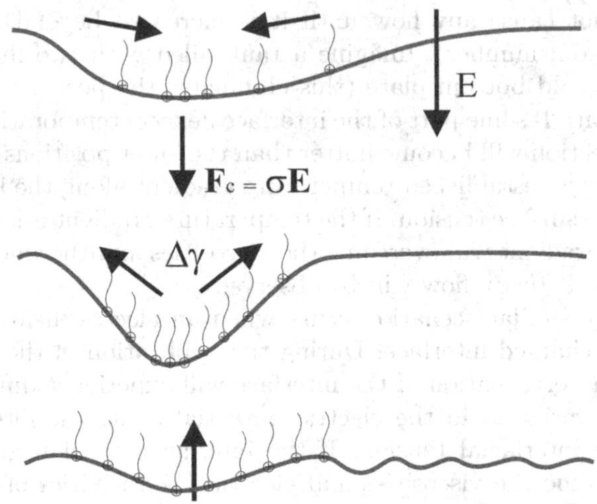

Fig. 10. Schematic of the periodic interfacial waves seen during larger vertical potential gradients and higher cationic surfactant concentrations. a The positively charged interface is initially pulled down, which causes the cationic surfactant to accumulate in the depression. b After some time the interfacial tension in the depression becomes significantly lower and c) the interface "snaps back" to a flatter shape

terfacial waves repeated every few minutes. At higher surfactant concentrations, the surfactant inhibits the surface motion [4]. This results in a much higher critical temperature needed to cause flow similar to Fig. 9. However, the positive charge on the interface causes the interface to be pulled down towards the negative terminal by the Coulombic force (Fig. 10). The depression in the interface now experiences a lower electric potential that attracts the positively charged surfactant molecules. As the surfactant concentration increases in the depressions, the local surface tension begins to decrease. Eventually, the higher surface tension at the higher surface regions snaps the interface back and causes the rippling waves seen in the experiment.

5 Conclusions

The purpose of this paper was to demonstrate the electrocapillary phenomenon for different systems. In particular, two distinct groups of experiments were given: one where a surface charge was induced at the fluid-fluid interface, and another where a non-zero surface charge existed in the absence of an applied field. It was shown that the electrocapillary flows for non-zero surface charge give rise to much stronger flows. When a non-zero surface charge is present, a direct analogy can be drawn between classical thermocapillary experiments and electrocapillary experiments. This was demonstrated using a tangential and perpendicular electric fields.

There are a number of important differences, though, between thermocapillarity and electrocapillarity. Among these are the fact that non-zero surface charges require that an ionic species be present at the interface. The adsorption of a surfactant on an interface necessarily requires one to consider the change in the chemical potential, or in other words, the solutal capillary stresses. This gives rise to additional phenomena such as the electro–elastic effect seen in the mercury–acid experiment. In addition, the ionic surfactants can desorb from the interface into the solution and, if the concentration is sufficiently high, form micelles in solution. Because of the number of existing and potential applications where electrocapillarity is present, this field requires further in-depth analysis.

References

1. R.J. Adria: Ann. Rev. Fluid Mech. **23**, 261, (1991)
2. C.D. Andereck, P.W. Colovas, M.M. Degen, Y.Y. Renardy: Int. J. Eng. Sci. **36**, 1451 (1998)
3. A.J. Bard, L.R. Faulkner: *Electrochemical Methods*, (Wiley, 1980)
4. J.C. Berg, A. Acrivos: Chem. Eng. Sci. **20**, 737 (1965)
5. K. Casson, D. Johnson: J. Coll. Int. Sci. **242**, 279 (2001)
6. K. Casson, D. Johnson: *to appear in* Phys. Fluids **242**, 279 (2001)
7. H.J. Cho, I.S. Kang, Y.C. Kweon, M.H. Kim: Int. J. Multi Flow **5**, 909 (1996)
8. J.W. Choi, C.H. Ahn: 'An Active Micromixer using Electrohydrodynamic (EHD) Convection'. In: *Solid State Sensor and Actuator Workshop*, (Hilton Head, SC 2000)
9. P.A. Davidson: Ann. Rev. Fluid Mech. **31**, 273 (1999)
10. S.H. Davis: Ann. Rev. Fluid Mech. **19**, 403 (1987)
11. R. Defay, I. Prigogine, A. Bellemans: *Surface Tension and Adsorption*, translated by D. H. Everett (Longmans, London 1966)
12. J.W. Ha, S.M. Yang: J. Coll. Int. Sci. **175**, 369 (1995)
13. D. Johnson, R. Narayanan: Phys. Rev. E **56**, 5462 (1997)
14. D. Johnson, R. Narayanan: J. Chaos **9**, 124 (1999)
15. A. Kitahara, A. Watanabe: *Electrical Phenomena at Interfaces: Fundamentals, Measurements, and Applications*, (Marcel Dekker, New York 1984)
16. Y.C. Kweon, M.H. Kim: Int. J. Multiphase Flow **24**, 145 (1998)
17. Y.C. Kweon, M.H. Kim: Int. J. Multiphase Flow **26**, 1351 (2000)
18. A.F. LaCamera, D P. Ziegler, R L. Kozarek: 'Magnetohydrodynamics in the Hall-Héroult Process, An Overview'. In: *Magnetohydrodynamics in Process Metallurgy*, (The Minerals, Metals, & Materials Soc. 1991)
19. V.G. Levich: *Physicochemical Hydrodynamics*, translated by Scripta Technica, Inc., (Englewood Cliffs, N.J., Prentice-Hall 1962)
20. R.J. McGreevy, R.S. Schechter: AIChE J. **37**, 169 (1991)

21. A.A. Nepomnyashchy, M.G. Velarde, P. Colinet: *Interfacial Phenomena and Convection*, (Chapman & Hall/CRC, Boca Raton, Florida 2002)
22. D.A. Saville: Ann. Rev. Fluid Mech. **29**, 27 (1997)
23. C.V. Sterling, L.E. Scriven: AIChE J. **5**, 514 (1959)
24. G.I. Taylor, J.R. Melcher: Ann. Rev. Fluid Mech. **1**, 111 (1969)

Direct Simulation of Drop Fragmentation under Simple Shear

Yuriko Renardy

Department of Mathematics, 460 McBryde Hall Virginia Tech, Blacksburg, VA 24061-0123, USA

Abstract. The numerical investigation of drop deformation and breakup under simple shear is a basis for theoretical studies of emulsification and mixing. A three-dimensional Couette flow without gravity is considered; the matrix liquid is bounded at the top and bottom by two parallel horizontal plates and the liquid drop is suspended there. The boundaries move horizontally, and generate a simple shear flow. The fluid and flow properties, such as viscosities, Reynolds number and plate separation, influence the subsequent evolution. The pinching off of daughter drops is investigated with the use of volume-of-fluid methods. The convergence of such methods, particularly with respect to drop size distributions, is evaluated, and a novel parabolic reconstruction of the interfacial tension force for the VOF method (PROST) is introduced.

1 Introduction

The study of dynamics of a drop in shear flow is of fundamental importance in dispersion science, and has evoked great interest, most notably since the experiments of G. I. Taylor [21,29]. These processes, which yield daughter drops, are paradigms of theoretical investigations into emulsification and mixing [19]. Emulsions arise in a wide range of industrial applications: in materials processing, waste treatment, and pharmaceuticals [3,16]. There has been considerable interest in incompatible polymer blends because of the need for recycling plastics [9,20]. In order to apply the emulsion technology, we need to control and manipulate its microstructure. A starting point is to consider the deformation and breakup behavior of single droplets in a well-defined flow field [7]. There are several broad categories for making emulsions; one category is shear mixing while adding one fluid to another [19]. The reader is referred to review articles [21,29] on drop breakup under a variety of flow conditions, and to recent experimental studies on breakup under simple shear [8,10,18,33].

In this paper, we focus on the numerical investigation of fragmentation with particular interest on the volumes of the daughter drops. Figure 1 shows a schematic of the initial condition for a numerical simulation. A drop of viscosity μ_d and density ρ_d is suspended in another liquid of viscosity μ_m and density ρ_m. In this article, we keep the densities the same ($\rho_d = \rho_m$) to simplify the investigation. There is an imposed constant shear rate $\dot{\gamma}$ (the top wall moves in the x-direction, and the bottom wall in the opposite direction). The undeformed radius is a, the plate separation is L_z, and the computational box has periodic boundary conditions in the x and y directions. There are seven dimensionless

Y. Renardy, Direct Simulation of Drop Fragmentation under Simple Shear, Lect. Notes Phys. **628**, 305–323 (2003)
http://www.springerlink.com/

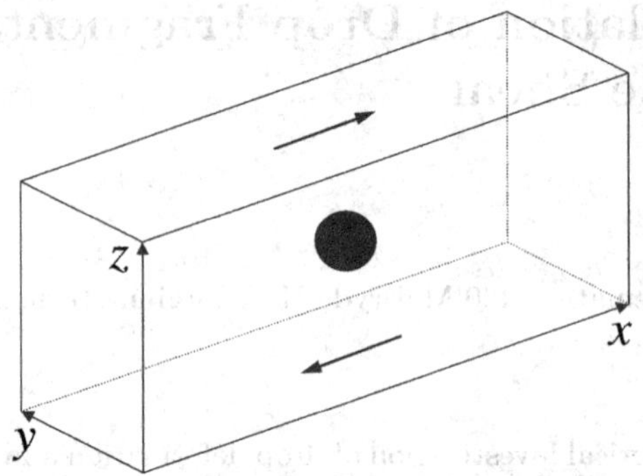

Fig. 1. Sketch of initial condition in our numerical simulation for three-dimensional drop deformation. A drop is suspended in a matrix liquid, and subjected to simple shear generated by the motion of the top and bottom walls

parameters: the viscosity ratio of the drop to matrix liquids

$$\lambda = \mu_d/\mu_m \;, \tag{1}$$

the capillary number,

$$Ca = \mu_m \dot{\gamma} a/\sigma \;, \tag{2}$$

the Reynolds number

$$Re = \rho_m \dot{\gamma} a^2/\mu_m \;, \tag{3}$$

the dimensionless plate separation L_z/a, and dimensionless spatial periodicities, L_x in the x and L_y in the y directions. We take the densities of the drop and matrix liquids to be the same and focus on the competition between viscous force, capillary force and inertia. The effect of changing other parameters, such as the viscosity ratio, is also interesting but is not touched in this article.

Taylor [30] gives a detailed description about his experiments with very viscous drops in Stokes flow. This is delved further in Sect. 5 of [5]. We summarize the main features here. Initially, the most noticeable motion is the elongation of the drop, stretched by viscous shear stress in the external flow. The matrix fluid undergoes a flow in the x-z plane with velocity

$$\begin{aligned} \dot{x} &= \dot{\gamma} z, \\ \dot{z} &= 0, \end{aligned} \tag{4}$$

where $\dot{\gamma}$ is a constant shear rate. We decompose this velocity field:

$$\begin{pmatrix} \dot{x} \\ \dot{z} \end{pmatrix} = \frac{\dot{\gamma}}{2}(\mathbf{S}+\mathbf{A})\begin{pmatrix} x \\ z \end{pmatrix}, \; \mathbf{S} = \begin{pmatrix} 0 & 1 \\ 1 & 0 \end{pmatrix}, \mathbf{A} = \begin{pmatrix} 0 & 1 \\ -1 & 0 \end{pmatrix}. \tag{5}$$

We denote $\mathbf{x} = \begin{pmatrix} x \\ z \end{pmatrix}$. The motion generated by the symmetric matrix $\mathbf{S}$ satisfies

$$\frac{d\mathbf{x}}{dt} = \mathbf{S}\mathbf{x} = \begin{pmatrix} z \\ x \end{pmatrix}. \tag{6}$$

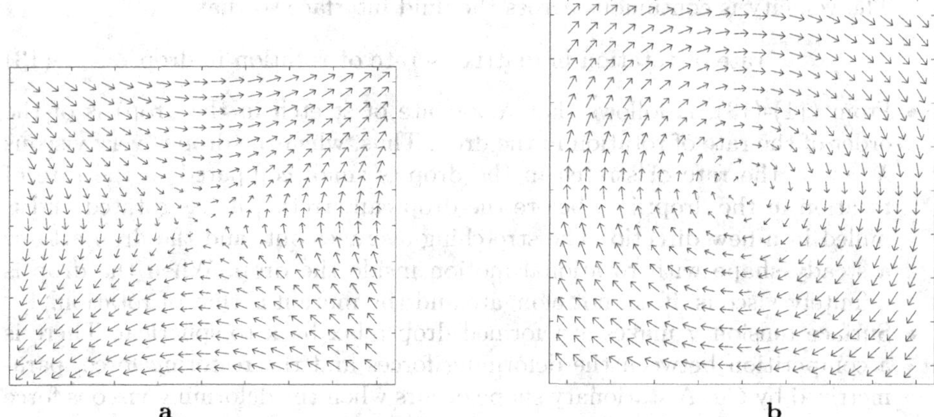

Fig. 2. In (5), the motion driven by **a S**, **b A**

The solution of this system is

$$c_1 e^t \begin{pmatrix} 1 \\ 1 \end{pmatrix} + c_2 e^{-t} \begin{pmatrix} 1 \\ -1 \end{pmatrix}. \tag{7}$$

Figure 2a shows the motion of the fluid, which is pure elongation (stretch), with the axis of elongation along $x = z$. The motion generated by the antisymmetric matrix **A** satisfies

$$\frac{d\mathbf{x}}{dt} = \mathbf{A}\mathbf{x} = \begin{pmatrix} z \\ -x \end{pmatrix}. \tag{8}$$

This is a conservative system with the solution

$$x^2 + z^2 = c, \tag{9}$$

where c is an integration constant. Figure 2b shows the motion which is pure rotation.

An order-of-magnitude estimate for drop stabilization vs breakup is as follows.

- First, shear stress is continuous across the drop/matrix interface, and this is the order of magnitude of viscous stress. Thus,

$$\mu_d \cdot \text{rate of elongation in the drop} \tag{10}$$
$$\sim \mu_m \cdot \text{ rate of elongation in the matrix liquid }, \tag{11}$$

  which yields that $\lambda \times$ (rate of elongation in the drop) is on the order of the rate of elongation in the matrix liquid.
- Secondly, the decomposition (5) shows that in the matrix liquid, the rate of rotation and rate of stretching have similar orders of magnitude:

$$\text{rate of rotation in matrix} \sim \text{rate of elongation in matrix.} \tag{12}$$

The velocity is continuous across the fluid interface so that

$$\text{rate of rotation in matrix} \sim \text{rate of rotation in drop.} \qquad (13)$$

- From (11)-(13), it follows that $\lambda \times$ (rate of stretch in the drop) is of the order of the rate of rotation in the drop. Thus, when the drop is very viscous $\lambda >> 1$, the rate of stretch in the drop is small compared to the rate of rotation in the drop: i.e., before the drop can stretch, it has rotated and is pulled in a new directio. The stretching averages out, and the drop attains a steady shape with rotational motion inside the drop. When the drop is infinitely viscous, it cannot elongate and the motion is that of rotation.
- Surface tension σ makes a deformed drop relax back to spherical. There is a competition between the deforming forces and the restoring force, parametrized by Ca. A stationary shape occurs when the deforming viscous force $\sim \mu\dot{\gamma}$ is much less than the restoring surface-tension force $\sim \sigma/a$ ($Ca << 1$). When $Ca >> 1$, the drop elongates with the flow. As Ca increases, the drop approaches the critical condition for breakup ($\lambda = 1$, Stokes flow):

$$Ca \sim 1. \qquad (14)$$

Figure 3 shows a steady state shape for Stokes flow, with swirling motion inside the drop.
- The *critical* capillary number is denoted Ca_c, above which a steady-state shape ceases to exist.

The sequence of events for fragmentation in laboratory experiments is described at length in [18]. Figures 4–5 show the main features.

- The drop evolves to an ellipsoidal shape, then to a *dumbbell* shape.
- The necks between the bulbs and the central portion of the drop continue to stretch and thin. There is vortical motion inside the bulbs.
- Next, the bulbs detach. These daughter drops reach steady state because their capillary numbers are a fraction of the critical value. This sequence is known as end-pinching. The first end-pinching results in the largest daughter drops.
- Experimental observations [18] of breakup all begin with a breakup of the elongative end pinching type. If the mother drop is large enough, the ends of the remaining portion will retract slightly due to surface tension, then *bulb up* and end-pinching repeats. If the capillary number is sufficiently high, end-pinching leaves behind a long cylindrical thread, where capillary wave breakups are observed. This results in a distribution of large satellite droplets, interspersed with small droplets.

In Stokes flow at $\lambda = 1$, the critical capillary number is approximately 0.43. Below this, the stationary shape is found. On the other hand, if we introduce inertia, then a drop can break up for $Ca < 0.43$ [15,22]. At each capillary number, an increase in the Reynolds number elongates the drop. There are three trends for the overall effect of inertia. These are illustrated in Fig. 6: velocity vector fields are shown in the $x - z$ cross-section for $Re = 1, 10, 60$.

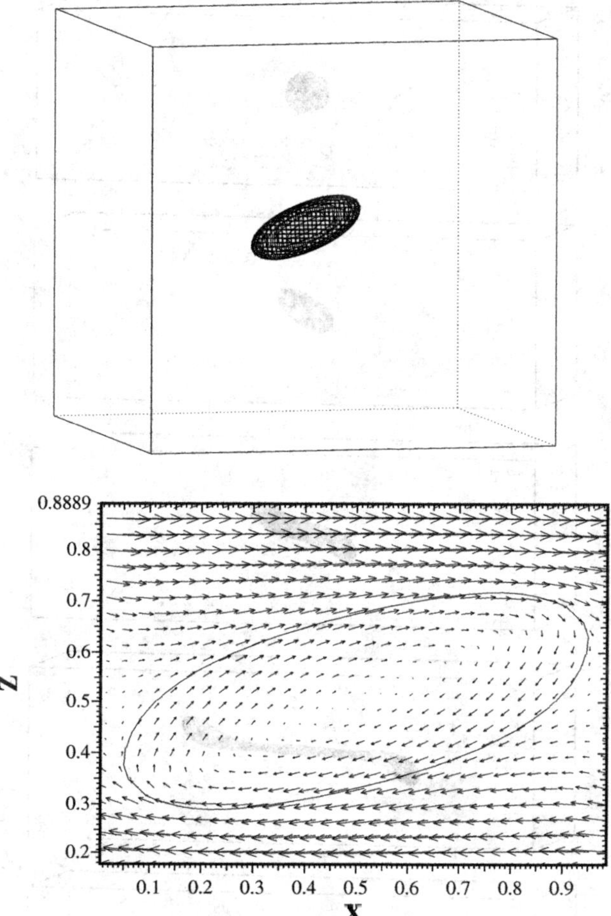

Fig. 3. Numerical simulation of drop evolution to steady drop shape, with velocity vector plot in the $x - z$ plane through the center of the drop. Stokes flow, $Ca = 0.35$, $\lambda = 1$ [23]. The algorithm is described in Sect. 3.2. This evolution has been checked to be within 1% of the results from the boundary integral code of [6]

- First, inertia rotates the drop, so that at higher Reynolds numbers, the steady states are more aligned toward the vertical than in Stokes flow and therefore the drop experiences greater shear.
- Secondly, in Stokes flow, the flow inside the drop consists of a single vortical swirl. When the Reynolds number increases, the velocity field bifurcates, with additional swirls.
- Thirdly, the length of the drop in steady states just below breakup shortens as inertia increases.
- The symmetry across the mid-plane of the steady state, evident in Stokes flow, is lost.

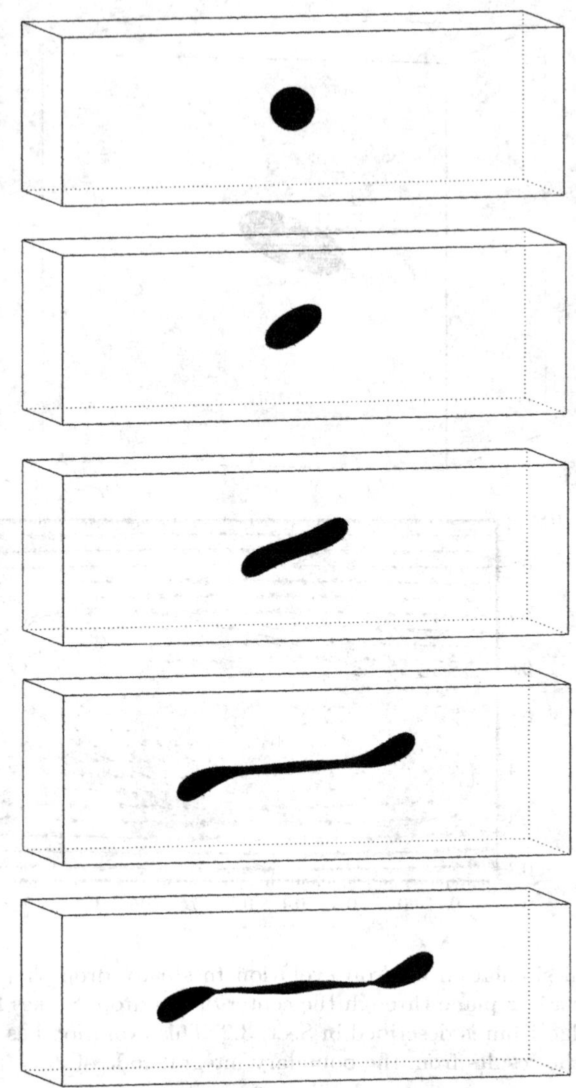

Fig. 4. Numerical simulation of fragmentation. $Re = 12, Ca = 0.175 = 1.14 Ca_c$, $\lambda = 1$. $t = 0, 1, 9, 19, 21, 22, 22.4, 24.8\dot{\gamma}^{-1}$. The VOF-PROST algorithm described in Sect. 3.2 is used

What is the mechanism for inertia-induced breakup? For large Reynolds numbers, the Reynolds stress is of order $\rho|\mathbf{v}|^2 \sim \rho\dot{\gamma}^2 a^2$. This is balanced by capillary stresses of order σ/a. The critical condition is, upon division by the viscous stress $\mu\dot{\gamma}$,

$$Re \sim 1/Ca. \tag{15}$$

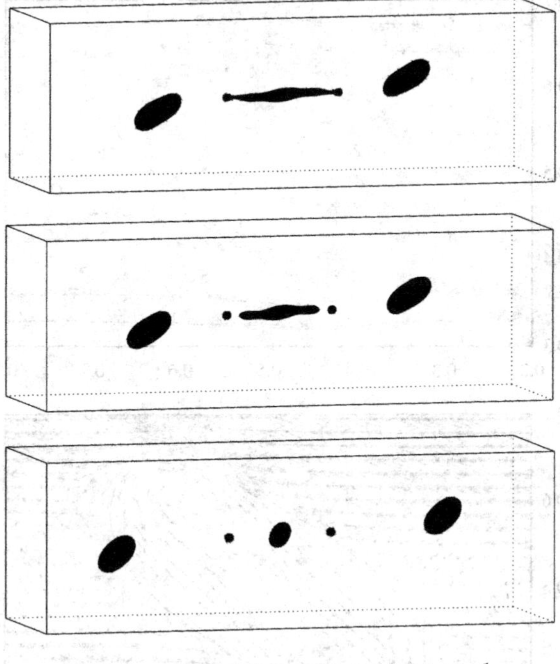

Fig. 5. continued. $t = 22, 22.4, 24.8\dot{\gamma}^{-1}$

In fact, the ratio of inertial to capillary forces is the Weber number

$$We = ReCa. \tag{16}$$

The inviscid limit law is that the critical Weber number $We_c \sim$ constant [24].

2 The Volume-of-Fluid Method: Try It and You'll Savor the (Finite) Difference

In the numerical treatment, we must answer three questions: (1) how do we represent the interface on a mesh? (2) how will the interface evolve in time? and (3) how should we apply the boundary conditions on the interface? The VOF method [1,2,11,14,28,32], provides a simple way of treating the topological changes of the interface. The algorithm is a finite-difference code on a Cartesian grid; Fig. 7 shows the cells. The MAC grid is used. This is a staggered grid for the unknowns; the pressure and the VOF function C are given at nodes in the center of each cell, and the velocities are given at the centers of the faces.

There are three components to the overall algorithm: a VOF method to track the interface, a projection method to solve the Navier-Stokes equations on the MAC grid, and a continuum method for modeling the interfacial tension force. We describe these components below.

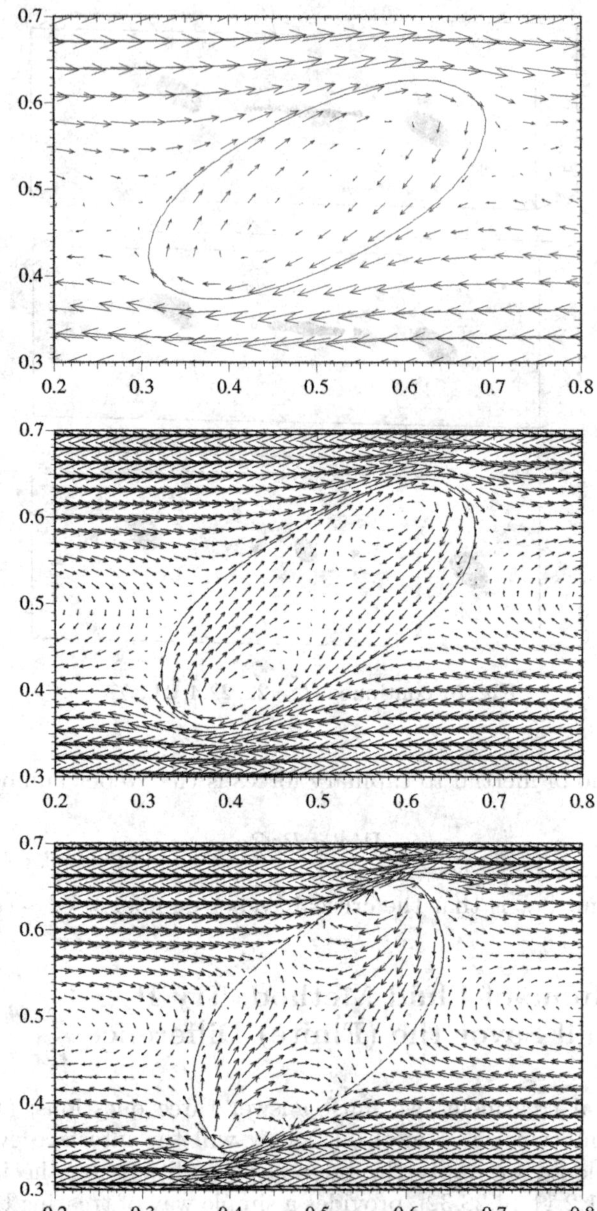

Fig. 6. Velocity vector plot for the cross-section through the center of the drop in the x-z plane, just below criticality. **a** $Re = 1, Ca = 0.27$, $L/a = 1.8$, $\theta = 25$ deg. **b** $Re = 10, Ca = 0.15$, $L/a = 1.9$, $\theta = 23$deg. **c** $Re = 60, Ca = 0.053$, $L/a = 1.52$, $\theta = 53$deg

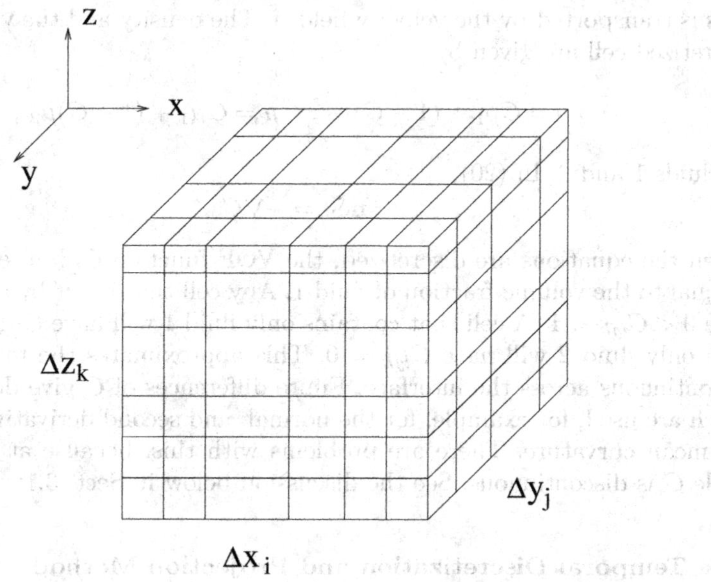

Fig. 7. Three dimensional Cartesian mesh with variable cell sizes

2.1 The Equations of Motion

The flow is incompressible:

$$\nabla.\mathbf{u} = 0 \,, \tag{17}$$

and governed by the Navier-Stokes equation:

$$\rho(\frac{\partial \mathbf{u}}{\partial t} + \mathbf{u} \cdot \nabla \mathbf{u}) = -\nabla p + \nabla \cdot \mu \mathbf{S} + \mathbf{F} \,, \tag{18}$$

where $\mathbf{S}$ is the viscous stress tensor:

$$\mathbf{S}_{ij} = (\frac{1}{2})(\frac{\partial u_j}{\partial x_i} + \frac{\partial u_i}{\partial x_j}). \tag{19}$$

At the fluid interface, velocity is continuous, shear stress is continuous, and the jump in the normal stress is balanced by interfacial tension force. These conditions appear in the VOF method as a body force over the cells which contain the interface. In the Continuous Surface Force (CSF) algorithm, [2,4,27], the body force $\mathbf{F}$ includes the interfacial tension force $\mathbf{F_s}$,

$$\mathbf{F_s} = \sigma \kappa \mathbf{n} \delta_S, \tag{20}$$

where σ denotes the coefficient of surface tension, κ the mean curvature, $\mathbf{n}$ the normal to the surface, and δ_S is a delta function concentrated on the interface.

Fluid interfaces are reconstructed from the values of a color (or VOF) function which represents the volume fraction of one of the fluids in each grid cell:

$$C(\mathbf{x}) = \begin{cases} 1 \text{ fluid } 1, \\ 0 \text{ fluid } 2. \end{cases} \tag{21}$$

This is transported by the velocity field $\mathbf{u}$. The density and the viscosity in each discretized cell are given by

$$\rho = C\rho_1 + (1-C)\rho_2 , \quad \mu = C\mu_1 + (1-C)\mu_2 , \tag{22}$$

for fluids 1 and 2. In (20),

$$\mathbf{n}\delta_S = -\nabla C. \tag{23}$$

When the equations are discretized, the VOF function C_{ijk} in each (ijk)th cell is equal to the volume fraction of fluid 1. Any cell that is cut by an interface will have $0 < C_{ijk} < 1$. A cell that contains only fluid 1 will have $C_{ijk} = 1$ and a cell with only fluid 2 will have $C_{ijk} = 0$. This approximates the function which is discontinuous across the interface. Finite differences of C give derivative values which are used, for example, for the normal, and second derivatives are used for the mean curvature. There are problems with this, because at the continuum levele C is discontinuous. See the discussion below in Sect. 3.1.

2.2 Temporal Discretization and Projection Method

The momentum equations are first solved for an approximate $\mathbf{u}^*$ without the pressure gradient, assuming that $\mathbf{u}^n$ is known:

$$\frac{(\mathbf{u}^* - \mathbf{u}^n)}{\Delta t} = -\mathbf{u}^n \cdot \nabla \mathbf{u}^n + (\frac{1}{\rho})(\nabla \cdot (\mu \mathbf{S}) + \mathbf{F})^n. \tag{24}$$

In general, the intermediate flow field $\mathbf{u}^*$ does not satisfy the incompressibility equation. It is corrected by the pressure

$$\frac{(\mathbf{u}^{n+1} - \mathbf{u}^*)}{\Delta t} = -(\frac{\nabla p}{\rho}), \tag{25}$$

in order to yield a divergence free velocity $\mathbf{u}^{n+1}$ at the next time step. The pressure field is not known in this equation, but by taking the divergence of this equation, it is found to satisfy a Poisson equation,

$$\nabla \cdot (\frac{\nabla p}{\rho}) = \frac{\nabla \cdot \mathbf{u}^*}{\Delta t}. \tag{26}$$

In the problems we address below, the boundary conditions for the velocity are periodicity and the Dirichlet condition.

For this explicit method, the time step should be less than the viscous time scale, $\rho h^2/\mu$, where h denotes the mesh size, ρ the density, μ the viscosity. Therefore, simulations of low Reynolds number flows typical of very viscous liquids, are subject to strict stability limitations and are expensive. In order to overcome this difficulty, an unconditionally stable semi-implicit scheme [13–15] has been incorporated. In comparison with the explicit scheme, the speed-up is twenty-fold or more in the simulations shown in this paper.

2.3 Advection of the Interface

Since we lose interface information when we represent the interface by a volume fraction field, the interface needs to be reconstructed approximately in each cell. To do this, the Piecewise Linear Interface Construction (PLIC) method is used. For a review of interface representation methods see [12].

The normal $\mathbf{n}$ to the interface is calculated in each cell, since this determines one unique linear interface with the volume fraction of the cell. The discrete gradient of C yields:

$$\mathbf{n} = \frac{\nabla^h C}{|\nabla^h C|} . \tag{27}$$

The second step of the VOF method is to evolve the volume fraction field C. The Lagrangian method is the natural choice for interface evolution. In this scheme (Fig. 8), once the interface is reconstructed, the velocity at the interface is interpolated linearly and then the new interface position is calculated via

$$\mathbf{x}^{n+1} = \mathbf{x}^n + \mathbf{u}(\Delta t). \tag{28}$$

2.4 Results

Numerical simulations for drop deformation under oscillatory shear is shown in Fig. 9. The figure gives a good comparison with experimental data. Here, the top and bottom walls oscillate with speeds $u = U\cos(\omega t)$ and $u = -U\cos(\omega t)$, respectively. The experimental data were taken by S. Wannaborworn [17] with certain polymeric fluids: PDMS as the drop and PIB as the matrix phase. The frequency is 0.3Hz ($\omega = 2\pi(0.3) = 1.88$ per second), and the percentage strain (i.e. the maximum relative displacement of the boundaries divided by their distance) is 250%. The initial drop diameter is 30.175 mm. The data are taken from a top view: The plots show are the major and minor axes of the deformed ellipsoid, respectively. The interfacial tension between these two polymers reported in the literature is 2.3-4 mN/m. The numerical simulations are performed for 4

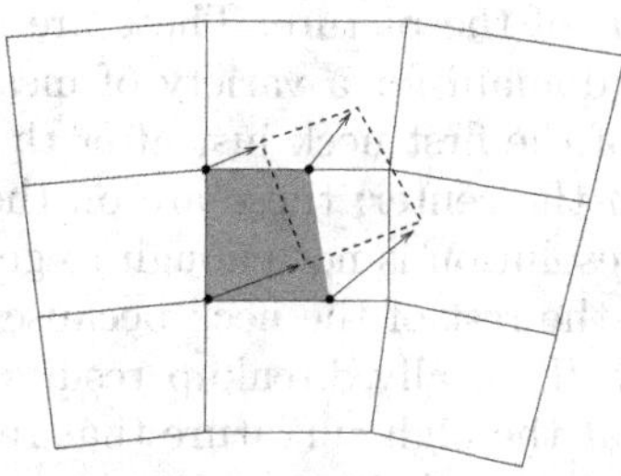

Fig. 8. The Lagrangian method illustrated for a 2D mesh. The shaded polygon represents the portion of the central cell occupied by the fluid. The broken line shows the polygon position after advection in the local velocity field (represented by arrows). The fluid is redistributed between neighboring cells, which are partially overlapped by the new polygon

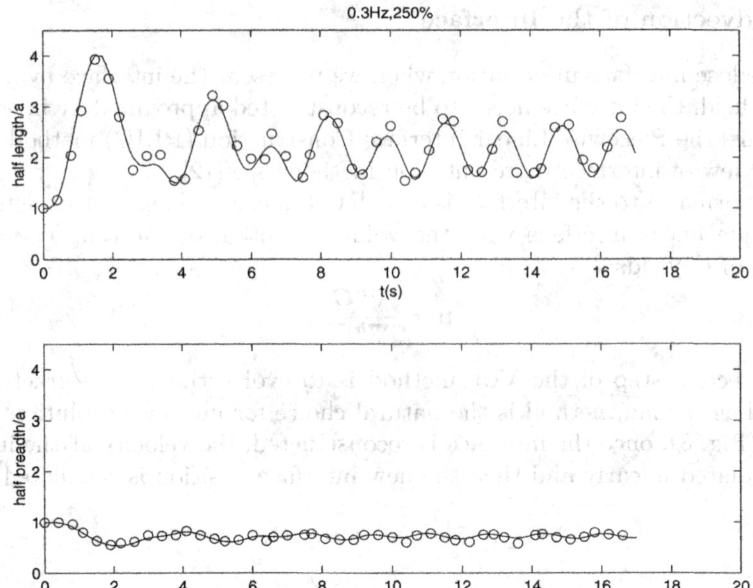

Fig. 9. Evolution of length and breadth of a drop undergoing oscillatory shear. *Circles* represent experimental data [17] and the *line* is the numerical simulation

mN/m, because the effect of varying surface tension over the interval of values is found to be small. The viscosity ratio is approximately 1 and the viscosity is 80 Pas.

The first daughters detach in a flow where the horizontal velocity is significant, as well as capillary effects which come into play to detach them. Simulations with mesh refinement on the volumes of the first daughters agree well. Next, we enter a regime where the capillary effects becomes more and more significant compared to the magnitudes of the velocity field. After the first daughter drops detach, the neck undergoes its fragmentation. Figure 10 shows the final states after fragmentation of the neck, which occurs after the first daughter drops have pinched off and moved out of the picture. These are snapshots taken from the top of the computational domain for a variety of mesh refinements. The small satellites which come off of the first neck just after the first daughters pinch off are the ones farthest from the center; these are on the scale of the mesh every time, so that the spatial resolution is not enough to get them correctly. This influences what happens to the rest of the neck because the evolution is sensitive to the volume that is left. Typically, breakup results in small moons between larger drops. It is clear that the high curvature that results at the first pinch-off degrades the subsequent numerical simulation.

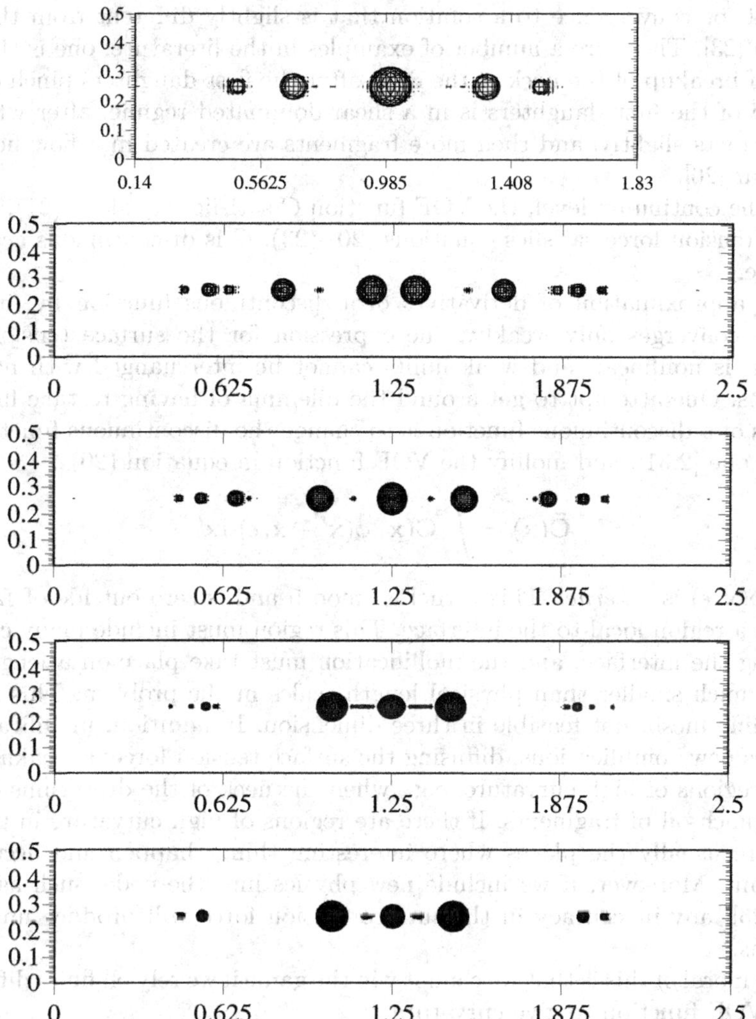

Fig. 10. $Re = 12, Ca = 0.175 = 1.14Ca_c$, top view. Fragmentation of the neck after the first daughter drops have detached and moved out of the picture. Mesh refinement from top down: $\Delta x = \Delta y = \Delta z = a/8$ $(t = 26\dot{\gamma}^{-1})$, $a/12$ $(t = 27\dot{\gamma}^{-1})$, $a/16$, $(t = 27\dot{\gamma}^{-1})$, $a/20$, $(t = 27\dot{\gamma}^{-1})$, $a/24$, $(t = 29\dot{\gamma}^{-1})$. VOF-CSS is used. Timesteps $\Delta t = 10^{-3}$

3 That's Incurable! (Unless You Devise a Sharp-Interface VOF Algorithm)

3.1 Calculation of Surface Tension Force from Finite Differences of the VOF Function

One weakness of the VOF-CSF algorithm is that when the capillary force is the dominant physical mechanism, there is lack of convergence with spatial refinement, or convergence to a solution that is slightly different from the exact solution [23]. There are a number of examples in the literature; one is the simulation of breakup of the neck of the drop after the first daughters pinch off. The creation of the first daughters is in a shear-dominated regime, after which the neck retracts slightly, and then more fragments are created in a flow field with less shear [26].

At the continuum level, the VOF function C satisfies equation (21), and the surface tension force satisfies equations (20)-(23). C is discontinuous across the interface.

The approximation of derivatives of a discontinous function by finite differences converges only weakly. The expression for the surface tension force, however, is nonlinear, and weak limits cannot be interchanged with nonlinear functions. One attempt to get around the dilemma of having to take finite differences of a discontinuous function is to change the discontinuous function to a smooth one [2,31], and mollify the VOF function in equation (20)

$$\tilde{\mathbf{C}}(\mathbf{x}) = \int_{\mathbf{\Omega}} \mathbf{C}(\mathbf{x}')\phi(\mathbf{x}' - \mathbf{x}, \epsilon)\mathbf{dx}' \,, \tag{29}$$

where $\phi(\mathbf{x}, \epsilon)$ is a *kernel.* This kernel is smooth and is zero outside of Ω, which denotes a region local to the interface. This region must include many cells surrounding the interface, and the mollification must take place on a length scale that is much smaller than physical length scales in the problem. This requires a very fine mesh, not feasible in three dimension. In addition, mollification introduces new complications, diffusing the surface tension force; a weakness that affects regions of high curvature, e.g., when the neck of the drop thins out just before pinch-off of fragments. If there are regions of high curvature in the flow, those are usually the places where interesting things happen and accuracy is important. Moreover, if we include new physics into the code, such as surfactants [25], any inaccuracy in the surface tension force will produce unphysical solutions.

The moral of this is that we cannot win the game if we rely on finite differences of the VOF function to give curvature.

Spurious Currents. A well-known example of inaccuracy in VOF schemes that rely on finite-differencing the VOF function is shown in this section. A spherical drop with non-zero interfacial tension is suspended in a second liquid of the same density and viscosity. The boundary conditions on the computational box are zero velocity. The initial velocity is zero. The exact solution for this problem is that the drop stay the same with zero velocity for all time.

The equations which matter in the numerical simulation are: $\mathbf{F_s} = \sigma\kappa\mathbf{n}\delta_S$, where $\mathbf{n}\delta_S = -\nabla C$, which is balanced by the pressure field:

$$\mathbf{F_s} = -\sigma\kappa\nabla C, \; \nabla p = -\mathbf{F_s}. \tag{30}$$

Therefore, $\mathbf{F_s}$ is a gradient when the interface is the exact solution, a sphere (constant κ). The surface tension force is cancelled by the pressure gradient;

at the discrete level, this holds if the same finite difference approximation is used for ∇C and ∇p. This is the case, and also in commercial codes. Where is the inaccuracy? Equation (30) shows that the crucial ingredient for avoiding spurious currents is an accurate approximation of the curvature (among other things).

Figure 11 shows the spurious currents from a VOF-CSF simulation, and these currents arise in all VOF methods prior to VOF-PROST. The magnitude and location of the currents remain the same when the mesh is refined. If the user is naive, then these simulations give the appearance of a spatially converged solution. The Ohnesorge number

$$Oh = (Ca/Re)^{1/2} = \frac{\mu^2}{\sigma \rho a} \tag{31}$$

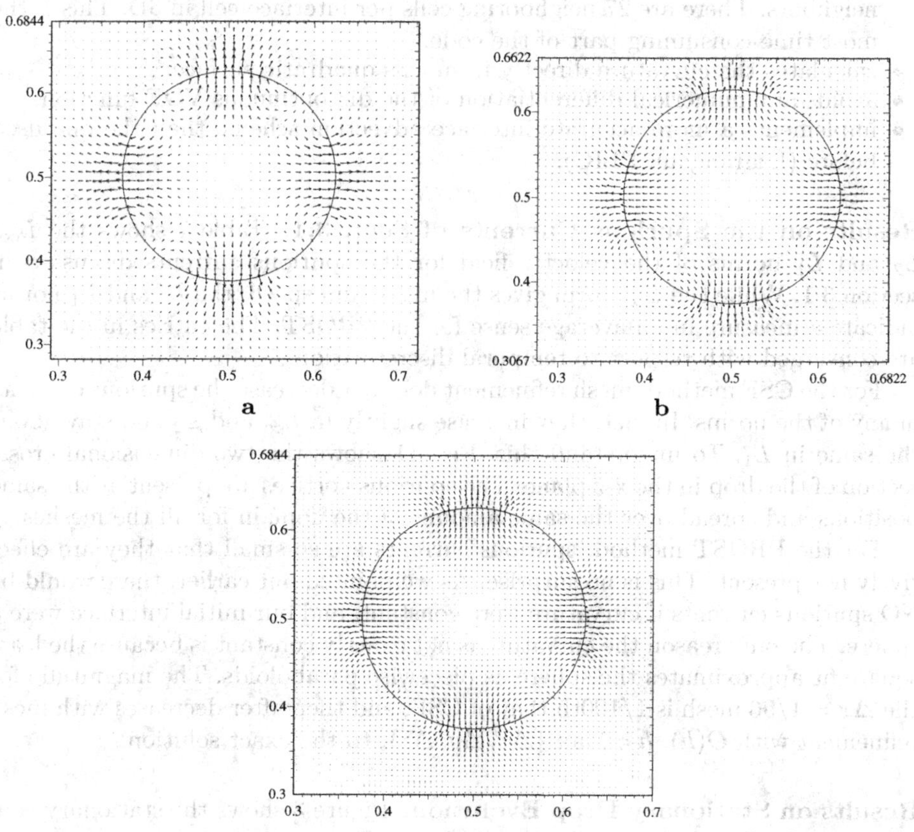

Fig. 11. CSF, velocity vector plot across centerline in the x-z plane at 200th timestep, $\Delta t = 10^{-5}$. These show the locations of the spurious currents with mesh refinement, $\Delta x = 1/96$ (**a**), 1/128 (**b**), 1/160 (**c**). The simulation does not converge to the exact solution (zero velocity) with spatial refinement

expresses the ratio of viscous effect to capillary and inertial effects. For instance, when the matrix liquid is a gas, and when surface tension is high, the Oh for the gas is small; spurious currents would be noticeable. Spurious currents are small when Oh is sufficiently large.

3.2 Calculation of Surface Tension Force with a Sharp-Interface Algorithm

To our knowledge, VOF-PROST (*P*roper *R*epresentation *O*f *S*urface *T*ension) is the first sharp-interface VOF code for 3D simulations. PROST is detailed in [23] and we summarize the main features here. If a cell is intersected by the interface, we call it an interface-cell.

PROST

- reconstructs the interface in each interface-cell from a least square fit of a paraboloid to the values of the color function in the given cell and its neighbors. There are 27 neighboring cells per interface-cell in 3D. This is the most time-consuming part of the code.
- calculates the curvature directly from this quadratic surface;
- avoids any numerical differentiation of the discontinuous VOF function;
- implements a more accurate interface advection scheme than the one used before (Lagrangian, PLIC).

Results on the Spurious Currents of Sect. 3.1. Table 1 shows the L_∞, L_2 and L_1 norms of the velocity field for the spurious currents discussed in section 3.1. While the L_∞ norm gives the maximum speed, the L_2 and L_1 norms indicate a measure in an average sense for the PROST. The entries in the table are converged with respect to temporal discretization.

For the CSF method, mesh refinement does not decrease the spurious currents in any of the norms. In fact, they increase slightly in L_∞ and L_2 and stay about the same in L_1. To understand this, Fig. 11 shows the two-dimensional cross-section of the drop in the x-z plane. The spurious vortices are present at the same positions and spread over the same amount of the domain for all the meshes.

For the PROST method, spurious currents are so small that they are effectively not present. This is no surprise. As we pointed out earlier, there would be NO spurious currents if curvature were constant, and our initial interface were a sphere. The only reason the curvature is not exactly constant is because the least square fit approximates the sphere as piecewise paraboloids. The magnitude for the $\Delta x = 1/96$ mesh is 1/100th that of CSF, and thereafter decreases with mesh refinement with $O(h)$, $h =$Max $\{\Delta x, \Delta y, \Delta z\}$, to the exact solution.

Results on Stationary Drop Evolution. Figure 3 shows the stationary configuration for Stokes flow at $Ca = 0.35$, $\lambda = 1$. Figure 12 compares the temporal evolution of its length/radius to steady state, for VOF-CSF, VOF-PROST and the boundary integral code of [5]. The CSF output remains 3% away from the solution of the boundary integral method, whereas the PROST output converges to it.

Table 1. Norms of velocity at 200th timestep, $\Delta t = 10^{-5}$

Δx	L_∞	L_2	L_1	method
1/96	0.00179982	0.00008403	0.00001473	CSF
1/128	0.00184090	0.00008542	0.00001539	
1/160	0.00189053	0.00008596	0.00001569	
1/192	0.00196880	0.00008627	0.00001569	
1/96	0.00002243	0.00000087	0.00000014	PROST
1/128	0.00001309	0.00000053	0.00000009	
1/160	0.00000954	0.00000041	0.00000007	
1/192	0.00000568	0.00000023	0.00000004	

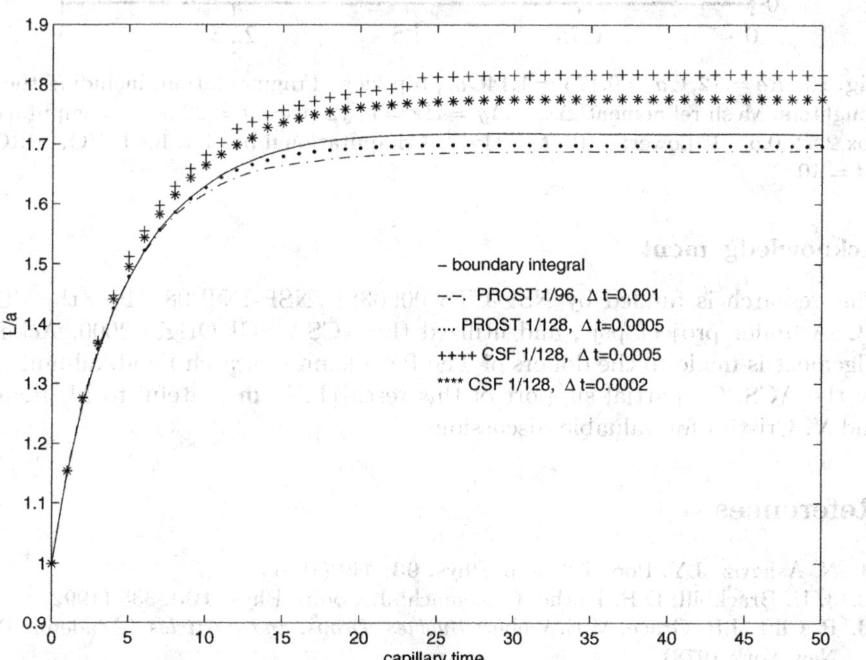

Fig. 12. Evolution of (half-length/initial-radius) for boundary integral method for $Ca = 0.35$, Stokes flow. Boundary integral method (line); VOF-PROST and VOF-CSF. [23]

Results on Fragmentation. Figure 13 shows a simulation for fragmentation for different mesh sizes (and different computational domain sizes). The flow parameters correspond to that of Fig. 10, but here there is convergence on the volumes of the fragments.

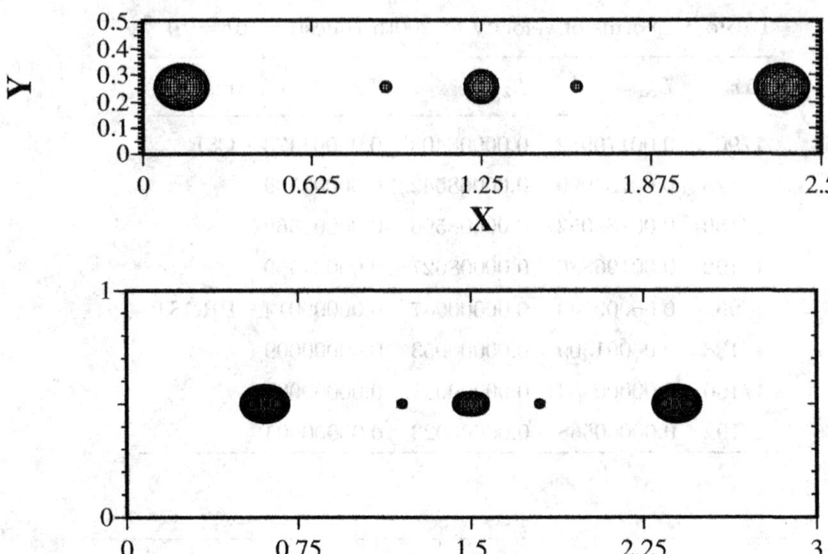

Fig. 13. $Re = 12, Ca = 0.175 = 1.14Ca_c$, top view. Fragmentation, including the first daughters. Mesh refinement $\Delta x = \Delta y = \Delta z = a/12$ (top), $t = 22.5\dot{\gamma}^{-1}$, computational box $2.5 \times 0.5 \times 1$. Lower: $a/16$, $t = 24\dot{\gamma}^{-1}$. Computational box $3 \times 1 \times 1$. VOF-PROST. $\Delta t = 10^{-3}$

Acknowledgement

This research is funded by NSF-CTS 0090381, NSF-INT 9815106, the Illinois NCSA under project psj , and utilized the NCSA SGI Origin 2000. Acknowledgement is made to the donors of The Petroleum Research Fund, administered by the ACS, for partial support of this research. I am grateful to M. Renardy and V. Cristini for valuable discussions.

References

1. N. Ashgriz, J.Y. Poo: J. Comp. Phys. **93**, 449 (1991)
2. J. U. Brackbill, D.B. Kothe, C. Zemach: J. Comp. Phys. **100**, 335 (1992)
3. R. Clift, J.R. Grace, M.E. Weber: *Bubbles, Drops, and Particles*, (Academic Press, New York 1978)
4. A.V. Coward, Y. Renardy, M. Renardy, J.R. Richards: J. Comp. Phys. **132**, 346 (1997)
5. V. Cristini: Drop dynamics in viscous flow. PhD thesis, Yale University, New Haven, CT (2000)
6. V. Cristini, J. Blawzdziewicz, M. Loewenberg: J. Comp. Phys. **x**, **168**, 445 (2001)
7. R.W. Flumerfelt: Ind. Eng. Chem. Fund. **11**, 312 (1972)
8. H.P. Grace: Chem. Eng. Commun. **14**, 225 (1982)
9. N. Grizzuti, O. Bifulco: Rheol. Acta **36**, 406 (1997)
10. S. Guido, F. Greco: Rheol. Acta **40**, 176 (2001)
11. C.W. Hirt, B.D. Nichols: J. Comp. Phys. **39**, 201 (1981)

12. D.B. Kothe, W.J. Rider: *Comments on modelling interfacial flows with volume-of-fluid methods*, (Los Alamos preprint, 1994)
13. J. Li, Y. Renardy: J. Fluid Mech. **391**, 123 (1999)
14. J. Li, Y. Renardy, M. Renardy: Phys. Fluids **10**, 3056 (1998)
15. J. Li, Y. Renardy, M. Renardy: Phys. Fluids **12**(2), 269 (2000)
16. K.J. Lissant: *Emulsions and emulsion technology, Part III*, (Marcel Dekker, 1984)
17. M.R. Mackley, S. Wannaborworn, P. Gao, F. Zhao: J. Microscopy and Analysis **69**, 25 (1999)
18. C.R. Marks: Drop breakup and deformation in sudden onset strong flows. PhD thesis, University of Maryland, College Park, MD (1998)
19. T. G. Mason, J. Bibette: Langmuir **13**, 4600 (1997)
20. M. Minale, P. Moldenaers, J. Mewis: Macromolecules **30**, 5470 (1997)
21. J.M. Rallison: Ann. Rev. Fluid Mech. **16**, 45 (1984)
22. Y. Renardy, V. Cristini: Phys. Fluids **13**(1), 7 (2001)
23. Y. Renardy, M. Renardy: J. Comp. Phys. **183**, 400 (2002)
24. Y. Renardy, V. Cristini: Phys. Fluids **13**(8), 2161 (2001)
25. Y. Renardy, M. Renardy, V. Cristini: Eur. J. Mech. B/ Fluids, **21**, 49 (2002)
26. Y. Renardy, V. Cristini, J. Li: Int. J. Mult. Flow, **28**, 1125 (2002)
27. W.J. Rider, D.B. Kothe: J. Comp. Phys **141**, 112 (1998)
28. R. Scardovelli, S. Zaleski: Ann. Rev. Fluid Mech. **31**, 567 (1999)
29. H.A. Stone: Ann. Rev. Fluid Mech. **26**, 65 (1994)
30. G.I. Taylor: Proc. R. Soc. A **146**, 501 (1934)
31. M. Williams, D. Kothe, E.G. Puckett: 'Accuracy and convergence of continuum surface-tension models'. In: *Fluid Dynamics at Interfaces*, ed. by W. Shyy, R. Narayanan, (Cambridge Univ. Press, Cambridge, UK 1999) pp. 294–305
32. D.L. Youngs: *An interface tracking method for a 3D Eulerian hydrodynamic code*, (Technical Report, 44/92/35 AWRE, 1984)
33. X. Zhao, J.L. Goveas: Langmuir **17**, 3788 (2001)

Dynamics, Stability and Solidification of an Emulsion under the Action of Thermocapillary Forces and Microgravity

V.V. Pukhnachov[1], O.V. Voinov[2], A.G. Petrova[3], E.N. Zhuravleva[4], and O.A. Gudz[5]

[1] Lavrentyev Institute of Hydrodynamics, Novosibirsk, Russia
[2] Institute of Mechanics of Multiphase Systems, Tyumen, Russia
[3] Altai State University, Barnaul, Russia
[4] Altai State Technological University, Barnaul, Russia
[5] Novosibirsk State University, Novosibirsk, Russia

Abstract. There are two reasons for understanding the nature of emulsion behavior under the conditions of microgravity. One of them is creating the scientific prerequisites for processing composites, where components have strongly differing densities, on board the orbiting station. The other one is the problem of purification of mixtures from gas and liquid inclusions under the conditions of microgravity. The possibilities of obtaining laboratory experiments with emulsions, where components differ in density, are very limited. That is why the role of mathematical modelling in solving the above problems is important.

Thanks to fundamental works of Hadamard [7], Rybczinski [21], and Young, Goldstein and Block [28] we know how a single droplet moves in a surrounding liquid under the action of gravity and thermocapillary forces. Our work is devoted to the effects of the joint action of drops or bubbles in irregularly heated liquids under the condition of microgravity.

Based upon the model of emulsion motion formulated in [19], we study the stability of the space-uniform state and a plane interface in an emulsion-pure liquid system. The phenomenological model of emulsion solidification is given. It is shown that there is a possibility of adjusting the material composition by means of changing the boundary conditions on temperature. The problem of hydrodynamical interaction of drop and crystallization fronts is given and solved for a general case of slightly changing external conditions.

1 Mathematical Model of Emulsion Motion

1.1 Introduction

The dynamics of multiphase media is one of the most rapidly developing subject areas. It has many applications to modern technologies. However, unlike advanced problems of mechanics of suspensions or gas-liquid mixtures (see [13], for example), the behaviour of emulsion motion has not been well investigated.

A classical example of an emulsion is ordinary milk. Another example is an emulsion of "oil drop - water". If our goal is to separate the components of emulsions, we could use centrifuges and sometimes the action of gravity would

V.V. Pukhnachov et al., Dynamics, Stability and Solidification of an Emulsion under the Action of Thermocapillary Forces and Microgravity, Lect. Notes Phys. **628**, 325–354 (2003)
http://www.springerlink.com/

be enough. The situation changes drastically when we consider microgravity conditions. Under these conditions, surface forces are comparable to bulk forces and can even dominate them. Among surface forces, thermocapillary forces play a special role, because they are very easy to control.

Thermocapillary effects can be used both for the purpose of cleaning liquids, metal melts or optic glass and for obtaining exotically composed materials of the "aluminium-lead" type, which are impossible to obtain on Earth. But there exist possible situations on Earth when the gravity effect is weakened — this occurs with emulsion motion where components have densities close to each other.

1.2 Governing Equations

Let the volume Ω contain a viscous incompressible liquid with drops of another viscous incompressible liquid immiscible with the first one. The number of drops is large enough that there exists an $r \ll diam\Omega$, such that every sphere of radius r contains a number of drops $n \gg 1$. It is supposed that the drops have a spherical form and equal radius R. We neglect Brownian motion as well as drop collision, junction and subdivision effects. We assume that the average distance l between the drops is such that, $R \ll l \ll diam\Omega$, and the concepts of mechanics of heterogeneous media [13] are applicable to the system.

It is supposed that the system is in local thermodynamic equilibrium. The relative phase motion is primarily due to the nonuniformity of the temperature field that produces the thermocapillary effect by virtue of the dependence of the coefficient of the surface tension σ on the temperature T. For simplicity, we suppose that this dependence is linear: $\sigma = \sigma_0 - \sigma_T(T - T_0)$, where σ_0, σ_T, and T_0 are some positive constants. In addition to the thermocapillary forces, the system is exposed to microgravity with constant acceleration. We suppose also that the volume concentration of the disperse phase c is small.

Let $\boldsymbol{u}$ and $\boldsymbol{v}$ stand for average velocities of the disperse and carrying phases correspondingly and p stand for pressure. Let us use the index d to denote the parameters of the disperse phase and m for the liquid matrix. Then the system of equations which describe the model is as follows:

$$\frac{\partial c}{\partial t} + div(c\boldsymbol{u}) = 0, \tag{1.1}$$

$$\frac{\partial(1-c)}{\partial t} + div((1-c)\boldsymbol{v}) = 0, \tag{1.2}$$

$$\varrho_d c\Big(\frac{\partial \boldsymbol{u}}{\partial t} + \boldsymbol{u}\cdot\nabla\boldsymbol{u}\Big) + \varrho_m(1-c)\Big(\frac{\partial \boldsymbol{v}}{\partial t} + \boldsymbol{v}\cdot\nabla\boldsymbol{v}\Big) = \tag{1.3}$$
$$= -\nabla p + div(\mu_m(1+Nc)(\nabla\boldsymbol{v} + (\nabla\boldsymbol{v})^*)) + \varrho_d c\boldsymbol{g} + \varrho_m(1-c)\boldsymbol{g},$$

$$\varrho_d\lambda_d c\Big(\frac{\partial T}{\partial t} + \boldsymbol{u}\cdot\nabla T\Big) + \varrho_m\lambda_m(1-c)\Big(\frac{\partial T}{\partial t} + \boldsymbol{v}\cdot\nabla T\Big) =$$
$$= div(k_m(1-Mc)\nabla T), \tag{1.4}$$

and

$$\boldsymbol{u} - \boldsymbol{v} = K\boldsymbol{g} + L\nabla T, \tag{1.5}$$

where ϱ is density, μ - dynamic viscosity, λ - heat capacity, k- thermal conductivity,

$$N = \frac{\mu_m + 5\mu_d/2}{\mu_m + \mu_d}, \quad K = \frac{2R^2(\varrho_d - \varrho_m)(\mu_m + \mu_d)}{3\mu_m(2\mu_m + 3\mu_d)},$$

$$M = \frac{3(k_m - k_d)}{2k_m + k_d}, \quad L = \frac{2Rk_m\sigma_T}{(2\mu_m + 3\mu_d)(2k_m + k_d)}. \tag{1.6}$$

Equations (1.1)–(1.5) were obtained in [19].

Equations (1.1) and (1.2) are the exact forms (within the framework of the heterogeneous media mechanics approach) of the mass conservation law for the disperse and carrying phases, respectively. Equations (1.3) and (1.4) are the asymptotically exact forms of the momentum and energy equations (we neglected the terms of the order c^2 in viscous stresses, and the diffusion heat flux). The dynamic viscosity coefficient in (1.3) is determined by the Taylor formula [22], whereas the thermal conductivity of the mixture in (1.4) is determined by Maxwell's equation [11]. We have also omitted the dissipation function in (1.4) taking into account the smallness of the arising velocities of the motion.

Equation (1.5) requires a special comment. This relation gives an approximate value of the relative phase velocity. In (1.5) we neglect the effect of apparent additional masses and Basset's hereditary force on account of the low Reynolds numbers. Applicability of (1.5) is restricted by small Reynolds numbers at relative phase velocity. The influence of the thermocapillary drift of a drop on the Reynolds number was studied in [4]. Equation (1.5) represents the superposition of the Hadamard-Rybczinski formula for the relative velocity of the drop moving under gravitation [7,21] and the Young-Goldstein-Block formula [28] for the drop in the nonuniform temperature field. In this connection it is necessary to mention [3] where the problem of creeping motion of a small drop with the gravity acceleration collinear to the temperature gradient at infinity was solved explicitly.

It is known [1], that for stationary thermocapillary motion of bubbles in mixtures their average velocity u relative to the liquid can differ from u_0 defined by Young-Goldstein-Block formula: $u = u_0(1 + O(c))$. In the case of drops the correction to u_0 depends upon the microstructure of medium. There is no universal formula similar to Taylor's one for the effective viscosity of emulsion or Maxwell's equation for the heat conductivity of the emulsion. On the other hand, the effects we consider in this work do not depend on the correction mentioned above. Therefore we don't discuss the concrete value of the concentration correction for u_0.

1.3 One-Dimensional Motion

Equations (1.1)–(1.5) is very complicated in view of its nonlinearity, high order, and mixed type. However, in the case of one-dimensional motion with plane waves it can be simplified drastically. (In this case we suppose that vectors $\boldsymbol{g}$

and ∇T are collinear and parallel to the axis x). In fact, (1.1) and (1.2) provide the integral

$$cu + (1-c)v = f(t) \tag{1.7}$$

where $f(t)$ is the mean volume velocity of the mixture. Now we can express u and v from (1.7) and the last equation of (1.5) in terms of c, T and f. As a result, we obtain a system to determine T and c :

$$\frac{\partial c}{\partial t} + \frac{\partial}{\partial x}\Big(\Big(c\Big(Kg + L\frac{\partial T}{\partial x}\Big) - f\Big)(1-c)\Big) = 0, \tag{1.8}$$

$$(\varrho_d\lambda_d c + \varrho_m\lambda_m(1-c))\Big(\frac{\partial T}{\partial t} + f\frac{\partial T}{\partial x}\Big)+$$
$$+(\varrho_d\lambda_d - \varrho_m\lambda_m)c(1-c)\Big(Kg + L\frac{\partial T}{\partial x}\Big)\frac{\partial T}{\partial x} = k_m\frac{\partial}{\partial x}\Big((1-Mc)\frac{\partial T}{\partial x}\Big) \tag{1.9}$$

Here $g = |\boldsymbol{g}|$.

Equation (1.8)–(1.9) must be completed with appropriate initial and boundary conditions. An additional boundary condition is necessary to determine the newly defined variable $f(t)$ (for example, $f = 0$ for motion possessing a plane of symmetry). A typical problem of such a kind was set up in [27]. The gas-liquid mixture is restricted by two parallel solid impermeable walls. The initial distributions of concentration and temperature are given. The heat flux at both walls is prescribed as a function of time. In addition, the impermeability condition for the liquid phase is satisfied at the boundaries of the flow domain. The solvability of the formulated problem is proved in [17].

The numerical results for one-dimensional motion of an emulsion and gas-liquid mixture were presented in [9,20].

2 Stability of the Space-Uniform State

2.1 Small Perturbations of the Space-Uniform State

The simplest solution of (1.1)–(1.5) corresponding to a uniform relative phase motion with constant concentration and with the temperature distribution given by a linear function of the Cartesian coordinate x and time is:

$$c = c_0,\ \boldsymbol{u} = \boldsymbol{u}_0,\ \boldsymbol{v} = \boldsymbol{v}_0,\ p = p_0 = \nabla p_0 \cdot \boldsymbol{x} \text{ and } \quad T = T_0 = \nabla T_0 \cdot \boldsymbol{x} + T_{0t}t \tag{2.1}$$

where $c_0, \nabla T_0 = const$ and $\nabla p_0 = (\varrho_d c_0 + \varrho_m(1-c_0))\boldsymbol{g} = const$.

The conditions of stability of the space-uniform state are necessary for the existence of the emulsion as a continuous media for a long time.

Due to Galilean invariance of the system one can assume that $\boldsymbol{v} = 0$. Then

$$T_{0t} = -\frac{\varrho_d\lambda_d c_0(K\boldsymbol{g} + L\nabla T_0)\cdot\nabla T_0}{\varrho_d\lambda_d c_0 + \varrho_m\lambda_m(1-c_0)} \quad \text{and} \quad \boldsymbol{u}_0 = K\boldsymbol{g} + L\nabla T_0.$$

Let us choose the coordinate system (Fig. 1), such that

$$\nabla T_0 = (G, 0, 0) \text{ and } \quad \boldsymbol{g} = (g_1, g_2, 0), \text{ then}$$

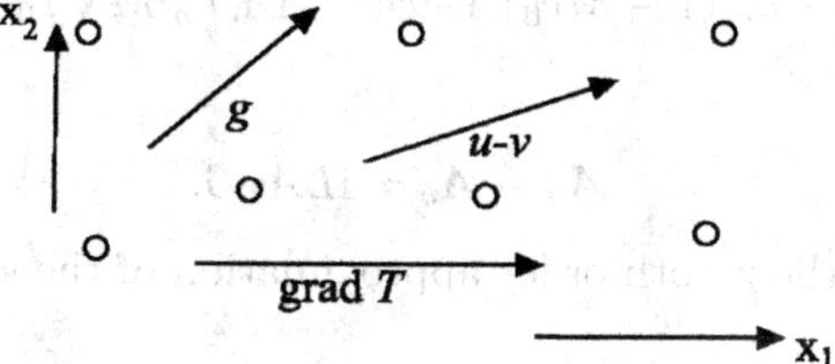

Fig. 1. Motion in chosen coordinate system

$$T_{0t} = -\frac{\varrho_d\lambda_d c_0(Kg_1 + LG)G}{\varrho_d\lambda_d c_0 + \varrho_m\lambda_m(1 - c_0)} \quad \text{and} \quad \boldsymbol{u}_0 = (Kg_1 + LG, Kg_2, 0). \tag{2.2}$$

Linearizing the problem, we obtain the following system for disturbances:

$$\frac{\partial c}{\partial t} + c_0 div\boldsymbol{u} + \boldsymbol{u}_0 \cdot \nabla c = 0, \qquad \frac{\partial(1-c)}{\partial t} + (1 - c_0)div\boldsymbol{v} = 0,$$

$$\varrho_d c_0\Big(\frac{\partial \boldsymbol{u}}{\partial t} + \boldsymbol{u}_0 \cdot \nabla\boldsymbol{u}\Big) + \varrho_m(1 - c_0)\frac{\partial \boldsymbol{v}}{\partial t} =$$

$$= -\nabla p + \mu_m(1 + c_0N)\Delta\boldsymbol{v} + (\varrho_d - \varrho_m)c\boldsymbol{g}, \tag{2.3}$$

$$\varrho_d\lambda_d c_0\Big(\frac{\partial T}{\partial t} + \boldsymbol{u}_0 \cdot \nabla T + \boldsymbol{u} \cdot \nabla T_0\Big) + \varrho_m\lambda_m(1 - c_0)\Big(\frac{\partial T}{\partial t} + \boldsymbol{v} \cdot \nabla T_0\Big) +$$

$$+\varrho_d\lambda_d c\Big(\frac{\partial T_0}{\partial t} + \boldsymbol{u}_0 \cdot \nabla T_0\Big) - \varrho_m\lambda_m c\frac{\partial T_0}{\partial t} = k_m(1 - Mc_0)\Delta T - k_m M\nabla T_0 \cdot \nabla c$$

and

$$\boldsymbol{u} - \boldsymbol{v} = L\nabla T.$$

We search for the solution of the linearized problem in the form

$$\boldsymbol{U} = \boldsymbol{A}_U exp(\alpha t + i\boldsymbol{\beta} \cdot \boldsymbol{x})$$

where $\boldsymbol{A}_U$ - is the constant vector , α is the growth exponent and $\boldsymbol{\beta}$ is the wave vector.

By substitution in (2.3), we obtain the system for determining α and the coefficients $\boldsymbol{A}_U$:

$$A_c\alpha + ic_0\boldsymbol{A}_u \cdot \boldsymbol{\beta} + iA_c\boldsymbol{u}_0 \cdot \boldsymbol{\beta} = 0,$$

$$-A_c\alpha + i(1 - c_0)\boldsymbol{A}_v \cdot \boldsymbol{\beta} = 0,$$

$$\varrho_d c_0\boldsymbol{A}_u(\alpha + i\boldsymbol{u}_0 \cdot \boldsymbol{\beta}) + \varrho_m(1 - c_0)\boldsymbol{A}_v\alpha =$$

$$= -iA_p\boldsymbol{\beta} - \mu_m(1 + Nc_0)\boldsymbol{A}_v|\boldsymbol{\beta}|^2 + A_c(\varrho_d - \varrho_m)\boldsymbol{g},$$

$$\varrho_d\lambda_d c_0 A_T(\alpha + i\boldsymbol{u}_0 \cdot \boldsymbol{\beta}) + \varrho_m\lambda_m(1 - c_0)A_T\alpha + \tag{2.4}$$

$$+\varrho_d\lambda_d c_0\boldsymbol{A}_u \cdot \nabla T + \varrho_m\lambda_m(1 - c_0)\boldsymbol{A}_v \cdot \nabla T_0 +$$

$$+A_c\varrho_d\lambda_d(T_{0t} + \boldsymbol{u}_0 \cdot \nabla T_0) - A_c\varrho_m\lambda_m T_{0t} =$$

$$= -k_m(1 - Mc_0)A_T|\boldsymbol{\beta}|^2 - \mathrm{i}A_c k_m M \nabla T_0 \cdot \boldsymbol{\beta},$$

and

$$\boldsymbol{A}_u - \boldsymbol{A}_v = \mathrm{i}LA_T\boldsymbol{\beta}.$$

The analysis of the zeroth order approximation of the system for the decompositions

$$\alpha = \alpha(0) + c_0\alpha'(0) + O(c_0^2) \text{ and } \quad \boldsymbol{A}_U = \boldsymbol{A}_U(0) + c_0\boldsymbol{A}'_U(0) + O(c_0^2),$$

shows that it contains five equations which involve $\alpha(0)$. Therefore the characteristic equation has five roots. The root corresponding to viscous disturbances is a multiple one (it is obtained from a vector equation). So we have three solutions, viz.,

Solution 1:

$$\alpha_1(0) = -\chi_m|\boldsymbol{\beta}|^2,$$

$$A_c = 0,\ A_p = 0,\ A_T \neq 0,\ \boldsymbol{A}_v = 0 \text{ and } \boldsymbol{A}_u = \mathrm{i}LA_T\boldsymbol{\beta}; \tag{2.5}$$

Solution 2:

$$\alpha_2(0) = -\nu_m|\boldsymbol{\beta}|^2,$$

$$A_c = 0,\ A_p = 0,\ A_T = \frac{\boldsymbol{A}_v \cdot \nabla T_0}{(\chi_m - \nu_m)|\boldsymbol{\beta}|^2},\ \boldsymbol{A}_v \cdot \boldsymbol{\beta} = 0,\ \boldsymbol{A}_v \neq 0, \tag{2.6}$$

$$\boldsymbol{A}_u \cdot \boldsymbol{\beta} = \mathrm{i}LA_T|\boldsymbol{\beta}|^2 \text{ and } \boldsymbol{A}_u \cdot \nabla T_0 + \mathrm{i}LA_T\nabla T_0 \cdot \boldsymbol{\beta};$$

and Solution 3:

$$\alpha_3(0) = -\mathrm{i}(K\boldsymbol{g} \cdot \boldsymbol{\beta} + L\nabla T_0 \cdot \boldsymbol{\beta}),$$

$$A_c \neq 0,\ A_p = \frac{\mathrm{i}A_c\rho_m}{|\boldsymbol{\beta}|^2}\Big(\mathrm{i}(\boldsymbol{u}_0 \cdot \boldsymbol{\beta})^2 - \nu_m\boldsymbol{u}_0 \cdot \boldsymbol{\beta}|\boldsymbol{\beta}|^2 - (\varrho_d/\varrho_m - 1)\boldsymbol{g} \cdot \boldsymbol{\beta}\Big),$$

$$\boldsymbol{A}_v \cdot \boldsymbol{\beta} = A_c\boldsymbol{u}_0 \cdot \boldsymbol{\beta},$$

$$\boldsymbol{A}_v \cdot \nabla T_0 = A_c\Big(-\boldsymbol{u}_0 \cdot \boldsymbol{\beta}\frac{\nabla T_0 \cdot \boldsymbol{\beta}}{|\boldsymbol{\beta}|^2} + (\varrho_d/\varrho_m - 1)\frac{1}{|\boldsymbol{\beta}|^2}\frac{(\boldsymbol{g} \cdot \boldsymbol{\beta})(\nabla T_0 \cdot \boldsymbol{\beta}) - \boldsymbol{g} \cdot \nabla T_0}{\mathrm{i}\boldsymbol{u}_0 \cdot \boldsymbol{\beta} - \nu_m|\boldsymbol{\beta}|^2}\Big),$$

$$A_T = \frac{A_c}{\mathrm{i}\boldsymbol{u}_0 \cdot \boldsymbol{\beta} - \chi_m|\boldsymbol{\beta}|^2}\Big(\frac{\varrho_d\lambda_d}{\varrho_m\lambda_m}\boldsymbol{u}_0 \cdot \nabla T_0 + \mathrm{i}\chi_m M\nabla T_0 \cdot \boldsymbol{\beta} - \frac{(\boldsymbol{u}_0 \cdot \boldsymbol{\beta})(\nabla T_0 \cdot \boldsymbol{\beta})}{|\boldsymbol{\beta}|^2} +$$

$$+(\varrho_d/\varrho_m - 1)\frac{1}{|\boldsymbol{\beta}|^2}\frac{(\boldsymbol{g} \cdot \boldsymbol{\beta})(\nabla T_0 \cdot \boldsymbol{\beta}) - \boldsymbol{g} \cdot \nabla T_0|\boldsymbol{\beta}|^2}{\mathrm{i}\boldsymbol{u}_0 \cdot \boldsymbol{\beta} - \nu_m|\boldsymbol{\beta}|^2}\Big),$$

$$\boldsymbol{A}_u \cdot \boldsymbol{\beta} = \mathrm{i}LA_T|\boldsymbol{\beta}|^2 - A_c\boldsymbol{u}_0 \cdot \boldsymbol{\beta} \text{ and } \quad \boldsymbol{A}_u \cdot \nabla T_0 = \mathrm{i}LA_T\nabla T_0 \cdot \boldsymbol{\beta} + \boldsymbol{A}_v \cdot \boldsymbol{\beta}.$$

Here $\chi_m = k_m/(\varrho_m\lambda_m)$ is a thermodiffusion coefficient and $\nu_m = \mu_m/\varrho_m$ is the kinematic viscosity of the carrying phase.

The first solution corresponds to the heat mode, the second to the viscous mode, and the third to the convective one. It is obvious from (2.5), (2.6) that the heat and viscous modes don't lead to instability. As for the convective mode, it is neutrally stable in the zeroth order approximation. The calculations show that

$$Re\alpha_3 = Lc_0 \cdot \Big[\chi_m|\boldsymbol{\beta}|^4((K\boldsymbol{g}\cdot\boldsymbol{\beta} + L\nabla T_0\cdot\boldsymbol{\beta})^2 + \nu_m^2|\boldsymbol{\beta}|^4)\times$$

$$\times(LG^2 + Kg_1G)\Big((1+M)\frac{\beta_1^2}{|\boldsymbol{\beta}|^2} - \frac{\varrho_d\lambda_d}{\varrho_m\lambda_m}\Big) + \Big(\frac{\varrho_d}{\varrho_m} - 1\Big)\Big(\frac{\beta_1^2}{|\boldsymbol{\beta}|^2} - 1\Big)\times$$

$$\times g_1G|\boldsymbol{\beta}|^2\Big(\chi_m\nu_m|\boldsymbol{\beta}|^4 - (K\boldsymbol{g}\cdot\boldsymbol{\beta} + L\nabla T_0\cdot\boldsymbol{\beta})^2\Big) + \chi_m(1+M)\times$$

$$\times K|\boldsymbol{\beta}|^2 g_2G\beta_1\beta_2\Big((K\boldsymbol{g}\cdot\boldsymbol{\beta} + L\nabla T_0\cdot\boldsymbol{\beta})^2 + \nu_m^2|\boldsymbol{\beta}|^4\Big) + \Big(\frac{\varrho_d}{\varrho_m} - 1\Big)\times$$

$$\times g_2G\beta_1\beta_2\Big(\nu_m\chi_m|\boldsymbol{\beta}|^4 - (K\boldsymbol{g}\cdot\boldsymbol{\beta} + L\nabla T_0\cdot\boldsymbol{\beta})^2\Big)\Big]/\Big[((K\boldsymbol{g}\cdot\boldsymbol{\beta}+$$

$$+L\nabla T_0\cdot\boldsymbol{\beta})^2 + \chi_m|\boldsymbol{\beta}|^4)((K\boldsymbol{g}\cdot\boldsymbol{\beta} + L\nabla T_0\cdot\boldsymbol{\beta})^2 + \nu_m^2|\boldsymbol{\beta}|^4)\Big] + O(c_0^2). \quad (2.7)$$

Therefore, the stability condition for the space-uniform state reduces to $Re\alpha_3 \leq 0$.

2.2 Analysis of Some Limiting Cases

For the study of $Re\alpha_3$ we pass on to dimensionless variables. Let $X = \chi_m/(LG)$ be the characteristic length, $\tau = \chi_m/(LG)^2$ – characteristic time, $V = LG$ – characteristic velocity, $\boldsymbol{l} = X\boldsymbol{\beta}$ – dimensionless wave vector, $l = |\boldsymbol{l}|$, $e_1 = l_1/l$, $e_2 = l_2/l$, $\gamma = \tau\alpha$ – dimensionless growth exponent of disturbances, $\varrho = \varrho_d/\varrho_m$ – densities ratio, $\mu = \mu_d/\mu_m$ – dynamic viscosity ratio, $\lambda = \lambda_d/\lambda_m$ – heat capacity ratio, $k = k_d/k_m$ – thermal conductivity coefficient ratio, θ – angle between vectors ∇T_0 and $\boldsymbol{g}$, so $g_1 = g cos\theta$, $g_2 = g sin\theta$, $g = |\boldsymbol{g}|$; $a = Kg/(LG)$ – dimensionless microacceleration, $Q = \nu_m/(XLG)$ – inverse Reynolds number, and $H = g\chi_m/(LG)^3$ – the "Galilean thermal number". In dimensionless variables (2.7) takes form

$$Re\gamma_3 = c_0\Big(l^2\Big((1+cos\theta)((1+M)e_1^2 - \varrho\lambda) + e_1e_2(1+M)a sin\theta\Big)\times$$

$$\times\Big((e_1(1+a cos\theta) + e_2 a sin\theta)^2 + Q^2l^2\Big) + \Big((e_1^2-1)cos\theta + e_1e_2 sin\theta\Big]\times$$

$$\times(\rho-1)H\Big(Ql^2 - (e_1(1+a cos\theta) + e_2 a sin\theta)^2\Big)\Big)/\Big(\Big(e_1(1+a cos\theta)+$$

$$+e_2 a sin\theta\Big)^2 + l^2\Big)\Big(\Big[e_1(1+a cos\theta) + e_2 a sin\theta\Big]^2 + Q^2l^2\Big)\Big) + O(c_0^2). \quad (2.8)$$

The main term in the right hand side of the last expression depends on 9 dimensionless variables, which can be divided into three groups: l, e_1, e_2 (space-uniform state disturbance characteristics); $\varrho\lambda, k$ (parameters determined by mechanical and thermal properties of the phases); $\theta, a, Q, (\varrho-1)H$ (these values depend also on the external forces). The parameter μ is not included in (2.8) explicitly. The parameter M in (2.8), is not independent. Let us recall its definition together with definitions of K and L, as these will be needed for the stability analysis:

$$M = \frac{3(k_m - k_d)}{2k_m + k_d} \equiv \frac{3(1-k)}{2+k},$$

$$K = \frac{2R^2(\varrho_d - \varrho_m)(\mu_m + \mu_d)}{3\mu_m(2\mu_m + 3\mu_d)} \quad \text{and} \quad L = \frac{2Rk_m\sigma_T}{(2\mu_m + 3\mu_d)(2k_m + k_d)}.$$

Note that constants M and K can have any sign while the parameter $L > 0$.

The condition $Re\gamma_3 = 0$ marks out a surface in the space of parameters, which divides the domains of stability and instability for the space-uniform state of the emulsion. It is difficult to give a description of this surface for the general case; however in some limiting cases sufficient or necessary conditions of stability can be found. Let us pass to the analysis of these conditions.

The case of weightlessness considered in [9] is the simplest one. In this case (2.8) can be simplified to the following:

$$Re\gamma_3 = \frac{l^2\Big[(1+M)e_1^2 - \varrho\lambda\Big]c_0}{l^2 + e_1^2} + O(c_0^2).$$

The stability condition with linear approximation with respect to c_0 becomes equivalent to

$$(1+M)e_1^2 - \varrho\lambda \leq 0. \tag{2.9}$$

It follows from (2.9) that the most dangerous for stability are one-dimensional disturbances for which $|e_1| = 1$. Inequality (2.9) in this case can be rewritten in the equivalent form

$$2(\varrho\lambda + k) + \varrho\lambda k \geq 5. \tag{2.10}$$

Condition (2.10) is obviously fulfilled if the parameter ϱ is large and the values λ and k are of the order unity. An example of such a situation is the emulsion "aluminium-lead" with aluminium as the carrying phase. We therefore have an emulsion with the stable state under the condition $g = 0$.

If the lead and aluminium are interchanged, (2.10) will be altered. It can be expected, that under the conditions close to weightlessness, similar emulsions will be destroyed under the action of thermocapillary forces. We also note, that (2.10) turns into an equality if $\varrho\lambda = k = 1$. In this case the emulsion is neutrally stable.

It can be supposed a priori that the condition of stability for the emulsion with the components of equal density (at $\varrho = 1$) coincides with (2.9), since this case is analogous to the case of weightlessness. This fact follows immediately from (2.8) with $a = 0$ and $\varrho = 1$.

Let us consider another limiting case: $\varrho \to 0$ together with $k \to 0$. This case describes well the behavior of gas-liquid mixtures [27]. Here one-dimensional disturbances can be studied entirely.

Considering (2.8) with $\varrho = k = 0$, $e_1 = 1$, $e_2 = 0$, we obtain the stability condition to the first approximation with respect to c_0:

$$1 + a cos\theta \leq 0.$$

Taking into account the definition of the parameter a, we can rewrite the last inequality in the form

$$LG + gK cos\theta \leq 0. \tag{2.11}$$

Note that for a gas-liquid mixture $K < 0$. Taking into account, that L, G are positive, that g is nonnegative, and using (2.11), we come to the conclusion that the stabilization of space-uniform distribution of bubbles in a temperature field with constant gradient by attaching microgravity is possible only if the angle between vectors ∇T_0 and $\boldsymbol{g}$ is acute. The maximum stabilizing effect is achieved when these vectors are parallel.

In the absence of gravity the uniform distribution of bubbles in the liquid under the constant gradient of temperature is absolutely unstable [27]. Let us rewrite the expression for $Re\alpha_3$ in dimensional variables:

$$Re\alpha_3 = \frac{5L^2G^2\beta^2\chi_m c_0}{2(L^2G^2 + \beta^2\chi_m^2)} + O(c_0^2),$$

where $\beta = |\boldsymbol{\beta}|$. Considering $Re\alpha_3$ as a function that contains the phase thermodiffusion coefficient χ_m, we can see that the maximum corresponds to χ_m close to LG/β. At the same time, $Re\alpha_3 \to 0$ when $\chi_m \to 0$, and $\chi_m \to \infty$.

The absolute instability of the uniform state of emulsion takes place also in the case when vectors ∇T_0 and $\boldsymbol{g}$ are collinear ($sin\theta = 0$), and the relative phase velocity is small: $1 + acos\theta \approx 0$ or $|a| \approx 1$. Let us consider for simplicity the limiting case $cos\theta = -1$, $a = 1$, then

$$Re\gamma_3 = \frac{(1 - e_1^2)(\varrho - 1)Hc_0}{Ql^2} + O(c_0^2). \tag{2.12}$$

The positivity of a means, that $K > 0$, i.e. $\varrho > 1$, but thus $Re\gamma_3 > 0$, if $|e_1| < 1$. Consequently the state under consideration is unstable with respect to three-dimensional disturbances. The analogous statement is also correct in the case, when $a = -1$.

Let us consider now the case, when the temperature gradient is perpendicular to the acceleration of gravity ($cos\theta = 0$). Here the sign of $Re\gamma_3$ in the first approximation with respect to c_0 coincides with the sign of the function

$$F = l^2\Big[(e_1 + e_2 a)^2 + Q^2 l^2\Big]\Big[(1 + M)e_1(e_1 + e_2 a) - \varrho\lambda\Big] + (\varrho - 1)He_1e_2\Big[Ql^2 - (e_1 + e_2 a)^2\Big]$$

which depends on six dimensionless parameters (instead of nine, as in the general case): l, Q, $\varrho\lambda$, $(1 + M)e_1$, $e_1 + e_2 a$, $(\varrho - 1)He_1e_2$. (Without loss of generality we can set $sin\theta = 1$, since the sign of e_2 is not fixed). It is easy to prove that this state of emulsion is unstable with respect to long-wave disturbances. Indeed, for small l the function F can be represented by

$$F = -(\varrho - 1)He_1e_2(e_1 + e_2 a)^2 + O(l^2).$$

In view of the arbitrariness of the sign of e_1e_2, the condition $F < 0$ cannot be satisfied uniformly for all e_1 and e_2 if $\varrho \neq 1$.

Similar reasoning leads to a conclusion about long-wave instability of the uniform state of emulsion with components of different densities, if microacceleration isn't collinear to the temperature gradient. In the case, when these vectors are collinear, stability with respect to long-wave disturbances takes place if $\varrho > 1$, when ∇T_0 and $\boldsymbol{g}$ are parallel, and if $\varrho < 1$, when the specified vectors are antiparallel.

As for short-wave disturbances, the situation with stability is more favourable. Passing in (2.8) to a limit at $l \to \infty$, we obtain with the accuracy up to terms of order c_0^2:

$$Re\gamma_3 = (1 + acos\theta)\Big[(1+M)e_1^2 - \rho\lambda\Big]c_0 + O(l^{-2}).$$

From this formula the condition of stability for the uniform state of emulsion at short-wave disturbances follows immediately:

$$(1 + acos\theta)\Big[(1+M)e_1^2 - \rho\lambda\Big] \leq 0$$

for all $e_1 \in [-1, 1]$. This condition is equivalent to (2.10) and

$$1 + acos\theta \geq 0.$$

The analysis carried out above leads to a conclusion that in the general case a uniform state of emulsion filling all space is unstable. The most dangerous disturbances are long-wave disturbances, and a destabilizing factor is the absence of colinearity of the temperature gradient and the microacceleration vector. One can hope that taking into account finite sizes of the domain occupied by flow leads to neutralization of the long-wave instability. Also the nonlinear dependence of surface tension coefficient on temperature and the influence of a chaotic movement of drops can ensure the stabilization of the uniform state of emulsion (as in [27], where this question was discussed for a gas-liquid mixture).

3 Discontinuous Solutions

3.1 Conditions on the Surface of Discontinuity

A peculiar property of the equations of motion for an emulsion is that the temperature, thermal flux, and the impulse flux on a surface of discontinuity are continuous, while the concentration and the velocities of both phases have a jump.

System (1.1)–(1.5) can be written down in divergent form as:

$$\frac{\partial c}{\partial t} + div(c\boldsymbol{u}) = 0, \tag{3.1}$$

$$div(c\boldsymbol{u} + (1-c)\boldsymbol{v}) = 0, \tag{3.2}$$

$$\frac{\partial}{\partial t}(\varrho_d c u_i + \varrho_m(1-c)v_i) + div(\varrho_d c u_i \boldsymbol{u} + \varrho_m(1-c)v_i\boldsymbol{v} + p\boldsymbol{e}_i -$$

$$-\mu_m(1+cN)\boldsymbol{A}_i) = (\varrho_d c + \varrho_m(1-c))g_i, \quad (3.3)$$

$$\frac{\partial}{\partial t}\Big(\varrho_d\lambda_d cT + \varrho_m\lambda_m(1-c)T\Big)+$$

$$+div(\varrho_d\lambda_d cT\boldsymbol{u} + \varrho_m\lambda_m(1-c)T\boldsymbol{v} - k_m(1-Mc)\nabla T) = 0 \quad (3.4)$$

and

$$\boldsymbol{u} - \boldsymbol{v} = K\boldsymbol{g} + L\nabla T \quad (3.5)$$

where $\boldsymbol{e}_i$ is the base vector of index i, A is the deformation velocities tensor, and $\boldsymbol{A}_i$ is its ith column.

Proceeding in the standard way [14], we obtain the following conditions on the surface of discontinuity:

$$[c(u_n - D)]_{1,2} = 0, \quad (3.6)$$

$$[cu_n + (1-c)v_n]_{1,2} = 0, \quad (3.7)$$

$$[\varrho_d c\boldsymbol{u}(u_n - D) + \varrho_m(1-c)\boldsymbol{v}(v_n - D) + p\boldsymbol{n} - \mu_m(1+cN)A\boldsymbol{n}]_{1,2} = 0, \quad (3.8)$$

$$[\varrho_d\lambda_d cT(u_n - D) + \varrho_m\lambda_m(1-c)T(v_n - D) - k_m(1-cM)\nabla T\cdot\boldsymbol{n}]_{1,2} = 0 \quad (3.9)$$

and

$$[\boldsymbol{u} - \boldsymbol{v} - L\nabla T]_{1,2} = 0 \quad (3.10)$$

where D is the surface speed and $\boldsymbol{n}$ is the unit normal to this surface, $g_n = \boldsymbol{g}\cdot\boldsymbol{n}$, $[f]_{1,2} = f_2 - f_1$. Here f_1 (f_2) is the limiting value of the expression on the left (right) of the interface.

The temperature must be continuous ($[T]_{1,2} = 0$) to avoid the possibility of indefinitely large heat fluxes. So by use of (3.6) and (3.7) the expression (3.9) can be simplified:

$$[k_m(1-cM)\nabla T\cdot\boldsymbol{n}]_{1,2} = 0.$$

Discontinuous solutions for the one-dimensional motion of gas-liquid mixtures were studied in [9].

3.2 Simplest Discontinuous Solution and Its Stability

The simplest discontinuous solution corresponds to the case when the plane surface of discontinuity separates a homogeneous emulsion and a pure liquid (carrying phase without inclusions). Also an emulsion moves under the action of microgravity and thermocapillary forces with linear temperature. The importance of the study of this special type of discontinuity is on account of applications to the purification of a liquid from a mixture. In this case condition (3.10) should be omitted, because (3.5) and its corollary (3.10) doesn't make sense for a pure liquid. Note, however, that the mathematical problem of stability of motion with a strong discontinuity and its stability doesn't lose sense.

For the stability analysis it is convenient to pass to the coordinate system which is connected with the discontinuity surface moving with constant speed. Let the plane $x = 0$ separate the emulsion and the pure liquid filling semispaces $x > 0$ and $x < 0$ respectively. It is obvious that we can take

$$\nabla T_0 = (Q_1, Q_2, 0).$$

From (3.6)—(3.9) and from the continuity of temperature it follows that the temperature gradient in emulsion and its concentration are tied to the relation

$$\Big((\varrho_d\lambda_d - \varrho_m\lambda_m)(1 - Mc_0) - M\varrho_m\lambda_m\Big)Q_1(1 - c_0)(LQ_1 + Kg_1) + \\ +\varrho_d\lambda_d Q_2(LQ_2 + Kg_2) = 0,$$

where g_i are the components of the vector $\boldsymbol{g}$. In the absence of gravity it leads to the condition

$$\Big((\varrho_d\lambda_d - \varrho_m\lambda_m)(1 - Mc_0) - M\varrho_m\lambda_m\Big)Q_1^2(1 - c_0) + \varrho_d\lambda_d Q_2^2 = 0, \tag{3.11}$$

which restricts the class of emulsions with discontinuous solutions. It is necessary that its characteristics satisfy the inequality

$$(\varrho_d\lambda_d - \varrho_m\lambda_m)(1 - Mc_0) - M\varrho_m\lambda_m \leq 0 \quad \text{where } c_0 \ll 1. \tag{3.12}$$

In particular for a gas-liquid mixture such solutions don't exist.

The solution for the linearized problem for perturbations are searched in the form:

$$\begin{aligned} S &= s\exp(i\gamma y + i\delta z - \alpha t) \quad \text{for the surface of discontinuity,} \\ \boldsymbol{U} &= \boldsymbol{F}(x)\exp(i\gamma y + i\delta z - \alpha t) \quad \text{in an emulsion,} \\ \boldsymbol{V} &= \boldsymbol{G}(x)\exp(i\gamma y + i\delta z - \alpha t) \quad \text{in a pure liquid} \end{aligned} \tag{3.13}$$

where α is a complex increase the constant, γ and δ are real wave numbers, s is a complex constant; components of $\boldsymbol{F}(x)$ are functions bounded at $x > 0$, and components of $\boldsymbol{G}(x)$ are bounded at $x < 0$.

The linear homogeneous boundary value problem for the perturbations represents an eigenvalue problem: non-trivial solutions exist only at certain values of α. Stability means that real parts of these values can only be positive.

A necessary condition of stability with respect to arbitrary perturbations of the form (3.13) requires

$$(Kg_1 + LQ_1) > 0 \text{ and } \quad LQ_1(M - 1) \leq Kg_1. \tag{3.14}$$

The stability of the space-uniform state of the emulsion, filling the whole space and moving under the action of thermocapillary forces with constant gradient of temperature and microgravity was studied in Sect. 2. In particular, for the case of weightlessness the following condition of stability with respect to one-dimensional perturbations was found:

$$1 + M < \frac{\varrho_d\lambda_d}{\varrho_m\lambda_m} \tag{3.15}$$

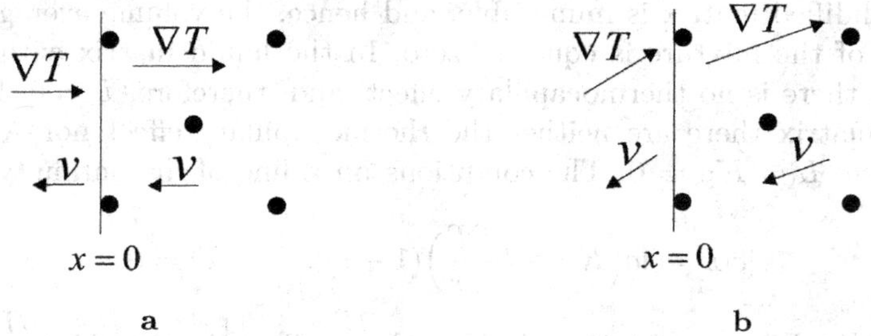

Fig. 2. The system emulsion-pure liquid with plane interface

Let us consider the problem of stability with respect to one-dimensional perturbations ($\gamma = \delta = 0$ in (3.13)) of the simplest discontinuous solutions in case $g = 0$. Note that if temperature gradients both in emulsion and in pure material are constant and directed perpendicularly to the interface (Fig. 2a), then (3.12) turns into

$$(\varrho_d \lambda_d - \varrho_m \lambda_m)(1 - Mc_0) = M \varrho_m \lambda_m, \tag{3.16}$$

and the concentration is uniquely determined by the parameters of emulsion. However, if we omit this requirement (Fig. 2b), we can consider emulsions, satisfying (3.12) with different small concentrations if the gradient of temperature is defined by (3.11).

The study of stability in case $g = 0$ with respect to one-dimensional perturbations allows us to conclude that the simplest discontinuous solution with $Q_1 > 0, Q_2 \neq 0$ (Fig. 2b) is stable if in addition to (3.15) characteristics of components satisfy the condition

$$M \leq 1,$$

which represents the corollary of (3.14) and which is caused by existence of the interface emulsion–pure liquid. The discontinuous solution with temperature gradient orthogonal to the interface (Fig. 2a) and concentration defined by (3.16) is stable when $Q_1 > 0, M \leq 1$. In fact, for both cases the stability is possible only if the values of M and $(\varrho_d \lambda_d - \varrho_m \lambda_m)/\varrho_m \lambda_m$ are very close.

In conclusion we note that in the limiting case $\varrho_d \lambda_d = \varrho_m \lambda_m$ and $M = 0$ a considered discontinuous solution exists for arbitrary $0 < c_0 \ll 1$, $Q_1 \neq 0,\ Q_2 = 0$ and is stable for $Q_1 > 0$.

4 Solidification of Emulsion

4.1 Formulation of the Problem

Let us consider one-dimensional motion. We will study the process of solidification under the following assumption: the process of solidification is described within the framework of the classical Stefan problem ignoring the jump in density during solidification.

The solidified matrix is immovable, and hence, the volume average velocity of motion of the mixture is equal to zero. In the liquid matrix with solidified inclusions, there is no thermocapillary effect, and, therefore, $L = 0$. Finally, in the solid matrix there are neither the thermocapillary effect nor Archimedes force, hence, $L = Kg = 0$. The conditions on a line of discontinuity have the form

$$[c]D = [c\Big(Kg + L\frac{\partial T}{\partial x}\Big)(1-c)], \qquad [T] = 0,$$

$$[U]D = [(\varrho_d\lambda_d - \varrho_m\lambda_m)c(1-c)\Big(Kg + L\frac{\partial T}{\partial x}\Big)T] - \Big[k_m(1-Mc)\frac{\partial T}{\partial x}\Big], \quad (4.1)$$

Here $[U]$ is the jump in internal energy, and D is the velocity of the phase boundary.

4.2 Additional Assumptions

The assumption of a low disperse-phase concentration gives ground for linearization of the concentration, and this is the main method of studying the problem in the present work. We introduce the "reduced concentration" $C = c/\varepsilon$, where ε is the maximum value of the initial concentration ($\varepsilon \ll 1$). Since thermal conduction as a first approximation ($\varepsilon = 0$) is determined by the parameters of the carrying phase (matrix), we assume that:

— The boundary of solidification of the matrix $x = s_m(t)$ is a Stefan boundary for the thermal problem, and the conditions of strong discontinuity for the temperature and concentration are satisfied on the boundary;

— The isotherm $T = T^d$, where T^d is the solidification temperature of the disperse phase, can be a strong discontinuity line only for the disperse-phase concentration [$x = s_d(t)$ is treated as the equation of this isotherm];

— The heat flux is directed towards the increasing x coordinate.

Let the solidification temperature of the matrix T^m be lower than the solidification temperature of the disperse phase T^d. The concentration of the solid disperse phase in the solidified matrix [the region $x < s_m(t)$] is denoted by $C^s(x,t)$, the concentration of the solid disperse phase in the liquid matrix (the region $s_m < x < s_d$) is denoted by $C^{sl}(x,t)$, and, finally, the concentration of the liquid disperse phase in the liquid matrix [the region $x > s_d(t)$] is denoted by $C^l(x,t)$. As a result of expansion in the small parameter ε, the problem of determining the functions of the first approximation $C^{sl}(x,t)$, $C^l(x,t)$, and $C^{sl}(x,t)$ takes the form

$$\begin{aligned}
&\frac{\partial}{\partial t}C^l + \frac{\partial}{\partial x}\Big(C^l(L\frac{\partial T^l}{\partial x} + Kg)\Big) = 0, && x > s_d(t);\\
&\frac{\partial}{\partial t}C^{sl} + \frac{\partial}{\partial x}(C^{sl}Kg) = 0, && s_m(t) < x < s_d(t);\\
&\frac{\partial}{\partial t}C^s = 0, && x < s_m(t); \quad (4.2)\\
&\frac{ds_d}{dt}(C^l - C^{sl}) = C^l\Big(L\frac{\partial T^l}{\partial x} + Kg\Big) - C^{sl}Kg, && x = s_d(t);
\end{aligned}$$

$$\frac{ds_d}{dt}(C^{sl} - C^s) = C^{sl} Kg, \qquad x = s_m(t).$$

The temperature $T^l(x,t)$ in the liquid mixture and the position of the free boundary $s_m(t)$ are tentatively obtained by solving the classical Stefan two-phase problem

$$\begin{aligned} &\varrho^l_m \lambda^l_m \frac{\partial T^l}{\partial t} = k^l_m \frac{\partial^2 T^l}{\partial x^2}, && x > s_m(t); \\ &\varrho^s_m \lambda^s_m \frac{\partial T^s}{\partial t} = k^s_m \frac{\partial^2 T^s}{\partial x^2}, && x < s_m(t); \\ &T^s = T^l = T^m, \quad \frac{ds}{dt}\gamma = k^s_m \frac{\partial T^s}{\partial x} - k^l_m \frac{\partial T^l}{\partial x}, && x = s_m(t), \end{aligned} \tag{4.3}$$

with some boundary and initial conditions, for example,

$$\begin{aligned} &T^s(0,t) = f^s(t) < T^m, \quad T^l_x(\infty,t) = 0, \quad s(0) = s_0; \\ &T^s(x,0) = \varphi^s(x) < T^m, \qquad x \in (0, s(t)); \\ &T^l(x,0) = \varphi^l(x) > T^m, \qquad x \in (s(t), \infty); \end{aligned} \tag{4.4}$$

(here γ is constant latent heat).

We now assume that the solidification temperature of the matrix T^m is higher than the solidification temperature of the disperse phase T^d. Since in the solidified matrix the disperse inclusions are immovable even if they have not solidified, the solidification boundary of the disperse phase $x = s_d(t)$ is no longer the line of concentration discontinuity. The problem of determining the disperse phase concentration in the solidified matrix $C^s(x,t)$ and the disperse phase concentration in the liquid matrix $C^l(x,t)$ consists of the equations

$$\frac{\partial}{\partial t} C^l + \frac{\partial}{\partial x}\Big(C^l \big(L\frac{\partial T^l}{\partial x} + Kg\big)\Big) = 0, \qquad x > s_m(t); \tag{4.5}$$

$$\frac{\partial}{\partial t} C^s = 0, \qquad x < s_m(t); \tag{4.6}$$

and condition

$$\frac{ds_m}{dt}(C^l - C^s) = C^l \Big(L\frac{\partial T^l}{\partial x} + Kg\Big), \qquad x = s_m(t). \tag{4.7}$$

The functions $T^l(x,t)$ and $s_m(t)$ are also solutions of the classical two-phase Stefan problem mentioned above.

For the systems described, we formulate the following two problems: the problem of determining the concentration distribution in the solidified part from a specified initial distribution of the concentrations $C^{sl}(x,0) = C^{sl}_0(x) > 0$ and $C^l(x,0) = C^l_0(x) > 0$ in the liquid matrix (called a direct problem), and the problem of determining the initial concentration distribution in the liquid matrix from a specified concentration distribution $C^s_0(x) \geq 0$ in the solidified

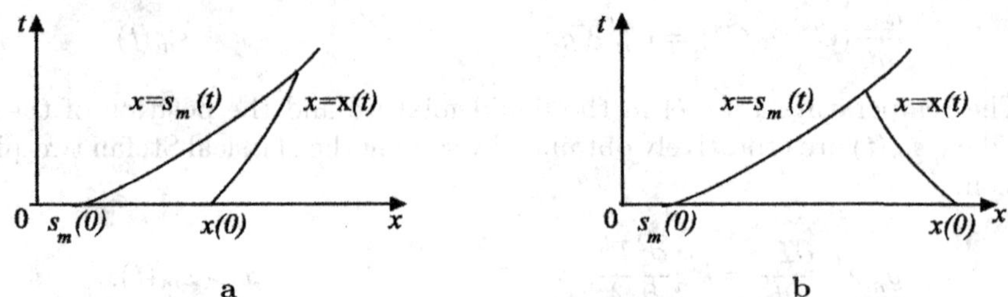

Fig. 3. slope of the characteristics

part (arbitrarily called an inverse problem). We search for a classical solution of the formulated problems with nonnegative restricted functions C^s, C^{sl}, and C^l. For simplicity, it is assumed that the functions $C_0^l(x)$ and $C_0^s(x)$ are specified on semi-infinite intervals.

Conditions of partial and complete displacement of the impurity

We are interested in the process of "oriented solidification," i.e., solutions of the Stefan problem that satisfy the inequality $ds_m/dt > 0$ for all values of time t. This requirement is easily ensured in terms of the input data of the Stefan problem [15].

We consider the relation of the disperse-phase concentrations on different sides of the boundary of solidification of the matrix in the problem with one front (the case $T^m > T^d$). Taking (4.7) into account, we see that a necessary condition for the unique solvability of the direct and inverse problems is the satisfaction of the inequality

$$\frac{ds_m}{dt} \geq L\frac{\partial T^l}{\partial x}(s_m(t),t) + Kg \text{ and } \quad t > 0. \tag{4.8}$$

It provides the "correct" slope of the characteristics and nonnegative sign of the concentrations on the solidification boundary. Inequality (4.8) implies that the velocity of transfer of the impurity in the liquid due to the thermocapillary effect and microgravity should not be higher than the velocity of motion of the solidification boundary. Then, in case $LT_x^l(s_m(t),t) + Kg \geq 0$ (Fig. 3a) we have $C^s(x) \leq C^l(x, s_m^{-1}(x))$. In case $LT_x^l(s_m(t),t) + Kg < 0$ (Fig. 3b) we obtain $C^s(x) > C^l(x, s_m^{-1}(x))$. In case of equality $\dot{s}_m(t) = LT_x^l(s_m(t),t) + Kg$, the direct problem can have only a trivial solution, and the inverse problem loses meaning.

Note, that the conditions of the "right" slope for characteristics of the two-front problem (4.2) are as follows:

$$\frac{ds_m}{dt} \geq Kg, \qquad \left(\frac{ds_d}{dt} - LT_x^l(s_d(t),t) - Kg\right) \cdot \left(\frac{ds_d}{dt} - Kg\right) \geq 0.$$

4.3 Solvability of the Problems and Some Special Solutions

Noting that the one-dimensional Stefan problem is well studied (for example [12]) it is possible to formulate the sufficient conditions for existence of bounded

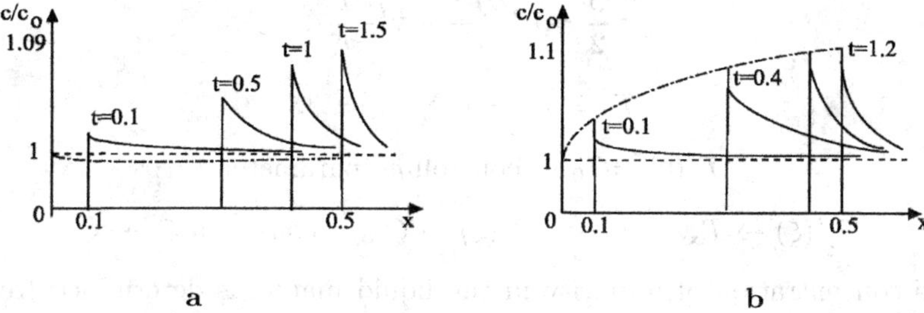

Fig. 4. The profiles of concentration for different time values

and nonnegative solution for direct and inverse problems at any time interval [16].

The exact solutions for (4.3), (4.5)–(4.7) with the special conditions of solidification (traveling wave with the solidification front equation $s(t) = Vt$, and self-similar with $s(t) = \beta\sqrt{t},\ g = 0$) are obtained and studied in [16].

The calculations were carried out on the example of the mixture aluminum–lead for the different values of Kg. As a test of the one-dimensional problem the exact solution of emulsion solidification with temperature regime such as a traveling wave was used. Fig. 4a shows the profiles of concentration for $Kg = 0$, Fig. 4b – for $Kg = -0.003$. The straight dotted lines stand for initial concentration, the curved dotted lines show the concentration in the solidified part at the last values of time and continuous lines represent concentration profiles at specific values of time.

The exact solution for the two-front problem is also obtained [16]. It should be noticed that the presence of the second line of the concentration discontinuity and the intermediate layer, representing the liquid matrix with solid inclusions, does not influence the qualitative picture of the solution behavior in the liquid emulsion and in the completely solidified part.

4.4 On the Problem of Control the Composition by Temperature

Let us now consider the one-dimensional linearized problem of solidification of emulsion in the case when the temperature of matrix solidification is higher than the temperature of solidification for impurities. It would then be interesting to investigate the possibility of obtaining the substance of given consistency by control of a temperature regime. We will study the problem of determination of the temperature value (controlling parameter) on the external boundary $x = 0$, which provides given constant concentration of impurity in the solidified part.

We will consider the self-similar variant of (4.3), (4.5)–(4.7) in the absence of gravity . Changing variable and using $\xi = x/\sqrt{t}$ we obtain

$$-\frac{\xi}{2}\frac{\partial T^s}{\partial \xi} = \chi^s \frac{\partial^2 T^s}{\partial \xi^2},\quad 0 < \xi < \beta, \tag{4.11}$$

$$-\frac{\xi}{2}\frac{\partial T^l}{\partial \xi} = \chi^l \frac{\partial^2 T^l}{\partial \xi^2},\quad \beta < \xi < \infty, \tag{4.12}$$

$$\gamma\frac{\beta}{2} = k^s\frac{\partial T^s}{\partial \xi} - k^l\frac{\partial T^l}{\partial \xi}, \tag{4.13}$$

$$T^l = T^s = T^m, \quad \xi = \beta. \tag{4.14}$$

$$T^s(0) = T_0 - \text{ controlling parameter.} \tag{4.15}$$

$$T^l(\xi) \to T_\infty = const, \; C^l(\xi) \to C_\infty = const, \quad \text{if } \xi \to \infty \tag{4.16}$$

The concentration of impurity in the liquid matrix is determined from the following equations:

$$\left(-\frac{\xi}{2} + L\frac{\partial T^l}{\partial \xi}\right)\frac{\partial C^l}{\partial \xi} + L\frac{\partial^2 T^l}{\partial \xi^2}C^l = 0, \beta < \xi < \infty \tag{4.17}$$

$$\frac{\beta}{2}(C^l - C^s) = LC^l\frac{\partial T^l}{\partial \xi}, \quad \xi = \beta. \tag{4.18}$$

and

$$C^s(x,T) = C^s,$$

where C^s is considered to be prescribed.

If we integrate (4.17) taking into account (4.16),(4.18) and make some transformations we obtain the following equation for finding β:

$$\frac{C^s}{C_\infty} = exp\left(\int\limits_1^\infty \frac{2LT_\infty exp(-\beta^2\eta^2/4\chi^l)/\eta \quad d\eta}{2LT_\infty e^{-\beta^2\eta^2/4\chi^l} - \beta^2\eta \int\limits_1^\infty e^{-\beta^2\eta^2/4\chi^l}d\eta}\right). \tag{4.19}$$

To prove the solvability of this equation let us study closely the behavior of the function

$$f(\beta) = \frac{\exp(-\beta^2/4\chi^l)}{\beta\int\limits_\beta^\infty \exp(-\xi^2/4\chi^l)d\xi} \tag{4.20}$$

as $\beta \to 0$ and as $\beta \to \infty$. It can be easily verified, that under the condition

$$\frac{LT_\infty}{\chi^l} < 1, \tag{4.21}$$

the equation $f(\beta) = 1/(2LT_\infty)$ has a unique root. Let us denote it by $\beta^\star$. The analysis of the behavior of function in (4.20) allows us to conclude that for $\forall\beta > \beta^\star$ the necessary condition for the "correct" slope of characteristics (4.8), which in our case has a form:

$$\frac{\beta}{2} > LT_\infty\frac{exp(-\beta^2/4\chi^l)}{\int\limits_\beta^\infty exp(-\xi^2/4\chi^l)d\xi} \tag{4.22}$$

is satisfied. Now let us analyze the right-hand part of (4.19) which we denote by $g(\beta)$ at $\beta \in (\beta^\star, \infty)$. It is possible to verify, that

$$\lim_{\beta \to \beta^\star} g(\beta) = 0 \text{ and } \lim_{\beta \to \infty} g(\beta) = 1.$$

So, the root of equation $C^s/C_\infty = g(\beta)$ exists under the condition

$$0 < \frac{C^s}{C_\infty} < 1. \tag{4.23}$$

The uniqueness of this root follows from $g(\beta)$ being a monotonic function, which is shown with the help of the analysis of the sign of $g'(\beta)$ in (4.21).

After β is determined, the formula for the temperature regime at $x = 0$

$$T_0 = \frac{\gamma \int\limits_0^\beta exp(-\xi^2/4\chi^s)d\xi}{k^s exp(-\beta^2/4\chi^s)} \left(-\frac{\beta}{2} - \frac{k^l T_\infty exp(-\beta^2/4\chi^l)}{\gamma \int\limits_\beta^\infty exp(-\xi^2/4\chi^l)d\xi} \right), \tag{4.24}$$

which provides the prescribed concentration C^s follows from the self-similar solution of (4.11)–(4.14).

Thus, if the input data satisfies (4.21),(4.23), then there exists the unique value (4.24) of a temperature on the external boundary $x = 0$, which corresponds to the prescribed distribution of the concentration in a solidified substance.

Remark 1. Note, that the dimensionless parameter LT_∞/χ^l does not exceed unity for most real cases, due to smallness of the coefficient L. In particular, for the emulsion "aluminium–lead" with lead as an impurity it has the order of 10^{-1} so the required condition (4.21) is satisfied.

Let us consider now the case when the initial concentration distribution is non-uniform, but the regime of solidification is still self-similar. Let's write out the characteristics for (4.5) at $\boldsymbol{g} = 0$:

$$dt = \frac{dx}{LT'_x} = \frac{dC}{-LT''_{xx}} \tag{4.25}$$

Here the derivatives of temperature are known from the self-similar solution. Integrating the first equation, we will obtain the family of characteristics:

$$x(t) = x(0) + \int\limits_0^t \frac{LBexp(-x^2/4\chi^l t)}{\sqrt{t}} dt, \tag{4.26}$$

where

$$B = \frac{T_\infty}{\int\limits_\beta^\infty exp(-\xi^2/4\chi^l)d\xi}.$$

Using the fact, that for any characteristic of family (4.26) the following equalities

$$\beta\sqrt{t} = s(t) < x(t) < x(t^\star) = \beta\sqrt{t^\star}$$

are correct, we get the estimation of behavior for $x(0)$:

$$\frac{x(0)}{\sqrt{t^\star}} = 0(1), \text{ at } t^\star \to \infty.$$

Integrating along the characteristics of the second of (4.25), we will obtain:

$$C^l(x(t^\star), t^\star) = C^l(x(0), 0) exp\Big(\int\limits_0^{t^\star} \frac{LBx}{2\chi^l t\sqrt{t}} exp(-x^2/4\chi^l t) dt\Big)$$

Considering the integral and making the substitution $\xi = x/\sqrt{t}$, in it, we obtain

$$\int\limits_0^{t^\star} \frac{LBx}{2\chi^l t\sqrt{t}} exp(-x^2/4\chi^l t) dt = \int\limits_\beta^\infty \frac{LB\xi exp(-\xi^2/4\chi^l)}{\chi^l \xi - 2\chi^l LB exp(-\xi^2/4\chi^l)} d\xi.$$

Therefore,

$$C^l(x(t^\star), t^\star) = C^l(x(0), 0) K(\beta), \tag{4.27}$$

where $K(\beta)$ does not depend on $t^\star$. It follows from passing in (4.27) to a limit at $t^\star \to \infty$, and also from using (4.7) on the solidification front, that the value of concentration on the solidification front from the side of solid part tends to the prescribed concentration.

Thus, for the self-similar regime of solidification under the conditions (4.21), (4.23) and for an arbitrary initial concentration distribution such, that $C^l(x, 0) \to C_\infty$ at $x \to \infty$, there exists the unique value of temperature at the point $x = 0$ such that the value of concentration on the solidification front from the side of the solid part tends to the prescribed concentration C^s at long times.

Remark 2. Note that the process under the consideration is meaningful for purification of mixtures, so the prescribed value of concentration in a solid part should be considerably less then the initial one. It is obvious however that the complete purification is impossible in the framework under consideration.

A numerical study carried out for the emulsion "aluminium-lead' with the lead as impurity has shown, that if the initial temperature differs from the melting temperatures for aluminium on 5 K, that for ten times lowering of impurity concentration in solid it is necessary to set $T_0 = 270.56$ K, thus $C^l(s(t), t)/C_\infty = 1.29$. For two times lowering of concentration: $T_0 = 264.74$ K, $C^l(s(t), t)/C_\infty = 1.22$.

5 Thermocapillary Drift of a Drop near the Surfaces of Phase Transition

We assume that liquid and solid phases of some substance is divided by a surface of phase transition. During solidification the surface is moving, absorbing a liquid phase. There is a liquid drop insoluble in the melted phase, moving under the action of thermocapillary forces in front of the phase transition surface. It is interesting to determine, how the drop influences the shape of a surface of phase transition. How does this surface change in time, if the distance between it and the drop decreases? How does this changing surface affect the velocity of thermocapillary drift of a drop? To answer these questions it is necessary to consider the problem of interaction of a drifting drop with the moving solidification front.

5.1 Statement of a Problem and Method of Solution

We accept the simplest model of movement of a surface of phase transition at solidification. Let surface S_0 of solidification move with respect to the solid substance. Let the speed v_0 be the speed of the surface. Let's write down the equation of energy on a surface S_0

$$\varrho_s \gamma_0 v_0 + k_1 \frac{\partial T_+}{\partial n} - k_s \frac{\partial T_-}{\partial n} = 0 \tag{5.1}$$

Here k_1, k_s are the heat conductivity coefficients for melt and solid correspondingly, γ_0 is a latent heat and ϱ_s - density of solid.

The temperature T_+ of a liquid and temperature T_- of a solid are equal and are constant along surface S_0:

$$T_+ = T_- = T_0 \tag{5.2}$$

The crystal is motionless, however on its surface the liquid velocity v differs from zero. This follows from the equation of mass conservation taking into account the jump of density at phase transition. The velocity v_+ of a liquid along the normal is proportional to the surface speed v_0:

$$\varrho_1 v_+ = (\varrho_1 - \varrho_s) v_0 \tag{5.3}$$

In most cases the density of the phases differ a little ($|\varrho_1 - \varrho_s| \ll \varrho_s$) and correspondingly the normal velocity v_+ of melted phase is small, $|v_+| \ll v_0$. It is necessary to take into account this velocity for adequate description of drop dynamics, when the drop's velocity is close to the velocity of the solidification front.

The following boundary conditions on S_0 for melted phase velocity correspond to (5.3):

$$\boldsymbol{v} \cdot \boldsymbol{n} = v_+ \text{ and } \quad \boldsymbol{v} - v_+ \boldsymbol{n} = 0 \tag{5.4}$$

At the normal direction to the solidification front, along axis x_1 the small spherical drop of radius R (Fig. 5) moves. The distance h from the center of the drop up to front is small enough in comparison with the sizes of system "melt-crystal" and is comparable to the small radius R. In this case the disturbances of temperature fields and velocities caused by a drop near the solidification front are important. At the same time, let $R/(2h)$ be a small parameter. Then the disturbances are small, but finite.

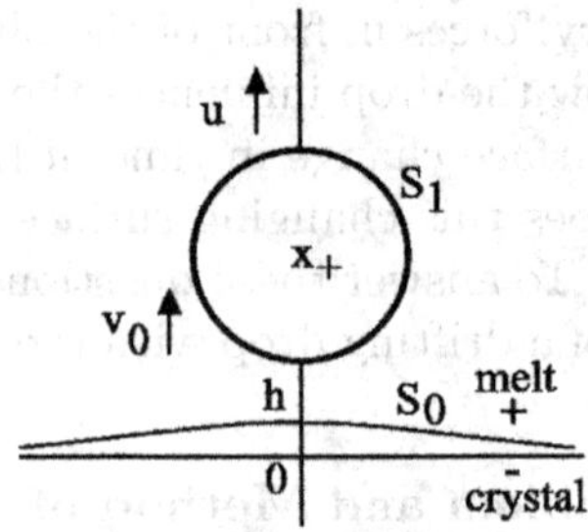

Fig. 5. Thermocapillary drift of a drop near the surface of phase transition

The temperature T changes negligibly on distance of the order of h. Hence, it is possible to neglect small changes of densities of the melted phase and crystal caused by nonhomogeneity of a field of temperature.

We assume, that the Peclet number $Pe = v_0 h/\chi_1 \ll 1$ is small (χ_1 is the coefficient of thermodiffusion of melt). The heat transfer is described by the quasistationary approximation and by the Laplace equation

$$\Delta T = 0 \tag{5.5}$$

In the absence of a drop the surface of phase transition is a plane $x_1 = 0$. The melted phase is at $x_1 > 0$ and the temperature T depends only on x_1. The center of the drop is on the point $\boldsymbol{x}_+ = (h, 0, 0)$.

At infinity the constant gradients of temperature N_+ and N_- ($N_\pm > 0$) are given by:

$$\nabla_1 T_+ \to N_+, \text{ at } x_1 > 0,\ |\boldsymbol{x}| \to \infty; \quad \nabla_1 T_- \to N_-, \text{ at } x_1 < 0,\ |\boldsymbol{x}| \to \infty \tag{5.6}$$

In the absence of a drop $T_\pm = T_0 + N_\pm x_1$ and (5.1) gives

$$\varrho_s \gamma_0 \bar{v}_0 = k_s N_- - k_1 N_+ \tag{5.7}$$

($\bar{v}_0$ is the undisturbed solidification front velocity). Under real conditions, gradients N_+ and N_- depend upon time. However interaction of a small drop with the solidification front is characterized by small time h/v_0. Therefore we assume, that N_+ and N_- are given constants.

There is another possible limiting case when at non-stationary external conditions the distance between the drop and the solidification front is big and could be compared with a container size. Taking into account the non-stationarity of the temperature field (without a drop) is necessary, and the influence of the drop motion on the dynamics and geometry of the front might be inessential. For such problems the dependence of distance between the drop and the front on time was studied for a special (self-similar) case of changing temperature field [2].

On the surface S_1 of the drop the heat flux and temperature are continuous:

$$k_1 \frac{\partial T_+}{\partial n} = k_2 \frac{\partial T_2}{\partial n}, \quad T_+ = T_2 \tag{5.8}$$

Index 2 corresponds to the area inside a drop. We determine a field of temperatures by :

$$T_{\pm} = T_{\pm}^{(0)} + T_{\pm}^{(1)} + \cdots \tag{5.9}$$

and

$$T_2 = T_2^{(0)} + T_2^{(1)} + \cdots \tag{5.9a}$$

Functions $T^{(0)}$, $T^{(1)}$ are harmonic in each of three domains, to which the bottom index $(+, -$ or $2)$ corresponds. The zero approximation $T^{(0)}$ includes the sum of the solution for the problem in the absence of a drop and the solution for the problem for a drop in the melt with temperature gradient in absence of a crystal:

$$T_{+}^{(0)} = T_0 + N_+ x_1 + b\nabla_1 \frac{1}{r_+}, \; b = \frac{k_2 - k_1}{2k_1 + k_2} R^3 N_+, \; r_+ = |\boldsymbol{x} - \boldsymbol{x}_+|, \tag{5.10}$$

$$T_{-}^{(0)} = T_0 + N_- x_1 \tag{5.11}$$

Disturbances $T_{\pm}^{(1)}$ are determined in absence of sphere S_1 from (5.2), (5.5) with (5.9)–(5.11).

Let's note a simple property of the solution. Equations (5.9)–(5.11) are equivalent to that the drop is in a nonuniform (external) field of temperatures

$$T' = T_0 + N_+ x_1 + T_{+}^{(1)} + \cdots, \tag{5.12}$$

T' - is a harmonic function at $x_1 > 0$ including the domain occupied by a drop; $T_+ - T'$ is harmonic outside of S_1, and $T_+ - T' \to 0$ at $r_+ \to \infty$.

We assume, that Reynolds number $|\boldsymbol{u} - \boldsymbol{v}_+|R/\nu \ll 1$ is small. The flow in the melt and drop is described by the Stokes equations

$$\mu \triangle \boldsymbol{v} = \boldsymbol{\nabla} p \text{ and } \quad \nabla \cdot \boldsymbol{v} = 0 \tag{5.13}$$

On a surface of a spherical drop S_1 the melted phase velocity is equal to velocity of flow in a drop, the normal velocity of a liquid is equal to normal velocity of the sphere:

$$\boldsymbol{v} \cdot \boldsymbol{n} = \boldsymbol{u} \cdot \boldsymbol{n} \text{ and } \quad \boldsymbol{v}_1 = \boldsymbol{v}_2 \tag{5.14}$$

The drop velocity $\boldsymbol{u}$ is unknown, $\boldsymbol{v}_\alpha$ is the velocity in the melted phase $(\alpha = 1)$, or in the drop $(\alpha = 2)$.

The drop moves under the action of thermocapillary forces. We assume that the surface tension σ depends linearly on temperature:

$$\sigma = \sigma_0 - \sigma_T (T - T_0); \quad \sigma_T > 0$$

Tangent stresses outside $\boldsymbol{P}_{\tau 1}$ and inside $\boldsymbol{P}_{\tau 2}$ of the drop are different because of the surface tension changes:

$$\boldsymbol{P}_{\tau 1} - \boldsymbol{P}_{\tau 2} = -\sigma_T [\boldsymbol{\nabla} T - (\boldsymbol{n} \cdot \boldsymbol{\nabla} T) \boldsymbol{n}]$$

$$\boldsymbol{P}_{\tau\alpha} = \boldsymbol{P}^\alpha \cdot \boldsymbol{n} - (\boldsymbol{n} \cdot \boldsymbol{P}^\alpha \cdot \boldsymbol{n})\boldsymbol{n}, \quad P^\alpha_{ij} = \mu_\alpha(\nabla_i v^\alpha_j + \nabla_j v^\alpha_i), \tag{5.15}$$

($\alpha = 1, 2;\ i, j = 1, 2, 3; \boldsymbol{n}$ is external normal). The sum total acting on caring liquid ($\alpha = 1$) from the side of a drop must be equal to zero:

$$\int_{S_1} \boldsymbol{P}^1 \boldsymbol{n} ds = 0$$

5.2 Thermocapillary Drift of a Drop in a Nonuniform Flow

The velocity $\boldsymbol{u}$ of a drop can be calculated on the basis of the exact solution of the problem of a spherical drop motion in an arbitrary nonuniform flow with variable velocity $\boldsymbol{v}'(\boldsymbol{x})$ and temperature $T'(\boldsymbol{x})$. The velocity is calculated by means of values $\boldsymbol{v}'(\boldsymbol{x}_+)$, $\Delta\boldsymbol{v}'(\boldsymbol{x}_+)$ and $\nabla T'(\boldsymbol{x}_+)$ at the center of the sphere [23,24]:

$$\boldsymbol{u} - \boldsymbol{v}' = L\nabla T' + \frac{\mu_2 R^2}{2(2\mu_1 + 3\mu_2)}\Delta\boldsymbol{v}' \text{ and } L = \frac{2R\sigma_T k_1}{(2k_1 + k_2)(2\mu_1 + 3\mu_2)} \tag{5.16}$$

The velocity $\boldsymbol{v}'$ satisfies the Stokes equations; the function $\boldsymbol{u} - \boldsymbol{v}'$ has no irregularities outside of the sphere S_1. In the case of an unbounded liquid it is possible to set a nonuniform flow by decomposition of the field of velocities (temperatures) in it's irregular points outside of the sphere S_1. However, if there exist the external boundaries of a flow, the velocity $\boldsymbol{v}'$ and temperature T' depend on boundary conditions on the sphere and can be known only approximately when its radius is small. In the problem under consideration they need to be determined

For a drop in uniform flow (5.16) corresponds to the Young-Goldstein-Block formula [28]. At large drop viscosity (5.16) approaches the Faxen formula [6,8] for dynamics of a solid particle. For the case, when the mass forces are essential, the formula for a drop velocity in nonuniform flow is also known [23]. Note that in (5.16) we can take into account the acceleration $\boldsymbol{g}$, as in (1.5). The formula (5.16) was used for the description of thermocapillary motion of the drops in a nonuniform heated emulsion [24]. This formula is valid for small Reynolds numbers. At large Reynolds numbers similar formulas for dynamics of gas bubble are known [25,26].

Equation (5.16) can be applied for many problems of thermocapillary motion for one or several drops in the presence of different boundaries. Influence of a solidification front (or another drop) on a drop is replaced by an external velocity field $\boldsymbol{v}'$ and external temperature field T'. The formula is effective when external velocity and temperature gradients change slightly with the drop diameter.

5.3 Determination of the Non-stationary Shape of the Melt-Solid Front

Without taking into account the interaction of a drop with the solidification front ($R/(2h) \to 0$) it follows from (5.4), (5.6), (5.13), (5.16) that

$$u_0 = LN_+ + v_+ \tag{5.17}$$

The front velocity concerning a drop is equal to

$$w = v_0 - u_0 = v_0 - v_+ - LN_+ = (\varrho_s/\varrho_1)v_0 - LN_+ \tag{5.18}$$

The velocity u is disturbed because of hydrodynamical and thermal interaction of the drop with the front.

The thermocapillary drift of a drop near an initially plane front ($x_1 = 0$) leads to disturbance of the front coordinate

$$x_1 = \delta(x_2, x_3, t), \quad \delta \ll h. \tag{5.19}$$

At a large distance from the axis of symmetry x_1 the front coincides with a plane: $\delta \to 0$, $x_2^2 + x_3^2 \to \infty$. Using the smallness of disturbance δ ($|\nabla\delta| \ll 1$), we will write down the following approximate expression for disturbances of the normal velocity of a surface:

$$v_0 - \bar{v}_0 = \frac{\partial \delta}{\partial t} \tag{5.20}$$

For the disturbed front (5.1) and (5.2) give boundary conditions at $x_1 = 0$ as follows:

$$-\varrho_s \gamma_0 \frac{\partial \delta}{\partial t} + k_s \nabla_1 T_-^{(1)} = k_1 \nabla_1 \nabla_1 \frac{b}{r_+} + k_1 \nabla_1 T_+^{(1)}, \quad r_+ = |\boldsymbol{x} - \boldsymbol{x}_+|$$

$$T_-^{(1)} + N_- \delta = 0 \text{ and } \quad T_+^{(1)} + b\nabla_1(1/r_+) + N_+ \delta = 0 \tag{5.21}$$

Functions $T_-^{(1)}$ and $T_+^{(1)}$ are harmonic at $x_1 < 0$ and $x_1 > 0$ respectively, and

$$T_\pm^{(1)} \to 0, \ |\boldsymbol{x}| \to \infty \tag{5.22}$$

The distance h of the drop from the undisturbed solidification front $x_1 = 0$ changes in time (according to zeroth order approximation) linearly as:

$$dh/dt = -w \le 0 \tag{5.23}$$

This is meaningful as long as $R/(2h) \ll 1$ is valid.

The non-stationary problem (5.5), (5.21) has the exact solution

$$T_-^{(1)} = E_- \nabla_1(1/r_+), \ r_+ = |\boldsymbol{x} - \boldsymbol{x}_+|, \tag{5.24}$$

$$T_+^{(1)} = E_+ \nabla_1(1/r_\star), \ r_\star = |\boldsymbol{x} - \boldsymbol{x}_\star|, \ \boldsymbol{x}_\star = -\boldsymbol{x}_+ \tag{5.25}$$

$$\delta(x_2, x_3, t) = A\nabla_1(1/r_+), \text{ at } x_1 = 0 \tag{5.26}$$

$$A = -\frac{2bk_1}{\rho_s \gamma_0 (w + v_0) + 2k_1 N_+}, \ E_+ = b + AN_+ \text{ and } E_- = -AN_-. \tag{5.27}$$

The point $\boldsymbol{x}_\star$ is symmetric with the point $\boldsymbol{x}_+$ with respect to the plane $x_1 = 0$, $\boldsymbol{x}_\star = (-h, 0, 0)$. According to (5.6), (5.26), (5.27) the coordinates of solidification surface S_0 are determined by formula

$$x_1 = \delta(x_2, x_3, t) = \frac{hR^3}{r^3} \cdot \frac{k_1 - k_2}{2k_1 + k_2} \cdot \frac{2k_1 N_+}{\rho_s \gamma_0 (w + v_0) + 2k_1 N_+}, \tag{5.28}$$

$$r = (h^2 + x_2^2 + x_3^2)^{1/2} \text{ and } dh/dt = -w = const.$$

here v_0 is undisturbed front velocity.

The disturbance δ of the solidification surface is caused by the thermal interaction of a drop with the phase front. According to (5.28), non-stationarity of waves on front is caused by the variable quantity $h(t)$. The velocity of the front and the drop enters the wave amplitude formula explicitly. Therefore for the same distance h there could be different wave profiles due to different values of w. The form of disturbances does not depend on time in coordinates x_2/h, x_3/h. The maximal amplitude grows as $1/h^2$ as the distance h of the drop from phase front decreases.

Equation (5.28) is not restricted by thermocapillary motion, the drop can move under action of gravity. Note that the most interesting case is, when the removal of heat from the front in a solid phase considerably exceeds the inflow of heat from the melted phase, $k_s N_- \gg k_1 N_+$.

For motion of a drop at a constant distance from the phase front the disturbance of this surface is limited and does not depend upon time. In this case the amplitude of disturbance is larger than in the case when its relative velocity differs from zero. With the growth of the velocity w the amplitude of the front disturbance decreases. The sign of curvature of the front is determined by the sign of difference between heat conductivity coefficients of the drop and the melt.

5.4 Velocity of a Drop in Thermocapillary Drift

In order to define a drop velocity it is necessary to find the velocity field in the liquid by solving the Stokes equations. The solution of similar problem of a drop close to the solid wall obtained earlier [5] with the help of series expansion in a special coordinate system. Note that problem considered could not be reduced to the problem of viscous interaction only, because it is necessary to take account of thermal interactions and non-stationary deformations of the solidification front.

At the small ratio $R/(2h)$ the velocity field in the melt can be given as

$$\boldsymbol{v} = \boldsymbol{v}^{(0)} + \boldsymbol{v}^{(1)} + \cdots, \quad \boldsymbol{v}^{(0)} = \boldsymbol{v}_+ + \frac{1}{2}(u - v_+)R^3 \nabla\nabla_1(1/r_+) \tag{5.29}$$

$$\boldsymbol{v}_+ = (v_+, 0, 0)$$

The disturbance $\boldsymbol{v}^{(1)}$ satisfies the Stokes equations in the domain $x_1 > 0$ subject to the conditions

$$\boldsymbol{v}^{(1)} = -\boldsymbol{v}^{(0)} + \boldsymbol{v}_+, \text{ at } x_1 = 0; \quad \boldsymbol{v}^{(1)} \to 0, \ |\boldsymbol{x}| \to \infty \tag{5.30}$$

The influence of the front shape disturbance on the velocity field is of the order $O(R^6/(2h)^6)$ and it can be neglected to the first approximation.

The velocity $\boldsymbol{v}'$ of nonuniform (external) flow in which the drop moves accordingly to (5.29) is equal to

$$\boldsymbol{v}' = \boldsymbol{v}_+ + \boldsymbol{v}^{(1)} \tag{5.31}$$

The melted phase velocity disturbance is determined from (5.13) and (5.30) as

$$v_i^{(1)} = -(u - v_+)\big(\nabla_1^2 I_{i1} + \frac{3}{2}\nabla_i\nabla_1\frac{1}{r_\star} + h\nabla_i\nabla_1^2\frac{1}{r_\star}\big)R^3 \quad (5.32),$$

where I_{i1} are the components of the Stokeslet velocity, oriented along an axis x_1,

$$I_{ij} = \frac{\delta_{ij}}{r_\star} + \frac{y_i y_j}{r_\star^3}, \quad \boldsymbol{y} = \boldsymbol{x} - \boldsymbol{x}_\star,$$

δ_{ij} – is Kronecker symbol. An equivalent to (5.32) is

$$v_i^{(1)} = (u - v_+)\{\frac{1}{2}\nabla_i\nabla_1\frac{1}{r_\star} + (x_1\nabla_i - \delta_{i1})\nabla_1^2\frac{1}{r_\star}\}R^3 \quad (5.33)$$

The operator $x_j\nabla_i - \delta_{ij}$, $j = 1$ corresponds to the Oberbeck formula [8,10], which gives the representation for components v_i of the velocity in Stokes flow via harmonic functions.

From (5.31) and (5.32) (or (5.33)) we determine the external velocity $\boldsymbol{v}\,'$. Then with the help of (5.11), (5.16), (5.25), (5.27) we find the drop velocity to an accuracy of $O(R^5/(2h)^5)$:

$$u = v_+ + (u_0 - v_+)\Big(1 - \frac{R^3}{2h^3} + \frac{R^3}{4h^3}\cdot\frac{k_2 - k_1}{2k_1 + k_2}\cdot\frac{\varrho_s\gamma_0(w + v_0)}{\varrho_s\gamma_0(w + v_0) + 2k_1 N_+}\Big) \quad (5.34)$$

(v_+ is determined in (5.3), $dh/dt = -w$). The first item in (5.34) and second item in parenthesis take account hydrodynamical interactions of the drop with the solidification front and the third item takes account the thermal interaction taking into consideration interaction of the drop with a non-stationary wave on the front. Equation (5.34) for the velocity has the important property that it is not dependent on material constants and on the values of temperature gradients from both sides of the phase transition surface. As a result of thermal and hydrodynamic interaction of a drop with the front it's velocity is always less than the velocity far from front, and dependence of velocity on the distance is monotonic. The solidification front "tends" to capture a new moving drop. If the drop in some moment of time came nearer to the front, it will come nearer to it also in the subsequent moments of time.

6 Discussion and Conclusion

1. The application range of the considered model is determined by the balance of gravitation and thermocapillary forces, which define the dynamics of a disperse particle. This balance is characterized by a dimensionless microacceleration a, introduced in the beginning of Sect. 2.2. If we accept typical values of material constants for a bubble in a water-spirit solution ($\varrho_m = 0.8\ g/cm^3$, $\sigma_T = 0.18\ g/(K\cdot s^2)$ and its radius $R = 0.1\ cm$, we will obtain that acceleration g and temperature gradient $G = 0.1\ K/cm$ give the same contribution to the drop velocity in case the acceleration g equals to $0.33\ cm/s$. This is 3000 times less than Earth acceleration. Note that on the orbital stations there could be obtained significantly smaller values of microacceleration.

2. The system of equations of the basic model (1.1)-(1.5) is not standard in comparison with the usual equations of mathematical physics, because it is of non-classical type together with high order and non-linearity. Therefore a correct statement of initial-boundary value problems for this system is of special interest. Thereupon the possibility of the reduction of a system of one-dimensional problems (motions with plane, cylindrical and spherical waves) to a system of only two differential equations seems especially valuable. However this system is non-classical — it is of a hyperbolic–parabolic type.

3. One of the most essential restrictions is the smallness of concentration. Equation (1.5) is based on additive contribution of mass forces and temperature to the resulting velocity. This is justified at small Peclet and Reynolds numbers.

4. The statement of a one-dimensional boundary value problem is based on a quasistationary form of the last equation in (1.5). Such an approach is similar to traditional quasistationary one, which is widely used in mechanics of continuum and in particular in drop dynamics [18].

5. Considered in Sect. 2 the problem about the stability of the space-uniform state of an emulsion didn't take into account the finiteness of dimensions of motion in the domain. It would be very interesting to investigate the influence of vessel walls on emulsion stability. In particular, to find the critical domain diameter, below which the long-wave instability is repressed, and to estimate the role of viscosity in a wall domain (we hope it would be stabilizing) would be interesting.

6. As far as we know, the mathematical statement of emulsion solidification problem was missing so far. The 4 model proposed in section uses a constant latent heat γ (see the last condition of (4.3)). This suggestion is in accordance to a linearization on small concentration c. Taking into account dependence of γ on concentration c could be interesting as well.

7. As shown in Sect. 5 the moving surface of solidification becomes curved, if a drop moves in a melt. The non-stationary disturbances of surface shape (5.28) are caused by thermal interaction of a drop with a solidification front. The curvature of a surface proceeds differently depending on the thermal conductivities of a drop and a melted phase.

8. The thermocapillary drift of a drop near a solidification front depends on viscous and thermal interactions of a drop with this front. The velocity of the thermocapillary motion is influenced by a non-steady disturbance of the front caused by a drop. The velocity of thermocapillary drift of a drop close to the front is always less than when it is far from it. Velocity (5.34) changes monotonically with distance from a drop front. Thus, the converging of drop and front accelerate with decreasing distance between them at any values of the parameters.

Acknowledgements

This work was supported by grant No 5-2000 of Siberian Division of Russian Academy of Science and grant No 00-15-96162 of the Council for support of leading Scientific School of the Russian Federation.

Authors are very grateful to Professor R. Narayanan for attention to their work and helpful advice, and to an anonymous reviewer for his valuable remarks.

References

1. A. Acrivos, D.J. Jeffrey, D.A. Saville: J. Fluid Mech. **212**, 95 (1990)
2. L.K. Antanovskii: Microgravity **2**, 103 (1998)
3. L.K. Antanovskii, B.K. Kopbosynov: J. Appl. Mech. and Techn. Phys. **2**, 59 (1986)
4. R. Balasubramaniam, R.S. Subramanian: Int. J. Multiphase Flow **22**, 593 (1996)
5. K.D. Barton, R.S. Subramanian: J. Colloid and Interface. Sci. **137**, 170 (1990)
6. H. Faxen: Arkiv. Mat. Astron. Och. Fys. **20**, No.8 (1927)
7. J. Hadamard: C. R. Acad. Sci. Paris **152**, 1735 (1911)
8. J. Happel, H. Brenner: *Low Reynolds Number Hydrodynamics*, (Prentice - Hall, 1960)
9. B.K. Kopbosynov: 'One-dimensional thermocapillary motion in the gas-liquid mixture'. In: *Dinamika Sploshnoi Sredy, vol.74*, ed by V.N. Monahov (Novosibirsk 1986) pp. 25–37 (in Russian)
10. H. Lamb: *Hydrodynamics*, (Cambridge University Press., London 1932)
11. J.C. Maxwell: *Treatise on Electricity and Magnetism* Vol.1 (Dover publ, London 1954)
12. A.M. Meirmanov: *The Stefan Problem*, (Walter De Gruyter, Berlin–New-York 1992)
13. R.I. Nigmatulin: *Dynamics of Multyphase Media*, (Hemisphere, 1991)
14. L.V. Ovsiannikov: *Lectures on the Bases of Gas Dynamics*, (Nauka, Moscow 1981)(in Russian)
15. A.G. Petrova: 'Monotonicity of the free boundary in two-phase Stefan problem'. In: *Dynamica Sploshnoi Sredy, vol.67*, ed. by S.N. Antontsev (Novosibirsk, 1984) pp. 97–99 (in Russian)
16. A.G. Petrova, V.V. Pukhnachov: J. Appl. Mech. and Tech. Phys. **40**, 128 (1999)
17. V.V. Pukhnachov: 'Two inverse problems of continuum mechanics'. In: *Ill-posed problems of mathematical physics and analysis*, (Nauka, Novosibirsk 1984) pp. 113–118 (in Russian)
18. V.V. Pukhnachov: J. Appl. Maths. and Mechs. **62**, 927 (1998)
19. V.V. Pukhnachov, O.V. Voinov: 'Mathematical model of motion of emulsion under effect of thermocapillary forces and microacceleration'. In: *Abstracts of Ninth European Symposium on Gravity Dependent Phenomena in Phisical Sciences*, (Berlin 1995) pp. 32–33
20. V.V. Pukhnachov, O.V. Voinov: 'Termocapillary motion in an emulsion'. In: *Third NASA Conference Fluid Physics in Microgravity*, (Cleveland 1996) pp. 337–342
21. M. Rybczynski: Intern. Acad. Polon. Sci. Cracovie, Ser. A **1**, 40 (1911)
22. G.I. Taylor: 'The viscosity of a fluid containing small drops of another fluid'. In: *Proceeding of the Royal Society*, A, vol. CXXXVIII (1932) pp. 41–48
23. O.V. Voinov: *Method for mathematical description of motion of multiple particles, drops or bubbles in a viscous liquid.* Transactions of VNIIBT **60** (Moscow 1985) pp. 3–15

24. O.V. Voinov: 'Hydrodynamic microwaves in two-phase medium'. In: *Joint X-th Europ. and VI-th Rus. Symp. on Phys. Sci. in Microgravity, vol.1*, ed. by V.S. Avduevski and V.I. Polezhaev, (Moscow 1997) pp. 176–180
25. O.V. Voinov: J. Fluid Mech. **348**, 247 (2001)
26. O.V. Voinov, A.G. Petrov: Sov. Physics Dokl. **210**, 1036 (1973)
27. O.V. Voinov, V.V. Pukhnachov: J.Appl. Mech. and Techn. Phys. **5**, 38 (1980)
28. N.O. Young, J.S. Goldstein, M.J. Block: J. Fluid Mech. **6**, 350 (1959)

Non-equilibrium Phase Change

S.P. Lin

Mechanical and Aeronautical Engineering Department, Clarkson University,
Potsdam, NY 13699-5725, USA

Abstract. A continuum theory of non-equilibrium transports of mass, momentum and energy accompanied by a phase change across an interface between two different fluids is developed. An exact solution of the governing differential system is obtained for the case of constant rate rapid evaporation or condensation due to differential heating or cooling across a flat interface in a microgravity environment. The results of the solution are used to demonstrate that the heat transfer to the interface is not only to change the phase of fluid but also to impart momentum and kinetic energy to the vapor and host dry gas mixture. The fraction of the heat transfer which is not used for the phase change is shown to be quite significant in some certain cases. An exact solution for the case of constant rate freezing or melting across a flat interface at microgravity is also obtained. The results of the solution is used to demonstrate how the motion of the liquid ahead of the moving interface is affected by the change of specific volume upon phase change.

1 Introduction

Rapid evaporation or condensation occurs in many applications including liquid fuel combustion [2,42] powder metallurgy [34], and film cooling and film coating [9,17,18,36]. The phase changes across a liquid-gas interface are usually observed to take place in temperature and mass concentration fields with steep gradients. Similar situations exist at liquid- solid interfaces where solidification or melting takes place. Solidification processes are encountered in crystal growth [3,5,6,12,15,28,32], nucleation [40], phase separation [38], photographic imaging [13], material processing [4], and environmental fluid mechanics [30]. It is commonly *assumed* that the Clapeyron equation, the thermodynamic equilibrium equation, can be adequately applied to describe the effect of phase change. The inadequacy of this assumption has been recognized [1,27,41], and alternatives have been pursued [35,37]. However, the correct mathematical boundary condition at an interface with phase change remains an unresolved question in physics.

Recently a constitutive equation for evaporation/condensation has been found to have important applications [9,10]. In these applications, the evaporation coefficient is treated as a free parameter. On the other hand, quantitative knowledge on the dependence of the evaporation coefficient on the rates of mass, momentum and heat transfer is necessary to assess the impact of phase change on the fluid dynamics of many important practical problems, including atomization [16,19–23,33,43]. Attempts to evaluate the evaporation coefficient on earth [27]

S.P. Lin., Non-equilibrium Phase Change, Lect. Notes Phys. **628**, 355–367 (2003)
http://www.springerlink.com/

have suffered from the lack of an exact theoretical solution due to the mathematical difficulty associated with gravity. This makes the comparison of tentative approximate solutions with experiments problematic. Moreover natural convection on earth masks the pure physical effect of evaporation and condensation. A theory which does not require the use of the evaporation coefficient is given in Sect. 2. An exact solution of the theoretically formulated problem is obtained in Sect. 3 for the problem of evaporation of a simple liquid into a host gas under weightless condition. Another exact solution for the problem of constant rate change of phase from liquid to solid is given in Sect. 4.

2 Formulation of the Problem

Consider two fluids separated by an interface. Fluid 2 below the interface is a pure liquid. The fluid above is a mixture of the host gas and the vapor of the liquid. Heat is transferred to the interface by conduction to effect the phase change. The fluids are assumed to be incompressible. The fields of velocity $\mathbf{V}$, temperature T and vapor concentration C associated with the phase change at the interface are governed by

$$\nabla \cdot \mathbf{V}_i = 0, \tag{1}$$

$$\rho_i \frac{D\mathbf{V}_i}{Dt} = \nabla \cdot \tau_i, \tag{2}$$

$$\frac{DT_i}{Dt} = \kappa_i \nabla^2 T_i, \tag{3}$$

$$\frac{DC_i}{Dt} = D_i \nabla^2 C_i, \tag{4}$$

Where the subscript i stands for the liquid or the gas mixture depending on $i = 2$ or 1, ρ_i is the density, D/Dt denotes the substantial derivative, and T is the temperature. τ is the stress tensor. κ and D are respectively the thermal and molecular diffusivity. Both κ and D are assumed to be constant. The assumptions that the material properties are constant, and the fluid is incompressible i.e. $D\rho_i/Dt = \partial\rho_i + \mathbf{V}_i \cdot \nabla\rho_i = 0$, are strictly speaking invalid, since the temperature and density fields vary in space and time in general. However the incurred errors are small for the particular solution to be obtained [24–26].

The boundary conditions can be obtained from applying the law of conservation of mass, the momentum principle, and the first law of thermodynamics to the mass in a flowing volume which contains the interface. The time derivative of the volume integral involved can be evaluated by the Reynolds transport theorem or the Leibniz rule. The two and three dimensional divergence theorems can then be applied to reduce all of the integrals involved to surface integrals. The sum of the integrands in the surface integrals for each physical principle then yields the

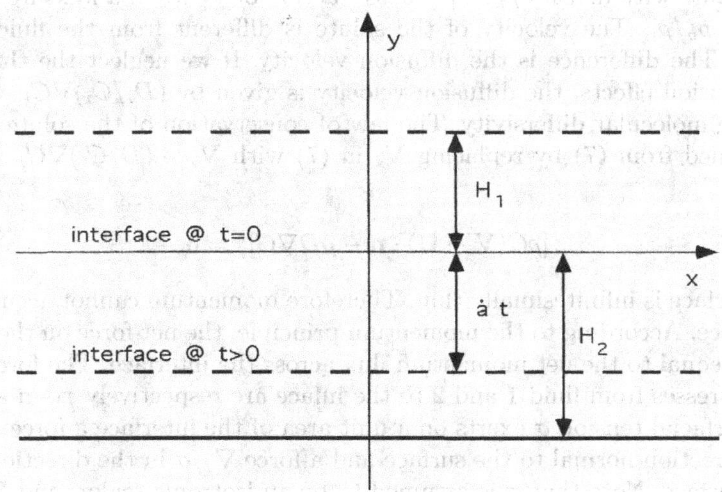

Fig. 1. Definition sketch

sought interfacial boundary conditions (C.F. Joseph and Renardy [11]). Here we apply the physical principles directly to the interface to obtain the conditions.

The interface is mathematically defined by

$$S = y - h(x, z, t) = 0. \tag{5}$$

Where (x, y, z) is the Cartesian coordinate, and h is the local instantaneous distance from the $x - z$ plane to the interface measured along the y-axis (cf. Fig. 1).

It follows from (5) that $DS/Dt = 0$, where D/Dt is the substantial derivative following the surface with its own velocity $\mathbf{U}$. Thus

$$\partial_t S + \mathbf{U} \cdot \nabla S = \partial_t h + \mathbf{U} \cdot \frac{\nabla S}{|\nabla S|} |\nabla S| = -\partial_t h + \mathbf{U}|\nabla S| \cdot \mathbf{n}, \tag{6}$$

where $\mathbf{n}$ is the unit normal vector at any location on S, i.e. $\mathbf{n} = \nabla S / |\nabla S|$. The kinematic condition (6) relates the interfacial velocity to the geometry of the interface.

The other kinematic condition is the conservation of mass across the interface. The fluid velocities relative to the moving interface from below and above are respectively $(\mathbf{V}_2 - \mathbf{U})$ and $(\mathbf{V}_1 - \mathbf{U})$. The mass absorption or desorption at the infinitesimally thin surface is not considered. Therefore mass flux (mass flow per unit area per unit time) from below $\rho_2(\mathbf{V}_2 - \mathbf{U}) \cdot \mathbf{n}$ and the mass flux from above $\rho_1(\mathbf{V}_1 - \mathbf{U}) \cdot \mathbf{n}$ above must add up to be zero,

$$\rho_1(\mathbf{V}_1 - \mathbf{U}) \cdot \mathbf{n} - \rho_2(\mathbf{V}_2 - \mathbf{U}) \cdot \mathbf{n} = [\rho(\mathbf{V} - \mathbf{U})]_2^1 \cdot \mathbf{n} = 0. \tag{7}$$

Note that the unit normal vector is defined to be positive if it is pointed from fluid 2 to fluid 1. The conservation of solute mass must also be observed. To simplify presentation, we demonstrate here with the case of a single species of dilute solute with density ρ_v. The mass fraction of solute in a mixture is defined as $C_i = \rho_v/\rho_i$. The velocity of the solute is different from the fluid mixture velocity. The difference is the diffusion velocity. If we neglect the thermo and baro-diffusion effects, the diffusion velocity is given by $(D_i/C_i)\nabla C_i$, where the D_i is the molecular differsivity. The law of conservation of the solute mass can be obtained from (7) by replacing $\mathbf{V}_i$ in (7) with $\mathbf{V}_i - (D/C_i)\nabla C_i$. It follows that

$$[\rho C(\mathbf{V} - \mathbf{U}) \cdot \mathbf{n} - \rho D \nabla C]_2^1 = 0. \tag{8}$$

The interface is infinitesimally thin. Therefore momentum cannot accumulate at the surface. According to the momentum principle, the net force on the interface must be equal to the net momentum flux across the interface. The force exerted by the stresses from fluid 1 and 2 to the inface are respectively $\tau_1 \cdot \mathbf{n}$ and $\tau_2 \cdot \mathbf{n}$. The interfacial tension σ exerts on a unit area of the interface a force $-\mathbf{n}\sigma\nabla \cdot \mathbf{n}$ in the direction normal to the surface and a force $\nabla_{//}\sigma$ in the direction tangent to the surface. Note that σ is assumed to be an isotropic scalor, and $\nabla_{//}$ is the gradient along the interface. The net momentum transfer per unit area per unit time across the interface is $\dot{m}(\mathbf{V}_1 - \mathbf{V}_2)$ where $\dot{m}$ is the mass flux

$$\dot{m} = \rho_1(\mathbf{V}_1 - \mathbf{U}) = \rho_2(\mathbf{V}_2 - \mathbf{U}) \tag{9}$$

It follows from the momentum principle that

$$[\rho \mathbf{V}(\mathbf{V} - \mathbf{U}) \cdot \mathbf{n} - \tau \cdot \mathbf{n}]_2^1 = \nabla_{//}\sigma - \mathbf{n}\sigma\nabla \cdot \mathbf{n} \tag{10}$$

The energy transfer across the interface must obey the first law of thermodynamics. The sum of the work done by all external forces on the interface and the net heat transfer and the net transfer of energy (including kinetic, potential and internal energy) across the same interface must result in the increase in the energy stored in the surface, according to the first law of thermodynamics. The work done by the surface tension per unit time on each unit surface area is

$$\mathbf{U} \cdot [\sigma(\nabla \cdot \mathbf{n})\mathbf{n}] \tag{11}$$

The net work done by the fluid stress per unit time per unit interface area is

$$[\mathbf{V} \cdot (\tau \cdot \mathbf{n})]_2^1 \tag{12}$$

The heat transfer to the interface by radiation will not be considered here. The net heat flux to the interface is given by the Fourier conduction

$$[k \nabla T \cdot \mathbf{n}]_2^1 \tag{13}$$

where k is the conductivity.
The net energy flux across the interface is

$$[\dot{m} f]_2^1 + [\rho D \nabla C \cdot \mathbf{n}(e_\alpha - e_\beta)]_2^1 \tag{14}$$

where f is the sum of the kinetic energy $1/2\ \mathbf{V}\cdot\mathbf{V}$, the potential energy, and the internal energy per unit mass. The last bracket in (14) arises from the internal energy flux due to the transport by diffusion of the changed phase, e_α and e_β being respectively the specific internal energy of the host component in fluid 1 and the evaporating phase. The sum of (11) - (14) must be equal to the time rate of increase in the surface energy per unit area of the interface as it moves with its velocity $\mathbf{U}$, i.e.

$$\frac{D\sigma}{Dt} = \mathbf{U}\cdot[\nabla_{//}\sigma - \sigma(\nabla\cdot\mathbf{n})\mathbf{n}] - [\dot{m}f - \mathbf{V}\cdot\tau\cdot\mathbf{n} + k\nabla T\cdot\mathbf{n} \\ -(\rho D\nabla C)(e_\alpha - e_\beta)\cdot\mathbf{n}]_2^1 \tag{15}$$

The field equations (1) - (4) must be supplemented by equations of state which relate material properties such as density to field variables such as pressure and temperature. In addition to the interfacial conditions (7), (8), (10) and (15), one must specify the temperature and the vapor concentration at the interface as well as the conditions at the boundary other than the interface. It is usually assumed that the gas and liquid phases are in thermal equilibrium, and the vapor concentration is prescribed [42] i.e.

$$T_1 = T_2$$
$$C_1 = C_1(T_1, P_{iv})$$

where P_{iv} is the vapor pressure above the interface. Then the interface cannot be in thermodynamic equilibrium, since in general $P_2 \neq P_1$ and thermodynamic equilibrium requires both thermal and mechanical equilibrium [14]. Thus the common use of the Gibbs-Thomson equation [29] which can be obtained from (10) under the thermodynamic equilibrium condition, i.e., $\mathbf{V} = 0$ and $\nabla\sigma = 0$, is no longer valid. On the other hand, there is no reason to disallow a temperature discontinuity [35]. Although on the molecular scale the temperature relaxation time is so short that the discontinuity is insignificant, on the macro-scale a discontinuous temperature jump across the interface may correspond more closely with the measured temperature difference. It may be reasonable to prescribe the temperatures on both sides of the interface, i.e.,

$$T_1 = T_1(P_1, \rho_1)$$
$$T_2 = T_2(P_2, \rho_2)$$

where P_2 may be taken as the saturation pressure. Then the vapor concentration cannot be prescribed, without using the nonequilibrium energy boundary condition (15). It is seen from this equation that the heat transfer is not only provided as the latent heat $(e_1 - e_2)$ to change the phase, but also to impart motion to the fluids through convection and diffusion.

3 Evaporation from a Liquid Layer

The above formulated theory will be applied to two different examples. The first is the rapid evaporation of a liquid layer into a host gas under weightless condition. The gas is treated as an ideal gas, and thus the equation of states for the gas phase is

$$\rho_1 = P_1 M / R T_1 \tag{16}$$

$$M = (P_{1v}/P_1) M_\beta + (1 - P_{1v}/P_1) M_\alpha \tag{17}$$

where R is the universal ideal gas constant, M is the molecular weight of the gas-vapor mixture, p_{1v} is the vapor pressure, and M_α and M_β are respectively, the molecular weight of the host gas and the vapor. The equation of state for the liquid is

$$\rho_2 = \rho_{20}[1 + \alpha(T_2 - T_{20})] \tag{18}$$

where ρ_{20} and T_{20} are respectively the liquid density and temperature at the interface, and α is the thermal expansion coefficient.

Here we seek an exact solution for the case of constant rate of evaporation. The rate depends on the relative rates of heat transfer from below and above the interface. Therefore the ratio r of heat transfer rates from either direction must be specified, i.e.

$$k_1 \nabla T_1 = r(-k_2 \nabla T_2) \tag{19}$$

When the interface is heated or cooled simultaneously from both sides r is positive. If one phase is heated but other is cooled, then r is negative. It should be pointed out that neither mechanical nor thermal equilibrium is assumed in the above formulation.

3.1 Exact Solution

A flat interface can be maintained with prescribed T_{10}, r_{10}, P_{20}, and C_{10} even at microgravity if the momentum, mass, and energy transport across the interface satisfy the governing equations with their boundary conditions exactly. Such an exact solution is given below

$$\mathbf{U} - \mathbf{j}\dot{a}, \qquad \mathbf{V}_1 = \mathbf{j}v_1, \tag{20}$$

$$T_1 = T_{10} + A\{\exp[\kappa_1 m^2 t + m(y - v_1 t)] - 1\}, \tag{21}$$

$$C_1 = C_{10} + B\{\exp[D_1 n^2 t + n(y - v_1 t)] - 1\}, \tag{22}$$

$$P_1 = P_{10}, \; \rho_1 = P_1 M/(R T_1), \tag{23}$$

$$\mathbf{V}_2 = 0, \; C_2 = 1, \tag{24}$$

$$T_2 - T_{20} - G[\exp(\kappa_2 l^2 t + l y) - 1], \tag{25}$$

$$P_2 = P_{20}, \tag{26}$$

$$l = -\frac{\dot{a}}{\kappa_2},\ m - \frac{v_0}{\kappa_1},\ n = \frac{v_0}{D_1},\ v_0 = v_1 - \dot{a}, \tag{27}$$

$$h = \dot{a}t, \tag{28}$$

$$\dot{a} = \pm\sqrt{\frac{\rho_{10}(P_{20} - P_{10})}{\rho_{20}(\rho_{20} - \rho_{10})}}, \tag{29}$$

$$v_1 = \frac{\rho_{10} - \rho_{20}}{\rho_{10}}\dot{a}, \tag{30}$$

$$B = C_{10} - 1, \tag{31}$$

$$A = rC_{p2}G/C_{p1}, \tag{32}$$

$$G = \frac{(\frac{v_1^2}{2} + \frac{v_1 P_{10}}{v_0 \rho_{10}}) + e(T_{10}) - e_2(T_{20}) + (C_{10} - 1)(e_\alpha(T_{10}) - e_\beta(T_{10}))}{(1+r)C_{p2}}, \tag{33}$$

where $\dot{a}$ is the constant rate of the interfacial displacement, the minus or plus sign in (29) is for evaporation or condensation, v_1 is the speed of the vapor-host gas mixture leaving the interface, and κ and C_p are respectively the thermal diffusivity and the constant pressure specific heat. It should be pointed out that the pressure and density in the expressions of $\dot{a}$ and v_1 are coupled with the latent heat and the gradients of temperature and vapor concentration at the interface through (15). The constant rate of evaporation cannot be achieved unless the heat flux Q and the mass flux J are held constant at the interface. These fluxes are given by

$$Q = -\rho_{10}C_{p1}\kappa_1\nabla T_{10} - \rho_{20}C_{p2}\kappa_2\nabla T_{20} = \rho_{20}(-\dot{a})(C_{p1}A + C_{p2}G), \tag{34}$$

$$J = -\rho_{10}D_1\nabla C_1 + \rho_{10}C_{10}v_0 = \rho_{20}\dot{a}. \tag{35}$$

By use of the relation

$$e_1(T_{10}) = C_{10}e_\beta(T_{10}) + (1 - C_{10})e_\alpha(T_{10}), \tag{36}$$

and (32) and (33), the heat flux can be written as

$$Q = \{[\frac{v_1^2}{2} + \frac{v_1 P_{10}}{v_0 \rho_{10}}] + [e_\beta(T_{20}) - e_2(T_{20})] + [e_\beta(T_{10}) - e_\beta(T_{20})]\}[\rho_{20}(-\dot{a})] \tag{37}$$

The first term in (37) i.e.

$$Q_m = (v_1^2/2 + v_1 P_{10}/v_0 \rho_{10})(-\rho_{20}\dot{a}) \tag{38}$$

represents the fraction of heat transfer which is used to provide the kinetic energy and the flow work for the convecting vapor-host gas mixture. The second term in (37) is the internal energy required for changing liquid to vapor at T_{20} (latent heat). The third term is the additional heat input required to change the vapor temperature from T_{20} to T_{10}. In the usual treatment only the second term in (37) is accounted for. It is seen that a constant rate of evaporation or condensation can be achieved at a liquid-gas interface under different temperature and pressure conditions as long as the net mass and heat transfer rate are specified by Eqs. (30) and (37). Note that Eqs. (35) and (37) are independent of r. Thus a given $\dot{a}$ can be obtained for given interfacial temperature and pressure jumps and a prescribed C_{10} (cf (29)) regardless of how the heat is transferred (i.e. regardless of the particular value of r) as long as Q and J are specified by (35) and (37).

In many practical problems, exact solution cannot be obtained. A constitutive equation is usually used to relate the mass rate of evaporation to the temperature and pressure changes across the interface. [9,10]

$$\rho_{10}C_1(\mathbf{V}_1 - \mathbf{U})\cdot\mathbf{n} - \rho_{10}D_1\nabla C_1\cdot\mathbf{n} = E(\frac{M_\beta}{2\pi R})^{1/2}(\frac{P_{2v}}{\sqrt{T_{20}}}\frac{P_{1v}}{\sqrt{T_{10}}}), \tag{39}$$

where E is the evaporation coefficient, and P_{1v} and P_{2v} are respectively the vapor pressure at T_{10} and T_{20}. The exact solution we have obtained can be substituted into (39) to evaluate E for practical use. P_{10} and P_{1v} in (39) can be obtained by applying the following ideal gas relations at the interface.

$$P_{1v} = P_1 C_1/[C_1 + (M_\beta/M_\alpha)(1 - C_1)], \tag{40}$$

$$\rho_1 = [p_{1v}M_\beta + (P_1 - P_{1v})M_\alpha]/RT_1, \tag{41}$$

To facilitate presentation of numerical results, we write (39) in dimensionless forms

$$E - \rho_r(\frac{M}{M_b})\frac{[-\dot{a}(\frac{2\pi M_b}{RT_{10}})^{1/2}]}{\{\frac{(P_{2v}/P_{10})}{T_r^{1/2}} - \frac{C_{10}}{[C_{10} + M_r(1 - C_1)]}\}}, \tag{42}$$

where

$$\rho_r = \frac{\rho_{20}}{\rho_{10}},\ T_r = \frac{T_{10}}{T_{20}},\ M_r = \frac{M_\beta}{M_\alpha},\ \frac{M}{M_\beta} = \frac{P_{1v}}{P_{10}} + \frac{1 - \frac{P_{1v}}{P_{10}}}{M_r}. \tag{43}$$

We also non-dimensionalize Q with

$$\rho_2 C_{p2} T_{20}(-\dot{a}),\ \text{i.e. } q = Q/\rho_2 C_{p2} T_{20}(-\dot{a}) \tag{44}$$

3.2 Numerical Results

The values of transport properties including the mass diffusivity of water vapor in air and specific heat and conductivity of water are taken from Raznjevic, [31], Haar et. al., [8] Edward et al. [7] and Tsederberg [39]. The ratio $v_1/\dot{a}$ as a function of M_r for given $T_{10} = 350°\text{C}$, $T_{20} = 200°\text{C}$, P_{20}=1.6 Mpa, $P_{20} - P_{10} = 10^{-7}$pa, are plotted in Fig. 2 for various water vapor concentrations C_{10}. It is seen that the direction of v_1 is always opposite to that of $\dot{a}$. For a dry

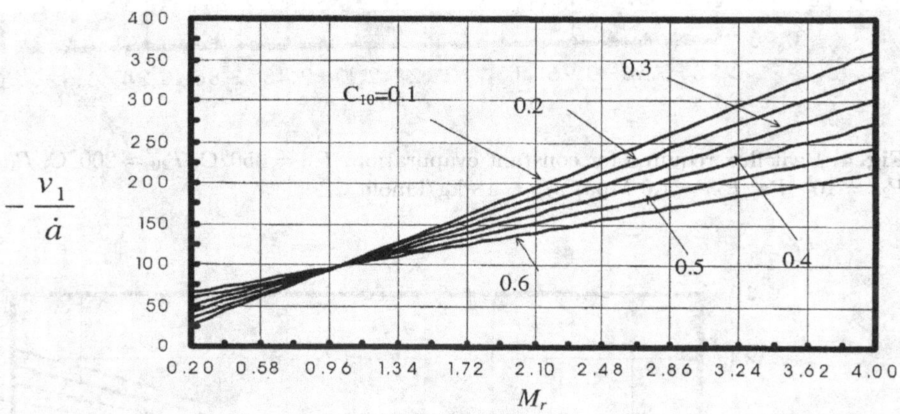

Fig. 2. Ratio of convective velocity to interfacial velocity. $T_{10} = 350°\text{C}$, $T_{20} = 200°\text{C}$, $P_{10} - P_{20} = 10^{-7}$Pa, $P_{20} = 1.6$ Mpa, $M_\beta = 18$ kg/kmole

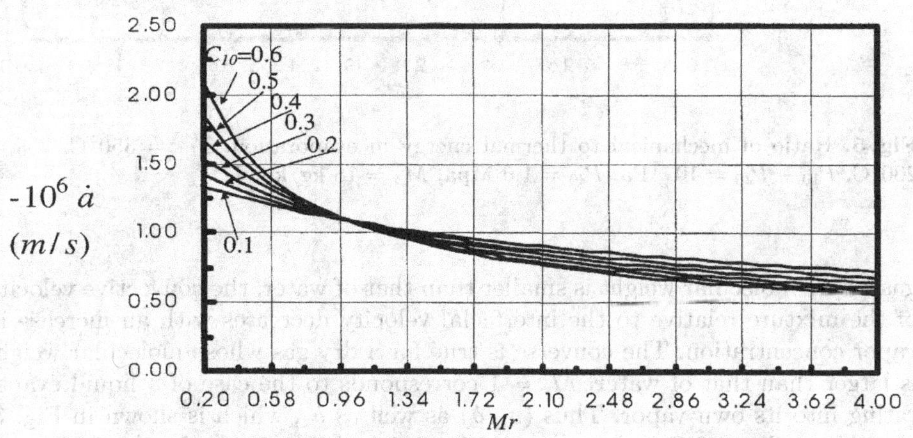

Fig. 3. Interfacial velocity. $T_{10} = 350°\text{C}$, $T_{20} = 200°\text{C}$, $P_{10} - P_{20} = 10^{-7}$Pa, $P_{20} =$ 1.6 Mpa, $M_\beta = 18$ kg/kmole

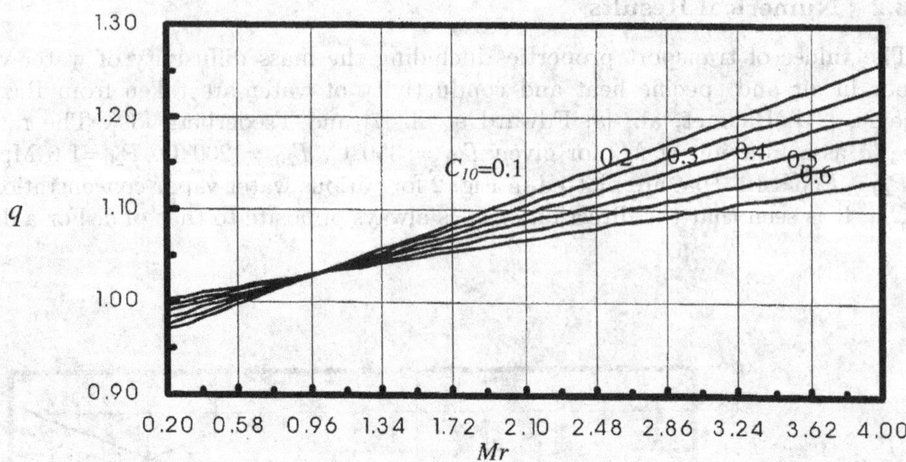

Fig. 4. Heat flux required for constant evaporation. $T_{10} = 350°\text{C}$, $T_{20} = 200°\text{C}$, $P_{10} - P_{20} = 10^{-7}\text{Pa}$, $P_{20} = 1.6$ Mpa, $M_\beta = 18$ kg/kmole

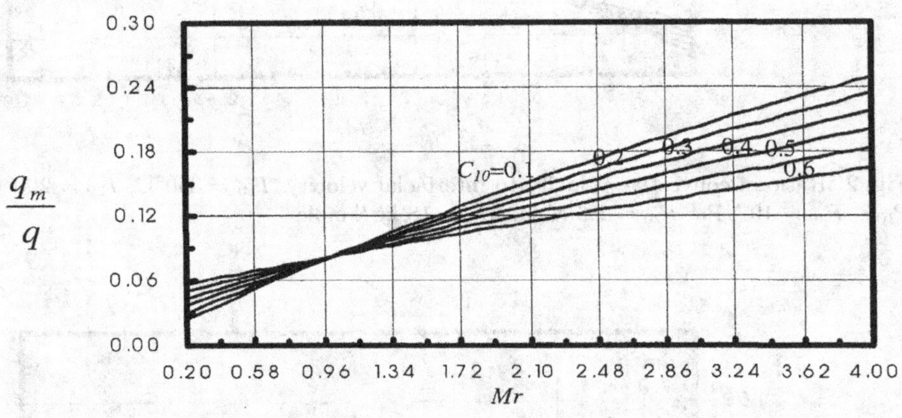

Fig. 5. Ratio of mechanical to thermal energy in evaporation. $T_{10} = 350°\text{C}$, $T_{20} = 200°\text{C}$, $P_{10} - P_{20} = 10^{-7}\text{Pa}$, $P_{20} = 1.6$ Mpa, $M_\beta = 18$ kg/kmole

gas whose molecular weight is smaller than that of water, the convective velocity of the mixture relative to the interfacial velocity decreases with an increase in vapor concentration. The converse is true for a dry gas whose molecular weight is larger than that of water. $M_r = 1$ corresponds to the case of a liquid evaporating into its own vapor. Thus $(v_1/\dot{a})$ as well as $\dot{a}$, which is shown in Fig. 3, are independent of C_{10} as expected. Figure 2 and Fig. 3 together enable one to calculate the corresponding convective velocity v_1. The heat transfer rates also depend on C_{10} as depicted in Figs. 4 and 5. Figure 5 shows that the fraction of

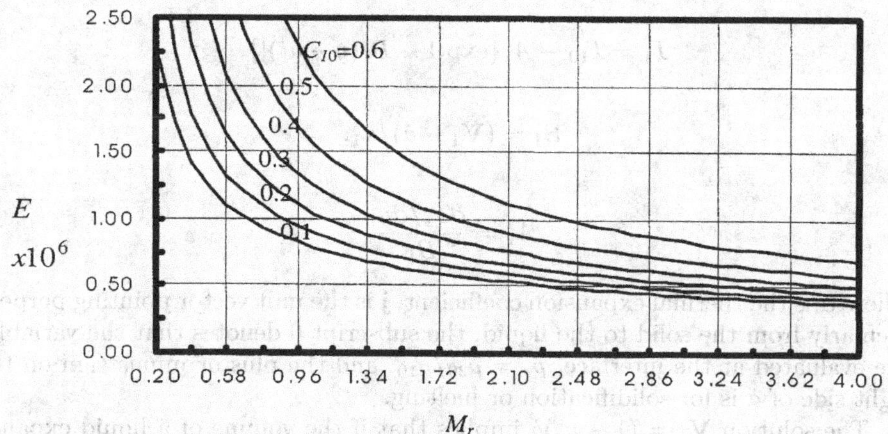

Fig. 6. Evaporation coefficient. $T_{10} = 350°\mathrm{C}$, $T_{20} = 200°\mathrm{C}$, $P_{10} - P_{20} = 10^{-7}\mathrm{Pa}$, $P_{20} = 1.6$ Mpa, $M_\beta = 18$ kg/kmole

heat transfer spent in producing mechanical energy may be significant, especially when the molecular weight of the liquid is larger than that of the dry gas. The evaporation coefficients in (39) corresponding to Figs. 2 and 3 are given in Fig. 6 which clearly indicates the functional dependence of E on other variables.

4 A Solidification Front

Consider the fields of temperature, pressure and density in a pure liquid ahead of its solidification front due to freezing. The frozen solid is considered as a rigid body. The governing equations given in Sect. 2 can be applied readily to the present problem. We need only to put $C_1 = C_2 = 1$, $P_2 = P_{20}$, $T_2 = T_{20}$, and $\mathbf{V}_2 = 0$ in these equations. If a flat solidification front is moving at a constant velocity $\dot{a}$ under weightless conditions, it is easily verified that the following solution satisfies the governing equations exactly.

$$\mathbf{U} = \mathbf{j}\dot{a},\ \dot{a} = \frac{\partial h}{\partial t} = \pm[P_{20} - P_{10}/\rho_{20}(\rho_r - 1)]^{1/2},$$

$$\rho_1 = \rho_{10}[1 - \alpha_1(T_1 - T_{10})],$$

$$\rho_2 = \rho_{20},$$

$$\mathbf{V}_1 = \mathbf{j}\mathrm{V}_1 = \mathbf{j}(1 - \rho_r)\dot{a},$$

$$\mathbf{V}_2 = 0,$$

$$T_2 = T_{20},$$

$$T_1 = T_{10} - A_1\{\exp[1 - B_1(y - \dot{a}t)]\},$$

$$B_1 = (\mathbf{V}_1 - \dot{a})/\kappa_1,$$

$$A_1 = \frac{\partial T_1/\partial y}{B_1},$$

where α is the thermal expansion coefficient, $\mathbf{j}$ is the unit vector pointing perpendicularly from the solid to the liquid, the subscript 0 denotes that the variables are evaluated at the interface, $\rho_r = \rho_{20}/\rho_{10}$, and the plus or minus sign on the right side of $\dot{a}$ is for solidification or melting.

The solution $\mathbf{V}_1 = (1 - \rho_r)\dot{a}$ implies that if the volume of a liquid expands upon solidification, such as in water, $\rho_r < 1$ and $\mathbf{V}_1 > 0$ upon freezing but $\mathbf{V}_1 < 0$ upon melting as it should. The expression of $\dot{a}$ also shows that for each assigned (P_{20}, T_{20}) and T_{10}, there is only one P_{10} which will allow the solidification to proceed at a constant interfacial velocity, since ρ_{10} and ρ_{20} are then fixed.

5 Conclusions

Examples have been given to demonstrate that the heat transfer to an interface undergoing a phase change is used not only to change the phase but also to impart kinetic energy of the fluid. The coupled effects of heat mass, and momentum are shown to be significant in the examples given. Therefore if the latent heat is defined to be the heat required to change the phase of a unit mass, then it cannot be treated as a material constant. It depends on the process of phase change under nonequilibrium thermodynamic conditions. Although a constant rate phase change across a flat interface is shown to be mathematically possible under nonequilibrium thermodynamic conditions, the stability of the process needs to be demonstrated.

References

1. T. Alty: Proc. Royal Soc. A **131**, 554 (1931)
2. C.J. Call, M. Kennedy: AIAA paper **93**, 09004 (1993)
3. B. Caroli, C. Calori, B. Roulet: J. Crystal Growth **66**, 575 (1984)
4. W. Chang, F. Cosandey, H. Hhan: Nanostructured Materials **2**, 2501 (1993)
5. A.A. Chernov: *Modern Crystallography*, III (Springer-Verlag, New York 1984)
6. S.R. Coriell, A.A. Chernov, B. Murray, G.B. McFadden: J. Crystal Growth **183**, 669 (1998)
7. W. Edward, J. Clarence, N. West, D. Ernest: *International Critical Tables of Numerical Data. Physics, Chemistry and Technology*, (McGraw-Hill, New York 1930)

8. L. Haar, J.S. Gallagher, J.S. Kell: *NBS/NRC Steam Tables: Thermodynamics and Transport Properties and Computer Programs for Vapor and Liquid States of Water in SI Units*, (Hemisphere Publishing Co., Washington, D.C. 1984)
9. S.W. Joo, S.H. Davis, S.G. Bankoff: J. Fluid Mech. **230**, 117 (1991)
10. S.W. Joo, S.H. Davis: 'Instabilities in Three-dimensional Heated Draining Film'. In: *Proceedings of IUTAM Symposium,, Nonlinear Instability of Nonparallel Flows*, ed. by S.P. Lin, W.R.C. Phillips, D.T. Valentine, (Springer-Verlag, New York 1994) pp. 425–435
11. D.D. Joseph, Y.Y. Rendardy: *Fundamentals of Two-Fluid Dynamics*, Vol. 1 (Springer-Verlag, New York 1992)
12. D.A. Kessler, J. Koplik, H. Levine: Advances in Physics **37**, 225 (1998)
13. A. Konstantinov, J. Panov, J. Malinowski: Photo Sci. **21**, 20 (1973)
14. L.D. Landau, E.M. Lifshitz: *Statistical Physics*, (Addision-Wesley, Reading, MA 1969)
15. J.S. Langer: Review of Modern Physics **52**, 1 (1980)
16. C.K. Law: Progress Energy Combustion Science **8**, 171 (1982)
17. S.P. Lin: J. Fluid Engineering **15**, 119 (1983)
18. S.P. Lin, C.Y. Wang: *Modeling Wavy Film Flows, in Encyclopedia of Fluid Mechanics.* Edited by N.P. Cheremisinoff, (Gulf Publishing 1986) **1**, 931
19. S.P. Lin, Z.W. Lian: Physics of Fluids, A **1**, 490 (1989)
20. S.P. Lin, Z.W. Lian: AIAA Journal **28**, 120 (1990)
21. S.P. Lin, E.A. Ibrahim: J. Fluid Mech. **218**, 614 (1990)
22. S.P. Lin, Z.W. Lian: Phys. of Fluids A **5**, 771 (1992)
23. S.P. Lin, D. Webb: Phys. Of Fluids **6**, 2671 (1994)
24. S.P. Lin, M. Hudman: Microgravity Science and Technology **8**, 163 (1995)
25. S.P. Lin, M. Hudman: J. Thermal Physics and Heat Trans. **10**, 497 (1996)
26. S.P. Lin, D.G. Xi: J. Jpn. Soc. Microgravity Appl. **15**, 1994 (1998)
27. J.R. Maa: Indust. Engng. Chem. Fund. **6**, 504 (1967)
28. W.W. Mullins, R.F. Sekerka: J. Appl. Phys. **35**, 444 (1964)
29. P. Pelce: *Dynamics of Curved Fronts*, (Academic Press, New York 1998)
30. V. Pot, C. Appert, A. Melayah, D.H. Rothman, S. Zaleski: J. De Physique II **10**, 1517 (1996)
31. K. Raznjevic: *Handbook of Thermodynamics Tables and Charts*, (Hemisphere Publishing Col., Washington, D.C. 1976)
32. T.P. Schulze, S.H. Davis: J. Crystal Growth **143**, 317 (1994)
33. W.A. Sirignano: Dynamics of Sprays – 1992 Freeman Scholar Lecture. J. Fluid Eng. 115, p. 345–378 (1993)
34. C.M. Sonsino, G. Schliper: Powder Metallurgy International **24**, 339 (1992)
35. Y. Sone, Y. Onishi: J. of Physical Soci. F Japan **44**, 1981 (1978)
36. S. Sreenivassan, S.P. Lin: Int. J. Heat Mass Transfer **21**, 1517 (1978)
37. H. Sugimoto, Y. Stone: Phys. Fluids A **4**, 419 (1992)
38. H. Tomita: Physica A. **204**, 693 (1994)
39. N.V. Tsederberg: *Thermal Conductivity of Gases and Liquids*, (MIT Press, Cambridge Massachusetts 1965)
40. Y. Viisanen, R. Strey: J. Chem. Phys. **101**, 7835 (1994)
41. G. Wylle, H.H. Wills: Royal Soc. **197**, 383 (1949)
42. Y.B. Zeldovich, G.I. Barenblatt, V.B. Librovich, G.M. Aakhviladze: *Combustion and Explosions G.M.*, (Consultant Bureau, New York 1985)
43. Z.W. Zhou, S.P. Lin: Phys. of Fluids A **4**, 277 (1992)

Lecture Notes in Physics

For information about Vols. 1–584
please contact your bookseller or Springer-Verlag
LNP Online archive: http://www.springerlink.com/series/lnp/

Vol.585: M. Lässig, A. Valleriani (Eds.), Biological Evolution and Statistical Physics.

Vol.586: Y. Auregan, A. Maurel, V. Pagneux, J.-F. Pinton (Eds.), Sound–Flow Interactions.

Vol.587: D. Heiss (Ed.), Fundamentals of Quantum Information. Quantum Computation, Communication, Decoherence and All That.

Vol.588: Y. Watanabe, S. Heun, G. Salviati, N. Yamamoto (Eds.), Nanoscale Spectroscopy and Its Applications to Semiconductor Research.

Vol.589: A. W. Guthmann, M. Georganopoulos, A. Marcowith, K. Manolakou (Eds.), Relativistic Flows in Astrophysics.

Vol.590: D. Benest, C. Froeschlé (Eds.), Singularities in Gravitational Systems. Applications to Chaotic Transport in the Solar System.

Vol.591: M. Beyer (Ed.), CP Violation in Particle, Nuclear and Astrophysics.

Vol.592: S. Cotsakis, L. Papantonopoulos (Eds.), Cosmological Crossroads. An Advanced Course in Mathematical, Physical and String Cosmology.

Vol.593: D. Shi, B. Aktaş, L. Pust, F. Mikhailov (Eds.), Nanostructured Magnetic Materials and Their Applications.

Vol.594: S. Odenbach (Ed.),Ferrofluids. Magnetical Controllable Fluids and Their Applications.

Vol.595: C. Berthier, L. P. Lévy, G. Martinez (Eds.), High Magnetic Fields. Applications in Condensed Matter Physics and Spectroscopy.

Vol.596: F. Scheck, H. Upmeier, W. Werner (Eds.), Noncommutative Geometry and the Standard Model of Elememtary Particle Physics.

Vol.597: P. Garbaczewski, R. Olkiewicz (Eds.), Dynamics of Dissipation.

Vol.598: K. Weiler (Ed.), Supernovae and Gamma-Ray Bursters.

Vol.599: J.P. Rozelot (Ed.), The Sun's Surface and Subsurface. Investigating Shape and Irradiance.

Vol.600: K. Mecke, D. Stoyan (Eds.), Morphology of Condensed Matter. Physcis and Geometry of Spatial Complex Systems.

Vol.601: F. Mezei, C. Pappas, T. Gutberlet (Eds.), Neutron Spin Echo Spectroscopy. Basics, Trends and Applications.

Vol.602: T. Dauxois, S. Ruffo, E. Arimondo (Eds.), Dynamics and Thermodynamics of Systems with Long Range Interactions.

Vol.603: C. Noce, A. Vecchione, M. Cuoco, A. Romano (Eds.), Ruthenate and Rutheno-Cuprate Materials. Superconductivity, Magnetism and Quantum Phase.

Vol.604: J. Frauendiener, H. Friedrich (Eds.), The Conformal Structure of Space-Time: Geometry, Analysis, Numerics.

Vol.605: G. Ciccotti, M. Mareschal, P. Nielaba (Eds.), Bridging Time Scales: Molecular Simulations for the Next Decade.

Vol.606: J.-U. Sommer, G. Reiter (Eds.), Polymer Crystallization. Obervations, Concepts and Interpretations.

Vol.607: R. Guzzi (Ed.), Exploring the Atmosphere by Remote Sensing Techniques.

Vol.608: F. Courbin, D. Minniti (Eds.), Gravitational Lensing:An Astrophysical Tool.

Vol.609: T. Henning (Ed.), Astromineralogy.

Vol.610: M. Ristig, K. Gernoth (Eds.), Particle Scattering, X-Ray Diffraction, and Microstructure of Solids and Liquids.

Vol.611: A. Buchleitner, K. Hornberger (Eds.), Coherent Evolution in Noisy Environments.

Vol.612: L. Klein, (Ed.), Energy Conversion and Particle Acceleration in the Solar Corona.

Vol.613: K. Porsezian, V.C. Kuriakose, (Eds.), Optical Solitons. Theoretical and Experimental Challenges.

Vol.614: E. Falgarone, T. Passot (Eds.), Turbulence and Magnetic Fields in Astrophysics.

Vol.615: J. Büchner, C.T. Dum, M. Scholer (Eds.), Space Plasma Simulation.

Vol.616: J. Trampetic, J. Wess (Eds.), Particle Physics in the New Millenium.

Vol.617: L. Fernández-Jambrina, L. M. González-Romero (Eds.), Current Trends in Relativistic Astrophysics, Theoretical, Numerical, Observational

Vol.618: M.D. Esposti, S. Graffi (Eds.), The Mathematical Aspects of Quantum Maps

Vol.619: H.M. Antia, A. Bhatnagar, P. Ulmschneider (Eds.), Lectures on Solar Physics

Vol.620: C. Fiolhais, F. Nogueira, M. Marques (Eds.), A Primer in Density Functional Theory

Vol.621: G. Rangarajan, M. Ding (Eds.), Processes with Long-Range Correlations

Vol.622: F. Benatti, R. Floreanini (Eds.), Irreversible Quantum Dynamics

Vol.623: M. Falcke, D. Malchow (Eds.), Understanding Calcium Dynamics, Experiments and Theory

Vol.624: T. Pöschel (Ed.), Granular Gas Dynamics

Vol.625: R. Pastor-Satorras, M. Rubi, A. Diaz-Guilera (Eds.), Statistical Mechanics of Complex Networks

Vol.626: G. Contopoulos, N. Voglis (Eds.), Galaxies and Chaos

Vol.627: S.G. Karshenboim, V.B. Smirnov (Eds.), Precision Physics of Simple Atomic Systems

Vol.628: R. Narayanan, D. Schwabe (Eds.), Interfacial Fluid Dynamics and Transport Processes

Printing: Strauss GmbH, Mörlenbach
Binding: Schäffer, Grünstadt

9783540405832